Healthcare Simulation Rese

Debra Nestel • Joshua Hui
Kevin Kunkler • Mark W. Scerbo
Aaron W. Calhoun
Editors

Healthcare Simulation Research

A Practical Guide

Editors
Debra Nestel
Monash Institute for Health and Clinical Education
Monash University
Clayton, VIC
Australia

Kevin Kunkler
Department of Surgery
University of Maryland, Baltimore
Baltimore, MD
USA

Aaron W. Calhoun
Department of Pediatrics
University of Louisville School of Medicine
Louisville, KY
USA

Joshua Hui
Emergency Medicine, Kaiser Permanente
Los Angeles Medical Center
Los Angeles, CA
USA

Mark W. Scerbo
Department of Psychology
Old Dominion University
Norfolk, VA
USA

ISBN 978-3-030-26836-7 ISBN 978-3-030-26837-4 (eBook)
https://doi.org/10.1007/978-3-030-26837-4

This Springer imprint is published by the registered company Springer Nature Switzerland AG
The registered company address is: Gewerbestrasse 11, 6330 Cham, Switzerland

Foreword

Every complex endeavor in the world depends heavily on research and scholarship to provide the basis for its theoretical underpinning, promote its practical development, determine its methods of implementation, and evaluate its results. While the overall community of practice will include many people who are the users of research and scholarship, it also must have a vigorous, and self-sustaining, set of people who actually conduct those activities. Many of our community in simulation in healthcare work in settings that value performing research and scholarship, such as universities or their affiliated hospitals, but only some of our members have the formal or experiential background to be seriously engaged in it. The Society for Simulation in Healthcare (SSH) has been sensitive for the need to provide some resources to those without such training or readily available mentorship. This book, *Healthcare Simulation Research: A Practical Guide*, is one such resource. But it's really meant for everyone. It contains many chapters (48!) on a large variety of topics covering a wide spectrum concerning research and scholarship in our field. Those new to techniques of research, development, and other elements of scholarship can find brief introductions to a host of aspects – many of which are basic and easy to fathom. Those farther along in their journey may still find these resources useful in their own work or in their teaching. They may also enjoy other sections that are about more advanced and occasionally arcane topics.

Although the coverage of the field of research and scholarship in simulation is encyclopedic, fortunately, each chapter is relatively brief – essentially an overview or introduction to the topic. Since nearly all of the topics in this book are each the subject of many (sometimes hundreds or thousands) of papers, presentations, book chapters, and whole books, this book neither intends to, nor can, substitute for the hard-won details of knowledge and skills described in this underlying literature. But for every one of the topics, the chapter provides a start.

To use some culinary metaphors, one could describe this book variously as a buffet (a large one – perhaps like on a cruise ship!), a tapas restaurant, or perhaps better yet tasting menus from a variety of skilled chefs (many small scrumptious dishes). To continue this metaphor, there are five celebrity executive chefs (editors) whose names will be well known to our community, as will most of the many dozens of chapter authors. I'm personally pleased to say that lots of them are my own long-time colleagues and friends.

Just as no one can eat a meal consisting of every dish created by dozens of master chefs (holiday dinners notwithstanding), this book is clearly not really intended to be read in one fell swoop. Instead, I expect that people will dip into it for some insights about things they've heard of but don't know much about as well as for a quick refresher about things they know about but may have forgotten. In a quote attributed to Mark Twain (even though he seems never to have said it this way), "It ain't what you don't know that gets you into trouble. It's what you know for sure that just ain't so." All of us can benefit from learning more in both categories.

Overall, I think that we collectively owe a debt of gratitude to the SSH, especially to its Research Committee, for commissioning this work and also – perhaps especially – to the many authors who worked hard to craft serious, short, useful "dishes" for the enjoyment and nourishment of us all.

David M. Gaba, MD
Associate Dean for Immersive and Simulation-Based Learning
Professor of Anesthesiology, Perioperative, and Pain Medicine, Stanford School of Medicine
Stanford, CA, USA
Staff Physician and Founder, and Co-Director
Patient Simulation Center at VA Palo Alto Health Care System
Palo Alto, CA, USA
Founding Editor in Chief, *Simulation in Healthcare*

Preface

This book is the product of an international community of leading scholars in healthcare simulation-based research. A diverse array of methodologies, designs, and theories are discussed as they apply to different aspects and phases of the research process, and the subject matter of each chapter has been carefully shaped to be accessible to a wide range of experience levels. Whether your interest is education, clinical outcomes, or assessment and whether you are a relative novice familiarizing yourself with this field or a seasoned researcher seeking information on advanced techniques or methodology outside of your usual scope of practice, this book has something in it for you. While we understand that a book cannot replace mentorship and experiential trial-and-error learning, each of us has wished that a textbook like this had been available when we embarked on our own journeys in healthcare simulation research. As the editors, we sincerely hope that this book will be a powerful resource for scholars and researchers of healthcare simulation, and we are deeply grateful to our colleagues who helped us bring it to fruition.

Clayton, VIC, Australia
Los Angeles, CA, USA
Baltimore, MD, USA
Norfolk, VA, USA
Louisville, KY, USA

Debra Nestel
Joshua Hui
Kevin Kunkler
Mark W. Scerbo
Aaron W. Calhoun

Contents

Contributors

Mark Adler, MD Feinberg School of Medicine, Department of Pediatrics (Emergency Medicine) and Medical Education, Northwestern University, Chicago, IL, USA

Ann & Robert H. Lurie Children's Hospital of Chicago, Chicago, IL, USA

Pamela Andreatta, EdD, PhD Uniformed Services, University of the Health Sciences, Bethesda, MD, USA

Tavis Apramian, MA, MSc, PhD Department of Family Medicine, McMaster University, Hamilton, ON, Canada

Michel Albert Audette, PhD, M. Eng, B. Eng. Department of Modeling, Simulation and Visualization Engineering, Old Dominion University, Norfolk, VA, USA

Marc Auerbach, MD, MSci Pediatrics and Emergency Medicine, Yale University School of Medicine, New Haven, CT, USA

Alexis Battista, PhD Graduate Programs in Health Professions Education, The Henry M Jackson Foundation for the Advancement of Military Medicine, Bethesda, MD, USA

Margaret Bearman, PhD Centre for Research in Assessment and Digital Learning (CRADLE), Deakin University, Docklands, VIC, Australia

Felicity Blackstock, PhD School of Science and Health, Western Sydney University, Campbelltown, NSW, Australia

John Boulet, PhD, MA, BSc Vice President, Research and Data Resources, Educational Commission for Foreign Graduates, Foundation for Advancement of International Medical Education and Research, Philadelphia, PA, USA

Birgitte Bruun, PhD Copenhagen Academy of Medical Education and Simulation (CAMES), Capital Region of Denmark, Center for Human Resources, Copenhagen, Denmark

Ryan Brydges, PhD The Wilson Centre, Toronto General Hospital, University Health Network and University of Toronto, Toronto, ON, Canada

The Allan Waters Family Simulation Centre, St. Michael's Hospital – Unity Health Toronto, Toronto, ON, Canada

Department of Medicine, University of Toronto, Toronto, ON, Canada

Aaron W. Calhoun, MD, FAAP, FSSH Department of Pediatrics, University of Louisville School of Medicine, Louisville, KY, USA

Todd P. Chang, MD, MAcM Department of Pediatric Emergency Medicine, Children's Hospital Los Angeles/University of Southern California, Los Angeles, CA, USA

Adam Cheng, MD Department of Pediatrics & Emergency Medicine, Albert Children's Hospital, University of Calgary, Calgary, AB, Canada

Jeffrey J. H. Cheung, MSc, PhD (c) The Wilson Centre, Toronto General Hospital, University Health Network and University of Toronto, Toronto, ON, Canada

Department of Medical Education, University of Illinois at Chicago College of Medicine, Chicago, IL, USA

Lou Clark PhD, MFA Hugh Downs School of Human Communication, Arizona State University, Tempe, AZ, USA

Andrew J. Collins, PhD, MSC, MA (Oxon); BA (Oxon) Department of Engineering Management and Systems Engineering, Old Dominion University, Norfolk, VA, USA

David A. Cook, MD, MHPE Mayo Clinic Multidisciplinary Simulation Center, Office of Applied Scholarship and Education Science, and Division of General Internal Medicine, Mayo Clinic College of Medicine and Science, Rochester, MN, USA

Peter Dieckmann, PhD, Dipl-Psych Copenhagen Academy of Medical Education and Simulation (CAMES), Capital Region of Denmark, Center for Human Resources, Copenhagen, Denmark

Department of Clinical Medicine, University of Copenhagen, Copenhagen, Denmark

Department of Quality and Health Technology, University of Stavanger, Stavanger, Norway

Walter J. Eppich, MD, PhD Feinberg School of Medicine, Department of Pediatrics, Division of Emergency Medicine, Ann & Robert H. Lurie Children's Hospital of Chicago, Northwestern University, Chicago, IL, USA

Rosemarie Fernandez, MD Department of Emergency Medicine and Center for Experiential Learning and Simulation, College of Medicine, University of Florida, Gainesville, FL, USA

Michael D. Fetters, MD, MPH, MA Department of Family Medicine, University of Michigan, Ann Arbor, MI, USA

Andree Gamble, MSN, RN Nursing Department, Holmesglen Institute, Southbank, VIC, Australia

Faculty of Medicine, Nursing & Health Sciences, Monash University, Clayton, VIC, Australia

James M. Gerard, MD Department of Pediatric Emergency Medicine, SSM Health Cardinal Glennon Children's Hospital/St. Louis University, St. Louis, MO, USA

Gregory E. Gilbert, EdD, MSPH, PStat® SigmaStats® Consulting, LLC, Charleston, SC, USA

Gerard J. Gormley, MB, BCh, BAO, MD, FRCGP, FHEA, DMH, PGCertMedEd Centre for Medical Education, Queen's University Belfast, Belfast, UK

Isabel T. Gross, MD, PhD, MPH Pediatrics and Emergency Medicine, Yale University School of Medicine, New Haven, CT, USA

Timothy C. Guetterman, PhD Creighton University, Omaha, NE, USA

Rose Hatala, MD, MSc, FRCP(C) Department of Medicine, St. Paul's Hospital, The University of British Columbia, Vancouver, BC, Canada

Joshua Hui, MD, MS, MHA Emergency Medicine, Kaiser Permanente, Los Angeles Medical Center, Los Angeles, CA, USA

Jennifer L. Johnston, MB, ChB, MPhil, PhD Centre for Medical Education, Queen's University Belfast, Belfast, UK

Suzan E. Kardong-Edgren, PhD, MS, BS Center for Medical Simulation, Boston, MA, USA

Grainne P. Kearney, MRCGP, MSc Centre for Medical Education, Queen's University Belfast, Belfast, UK

Michelle A. Kelly, BSc, MN, PhD Simulation and Practice, School of Nursing, Midwifery and Paramedicine, Curtin University, Perth, WA, Australia

David O. Kessler, MD, MSc Vagelos College of Physicians & Surgeons, Department of Emergency Medicine, Columbia University, New York, NY, USA

Abigail W. Konopasky, PhD Graduate Programs in Health Professions Education, The Henry M. Jackson Foundation for the Advancement of Military Medicine, Bethesda, MD, USA

Kevin Kunkler, MD School of Medicine – Medical Education, Texas Christian University and University of North Texas Health Science Center, Fort Worth, MD, USA

Saadi Lahlou, ENSAE, PhD, HDR Department of Psychological and Behavioural Science, London School of Economics and Political Science, London, UK

James F. Leathrum, PhD, MS (Duke University), BS (Lehigh University) Department of Modeling, Simulation and Visualization Engineering, Old Dominion University, Norfolk, VA, USA

Matthew Lineberry, PhD Zamierowski Institute for Experiential Learning, University of Kansas Medical Center and Health System, Kansas City, KS, USA

Karen Livesay, PhD, MPET, GDBA, GCTertEd, RN, RM School of Health and Biomedical Sciences, RMIT University, Melbourne, VIC, Australia

Joseph O. Lopreiato, MD, MPH Uniformed Services University of the Health Sciences, Bethesda, MD, USA

Karen Mangold, MD, Med Feinberg School of Medicine, Department of Pediatrics (Emergency Medicine) and Medical Education, Northwestern University, Chicago, IL, USA

Ann & Robert H. Lurie Children's Hospital of Chicago, Chicago, IL, USA

William C. McGaghie, PhD Feinberg School of Medicine, Department of Medical Education and Northwestern Simulation, Northwestern University, Chicago, IL, USA

Lisa McKenna, PhD, MEdSt, RN, RM School of Nursing and Midwifery, La Trobe University, Melbourne, VIC, Australia

Nancy McNaughton, MEd, PhD Institute of Health Policy, Management and Evaluation, University of Toronto, Toronto, ON, Canada

Michael J. Meguerdichian, MD, MHP-Ed Department of Emergency Medicine, Harlem Hospital Center, New York, NY, USA

Roland R. Mielke, PhD Electrical Engineering, University of Wisconsin-Madison, Madison, WI, USA

Department of Modeling, Simulation and Visualization Engineering, Old Dominion University, Norfolk, VA, USA

Sharon Muret-Wagstaff, PhD, MPA Department of Surgery, Emory University School of Medicine, Atlanta, GA, USA

David J. Murray, MD Department of Anesthesiology, Washington University School of Medicine, St Louis, MO, USA

Vinay M. Nadkarni, MD, MS Department of Anesthesiology, Critical Care and Pediatrics, The Children's Hospital of Philadelphia, Perelman School of Medicine, University of Pennsylvania, Philadelphia, PA, USA

Debra Nestel, PhD, FSSH Monash Institute for Health and Clinical Education, Monash University, Clayton, VIC, Australia

Austin Hospital, Department of Surgery, Melbourne Medical School, Faculty of Medicine, Dentistry & Health Sciences, University of Melbourne, Heidelberg, VIC, Australia

Catherine F. Nicholas, EdD, MS, PA Clinical Simulation Laboratory, University of Vermont, Burlington, VT, USA

Francisco M. Olmos-Vega, MD, MHPE, PhD Department of Anesthesiology, Faculty of Medicine, Pontificia Universidad Javeriana, Bogotá, Colombia

Anesthesiology Department, Hospital Universitario San Ignacio, Bogotá, Colombia

Stephanie O'Regan, BNurs, MSHM, MHSc(Ed) Sydney Clinical Skills and Simulation Centre, Northern Sydney Local Health District, St Leonards, NSW, Australia

Faculty Medicine, Nursing and Health Sciences, Monash University, Clayton, VIC, Australia

Miguel A. Padilla, BA, MAE, PhD Department of Psychology, Old Dominion University, Norfolk, VA, USA

Mary D. Patterson, MD, MEd Department of Emergency Medicine and Center for Experiential Learning and Simulation, College of Medicine, University of Florida, Gainesville, FL, USA

Shawna J. Perry, MD, FACEP Department of Emergency Medicine, University of Florida College of Medicine – Jacksonville, Jacksonville, FL, USA

Emil R. Petrusa, PhD Department of Surgery and Learning Lab (Simulation Center), Harvard School of Medicine, Massachusetts General Hospital, Boston, MA, USA

Gabriel B. Reedy, MEd, PhD CPsychol Faculty of Life Sciences and Medicine, King's College London, London, UK

Jill S. Sanko, PhD, MS, ARNP, CHSE-A, FSSH School of Nursing and Health Studies, University of Miami, Coral Gables, FL, USA

Mark W. Scerbo, PhD Department of Psychology, Old Dominion University, Norfolk, VA, USA

Joshua M. Sherman, MD Regional Medical Director, PM Pediatrics, Los Angeles, CA, USA

Jayne Smitten, PhD, Med, CHSE-A College of Health and Society, Experiential Simulation Learning Center, Hawai'i Pacific University, Honolulu, HI, USA

Jessica Stokes-Parish, BN, MN (Adv. Prac.), GCCC(ICU), PhD (c) School of Medicine & Public Health, University of Newcastle, Callaghan, NSW, Australia

Jill Stow, RN, BEdSt, MN, PhD School of Nursing and Midwifery, Monash University, Melbourne, VIC, Australia

Karen Szauter, MD Department of Internal Medicine, Department of Educational Affairs, University of Texas Medical Branch, Galveston, TX, USA

Jo Tai, MBBS, BMedSc, PhD Centre for Research in Assessment and Digital Learning (CRADLE), Deakin University, Melbourne, VIC, Australia

Pim W. Teunissen, MD, PhD Faculty of Health Medicine and Life Sciences (FHML), Department of Educational Development and Research, School of Health Professions Education (SHE), Maastricht University, Maastricht, The Netherlands

Department of Obstetrics and Gynaecology, VU University Medical Center, Amsterdam, The Netherlands

Brent Thoma, MD, MA, MSc Department of Emergency Medicine, University of Saskatchewan, Saskatoon, SK, Canada

Christopher J. Watling, MD, MMEd, PhD, FRCP© Departments of Clinical Neurological Sciences and Oncology, Office of Postgraduate Medical Education, Schulich School of Medicine and Dentistry, Western University, London, ON, Canada

Sharon Marie Weldon, BSc, MSc, PhD Department of Education and Health, University of Greenwich, London, UK

Travis Whitfill, MPH Pediatrics and Emergency Medicine, Yale University School of Medicine, New Haven, CT, USA

Cylie M. Williams, BAppSc, MHlthSc, PhD School of Primary and Allied Health Care, Monash University, Frankston, VIC, Australia

Michelle H. Yoon, PhD University of Colorado, Superior, CO, USA

Part I

Introduction to Healthcare Simulation Research

Developing Expertise in Healthcare Simulation Research

1

Debra Nestel, Joshua Hui, Kevin Kunkler, Mark W. Scerbo, and Aaron W. Calhoun

Overview
This book is the product of an international community of scholars in healthcare simulation research. Although the book has a strong focus on simulation as an educational method, the contents reflect wider applications of simulation. The book covers a broad range of approaches to research design. It is written for anyone embarking on research in healthcare simulation for the first time, or considering the use of a technique or method outside their usual practice. In this chapter, we share the origins of the book, an orientation to each part of the book, some biographical information on the editors and contributors, finishing with our own tips on developing research expertise in healthcare simulation.

D. Nestel (✉)
Monash Institute for Health and Clinical Education, Monash University, Clayton, VIC, Australia

Austin Hospital, Department of Surgery, Melbourne Medical School, Faculty of Medicine, Dentistry & Health Sciences, University of Melbourne, Heidelberg, VIC, Australia
e-mail: debra.nestel@monash.edu; dnestel@unimelb.edu.au

J. Hui
Emergency Medicine, Kaiser Permanente, Los Angeles Medical Center, Los Angeles, CA, USA

K. Kunkler
School of Medicine – Medical Education, Texas Christian University and University of North Texas Health Science Center, Fort Worth, TX, USA
e-mail: k.kunkler@tcu.edu

M. W. Scerbo
Department of Psychology, Old Dominion University, Norfolk, VA, USA
e-mail: mscerbo@odu.edu

A. W. Calhoun
Department of Pediatrics, University of Louisville School of Medicine, Louisville, KY, USA
e-mail: Aaron.calhoun@louisville.edu

Introduction

This book is the product of an international community of scholars in healthcare simulation research. Although the book has a strong focus on simulation as an educational method, the contents reflect wider applications of simulation. The book covers a broad range of approaches to research design. It is written for anyone embarking on research in healthcare simulation for the first time, or considering the use of a technique or method outside their usual practice. It offers guidance on developing research expertise.

Why a Book on Healthcare Simulation Research?

As editors, we have held the roles of chair, co-chair, and vice chair of the Society for Simulation in Healthcare (SSH) Research Committee. This international professional association has championed scholarship in healthcare simulation since its inception in 1994. The Research Committee was formed in 2005, and our leadership roles commenced in 2011. In these roles, we have sought to support the global healthcare simulation community as it undertakes research and scholarship activities. This is perhaps seen most clearly at the annual International Meeting for Simulation in Healthcare (IMSH) , the major event in the SSH calendar, at which the demand for guidance in research remains high. Each year the committee oversees the review process for conference research abstracts as well as the competitive bidding for SSH-based research funding. These experiences bring the importance of clarity in approaches to healthcare simulation research into sharp focus.

While many members of the SSH have clinical research experience, this does not always translate easily to education-focused or other research areas in healthcare simulation. In order to support members' requests, the Research Committee has undertaken several initiatives, and this edited book is one example of the growing base of resources.

D. Nestel et al. (eds.), *Healthcare Simulation Research*, https://doi.org/10.1007/978-3-030-26837-4_1

How Is the Book Organized?

The book comprises seven parts. In Part I, *Introduction to Healthcare Simulation Research*, we orient readers to the healthcare simulation research. We begin by documenting contemporary history of healthcare simulation with reflections from three editors-in-chief of healthcare simulation-focused journals (Chap. 2). Battista et al. offer examples of programs of research illustrating how established researchers have built their research practices (Chap. 3). In Chap. 4, Cheung et al. offer guidance on getting started in research, of identifying a problem worthy of study, of locating it in the literature and framing the research question with hints of direction of study. We then have two chapters that provide overviews of specific simulation modalities – serious gaming and virtual reality (Chap. 5) and computational modeling and system level research (Chap. 6).

In part II, *Finding and Making Use of Existing Literature*, we have two chapters. It may seem obvious to state that it is essential that we identify and acknowledge what is known on our research topics of interest. Enthusiasm to get started in research may curb a thorough search for established knowledge. However, we are reminded by Kessler et al., on the importance of identifying what is already known on our research area of interest and achieving this through a thorough search and review of literature (Chap. 7). We then learn from Cook as he shares his extensive expertise of systematic reviews in Chap. 8.

Qualitative research approaches are offered in Part III. The twelve chapters cover some key elements of qualitative research and where possible applied to healthcare simulation research. Chapter 9 outlines some fundamental concepts in qualitative research as well as orienting readers to the part. Bearman (Chap. 10) and Smitten (Chap. 11) continue the exploration of key concepts. We then shift gears to considering methods that are commonly used in qualitative research. Eppich et al. cover in-depth interviews (Chap. 12), McNaughton and Clark on focus groups (Chap. 13), observational methods from Bruun and Dieckmann (Chap. 14), and, other visual methods from Dieckmann and Lahlou (Chap. 15). Although not specifically qualitative in focus, Kelly and Tai share approaches to using survey and other textual data (Chap. 16). In Chap. 17, from Nicholas et al. we are guided through key components of data transcription and management. The three remaining chapters move to the next phase of research – analysis of data. Eppich et al. outlines Grounded Theory, (Chap. 18), Gormley et al. thematic and content analysis (Chap. 19), and McKenna et al. conversation, discourse and hermeneutic analysis (Chap. 20).

Part IV contains ten chapters on quantitative research approaches. This section opens with an introduction by Calhoun, Hui, and Scerbo (Chap. 21) that addresses important concepts, as well as common pitfalls, in quantitative research methods as applied to simulation. This is immediately followed by a deeper exploration by the same authors of the role played by theory and theoretical constructs in the formulation and testing of hypotheses (Chap. 22). The section next turns to an overview of quantitative study design by Mangold and Adler (Chap. 23), and a discussion by Andreatta of variables and outcome measures (Chap. 24). Gathering these data often requires valid and reliable assessment tools, the use of which are addressed in the next two chapters. Boulet and Murray discuss issues with tool design and selection (Chap. 25) while Hatala and Cook explore in depth the important concepts of validity and reliability (Chap. 26). The section ends with four chapters that address statistical reasoning and analysis. Lineberry and Cook begin the discussion by unpacking key statistical concepts and terminology and highlighting the importance of clear, open interaction between primary investigators and statisticians (Chap. 27). This is followed by a deeper exploration of more complex statistical issues in the following three chapters, including discussions of non-parametric statistics by Gilbert (Chap. 28); p-values, power and sample-size issues by Petrusa (Chap. 29); and advanced analytical methods such as hierarchical linear models and generalizability theory by Padilla (Chap. 30).

Part V consists of two chapters addressing mixed methods research. The mixed methods approach seeks to combine aspects of both qualitative and quantitative methods in order to more holistically address research questions. The section begins with a conceptual introduction to mixed methods by Guetterman and Fetters (Chap. 31) that explores various study design considerations using relevant examples from the literature. Next, Sanko and Battista (Chap. 32) describe various data types that can be incorporated into the research process, complementing the discussions of the preceding three sections.

Although it is important to learn about research design, successful research practice has many other considerations. In Part VI there are eleven chapters covering *Professional Practices in Healthcare Simulation Research*. It begins with a chapter from Kunkler on writing a research proposal (Chap. 33) and is followed by a detailed discussion of ethical issues associated with healthcare simulation research from Reedy et al. (Chap. 34). The next three chapters continue the theme of careful preparation in research. In Chap. 35 we learn from Muret-Wagstaff and Lopreiato about developing a strategy for your research, from Kunkler approaches to identifying and applying for funding (Chap. 36) and from Patterson et al., analysis of a research grant (Chap. 37). This chapter draws much of the preceding content into an exemplar. In Chap. 38, Whitfill et al. offer practical guidance on setting up and maintaining multi-site studies. Nestel et al. describe elements of research supervision focusing on graduate student research supervision (Chap. 39). Training in formal research project management often does not appear in books on research practices, but is another of the important professional aspects of healthcare simulation research. Williams and Blackstock summarize contemporary approaches to project management (Chap. 40). The next two chapters consider different facets of disseminating research. From Cheng et al., a diverse range of dissemination activities (Chap. 41) and from McGaghie, guidance in writing for publication in peer reviewed journals (Chap. 42). The section finishes with a chapter on peer review from Nestel et al., an essential role in the development of scholarship in our field (Chap. 43).

The final part of the book has a strong practical and experiential theme. The first chapter by Bearman et al., describe what they call the social dimensions of research (Chap. 44). Chapter 45 is from O'Regan in which she shares how she identified the 'research conversations' she wanted to join by conducting systematic review on the role of observers in simulation education. In Chap. 46, Weldon, a nurse describes her experiences of becoming a qualitative researcher working collaboratively with a social scientist in studies set in the operating theatre. Gilbert and Calhoun offer an account of their quantitative research (Chap. 47). Finally, in Chap. 48, Stokes-Parrish shares her experience as a doctoral student in the peer review process.

Who Are the Editors?

Debra Nestel was co-chair of the Research Committee (2014–2015), SSH. Now based in Australia, she is a professor of healthcare simulation at Monash University and professor of surgical education at the University of Melbourne. She has held extended academic appointments at the University of Hong Kong and Imperial College London. Debra's first degree (BA, Monash University) was in sociology and her PhD was a mixed methods study of educational methods to supporting the development of patient-centred communication skills in medical students, doctors and dentists in Hong Kong (University of Hong Kong). Debra has led a national faculty development program for simulation educators, created a virtual network for simulated participants (www.simulatedpatientnetwork.org), was founding Editor-in-Chief of the open access journal – *Advances in Simulation* and is the new Editor-in-chief of *BMJ Simulation and Technology Enhanced Learning*. Debra's current research program is mainly qualitative in design with a strong interest in simulated participant methodology.

Mark Scerbo was Vice Chair of the Research Committee from 2013 to 2014. He is presently professor of human factors psychology at Old Dominion University and adjunct professor of health professions at Eastern Virginia Medical School in Norfolk, VA, USA. Mark has over 35 years of experience using simulation to research and design systems that improve user performance in academic, military, and industrial work environments. Within healthcare, he has conducted research in emergency medicine, family medicine, interventional radiology, nursing, obstetrics and gynecology, oncology, pediatrics, surgery, as well as with physician's assistants and standardized patients. In addition to healthcare simulation, he is involved with the training and education of simulation professionals and is past Chair of the Old Dominion University Modeling & Simulation Steering Committee that manages and guides the pedagogical concerns of modeling and simulation across the university, including modeling and simulation certificate programs in business, computer science, education, health sciences, and human factors psychology. Mark served as the SSH Chair, Second Research Summit, Beyond our Boundaries, in 2017, and currently serves as Editor-in-Chief of *Simulation in Healthcare*.

Kevin Kunkler is currently the Executive Director for Simulation Education, Innovation and Research at the Texas Christian University and University of North Texas Health Science Center School of Medicine. He is also a member of the faculty serving as Professor. Kevin served as Vice Chair of the Research Committee from 2015 to 2016. Previously, Kevin worked at the University of Maryland, School of Medicine, Department of Surgery and was loaned to the United States Army Medical Research and Materiel Command (MRMC) at Fort Detrick. At MRMC, he was with the Telemedicine and Advanced Technology Research Center for 3 years and then served four and half years as Portfolio Manager and Chair of the Joint Program Committee for the simulation portion of the Medical Simulation and Information Sciences Research Program. Kevin completed his medical degree from the Indiana University School of Medicine and his masters within the science of regulatory affairs from the Johns Hopkins – Kreiger School of Arts and Sciences.

Joshua Hui is the Past President, Society of Academic Emergency Medicine (SAEM) Simulation Academy of which the members are academic emergency physicians with a focus on simulation-based endeavors. Joshua launched a novice research grant for simulation-based study during his tenure and subsequently served as the co-chair for 2017 SAEM Consensus Conference on simulation at systems levels. At SSH, Joshua has been chair of the Research Committee (2013–2014) and Scientific Committee of IMSH. Joshua launched the SSH Novice Research Grant. He also served as the reviewer of simulation-based research grant applications submitted to the Joint Program Committee 1 – Medical Simulation and Training Technologies under the Department of Defense. For the American College of Emergency Medicine California Chapter, he served as Chair of its Annual Assembly from 2013 to 2015. He has also served on the advisory board of Hong Kong Hospital Authority Accident and Emergency Training Centre. Joshua received a scroll of commendation from the Los Angeles County Board of Supervisors in 2012 and the Best Implemented Patient Safety Project Award in 2011 for his simulation endeavors in Los Angeles County. Joshua was selected as the 2015 Education Award recipient from American College of Emergency Medicine California Chapter. Chronologically, Joshua was awarded bachelor degrees in neuroscience and psychobiology from UCLA, a medical degree from UCLA, a master in clinical research from UCLA as well as a master in healthcare administration and interprofessional leadership from UC San Francisco.

Aaron Calhoun is a tenured associate professor in the Department of Pediatrics at the University of Louisville and an attending physician in the Just for Kids Critical Care Center at Norton Children's Hospital. Aaron received his B.A. in biology with a minor in sociology at Washington and Jefferson College and his M.D. from Johns Hopkins University School of Medicine. Aaron has also completed a residency in general pediatrics at Northwestern University Feinberg School of Medicine/Children's Memorial Hospital, and a fellowship in pediatric critical care medicine at the Harvard University School of Medicine/ Children's Hospital of Boston. Aaron is the current director of the Simulation for Pediatric Assessment, Resuscitation, and Communication (SPARC) program at Norton Children's Hospital, and has significant experience in simulation-based healthcare education and simulation research. Primary research foci include simulation-based assessment, in-situ simulation modalities, and the psychological and ethical issues surrounding challenging healthcare simulations. Aaron served in the past as the Scientific Content Chair for IMSH and currently chairs the SSH Research Committee. He also serves as co-chair of the International Network for Simulation-based Pediatric Innovation, Research, and Education (INSPIRE), an associate editor for the journal *Simulation in Healthcare*, and is a founding member of the International Simulation Data Registry (ISDR).

Who Are the Authors?

There are 78 contributors to the book working in six countries (Australia, Canada, Denmark, The Netherlands, United Kingdom and the United States) with multiple roles as simulation practitioners, clinicians, researchers and other specialist roles. They have each developed expertise in healthcare simulation research and have generously shared their knowledge.

Developing Research Expertise

We know that knowledge alone will not result in the development of expertise. In Box 1.1, we share tips on how we have developed and made use of knowledge in our different research trajectories. Some of the ideas are overlapping and come from a virtual conversation on developing research expertise.

Box 1.1 Tips on developing research expertise in healthcare simulation

- "Although it's important to be part of a research community, undertaking courses on research methods can be really important to get the fundamentals established."
- "Join a journal club."
- "Read, think, discuss, do, reflect, read, think, discuss, do, reflect …"
- "Air your ideas to different audiences, get used to summarizing your research, of framing and reframing it to make it meaningful."
- "Don't think of each study you engage in as an isolated event but instead as a part of a larger conversation within the simulation research community."
- "Consider how each individual study you perform might lead to a fruitful program of research."
- "Attend conferences, professional meetings etc."
- "Go to sessions at conferences that are outside your usual interests."
- "Read different journals. Tell someone about what you're reading."
- "Be open to new ways of thinking and doing."
- "Try not repeat what has been published. Try to ask novel research questions or at least perform a *better* study with the limitations of previous studies in mind."
- "Be curious. Ask questions. Search the literature to see how others have (or have not) tried to answer your questions."
- "Write to authors and ask them about their research, what they've learned, and what they still want to know."
- "Ask authors how they would have done their study differently if given the opportunity to do it again."

- "Look outside your own specialty. How do other domains deal with issues similar to ones that concern you?"
- "Keep a reflexive diary."
- "Read your published papers – again."
- "Identify researchers whose work you enjoy and follow them on social media and research networks – it helps you to keep up with what they're doing, where and with whom."
- "Pursue formal training in research"
- "Seek a mentor and a sponsor."
- "Seek knowledge *and* understanding."

Closing

We hope that this book will make an important contribution to the resources of the healthcare simulation community. We believe we would all have benefited from having access to a resource like this when we each started research in healthcare simulation. We are grateful to our colleagues around the world for their generosity in sharing their knowledge and experience. It has also been a privilege to build our own research practice networks through editing this book. We hope that you will enjoy the offerings as much as we have in the process of developing this book.

A Contemporary History of Healthcare Simulation Research

2

Debra Nestel, Mark W. Scerbo, and Suzan E. Kardong-Edgren

Overview
This chapter reviews the major developments and milestones in simulation research over the last 20 years. While we acknowledge that simulation has many applications outside education, our focus in this chapter is on documenting contemporary history with a strong education focus. We first outline major developments in medicine and nursing. We consider different approaches to research. We note the importance of the role of professional societies and associations in the dissemination of healthcare simulation research.

Practice Points

- Research surrounding healthcare simulation began to appear in the 1990s, but started to increase dramatically in the mid-2000s.
- The evolution of healthcare simulation research has been propelled by several important milestones and events including the development of simulation societies and associations and peer reviewed journals.
- Research paradigms – qualitative, mixed methods and quantitative – all have potential value in healthcare simulation research.
- In healthcare simulation, researchers and their audiences are diverse and include simulation practitioners, health and social care professionals and educators, psychologists, sociologists, biomedical scientists, engineers, information technologists, economists, programme evaluators, policy makers and others.

Introduction

Healthcare simulation education has a long and at times ancient history [1], however, scholarly research on the topic has only appeared more recently. In 1902, The BMJ published an article in which the author called for "Future research … to determine the role of advanced educational techniques, including the use of simulators, in facilitating bronchoscopy education [2]." Owen (2016) notes how the first half of the twentieth century was the "dark ages" in healthcare simulation and it was only in the latter part of the twentieth century that healthcare simulation was "rediscovered" [1]. It is from this time that we describe the contemporary history of healthcare simulation research. It is really only in the last 30 years that research with and about simulation has grown, and this growth has been exponential. A PubMed search using the terms: simulation and patient safety, simulation and healthcare, and human patient simulation between 1980 and 2018, demonstrates the dramatic growth in simulation publications (see Fig. 2.1).

Research on healthcare simulation has been diverse with respect to intent, simulation modality and context. It has been descriptive, experimental, evaluative, explanatory and exploratory, meaning the methodologies and methods have drawn from quantitative, qualitative and mixed methods

D. Nestel (✉)
Monash Institute for Health and Clinical Education, Monash University, Clayton, VIC, Australia

Austin Hospital, Department of Surgery, Melbourne Medical School, Faculty of Medicine, Dentistry & Health Sciences, University of Melbourne, Heidelberg, VIC, Australia
e-mail: debra.nestel@monash.edu; dnestel@unimelb.edu.au

M. W. Scerbo
Department of Psychology, Old Dominion University, Norfolk, VA, USA
e-mail: mscerbo@odu.edu

S. E. Kardong-Edgren
Center for Medical Simulation, Boston, MA, USA
e-mail: skardongedgren@harvardmedsim.org

D. Nestel et al. (eds.), *Healthcare Simulation Research*, https://doi.org/10.1007/978-3-030-26837-4_2

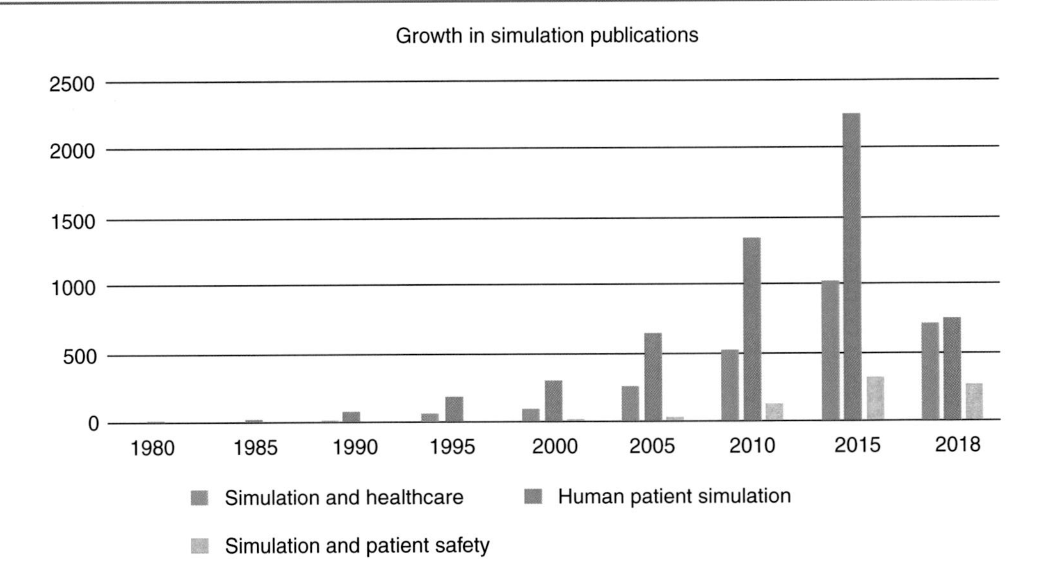

Fig. 2.1 Growth in healthcare simulation publications, 1980–2018

research approaches. Researchers and their audiences are also diverse and include simulation practitioners, health and social care professionals and educators, psychologists, sociologists, biomedical scientists, engineers, information technologists, economists, programme evaluators, policy makers and others [3]. While we acknowledge that simulation has many applications outside education, our focus in this chapter is on documenting contemporary history with a strong education focus. We first outline major developments in medicine and nursing. We consider different approaches to research. We note the importance of the role of professional societies and associations in the dissemination of healthcare simulation research.

Major Developments: Medicine

Even in the early 2000s, simulation in healthcare was viewed as a novelty by many. Over the course of the decade, however, there was a paradigmatic shift toward viewing simulation as an essential method for training and education. Several critical articles were published offering empirical evidence of the benefits of simulation training. In the late 1990s, Gaba and colleagues reported on the beneficial effects of simulation training in anesthesiology [4, 5]. In 2002, Seymour and colleagues published the first double-blind experiment comparing a traditional apprenticeship training approach to laparoscopic surgery with training on a virtual reality simulator [6]. Their results showed that residents who trained on the simulator needed 30% less time to perform a genuine procedure than those trained according to the traditional method and were also less likely to injure the patient. Then, in 2005, Issenberg and colleagues published a systematic review of the literature from 1969 to 2003 and concluded that 'high-fidelity' (manikin) medical simulation-based education was an effective method that complemented education in patient care settings, but that more rigorous research was still needed [7]. This review was repeated in 2010, and the authors noted advances from the earlier study [8]. It is valuable to report their findings since they reflect the focus of research to that time and have influenced what followed. The "features and best practices of simulation-based medical education" reported were: (i) feedback; (ii) deliberate practice; (iii) curriculum integration; (iv) outcome measurement; (v) simulation fidelity; (vi) skill acquisition and maintenance; (vii) mastery learning; (viii) transfer to practice; (ix) team training; (x) high-stakes testing; (xi) instructor training, and (xii) educational and professional context [8].

Perhaps equally important, several key leaders in medicine began to embrace the need to shift away from traditional approaches to training and education in favor of evidence-based alternatives that decreased the risk to patients [9–11]. In 2003, Ziv and colleagues argued that simulation-based training in healthcare had reached the point of becoming an ethical imperative [12].

Major Research Developments: Nursing

In 2005, the National League for Nursing (NLN) and Laerdal Medical jointly funded Jeffries and Rizzolo to develop simulation for nursing education in the USA. This work resulted in the first multisite nursing study in simulation and produced a framework which drove much future nursing research [13]. This was followed in 2015 with a more developed NLN Jeffries Simulation Theory [14]. In 2011, the Boards of Nursing in the USA pressed their National Council of State Boards of Nursing to provide evidence for the use of simulation in nursing education. This resulted in a cohort study of 600+ students in 10 schools of nursing around the USA over 2 years [15]. Results indicated that the substitution of up to 50% of traditional clinical time with high quality simulation using the INACSL Standards of Best Practice, did not interfere with students' abilities to pass the final certification exam, the NCLEX. Hospital educators and charge nurses who hired those graduates in the first 6 months post-graduation could not distinguish their performance from other new graduates [15].

Focus of Contemporary Research

This book explores different research approaches – qualitative, mixed methods and quantitative. All are present in contemporary research. McGaghie et al. argue for translational research in healthcare simulation [16]. This is the *bench to bedside* notion associated with biomedical and clinical sciences. The multiple levels from T1 (e.g. research that measures performance during simulation scenario), T2 (e.g. performance in clinical settings) and T3 (e.g. economic evaluations and sustainability) [17] all need investigation. We see many examples of research at T1 & T2 levels and increasing interest in T3.

Writing from a broader perspective than simulation, Regehr wrote of the need to re-orient two of the dominant discourses in health professions' education research: (i) from the imperative of proof to one of understanding, and (ii) from the imperative of simplicity to one of representing complexity well [18]. In an editorial of a new simulation journal, Nestel argued that his words resonated with the importance of valuing research that seeks understanding of the complex practice of simulation-based education [3].

The Role of Professional Societies in Healthcare Simulation Research

Late in the twentieth century, professional societies dedicated solely to healthcare simulation began to emerge. The Society in Europe for Simulation Applied to Medicine (SESAM) was established in 1994 and shortly thereafter the Society for Medical Simulation (later renamed the Society for Simulation in Healthcare; SSH), was established in the United States. The International Nursing Association for Clinical Simulation in Nursing (INACSL) was incorporated in 2003. Numerous organizations have emerged since then serving special niches within healthcare (e.g. International Pediatric Simulation Society – IPSS etc.), different simulation modalities (e.g. Association of Standardized Patient Educators – ASPE, for educators working with simulated participants), different countries (e.g. national societies), or geographical regions (e.g. California Simulation Alliance, Victorian Simulation Alliance etc.).

In 2006, SSH published *Simulation in Healthcare* and the INACSL began publication of *Clinical Simulation in Nursing*, the first two peer-reviewed journals dedicated solely to simulation. Since then, additional simulation journals have emerged including, *Advances in Simulation* and *BMJ Simulation & Technology Enhanced Learning*. Both of these journals are associated with professional societies. Other journals that address modelling and simulation more broadly have also begun to dedicate sections to healthcare simulation technology and systems (e.g., *Simulation*). Most of these professional societies and associations provide at least annual events in which research can be shared (See Chap. 41).

Standards of Simulation Practice

An important contribution to the healthcare simulation community has been the development of standards for simulation performance first published by the INACSL organization in 2010 [19]. The standards incorporated the then "best evidence" to provide guidance in the performance of high quality simulation education. The INACSL Standards for Best Practice: Simulation℠ are updated on a recurring cycle and are available freely to all (https://www.inacsl.org/inacsl-standards-of-best-practice-simulation/). Similarly, the ASPE have published standards for best practices for educators working with simulated participants [20]. Linked with the INACSL standards, the ASPE standards are based on research evidence in the discipline of simulated participant methodology.

Research Summits

Several professional societies and associations have held research summits and/or established research agendas. Nestel and Kelly have documented this history [21]. In 2006, the Society for Academic Emergency Medicine (SAEM) Simulation Task Force [22]. Issenberg and colleagues reported an Utsein-style meeting designed to establish a research agenda for simulation-based healthcare education [23]. In 2011, SSH held its first Research Summit bringing together experts from a wide range of professions and disciplines to review and discuss the current state of research in healthcare simulation and establish an agenda for future research [24]. Topics addressed at the Summit included: procedural skills, team training, system design, human and systems performance, instructional design and pedagogy, translational science and patient outcomes, research methods, debriefing, simulation-based assessment and regulation of professionals, and reporting inquiry in simulation. The Summit reaffirmed that research surrounding healthcare simulation had grown enormously. Although this increased research activity is certainly welcome, the reporting practices in the scholarly literature varied widely. Stefanidis et al.

(2012) report research priorities in surgical simulation for the twenty-first century using a Delphi study with members of the US-based Association for Surgical Education [25]. In 2013, the Australian Society for Simulation in Healthcare established a research agenda [21]. And, reported in 2014–2015, the International Network for Simulation-based Paediatric Innovation, Research, and Education (INSPIRE), brought together two research networks with the vision "to bring together all individuals working in paediatrics simulation-based research to shape and mould the future of paediatrics simulation research by answering important questions pertaining to resuscitation, technical skills, behavioural skills, debriefing and simulation-based education" [26]. These broad ranging initiatives all sit within professional societies and networks.

Research Reporting Standards for Simulation-Based Research

Several guidelines have been established to bring more uniformity to reporting research practices in medicine and other scientific disciplines fields, such as the Consolidated Standards of Reporting Trials (CONSORT) Statement for randomized trials and the Strengthening the Reporting of Observational Studies in Epidemiology (STROBE) Statement for observational studies. In 2015, a consensus conference was held to review the CONSORT and STROBE guidelines and introduce extensions aimed at simulation-based research. These modified guidelines represent an important step forward in standardizing and improving the reporting practices of healthcare simulation research. They were endorsed by four healthcare simulation journals; *Advances in Simulation, BMJ Simulation & Technology Enhanced Learning, Clinical Simulation in Nursing,* and *Simulation in Healthcare*; and appeared in the first joint publication among these journals (See Chap. 42) [27].

Recent Trends in Healthcare Simulation Research

In 2004, Gaba proposed eleven dimensions to describe the breadth of healthcare simulation at that point in time [28]. Scerbo and Anderson later organized those dimensions into three higher-level categories [29]. The first category describes the goals for using simulation (its purpose, healthcare domain, knowledge, skills, and attitudes addressed, and patient age). The second category addresses user characteristics (unit of participation, experience level, healthcare discipline of personnel, education, training, assessment, rehearsal, or research). The third category concerns the method of implementation (type of simulation or technology, site of event., the level of participation from passive to immersive, and the type of feedback given).

Several recently published articles confirm this broad scope of healthcare simulation research. Scerbo offered a picture of the breadth of research published in *Simulation in Healthcare* between 2013 and 2015 [30]. Regarding topic areas, articles on assessment, education/training, and technology accounted for almost two thirds of the publications. Another 10% of the articles addressed validation, teams, human factors issues, simulation theory, and patient safety. Articles on medical knowledge, patient outcomes, and patient care made up only 6% of the content. Articles addressing different clinical specialties revealed that most of the content came from anesthesiology, emergency medicine, general medicine, surgery, nursing, pediatrics, and obstetrics and gynecology. Three quarters of the articles addressed practicing clinicians and residents with a smaller minority focused on students or expertise at multiple levels. About half of the articles addressed research with mannequin or physical model simulators. Research with standardized (simulated) patients, virtual reality, hybrid systems, or multiple formats made up the remainder of the content. Scerbo concluded that much of the research published in the journal during that period focused on how to use simulation for training and assessment, how to improve the simulation experience for learners, and how to develop and evaluate new simulation systems. He also suggested that publications tended to come from clinical areas where simulation systems are more plentiful and have longer histories.

Nestel (2017) thematically analysed articles published in *Simulation in Healthcare* as *editorials* [31]. This is an indirect way of making meaning of contemporary healthcare simulation research. The five themes were:

1. "Embedding" simulation (Research that sought ways to embed simulation in medical and other curricula, in healthcare organisations such that simulation is part of education and training across professional practice trajectories);
2. Simulation responding to clinical practice (Research that addressed to elements of clinical practice that required improvements such as handoff, sepsis guidelines, etc.);
3. Educational considerations for simulation (Research that addresses ideas such as the relationship of realism to learning, the importance of creating psychological safety for participants, exploring debriefing approaches etc.);
4. Research practices (Research that considers methods and methodologies especially important to healthcare simulation); and,
5. Communicating leadership and scholarship about the community (This theme addressed ideas offered in editorials that were of interest to the simulation community such as language preferences etc.)

In nursing education, three major research reviews of simulation were published in the last 4 years [32–34]. Findings from these reviews indicated incremental improvements in research rigor over time but equivocal results overall. They also indicated the realities of educational research, a continued lack of funding, many one-group posttest designs, an abundance of self-report measures unaccompanied by objective measures, a lack of trained evaluators, inconsistent use of terminology, and a lack of adherence to standardized reporting guidelines [32–34]. In 2018, both Mariani et al. [35] and Cant et al. [32] evaluated research articles published in *Clinical Simulation in Nursing* for research rigour using the Simulation Research Rubric [36] and/or the Medical Education Research Study Quality Instrument [37]. The ratings from both evaluations showed the research to be of moderate to high quality. In summary, research in nursing is thriving and improving in rigor but continues to be underfunded. More multisite studies using reliable and valid instruments are needed. The INACSL publishes a research priorities needs list which can be found on its website (https://member.inacsl.org/i4a/pages/index.cfm?pageID=3545).

Another way to view the breadth and trends of healthcare simulation research is to examine what gets cited in the literature. Recently, Walsh and colleagues offered a bibliometric review of the 100 most cited articles in healthcare simulation [38]. They searched in Scopus and the Web of Science databases (Clarivate Analytics, Philadelphia, PA) in 2017, but compiled their list based on the Scopus search. The found that there were very few citations until about 2005. In fact, of their top 100 articles, citations did not exceed 10 per year until 2005. As might be expected review articles received the most citations followed by articles on interventions and tool development. Regarding topics and discipline, the most cited articles addressed clinical competence and quality of care, but those citations were limited to just six articles on their list. Other topics that were cited most frequently were medical training/education, surgery, primary care, oncology, anesthesiology, and doctor-patient communication. Articles addressing technical skills or the combination of technical and so-called ‘non-technical’ skills were cited more often than non-technical skills alone. Also, articles addressing physical and virtual reality part-task training systems and standardized or simulated patients were cited more frequently than other forms of simulators.

Closing

In his 2004 article, Gaba offered two different predictions for the future [28]. One path was pessimistic where he cautioned that interest in simulation within the medical community could wane. The other path was much more optimistic where he saw simulation training in healthcare becoming a requirement and a driving force behind changes to healthcare curricula. He also envisioned a public that demanded levels of safety in healthcare comparable to those in aviation and regulatory agencies that required simulation-based standards for training and evidence for devices gathered in trials using simulation.

Today, one could argue that we are closer to Gaba's optimistic view. There is no doubt that simulation has begun transforming healthcare training and education, but there is still a way to go. Healthcare research is increasing in importance in the scholarly literature. The articles at the top of Walsh et al.'s list of most cited papers exceed 1000 citations. New scholarly journals addressing special areas of healthcare simulation continue to emerge. However, this growth is certainly not uniform across the 11 dimensions that Gaba described 15 years ago. There are clinical specialties that are still underrepresented in the simulation literature. The promise of some forms of simulation technology have still not been realized. Translational studies showing direct benefits of simulation training on patient outcomes are still few and far between.

Collectively, these gaps in the research paint a picture of a discipline that is still evolving and volatile. Clearly, there is a lot of work to be done, but this is a picture of a research landscape that is ripe with opportunity for inquisitive minds. We hope that the research methods and tools described in this book provide a sturdy canvas for investigators to contribute to the bigger picture.

References

1. Owen H. Simulation in healthcare education: an extensive history. Cham: Springer; 2016.
2. Killian G. On direct endoscopy of the upper air passages and oesophagus; its diagnostic and therapeutic value in the search for and removal of foreign bodies. Br Med J. 1902;2:560–71.
3. Nestel D. Open access publishing in health and social care simulation research – advances in simulation. Adv Simul. 2016;1:2.
4. Gaba DM. Improving anesthesiologists' performance by simulating reality. Anesthesiology. 1992;76(4):491–4.
5. Gaba DM, Howard SK, Flanagan B, Smith BE, Fish KJ, Botney R. Assessment of clinical performance during simulated crises using both technical and behavioral ratings. Anesthesiology. 1998;89(1):8–18.
6. Seymour NE, Gallagher AG, Roman AA, O'Brien MK, Bansal VK, Andersen DK, Satava RM. Virtual reality training improves operating room performance: results of a randomized, double-blinded study. Ann Surg. 2002;236:458–64.
7. Issenberg SB, Mcgaghie WC, Petrusa ER, Lee Gordon D, Scalese RJ. Features and uses of high-fidelity medical simulations that lead to effective learning: a BEME systematic review. Med Teach. 2005;27(1):10–28.
8. McGaghie WC, et al. A critical review of simulation-based medical education research: 2003–2009. Med Educ. 2010;44(1):50–63.
9. Gould D, Patel A, Becker G, Connors B, Cardella J, Dawson S, Glaiberman C, Kessel D, Lee M, Lewandowski W, Phillips R,

Reekers J, Sacks D, Sapoval MK, Scerbo M. SIR/RSNA/CIRSE joint medical simulation task force strategic plan executive summary. J Vasc Interv Radiol. 2007;18:953–5.
10. Healy GB. The college should be instrumental in adapting simulators to education. Bull Am Col Surg. 2002;87(11):10–1.
11. Reznick RK, MacRae H. Teaching surgical skills—changes in the wind. New Engl J Med. 2006 Dec 21;355(25):2664–9.
12. Ziv A, Wolpe PR, Small SD, Glick S. Simulation-based medical education: an ethical imperative. Acad Med. 2003 Aug 1;78(8):783–8.
13. Jeffries PR. A framework for designing, implementing, and evaluating: simulations used as teaching strategies in nursing. Nurs Ed Per. 2005;26(2):96–103.
14. Jeffries PR. The NLN Jeffries simulation theory. Philadelphia: Wolters Kluwer; 2015.
15. Hayden J, Smiley R, Alexander MA, Kardong-Edgren S, Jeffries P. The NCSBN national simulation study: a longitudinal, randomized, controlled study replacing clinical hours with simulation in prelicensure nursing education. J Nurs Regul. 2014;5(2 Supplement):S1–S64.
16. McGaghie WC, et al. Evaluating the impact of simulation on translational patient outcomes. Simul Healthc. 2011;6(Suppl):S42–7.
17. Gaba D. Expert's Corner: Research in Healthcare Simulation. In: Palaganas J, et al., editors. Defining excellence in simulation programs. Philadelphia: Wolters Kluwer; 2015. p. 607.
18. Regehr G. It's NOT rocket science: rethinking our metaphors for research in health professions education. Med Educ. 2010;44:31–9.
19. INACSL standards of best practice: simulation. Retrieved on 20 November 2018 on the WWW from https://www.inacsl.org/inacsl-standards-of-best-practice-simulation/history-of-the-inacsl-standards-of-best-practice-simulation/
20. Lewis K, et al. The Association of Standardized Patient Educators (ASPE) Standards of Best Practice (SOBP). Adv Simul. 2017;2:10.
21. Nestel D, Kelly M. Strategies for research in healthcare simulation. In: Nestel D, et al., editors. Healthcare simulation education: evidence, theory and practice. West Sussex: Wiley; 2018. p. 37–44.
22. Bond WF, et al. The use of simulation in emergency medicine: a research agenda. Acad Emerg Med. 2007;14(4):353–63.
23. Issenberg SB, et al. Setting a research agenda for simulation-based healthcare education: a synthesis of the outcome from an Utstein style meeting. Simul Healthc. 2011;6(3):155–67.
24. Dieckmann P, Phero JC, Issenberg SB, Kardong-Edgren S, Østergaard D, Ringsted C. The first research consensus summit of the society for simulation in healthcare: conduction and a synthesis of the results. Simul Healthc. 2011;6(Suppl):S1–9.
25. Stefanidis D, et al. Research priorities in surgical simulation for the 21st century. Am J Surg. 2012;203(1):49–53.
26. INSPIRE. INSPIRE network report 2014–2015. 2015.
27. Cheng A, et al. Reporting guidelines for health care simulation research: extensions to the CONSORT and STROBE statements. Adv Simul. 2016;1:25.
28. Gaba DM. The future vision of simulation in health care. Qual Saf Health Care. 2004;13(Sup. 1):i2–i10.
29. Scerbo MW, Anderson BL. Medical simulation. In: Carayon P, editor. Handbook of human factors and ergonomics in health care and patient safety. 2nd ed. Boca Raton: CRC Press; 2012. p. 557–71.
30. Scerbo MW. Simulation in healthcare: Growin' up. Simul Healthc. 2016;11:232–5.
31. Nestel D. Ten years of simulation in healthcare: a thematic analysis of editorials. Simul Healthc. 2017;12:326–31.
32. Cant RP, Levett-Jones T, James A. Do simulation studies measure up? A simulation study quality review. Clin Sim in Nurs. 2018;21:23–39.
33. Cantrell MA, Franklin A, Leighton K, Carlson A. The evidence in simulation-based learning experiences in nursing education and practice: an umbrella review. Clin Sim in Nurs. 2016;13:634–67.
34. Doolen J, Mariani B, Atz T, Horsely TL, O'Rourke J, McAfee K. High-fidelity simulation in undergraduate nursing education: a review of simulation reviews. Clin Sim in Nurs. 2016;12:290–302.
35. Mariani B, Fey MK, Gloe D. The simulation research rubric: a pilot study evaluating published simulation studies. Clin Sim in Nurs. 2018;22:1–4.
36. Fey MK, Gloe D, Mariani B. Assessing the quality of simulation-based research articles: a rating rubric. Clin Sim in Nurs. 2015;11:496–504.
37. Reed DA, Cook DA, Beckman TJ, Levind RB, Kern DE, Wright T. Association between funding and quality of published medical education research. JAMA. 2007;298:1002–9.
38. Walsh C, Lydon S, Byrne D, Madden C, Fox S, O'Connor P. The 100 most cited papers on healthcare simulation: a bibliometric review. Simul Healthc. 2018;13:211–20.

Programs of Research in Healthcare Simulation

3

Alexis Battista, Abigail W. Konopasky, and Michelle H. Yoon

Overview
In this chapter, we outline a working definition of what a *program of research* is and describe some of the key components necessary for pursuing a program of research. We next highlight select programs of research within healthcare simulation, highlighting differing ways in which a program of research may arise (e.g., personal or organizational interests, research collaborations) and how programs grow and change as they mature. In keeping with the goals of this text, this chapter is primarily intended for individuals who are newly engaging in or are considering developing a program of research in healthcare simulation.

Practice Points/Highlights

- A program of research can be defined as a purposeful strategy for pursuing a coherent and connected line of inquiry.
- Programs of research can be viewed on a continuum – ranging from those programs just starting out to those that have grown and matured over time.
- The core components of a program of research are a central focus and flexible plan, committed researchers, appropriately selected research methods, and a web of supporting resources, such as space, materials, training opportunities, operational support, funding streams, and partnering groups or organizations.
- Programs of research may be derived through a variety of sources, including personal or institutional interests, accreditation body interests or guidance or research collaborations.

Introduction

Individuals working in healthcare simulation tend to be flexible, innovative, and focused – it is part and parcel of a growing and ever evolving field like simulation – but it may be difficult for them to find time and resources to purposefully pursue a stable research focus amid changing needs and demands. Yet it is precisely a *program of research* that can help build and sustain individuals, programs, and organizations.

In describing programs of research in this chapter, we draw from a rich tradition of varying definitions, from a sustained research enterprise with one or more components [1] to the development of a coherent group of research findings [2] to a series of connected studies that benefit the public welfare [3]. Drawing on these key ideas, we define a program of research as: *a purposeful strategy for pursuing a coherent and connected line of inquiry* [2, 3].

In this chapter, we begin by describing some of the key components necessary for pursuing a program of research. We next highlight select programs of research within healthcare simulation, highlighting differing ways in which a program of research may arise (e.g., personal or organizational interests, research collaboration) and how programs grow and change as they mature.

A. Battista (✉) · A. W. Konopasky
Graduate Programs in Health Professions Education, The Henry M. Jackson Foundation for the Advancement of Military Medicine, Bethesda, MD, USA
e-mail: alexis.battista.ctr@usuhs.edu; abigail.konopasky.ctr@usuhs.ed

M. H. Yoon
University of Colorado, Superior, CO, USA

D. Nestel et al. (eds.), *Healthcare Simulation Research*, https://doi.org/10.1007/978-3-030-26837-4_3

Table 3.1 Components of programs of research

Author	Year	Critical components
Sandelowski	1997	Careful researcher *planning*; theoretical *connection* among studies; goals related to broader *social good* [3]
Parse	2009	Discernable *patterns* in a researcher's line of inquiry [2]
Morse	2010	Large-scale programmatic *aim*; *self-contained* but *interconnected* projects [4]
SSHRC	2013	*Resources* (people and funding) to support quality work; *connections* across research communities; positive *impact* on society [1]
Taylor and Gibbs	2015	*Focus* on a real-world topic; formal and informal *support* and *collaboration*; institutional *resources* (e.g., library access, equipment, staff time); research team *training* [5]
Beck	2015	Systematic *planning*; addressing a *knowledge* gap that drives methods choices; *self-contained* studies that *build on* each other [6]

Key Components of Programs of Research

Across this body of literature, programs of research tend to have several core components, as Table 3.1 evidences: (a) a central focus and a flexible plan for pursuing that focus, (b) a team of researchers committed to the focus, (c) research methods for approaching questions related to the focus, and (d) a web of resources that supports the first three components. We touch on each component of the model below.

A central focus and flexible plan. What distinguishes a group of research projects in healthcare simulation from a program of research is a central area of focus. A central focus – on an assessment or treatment goal, on social needs or the social good, on a gap in the literature, on a new or poorly understood phenomenon, or on other real-world problems – is the main driver of a research program. For example, the National State Boards of Nursing program of research seeks to understand the use and role of simulation in pre-licensure nursing education. They first examined how schools of nursing utilized simulation and later considering whether simulation could be used in lieu of clinical time under specific circumstances without having a detrimental impact on board passage rates or readiness for transition to practice [7, 8].

Additionally, the plan for pursuing a focus within a program of research must be flexible. In order to reach program goals, team members must be ready to change plans when (not if!) the situation (e.g., funding, staffing, local program demands) changes. This flexibility is particularly important when pursuing a new area of research (or research on an existing topic in a new context, as is true of much simulation research), where unexpected findings may alter the original plan.

A team of researchers and practitioners committed to the focus. Programs of research are most often carried out by teams of researchers and practitioners. Frequently, these team members may not share the same approaches to research (e.g., quantitative versus qualitative versus mixed methods) and often have different professional training (e.g., clinician, psychologist, psychometrician) but they do have a shared commitment to the focus of the research. Often this allows research program leadership to broaden or strengthen the original team's networks, bringing in specialists with expertise in research methods, clinical practice, or simulation; or connecting with groups in other institutions. A clearly articulated focus for the program helps the team stay true to the larger goals while allowing for innovation and growth.

Methods for data collection and analysis appropriate for the focus. Which data to collect and how to collect and analyze it are all critical research design decisions. Teams often need to incorporate new methods in order to maintain their research focus, perhaps even developing new methodological or simulation tools. The relative novelty and flexibility of the simulation context allows teams to try out a variety of approaches to gathering and analyzing data (e.g., simulator outcome data, video analysis, written or oral assessments), but these choices must be made with the research focus in mind. For instance, if the focus is on improving team leadership skills during resuscitation efforts, an analysis of interactions among participants and clinical team members might be appropriate to determine which leadership skills individuals need to improve; however, future efforts to examine if a newly designed intervention improves those leadership skills may be better measured by using an Objective Structured Clinical Exam (OSCE).

Growing Web of Resources

Developing a program of research is an *emergent* process, meaning that, while research teams do make plans for upcoming studies, these plans change as findings from each successive study are considered and resources shift. Thus, the key components of a program of research are supported by an ever-growing web of resources: training and available time of team members, space and materials, access to technology, funding internal and external to the institution, professional organizations in research and simulation, and community connections. The model in Fig. 3.1 emphasizes the interconnectedness of the focus and plan, team of researchers, and research methods, all supported by a web of resources that help researchers carry out their efforts.

Building the infrastructure of that web is critical to the long-term success of a program of research in simulation. Early on, this may mean a loosely connected group of self-contained projects across different institutions with the same focus. These individual studies will most likely draw mainly

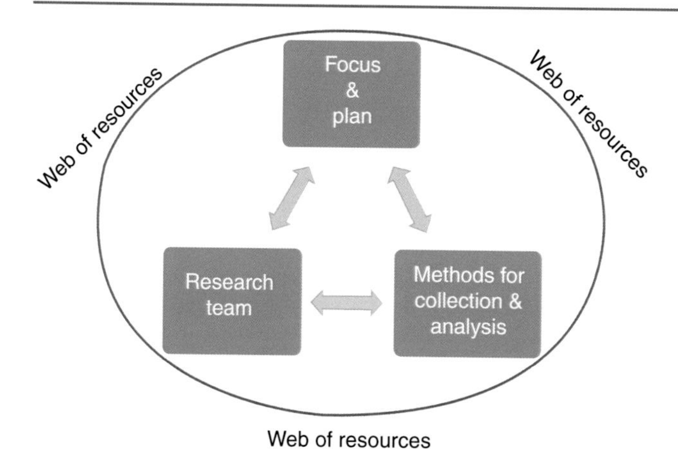

Fig. 3.1 A model for creating programs of research in simulation

on the resources at their local institutions and shared resources in regional and local organizations. As programs grow – and, with effort, time, and luck receiving funding – the infrastructure may formalize or centralize so that study teams are working together in one or two institutions or organizations. At this stage, institutions may become more actively involved, perhaps promoting the focus of the project as one of their core missions. Wherever a program of research stands, team members must consider what level of research (number, size, and type of studies and how interconnected they are) is *sustainable* given the available resources.

In addition to developing a web of resources, programs of research are *reflexive*, meaning they are also responsive to numerous driving forces that further shape future research efforts. These driving forces can range from the long-time research interests of individual investigators to the needs of institutions to the commitments of accreditation bodies. The examples of programs of research in simulation below highlight this range. Simulation researchers like Hunt, Draycott, and Brydges, all of whose research is discussed below, are deeply committed to the work as individuals, but they draw on other sources like accreditation bodies' desire for high-quality and safe educational opportunities, local organizations' needs for improving the quality and safety of patient care, and a growing community of researchers seeking to explore the unique opportunities presented by the simulation context. Recognizing – and drawing from – these driving forces can help simulation researchers formulate and grow a sustainable program of research.

Programs of Research in Healthcare Simulation

Simulation-based research (SBR) offers numerous examples of programs of research with the above components: a focus shared by a diverse team that flexibly draws from a variety of methods and is supported by a web of resources to address real-world clinical issues.

For example, Hunt sought to improve healthcare provider performance and management of pediatric resuscitation events (e.g., cardio-pulmonary, trauma resuscitation) in the clinical setting. To achieve this larger goal, Hunt and her team conducted a series of interconnected studies utilizing simulations to study healthcare professionals' behaviors and actions [9, 10]. As Hunt and colleagues' research program evolved, they also used simulation as an educational strategy to improve resident management of cardiopulmonary arrest [11, 12]. Hunt and colleagues have also employed simulations to develop, test and refine evaluation and assessment tools used for studying resuscitation events based in the clinical setting (Personal Communication with E. Hunt, 2018).

Over time, as Hunt and colleagues' research program matured, their efforts played a contributing role in the formation of the International Network for Simulation-based Pediatric Innovation, Research and Education (INSPIRE) research program, discussed later in this chapter. According to Cheng and colleagues, by forming the INSPIRE collaborative, the research team was enhanced by researchers across diverse fields, such as human factors engineering [13]. Additionally, by forming INSPIRE, their web of resources was enhanced, including "building capacity for the acquisition of grant funding and maintenance of multiple ongoing projects" [13].

In another example, Draycott's program of research seeks to improve multidisciplinary teams' care for mothers and newborns – a real world problem! Towards this focus, Draycott's efforts include a series of studies that build on each other, including those that describe the development and implementation of simulation-based learning activities, improvements in simulator design and the development of a dashboard used to track the impact of training on patient care. For example, in the late 1990s Draycott noted that there were few training programs that could easily accommodate multi-professional teams learning about responding to obstetric emergency situations (e.g., midwives, doctors, ancillary staff) [14]. Given this, Draycott and colleagues developed and implemented courses that included 'fire drills' to improve response to preeclampsia [14]. They further realized and developed a simulator that could support the training needs of multidisciplinary teams that could also provide force feedback measures, such as delivery force [15]. Subsequently, Draycott and colleagues also sought to measure and evaluate the impact of their training programs on the outcomes of mothers and infants in the clinical setting [16].

Another program of research highlighting a discernable pattern of research efforts is Brydges' program of research focusing on exploring how the healthcare professional's behaviors are influenced by training activities. To achieve this goal, Brydges and colleagues conduct studies that examine how individuals manage and direct their learning and strategies for optimizing the simulation-based practice environment. Brydges and colleagues' studies are methodologically diverse and include systematic reviews examining the efficacy of simulation-based instructional design [17] and qualitative, quantitative and mixed-methods studies. Furthermore, many of these studies are theoretically connected, often drawing from the social cognitive theory of self-regulated learning theory [18] to examine effective ways to structure clinical skills practice [19, 20].

Although these examples represent selected programs of research in healthcare simulation, they exemplify many of the key characteristics outlined earlier in this chapter, including a focus on real-world problems, being goal oriented rather than methodologically focused, representing diverse research teams, and drawing in networks of resources to continue and expand the work. Additionally, although these examples demonstrate mature programs of research they also highlight how an individual's own research interests can evolve and grow over time.

Contributions of research programs and priorities guided by accrediting agencies. In addition to local and historical factors, accrediting agencies and bodies also direct and influence programs of research. For example, The National Council of State Boards of Nursing (NCSBN) conducted a series of studies aimed at developing guidelines and policy for the use of simulation in nursing education in the United States. The first phase of this program of research initially examined how nursing schools were using simulation through a survey completed by 1060 pre-licensure nursing programs in the United States [7]. The findings from this survey led to a second phase which included a longitudinal randomized controlled trial to determine if simulations and simulation-based learning (SBL) could replace 25–50% of clinical rotations, while not having a detrimental effect on commonly used outcome measures (e.g., knowledge assessments, clinical competency ratings, board pass rates) [8]. The third phase followed student participants as they transitioned to the workplace to determine the longer-term impact of substituting simulations for clinical time. This effort resulted in regulatory recommendations for the use of simulation in lieu of clinical rotations and guidelines for developing, implementing and supporting high-quality simulation for nursing education [21].

Contributions of research programs and priorities set by research consortiums and collaboratives. Programs of research have also been constructed through the formation of research consortia and collaboratives. For example, the International Network for Simulation-based Pediatric Innovation, Research and Education (INSPIRE) was formed in 2011 to facilitate multicenter, collaborative simulation-based research with the aim of developing a community of practice for simulation researchers; as of 2017 it has 268-member organizations and 688 multidisciplinary individual members worldwide [13]. In addition to supporting and providing guidance for research priorities, the group also provides support for members through meetings, conferences and mentoring to name a few.

Conclusions

In this chapter we have described several key components of programs of research (i.e., planning around a central focus, a committed team, flexible and emergent methods, and a web of resources) and provided examples of programs of research within the field of healthcare simulation, including some that are coordinated through collaboratives or professional organizations. We have also discussed how these select programs of research have evolved and matured over time, highlighting how programs of research can be viewed on a continuum from their early stages to maturity. In the chapters that follow, this text will help you take the next steps in developing your own program of research (see, for example, Chap. 4, Choosing your Research Topic), help you explore diverse research methods (i.e., qualitative, quantitative, mixed-methods) that can help you achieve your research goals, and offer strategies for conducting multi-site studies (Chap. 39).

References

1. Social Sciences and Humanities Research Council of Canada. Framing our direction [Internet]. Available from: http://www.sshrc-crsh.gc.ca/about-au_sujet/publications/framing_our_direction_e.pdf. Cited 29 Dec 2017.
2. Rizzo Parse R. Knowledge development and programs of research. Nurs Sci Q [Internet]. 2009;22(1):5–6. Available from: http://journals.sagepub.com/doi/10.1177/0894318408327291. Cited 14 May 2018.
3. Sandelowski MJ. Completing a qualitative project [Internet]. Janice M. Morse, editor. Thousand Oaks: SAGE Publications; 1997. p. 211–26. Cited 14 May 2018.
4. Morse JM. What happened to research programs? Qual Health Res [Internet]. 2010;20(2):147. Available from: http://journals.sagepub.com/doi/10.1177/1049732309356288. Cited 14 May 2018.
5. Cleland J, Durning SJ. Researching medical education. John Wiley & Sons; 2015.
6. Tatano Beck C. Developing a program of research in nursing [Internet]. 1st ed: Springer Publishing Company; 2015. Available from: http://www.springerpub.com/developing-a-program-of-research-in-nursing.html. Cited 14 May 2018.
7. Hayden J. Use of simulation in nursing education: national survey results. J Nurs Regul [Internet]. 2010;1(3):52–7. Available from: https://www.sciencedirect.com/science/article/pii/S2155825615303355. Cited 14 May 2018.
8. Hayden J, Smiley R,... MA-J of N, 2014 undefined. Supplement: the NCSBN national simulation study: a longitudinal, randomized, controlled study replacing clinical hours with simulation in preli-

censure. mtcahn.org [Internet]; Available from: http://mtcahn.org/wp-content/uploads/2015/12/JNR_Simulation_Supplement-2015.pdf. Cited 14 May 2018.
9. Shilkofski NA, Nelson KL, Hunt EA. Recognition and treatment of unstable supraventricular Tachycardia by pediatric residents in a simulation scenario. Simul Healthc J Soc Simul Healthc [Internet]. 2008;3(1):4–9. Available from: https://insights.ovid.com/crossref?an=01266021-200800310-00002. Cited 19 May 2018.
10. Hunt E, Walker A, Shaffner D, Pediatrics MM. Simulation of in-hospital pediatric medical emergencies and cardiopulmonary arrests: highlighting the importance of the first 5 minutes. Am Acad Pediatr [Internet]. 2008 undefined. Available from: http://pediatrics.aappublications.org/content/121/1/e34.short. Cited 14 May 2018.
11. Nelson KL, Shilkofski NA, Haggerty JA, Saliski M, Hunt EA. The use of cognitive aids during simulated pediatric cardiopulmonary arrests. Simul Healthc J Soc Simul Healthc [Internet]. 2008;3(3):138–45. Available from: https://insights.ovid.com/crossref?an=01266021-200800330-00003. Cited 14 May 2018.
12. Hunt EA, Duval-Arnould JM, Nelson-McMillan KL, Bradshaw JH, Diener-West M, Perretta JS, et al. Pediatric resident resuscitation skills improve after "rapid cycle deliberate practice" training. Theatr Res Int. 2014;85(7):945–51. Available from: http://www.ncbi.nlm.nih.gov/pubmed/24607871. Cited 14 May 2018.
13. Cheng A, Auerbach M, Calhoun A, Mackinnon R, Chang TP, Nadkarni V, et al. Building a community of practice for researchers: the international network for simulation-based pediatric innovation, research and education. Simul Healthc [Internet]. 2017. Available from: http://www.ncbi.nlm.nih.gov/pubmed/29117090. Cited 14 May 2018.
14. Draycott T, Broad G, Chidley K. The development of an eclampsia box and fire drill. Br J Midwifery [Internet]. 2000;8(1):26–30. Available from: http://www.magonlinelibrary.com/doi/10.12968/bjom.2000.8.1.8195. Cited 14 May 2018.
15. Crofts JF, Attilakos G, Read M, Sibanda T, Draycott TJ. Shoulder dystocia training using a new birth training mannequin. BJOG An Int J Obstet Gynaecol [Internet]. 2005;112(7):997–9. Available from: http://doi.wiley.com/10.1111/j.1471-0528.2005.00559.x. Cited 14 May 2018.
16. Draycott TJ, Crofts JF, Ash JP, Wilson LV, Yard E, Sibanda T, et al. Improving neonatal outcome through practical shoulder dystocia training. Obstet Gynecol [Internet]. 2008;112(1):14–20. Available from: http://content.wkhealth.com/linkback/openurl?sid=WKPTLP:landingpage&an=00006250-200807000-00006. Cited 14 May 2018.
17. Cook DA, Hamstra SJ, Brydges R, Zendejas B, Szostek JH, Wang AT, et al. Comparative effectiveness of instructional design features in simulation-based education: systematic review and meta-analysis. Med Teach [Internet]. 2013;35(1):e867–98. Available from: http://www.tandfonline.com/doi/full/10.3109/0142159X.2012.714886. Cited 14 May 2018.
18. Brydges R, Manzone J, Shanks D, Hatala R, Hamstra SJ, Zendejas B, et al. Self-regulated learning in simulation-based training: a systematic review and meta-analysis. Med Educ [Internet]. 2015;49(4):368–78. Available from: http://doi.wiley.com/10.1111/medu.12649. Cited 14 May 2018.
19. Brydges R, Carnahan H, Safir O, Dubrowski A. How effective is self-guided learning of clinical technical skills? It's all about process. Med Educ [Internet]. 2009;43(6):507–15. Available from: http://doi.wiley.com/10.1111/j.1365-2923.2009.03329.x. Cited 14 May 2018.
20. Brydges R, Nair P, Ma I, Shanks D, Hatala R. Directed self-regulated learning versus instructor-regulated learning in simulation training. Med Educ [Internet]. 2012;46(7):648–56. Available from: http://doi.wiley.com/10.1111/j.1365-2923.2012.04268.x. Cited 14 May 2018.
21. Alexander M, Durham CF, Hooper JI, Jeffries PR, Goldman N, Kardong-Edgren S"S", et al. NCSBN simulation guidelines for prelicensure nursing programs. J Nurs Regul [Internet]. 2015;6(3):39–42. Available from: https://www.sciencedirect.com/science/article/pii/S2155825615307833. Cited 14 May 2018.

Starting Your Research Project: From Problem to Theory to Question

4

Jeffrey J. H. Cheung, Tavis Apramian, and Ryan Brydges

Overview
This chapter represents a guide for how to start a meaningful research project in simulation-based education. Rather than a checklist for conducting research, we introduce how to begin your foray into research, and how to think like a principal investigator. Conducting high quality research takes more time and assistance than most expect. For novice researchers reading this chapter, we suggest you first aim to produce impactful research with the help of your collaborators, as a co-investigator, before taking on the role of principal investigator. We emphasize use of theory to guide your research project and to lay the foundations for a research career with longevity. We argue that focusing on theory will help distill your research problems into research questions that align with established methodologies and methods to provide meaningful answers. We believe strongly that using theory ensures findings from a single research project can transcend their original context, providing meaningful insights to researchers and educators writ large. We also acknowledge the challenges of using theory, and note the need to develop a strong, well-rounded research team with the requisite resources, time, and expertise (in theory and clinical education).

Practice Points

- Start with a problem you are curious about, identify gaps in the literature, and find a hook that articulates the importance of filling this gap.
- Think about conducting research that goes beyond filling local educational gaps, and allows you to join existing conversations rather than starting your own.
- Develop a theoretical understanding to help transform your problem into a meaningful research question, the answer to which will be more likely to benefit you and other scholars.
- Once you choose your theory, the appropriate methodology and methods will follow from that theory's accompanying research tradition.
- Remember that research and science is a team sport. Find other researchers who have the expertise, time, and interest to help complete your project and build it into a research program.

Introduction

In academic healthcare, there is a constant pressure to produce – research articles, abstracts, simulation scenarios, curriculum plans, lesson plans, assessment tools and more. Most often, the aim is to fill gaps; gaps in curriculum, administration, assessment, and evaluation. "We don't have a ________ for ________, why don't you build one?" (Feel free to fill in the blanks with whatever may be relevant to your context; we're confident you can!) Though it is tempting to succumb to the pressures of production and to try to 'make a research project' out of a current educational proj-

J. J. H. Cheung (✉)
The Wilson Centre, Toronto General Hospital, University Health Network and University of Toronto, Toronto, ON, Canada

Department of Medical Education, University of Illinois at Chicago College of Medicine, Chicago, IL, USA

T. Apramian
Department of Family Medicine, McMaster University, Hamilton, ON, Canada
e-mail: tavis.apramian@schulich.uwo.ca

R. Brydges
The Wilson Centre, Toronto General Hospital, University Health Network and University of Toronto, Toronto, ON, Canada

The Allan Waters Family Simulation Centre, St. Michael's Hospital – Unity Health Toronto, Toronto, ON, Canada

Department of Medicine, University of Toronto, Toronto, ON, Canada
e-mail: ryan.brydges@utoronto.ca

D. Nestel et al. (eds.), *Healthcare Simulation Research*, https://doi.org/10.1007/978-3-030-26837-4_4

ect, we implore you to resist this urge and think instead about conducting quality research. Production for production's sake does not make for a quality research project. Helping to fill your institution's local gaps is important work, yet it does not always translate well into meaningful evidence for others outside your specific context. We argue that meaningful research requires more than filling gaps, and can only be achieved when theory is woven into the fabric of the research process from start to finish.

We wrote this chapter as a guide for novice and experienced researchers starting a research project in the field of healthcare simulation. To begin, we present the problem-gap-hook heuristic [1], as a structure for thinking through the process of telling and uncovering one's research story. We then focus on how theory can guide us in starting, refining, and following-up on research projects. We end with pragmatic considerations when conducting research using healthcare simulation.

A Problem, a Gap, and a Hook

The Problem-Gap-Hook heuristic is a writing tool designed to refine arguments into a compelling narrative format [1]. Many research projects begin with a *problem*. We encourage you to think about something that makes you curious, or irks you, something that typically arises from seeing patients receiving care, or from seeing learners acquiring clinical skills. Identifying your problem requires observation and reflection. An assessor may observe that trainee performance in a simulation decreases when the assessment stakes are high. A graduate student may notice few trainees are using the 24-hour simulation room. An educator may encounter resistance to feedback following a simulation session. Finding a personally meaningful research problem usually opens a dam of downstream questions: Why does this problem exist? Has someone solved this problem before? Is this a solvable problem?

With your initial problem identified, you must then appraise the literature related to your problem. Have others noticed what you have noticed? Where else has this been studied? What is the language used to describe the problem? This appraisal helps establish the boundaries of current understanding, and uncover any *gaps* in how the problem has been studied. To ensure this phase is comprehensive, we recommend working with librarians to help refine your search strategies, and with informed colleagues to help identify key researchers and articles.

Not all problems are good research problems; good research problems need a *meaningful* gap—they need a *hook*. The rationale that "no one has studied this before in this context" is necessary, but not sufficient. A good hook addresses why the findings of the research matter, and why the gap needs to be filled. The hook relates the research problem and the potential findings to peoples' lives through greater insights about an issue, patient outcomes, potential educational policy changes, social costs, or other tangible impacts. You must consider how the findings of your research will hook into meaningful conversations in the literature.

Like all heuristics, "problem-gap-hook" is useful but limited. The framing it provides can help you overcome the inertia accompanying the start of any research project. Beyond framing, you'll need a rich body of knowledge and understanding that comes only with time and reflection. For those looking to act as principal investigators on a research project, this is essential. For novice researchers, we recommend collaborating with more knowledgeable peers and research mentors who can help direct the project and also provide an opportunity for you to build up knowledge and understanding of the research area. Next, we focus on the importance of theory for shaping and refining research problems, and for helping establish lines of inquiry with longevity beyond your initial research project.

Using Theory to Refine Your Problem

What Is Theory?

Theory can be thought of as a way of viewing the world. Theory can illuminate gaps in understanding and help organize our thoughts and actions according to predictions or explanations of how the world seems to work. There are various schools of thought about what constitutes a theory. These approaches to theory stem from the various perspectives on the nature of knowledge and what can and cannot be known (e.g., epistemology and ontology). We will not discuss these nuances in this chapter. Instead, we highlight the value of theory for scientific inquiry in healthcare education, the importance of using theory from the very beginning of your research project, and the investments required to develop your theoretical lens. Beyond our introduction here, other articles provide overviews of theory and its value in health professions education research [2–6]. See also Chap. 23.

Why Use Theory?

Through theory, your initial observations and reflections can be distilled into a meaningful problem that demands resolution, a gap yes, but more specifically, a research question. Reviewing what the literature says about your problem will provide some guidance, but on its own is insufficient. For example, you might want to study the problem of delivering effective feedback during a simulation debrief, however, such a problem can produce an infinite number of potential gaps and research questions. Homing in on a specific and impactful research question requires recognizing that there may be more to your problem than you considered at first glance. Within the feedback and simulation debriefing processes lies a world of phenomena, or constructs, that transcend your specific setting and observations.

Actively searching for your theory will reveal ways to approach your question from disciplinary perspectives you might otherwise miss. Exploring different theoretical lenses will expose you to numerous 'constructs' you might study. For example, viewing the debriefing example with a cognitive psychology theory lens, you might be inclined to examine how feedback impacts learning at the level of trainees' cognitive processes, such as memory, decision-making, or emotion. Depending on the specific theory you choose, you might then consider how aspects of an instructor's feedback (e.g., credibility, experience level, relationship to the trainee) impact those cognitive processes, and ultimately, trainees' learning. In any of these possible scenarios, theory guides your interpretation and study of a problem by providing a structure and language to identify, define, and operationalize constructs. In doing so, theory connects your problem to established hypotheses and philosophies, making constructs visible and providing lines of inquiry that can inform research and understanding beyond the confines of your specific problem context. These features make theory the most powerful tool in a researcher's toolkit!

Further, theory helps tie your research findings to foundations that both you and other researchers can build from; a common set of problems, gaps and hooks. When communicating your work, theory also provides a common language for conversation. Too often, atheoretical research projects create disarray in the literature. Authors may be speaking into the ether, unaware of how research findings from their specific context may tie to another researcher's findings from another context. Viewed in this light, theory is a unifying force for scientific communication and the collective building of new knowledge.

Choosing a Theory

For the uninitiated, choosing a theory to guide your research can be the most challenging aspect of research, as there is no textbook listing the appropriate theories for your research problem. You will need to invest time and hard work initially, which will pay dividends, saving you time and headaches in the future. A good starting point is your literature review about your problem, where you can note the theories other authors used to illuminate and fill previous gaps. These theories may be explicit, and cited within the text, or they may be implicit, requiring you to trawl through the sea bed of references or even read between the lines.

Hodges and Kuper [5] provide some tips for choosing a theory, prompting researchers to consider the scope and school of a theory. For scope, they suggest thinking about whether you want to study individuals' attitudes, thoughts or behaviours (i.e., micro level), the interactions between two or more people (i.e., meso level), or social structures and institutions (i.e., macro level). For school, they prompt thinking about how your chosen scope might best be studied. Studying the micro level might require 'bioscience theories' (e.g., neurophysiological theory to study stress responses), the meso level might require 'learning theories' (e.g., situated learning theory to study how healthcare professionals learn through collaboration), and the macro level might require 'sociocultural theories' (e.g., critical theory to understand how power hierarchies affect workplace based learning). Though the distinctions between these clusters of theory can and do overlap, thinking about the scope and school of your problem will help frame your research project by characterizing your theory, and narrowing your choice of theories substantively.

Let's work with the problem of the underused 24-hour simulation room for learning bedside procedures. To begin, you may read about simulation-based procedural skills training, independent practice, and learning outcomes of individuals, situating your scope at the micro level. You would find several articles describing the shortcomings of self-learning, arguing for the need of instructor guidance. By sifting through the reference sections, you find the roots of the argument come from the field of educational psychology, situating your potential choice of theory within the school of what Hodges and Kuper [5] labeled bioscience theories. That literature describes the challenges arising for learners without 'direct instruction' (i.e., education organized and delivered by an instructor who knows the material), and thus provides specific hypotheses for why self-learning may yield poorer learning outcomes. However, your search would also find researchers using terms like "self-regulated learning" and "discovery learning" to argue that self-learning can be effective, when designed well. A-ha! Perhaps all is not lost, perhaps you can incorporate these educational principles into the design of your simulation room to encourage more effective learning outcomes. Here your problem has shifted, you now have a rich evidence base and set of theories to help you think about optimizing the learning outcomes of your trainees' independent learning. This example highlights the diverse ways of thinking about your problem you are likely to encounter when searching for "your" theory.

Building "your" Theoretical Lens

Like choosing your research problem, choosing your theory is often a matter of personal preference. The way you view and interpret your research problem will be shaped by your own personal theories of how the world works, which will influence the theories you find intriguing for studying the constructs underlying your problem. Those studying simulation-based procedural skills will likely find theories of motor skill acquisition attractive, while those studying the debriefing process may be drawn to theories of communication. There is no "right" or "wrong" theory, only the theory that most appropriately helps you breathe life into your ambitions to study your research problem. Often the theory that researchers choose is one they have inherited from their supervisors or trusted colleagues. But making that decision—and, more importantly, convincingly defending it—requires a broad survey and many attempts at finding a fit.

At times, the lenses through which you see, operate, and question the world can be implicit. Everyone operates through theory, often without knowing it, or thinking about it this way. Every day, people filter their experiences from the world through a theoretical lens crafted by their individual histories. Experience from the workplace often serves as a core to your personal theories. Personal theories however, can only take your research so far. For your research to connect with others, you will need to do the difficult work of excavating your implicit assumptions, and then contrasting your personal theories to established theories.

Though an essential part of producing meaningful research, a theory alone is inadequate. A deep and rich understanding of the clinical context and problems plaguing clinical training and education also requires investment in observation and reflection. The key challenge lies in coordinating and aligning the language between the world of theory, and your (or your colleagues') tacit clinical knowledge and experience. That intersection, which has been referred to as Pasteur's Quadrant [7], is where researchers work to optimize the theoretical and applied contributions of their work. This use-inspired approach to basic research aims to yield practical findings that also contribute to fundamental understanding (e.g., Pasteur's discovery of pasteurization advanced fundamental understanding of microbiology). Such research contrasts with pure basic research pursuing only fundamental understanding, and pure applied research pursuing only utility. We believe research aiming for Pasteur's Quadrant has great potential for advancing both theoretical understanding and pragmatic needs for healthcare education, simulation, and professional practice.

From Problem to Question and Back Again

A look through your new theory view-finder will help translate your problem into a research question, and it may also encourage you to reimagine the problem altogether. What may have originally struck you as a problem of poor self-learning in the case of 24-hour simulation rooms might be an issue of faculty not role-modeling the behaviour for their trainees. Further theoretical exploration may highlight the potential benefits of pairing trainees with their peers to capitalize on the benefits of peer-assisted learning. You might be prompted to consider how trainees form their professional identity, and how simulation training might be designed to affect that process. These possibilities arise when you think about the world using theory. Hence, as you consider multiple theories, you will likely experience discomfort and uncertainty through this necessarily iterative process.

Refining Your Problem into a Question

Informed by theory, and having chosen your problem, you are poised to articulate a research question. Researchers with different theoretical lenses will arrive at different research questions even when exploring similar problem contexts. For example, informed by the importance of feedback in the acquisition of psychomotor skills, a researcher may ask "what is the relationship between the level of supervision available (i.e., how much feedback students receive), and learning of invasive procedures in a 24-hour simulation room?"—a question that is ripe for an experimental study design. Alternatively, informed by theories of learner autonomy and professional identity, a second researcher may ask "how do trainees conceptualize independent learning when using a 24-hour simulation room?"—a question ripe for qualitative inquiry. The corresponding methods and formatting for both questions depends on the theory's research tradition. For a summary of how different research traditions are associated with different research questions, see the guide from Ringsted et al [8].

Unfortunately, there are no hard and fast rules for what constitutes a 'good' research question. More important than following guidelines for asking a research question, we emphasize focusing on how your project contributes understanding to the theoretical undercurrent of constructs and phenomena. Put another way, be sure to always ask yourself whether your research question addresses a generalizable problem with value beyond your own local problem context (am I joining and contributing to a conversation, or am I speaking to myself?).

Theory Before Methods

When your question becomes more specific and concrete, the methodology and methods to address it will begin falling into place. Always choose your theory, and format your question, before you choose your methods. Using methods without theory, you run the risk of producing meaningless data, or at best, data you will find challenging to interpret. Some exploratory studies are needed to help sensitize understanding, but we have seen many colleagues become lost in a sea of unmoored data. Indeed, studies using research methods without theoretical grounding often end up as islands without theory to map where they lie with respect to other work. For example, you may find success at your institution by developing a means to improve use of the 24-hour simulation room, but if your method for doing so does not have a theoretical frame it becomes challenging to identify the ingredients of the intervention that led to your improved outcome, and others will likely find it difficult to replicate the same effects at their institutions. Did your intervention work because your instructors had great rapport with the trainees? Because your institutional culture prizes self-regulated learning? Perhaps the blueprinting of your local curriculum played a role? Or maybe the proximity of the simulation room to where trainees had their meal breaks? These factors cannot be accounted for without theory. We strongly recommend you avoid building islands: plan your course ahead of your methods by building a theoretically grounded question. We summarize our views on the research journey with and without theory in Fig. 4.1.

Thinking Programmatically

Your theoretical lens also helps set you on course to lead a productive program of research. A common experience in the early stages of a researcher's career involves conducting studies to answer disconnected problems specific to one's local context, which produces several disconnected islands. By contrast, programmatic research involves iterative flow from problem to theory to question. You use theory to shape your problem into an initial question, which also shapes your research project, methodology, and methods. Once you complete your project, you will use theory to interpret your results, and to yield more research questions. These research questions then must be contextualized by relating them to the original problem in a meaningful way (the *hook*, i.e., why does this matter?). The previous chapter by Battista et al. elaborates on programs of research and provide examples using researchers in healthcare simulation.

Pragmatics of Starting Your Research Project

As we have noted repeatedly, you must commit to reading widely across the healthcare simulation, health professions education, and disciplinary (e.g., educational psychology, social sciences) literatures. Such searching and reading can feel unproductive at times, so you will need to find informed colleagues and effective collaborators to bounce your ideas off, for guidance and feedback. To find them, we suggest attending courses on research skills offered at your local institutions, national organizations, or at international conferences (e.g., the Research Advanced Skills in Medical Education offered at the Association for Medical Education in Europe).

A more structured, and time-consuming, approach is to pursue formal graduate training. Such intensive training optimizes your time to read, think, question, and develop your theoretical lens within a community of supportive and knowledgeable scholars. Encountering many theories will teach you the blind spots of different lenses, and help you select the most appropriate theories to guide your project, and shape your research question. This theoretical nimbleness comes with time, and by some accounts, requires the ambiguity and 'productive wallowing' that only specialized training provides.

Either through your colleagues, or formal training, pay close attention to the common language used in your chosen area: What terms and theories are others using? Using appropriate language will help you join a conversation, and avoid contributing to the currently splintered healthcare simulation research literature. Further, by becoming nimble with your language, you can modify your *hook* for different audiences, which helps when presenting your work at conferences, meetings, and funding agencies.

Before starting your project, conduct a full audit to assess the resources and skills required to complete it successfully. Here, you must ask yourself tough questions like: Who will coordinate the project and keep it moving forward? How will the data be collected, and by whom? How will you access the

Research Project *with* Theory

Read broadly to identify gaps in literature about your problem and how various theories attempt to fill these gaps

Using theory, refine your problem into a question informed by the methodology and methods associated with your theory

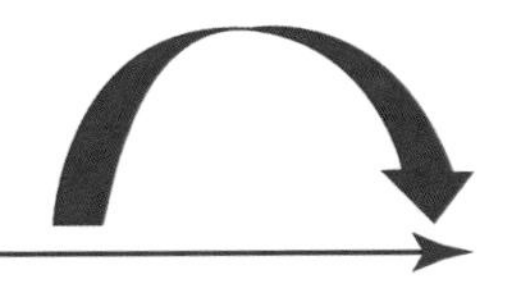

Problem

Observation or challenge of personal interest

Research Question

Theory-grounded question with a clear methodology and method that follow the theory's research tradition

Continue reading about specific theories to build "your" theoretical lens

Potential Outcomes

- You will join a conversation with others who have tried to address similar problems to your own. Your data and results will contribute to a general understanding that will help other researchers and educators in different contexts.
- With theory weaved throughout your research project, your findings will highlight new avenues ripe for research.

Research Project *without* Theory

Without a literature review, you may miss other researchers' answers to your problem

Without a theoretical lens your are gambling that your methodology and method will produce data from which meaningful conclusions can be drawn – "Let's do a study and see what happens"

Problem

Research Question

Observation or challenge of personal interest

Without investing in building your theoretical lens, you will be challenged in formulating a question, designing and executing a study, interpreting the data, and understanding where to go next

"Intuitive leap" to fill a locally specific gap in understanding or educational programming

Potential Outcomes

- You may end up only speaking to yourself. Your data and results may not contribute to a general understanding that helps other researchers and educators in different contexts.
- Even adding theory post-hoc, your interpretation of your results is speculative at best, especially if the data gathered are inappropriate for the scope and school of theory you applied.

Fig. 4.1 The research journey with and without theory

simulators, students, staff? Do you need funding for equipment, software, or to remunerate raters, statisticians, and participants? Does the project require ethics approval? And so on. Choose a project lead with the time, interest, and skills to make it happen, a step that is often over-looked, or assumed to be obvious.

Another key area to focus on is ethical research practice, including obtaining research ethics approval and everyday ethics of research. At an early stage, we recommend reviewing authorship requirements and discussing each team member's roles, responsibilities, and promised position for authorship, which helps to avoid unpleasant conversations later. Most academic centres will have further resources (often available online and through institutionally mandated training) that provide more detailed guidance on the matter of research ethics.

Conclusion

We do not know of a sure-fire recipe for a high quality research project, though we do advise that you seek out problems nagging you to constantly ask this simple question: Why? Our experience as researchers in healthcare simulation prompts us to note that a major challenge facing our community (and the broader field of health professions education research) is a lack of theory-oriented research. Unfortunately, a minority of articles in the simulation literature involve authors engaging constructs theoretically and programmatically, beyond their specific simulator, clinical skill, setting, or geography. We hope you feel motivated to help improve the quality, integrity, and rigour of research in healthcare simulation. Remember, theory comes first, and with it your research question, methodology, and finally the methods for your specific project. We believe strongly in the power of reading widely, and in allowing ourselves to be curious, initially promiscuous with theory, intellectually adventurous, and academically collaborative. Temper your enthusiasm to "just publish", with a well-planned and well-executed contribution to our community. When bogged down deep in the trenches of a literature review or venturing down a seemingly endless rabbit hole of theory, try to remember these words of wisdom: *A well-planned project will feel like it takes twice as long to complete, while a poorly planned project will actually take three times as long.*

References

1. Lingard L. The writer's craft. Perspect Med Educ. 2015;4(2):79–80. https://doi.org/10.1007/s40037-015-0176-x.
2. Bearman M, Nestel D, McNaughton N. Theories informing healthcare simulation practice. Healthcare Simul Educ. 2017; https://doi.org/10.1002/9781119061656.ch2.
3. Bordage G. Conceptual frameworks to illuminate and magnify. Med Educ. 2009;43(4):312–9. https://doi.org/10.1111/j.1365-2923.2009.03295.x.
4. Bordage G, Lineberry M, Yudkowsky R. Conceptual frameworks to guide research and development in health professions education. Acad Med. 2016;1 https://doi.org/10.1097/ACM.0000000000001409.
5. Hodges BD, Kuper A. Theory and practice in the design and conduct of graduate medical education. Acad Med. 2012;87(1):25–33. https://doi.org/10.1097/ACM.0b013e318238e069.
6. Nestel D, Bearman M. Theory and simulation-based education: definitions, worldviews and applications. Clin Simul Nurs. 2015;11(8):349–54. https://doi.org/10.1016/j.ecns.2015.05.013.
7. Stokes DE. Pasteur's quadrant: basic science and technological innovation. Washington, DC: Brookings Institution Press; 1997.
8. Ringsted C, Hodges B, Scherpbier A. 'The research compass': an introduction to research in medical education. AMEE guide no. 56. Med Teach. 2011;33(9):695–709.

Overview of Serious Gaming and Virtual Reality

5

Todd P. Chang, Joshua M. Sherman, and James M. Gerard

Overview
Serious Games and Virtual Reality (VR) have been accelerating in their quality and ubiquity within healthcare simulation, and the variety of technological innovations is outpacing the healthcare research community's ability to evaluate their effects as an intervention or their utility in simulating an environment for research. This chapter seeks to highlight unique advantages and challenges when using serious games or VR for healthcare research that are different than those encountered with other simulation modalities such as manikins, simulated/standardized patients etc. First, we define the terminology surrounding the concept of serious games and VR, including the advantages and disadvantages for their utility in answering important healthcare research questions. Second, we provide insight into optimal models of research that are suited for serious games or VR. Finally, we describe the development process for researchers to integrate research methodologies during the development phase.

Practice Points

- Screen-based Simulation (SBS) consists of any digital simulation using a computer or mobile device screen or a virtual reality headset.
- Serious Games and VR have distinct advantages and disadvantages over manikin-based simulation for both development, implementation, and data collection for research.
- Data collection in serious games and VR must be built-in during the development process for the software.

T. P. Chang (✉)
Department of Pediatric Emergency Medicine, Children's Hospital Los Angeles/University of Southern California, Los Angeles, CA, USA
e-mail: tochang@chla.usc.edu

J. M. Sherman
Regional Medical Director, PM Pediatrics, Los Angeles, CA, USA

J. M. Gerard
Department of Pediatric Emergency Medicine, SSM Health Cardinal Glennon Children's Hospital/St. Louis University, St. Louis, MO, USA
e-mail: gerardjm@slu.edu

Introduction

While **serious games** can be defined in a variety of ways, they can be best described as games that educate, train, and inform, for purposes other than mere entertainment [1]. Serious games can be applied to a broad spectrum of application areas, (e.g., military, government, educational, corporate, and healthcare). Many attempts have been made at defining what constitutes games – to understand how they work to facilitate learning. Specific attributes define a simulation as a serious game, which include a taxonomy of concepts described by Bedwell et al.: assessment, conflict, control, environment, rules, goals, fantasy, and immersion [2]. Not all serious games require a screen or electricity, as board and card games that facilitate learning can also be a form of serious game.

Virtual Reality (VR) is an artificial reality which is experienced through sensory stimuli, such as sight and sound, provided by a computer in which one's actions determines what happens next in the environment. VR is constantly changing with the exponential growth of technology. In the past, VR described an environment or situation through the eyes of a computerized avatar such as a first-person video game. However, as hardware technology improved, types and opportunities for healthcare VR have also advanced. Of note, VR differs from Augmented Reality in which digital imagery, text, or characters are superimposed onto a display of an individual's real environment. This contrasts with VR's

D. Nestel et al. (eds.), *Healthcare Simulation Research*, https://doi.org/10.1007/978-3-030-26837-4_5

ability to shut out the real environment entirely to allow a completely immersive experience. Augmented Reality will not be discussed in the chapter.

Screen-based simulation (SBS) is a form of simulation in which one or more scenarios are presented through a digital screen surface [3]. This can include virtual patients, virtual worlds, and virtual trainers. The user interacts with a game selecting the next step or test from selection menus. As with other forms of simulation, SBS provides the user with a safe place for experiential learning and assessment. SBS includes serious games and VR, but not all SBS require game elements or game mechanics. Examples are shown in Figs. 5.1, 5.2 and 5.3.

3D VR or Head mounted VR refers to the use of a goggle/headset type device such as the Oculus Rift (Oculus VR, LLC, Menlo Park, CA), HTC Vive (HTC Corporation, Xindian City, Taipei), Gear VR (Samsung, Ridgefield Park, NJ), or Google Cardboard (Google, Mountainview, CA) to create a fully immersive 360° environment that substitutes one's audiovisual reality with a virtual environment. Many of these definitions can refer to the same product. A serious game may use a VR headset, though not all VR experiences are games. Examples of these VR simulations are shown in Figs. 5.3 and 5.4.

Advantages and Disadvantages

Screen-based simulation (SBS) has five main advantages over other forms of simulation; all of which can be useful to healthcare researchers. These advantages as noted in the literature are: standardization, portability, distribution, asynchrony, and data tracking [4, 5]. Because SBS is basically a predetermined computer algorithm; by definition, it is standardized for each user. Although modifications can be made to accommodate different levels of player expertise, each user at the same level will experience the same simulation with the same options. Portability and distribution are similar concepts. SBS can use mobile devices, tablets, laptops, or VR headsets, which are

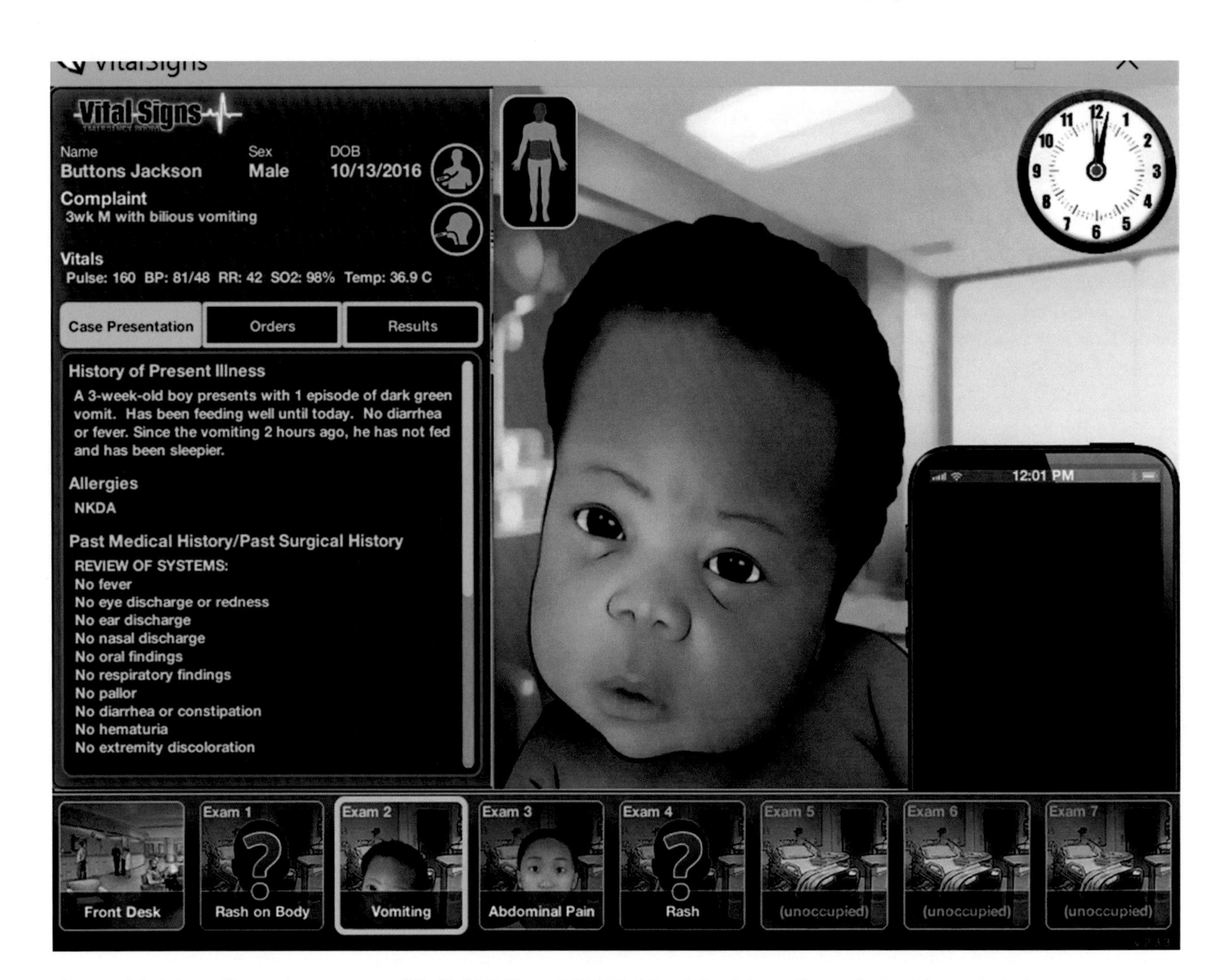

Fig. 5.1 Vital signs. (Screenshots courtesy of Dr. Todd P Chang, MD MAcM, and BreakAway Games, Ltd., with permission)

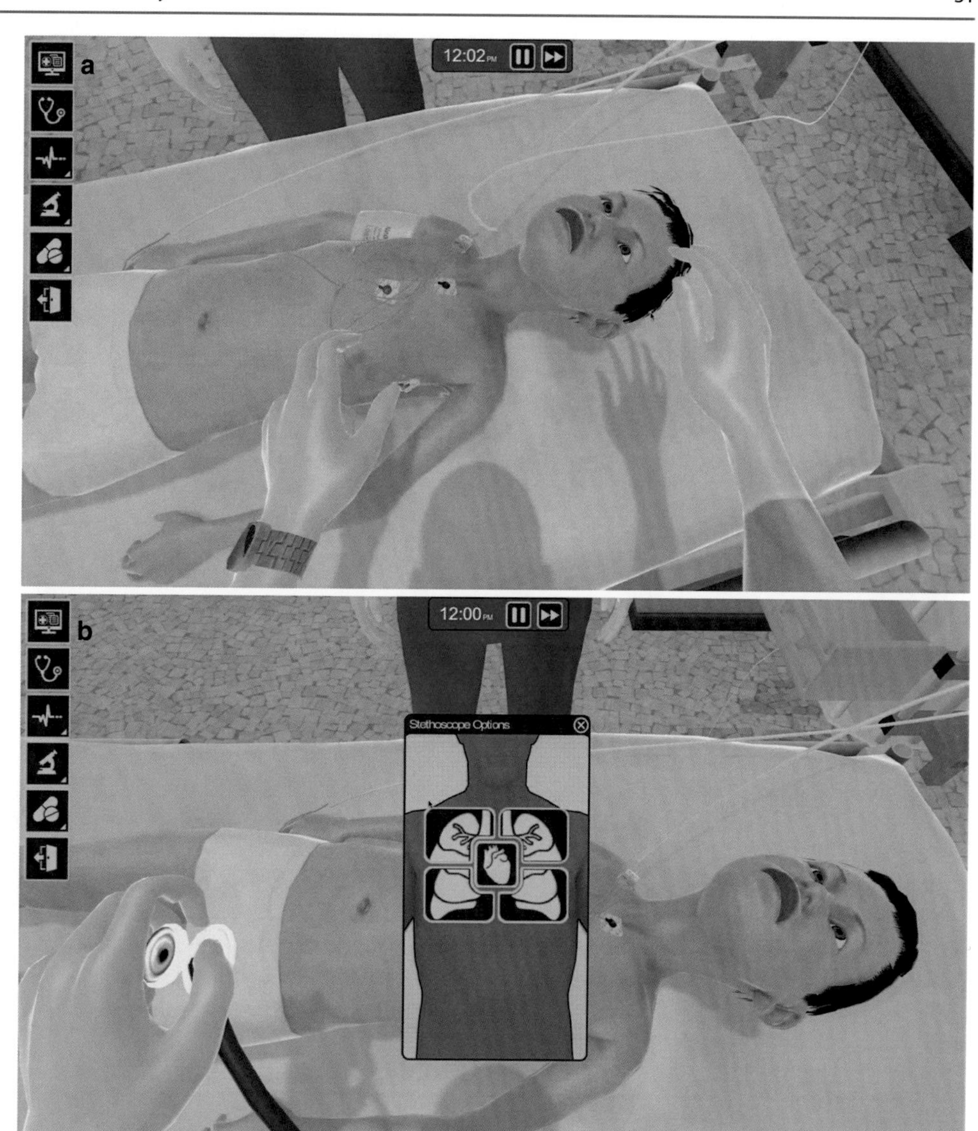

Fig. 5.2 (**a**, **b**) Pediatric resuscitation simulator. (Screenshots courtesy of Dr. James Gerard, MD, and BreakAway Games, Ltd., with permission)

common items. *Portability* refers to the ability to move the hardware or proprietary devices easily across healthcare arenas or institutions, resulting in reduced equipment and travel. Similarly, *distribution* is the ability of the software to be replicated or copied across hardware (such as a flash drive) and online, whether through a proprietary network or the world wide web. The combination of portability and distribution reduces many of the barriers in conducting multi-center trials, for example, when compared to manikin-based simulation. Both portability and distribution allow for *asynchrony*, which is the use of the simulation without a facilitator or instructor immediately present. As an example, manikin-based simulation (MBS) requires a technician, confederates, and/or a facilitator for debriefing at the time of the simulation, which would be considered *synchronous*. While facilitator-led debriefing is standard and common in MBS, there is no such standard with SBS, because it can be completed asynchronously on one's own. There may be a benefit to having a facilitator or briefer/debriefer available synchronously (live) either physically next to the subject, or communicating to the subject remotely. Alternatively, the debriefing can happen at a different time and location, so the users can practice and improve at their own pace and when most convenient for them. Of note, a fundamental question for SBS, serious games, and VR, is the optimal structure and format for debriefing in this relatively new modality of simulation.

SBS also has built-in *data tracking*; all user actions - whether input through a keyboard, mouse, controller, or VR head movement - are documented by the software with very precise timestamps. The variety of data tracked can be massive, and the researcher is advised to pick out the most meaningful data to answer their research questions, rather than to request all data. These performance data can easily and objectively be tracked and stored for either real-time or future review and assessment.

3D head mounted VR has the same advantages as SBS, with the addition of full 360° immersion, and shows promise

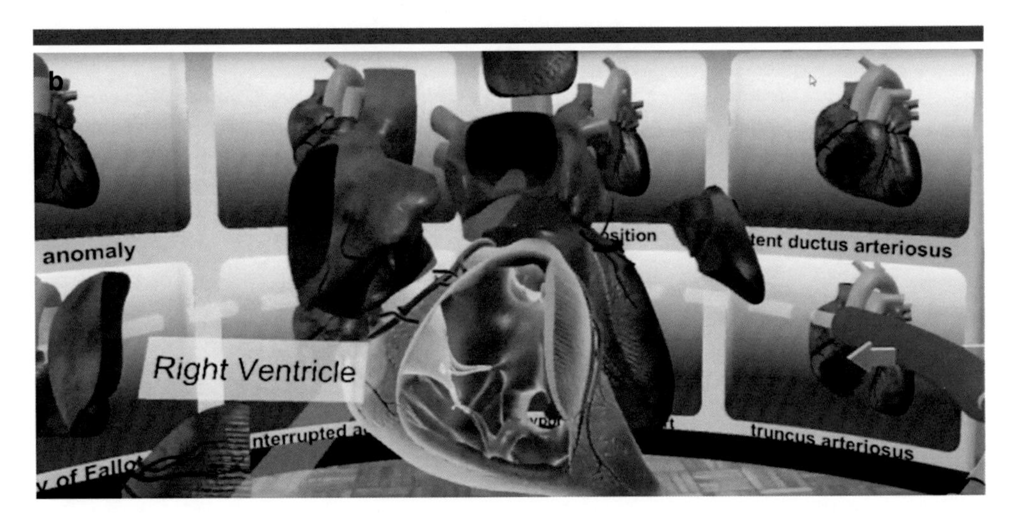

Fig. 5.3 (**a**, **b**) Stanford heart project. (Screenshots courtesy of Dr. David Axelrod, MD, and the Lighthaus Inc., with funding support from The Betty Irene Moore Children's Heart Center at Stanford and Oculus from FaceBook, with permission)

for the removal of potentially distracting selection menus due to the nature of its interactivity. Most SBS use drop-down menus for item selection and scenario advancement. In VR, these selections and interactions can be done with realistic movements, such as pointing, grabbing, or simply staring at an item of choice. For example, a virtual crash cart can have drawers that the player physically pulls open to reveal medical choices. In other words, a well-crafted VR environment allows the user to select an item that is in their virtual environment without needing a drop-down menu. In addition, 3D head mounted VR can track gaze patterns automatically within the hardware. Researchers wishing to incorporate gaze tracking as a measure of situational awareness or of attention and focus, can do so more readily within a 3D head mounted VR environment.

Disadvantages

The major disadvantages of SBS relate to inherent technological limitations and the concept of selective fidelity. Selective fidelity refers to the limitations of SBS in maxi-

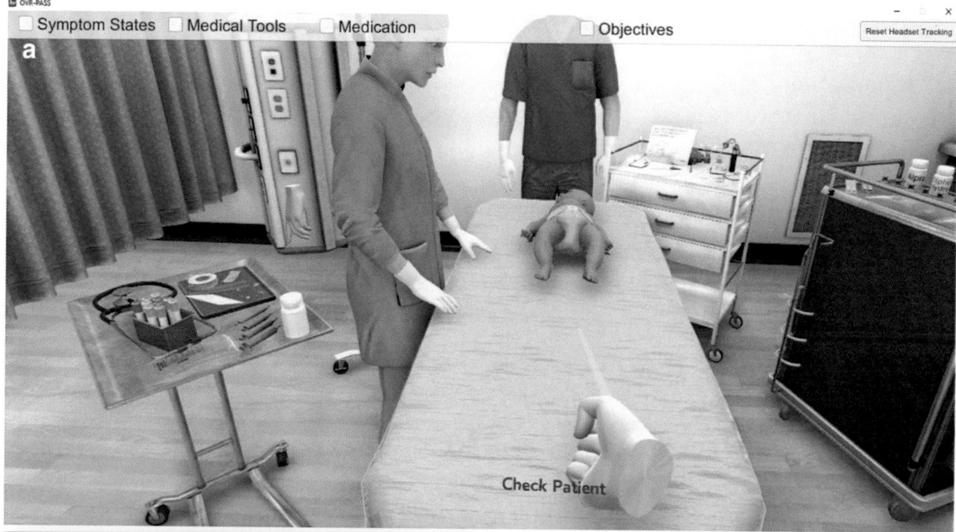

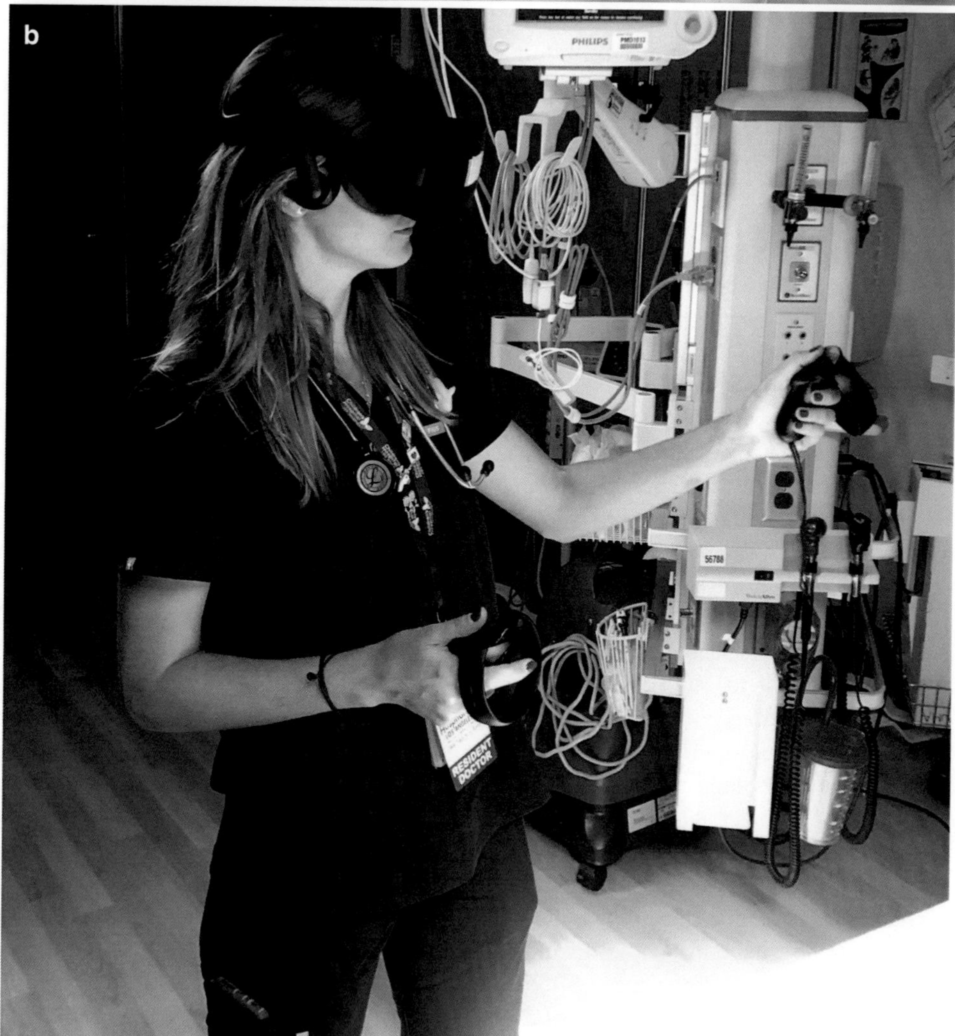

Fig. 5.4 (**a**, **b**) Oculus CHLA virtual reality project. (Screenshots courtesy of Dr. Joshua Sherman, MD, Dr. Todd P Chang, MD MAcM, a.i.Solve, Ltd., BioFlightVR, and Oculus from FaceBook, with permission)

mizing different facets of fidelity: physical, functional, and psychological [6]. VR and SBS can have incredible visual and even auditory realism, but haptic realism is still in its infancy. In other words, SBS technology is *limited* in portions of physical fidelity it can provide. Providing realistic sensations of touch of healthcare instruments, and particularly the human body, is still a formidable challenge. The potential lack of haptics in SBS and 3D head mounted VR is a significant limitation compared to MBS, currently making it less ideal for procedural training, practice and assessment. But with the advancement of technology that could very well be mitigated in the relatively near future. The limitation of using a screen in SBS and serious games that are screen-based simulations may limit the degree of immersion and thus affect psychological fidelity. With 3D head mounted VR, the level of immersion has improved significantly given the 360° and interactive nature of the VR experience. Finally, the use of multiple menus, drop-downs, and computer-based interactions may also affect functional fidelity, such that the interactions within the SBS feel artificial or tedious.

Development of SBS and 3D head mounted VR both come with a high front-end cost and long development time, which can often be a rate-limiting step in a research study. The subject matter experts and researchers must work together with the developers and coders, which is costly and time-consuming. Contrast this with manikin-based simulation: once the manikin is purchased the researcher can immediately begin scenarios without engineering skill. Even though space and human resources is a cost and a concern for manikins, the skilled simulationist can work around it. With SBS, if there is no module or product, there is nothing to work with at all [3, 4]. In essence, the quality of the research depends wholly on the developers; even when contractual agreements are present (a necessity when doing research in serious games or VR), and the final product may be different than that which the research envisioned because of funding or timeline limitations.

As with most technology, SBS and 3D head mounted VR are subject to technological problems such as glitches, slowing, and even complete blackout. While manikins also rely on computerized parts and connections, a shutdown may be mitigated with other manikins, retooling, or modifications to the scenarios. Internet connectivity is also an issue, as a serious game or VR that relies on wi-fi will be completely useless without. Multi-player games or SBS require particularly robust connectivity. With traditional SBS prior to 3D head mounted VR, there has been a concern over the limit of functional fidelity given the 2D nature of the simulation and lack of full immersion.

Models of Research

As with other methods of simulation, there are two main types of simulation-based research related to games and VR: (1) research that assesses the efficacy of the VR simulation as a training methodology (e.g., simulation as the subject), and (2) research where the VR simulation is used as an investigative methodology (e.g., simulation as the environment) [7].

Once developed, initial studies should be conducted to collect evidence that supports the validity of the game or VR when used as an assessment and/or training tool by the target audience. Non-comparative studies are useful for assessing factors such as the content, internal structure, and discriminant ability of the game and game scores [8]. Further details are explored in Chap. 26. We draw a distinction between evaluating the technology itself (as a validity trial), which is different than evaluating the educational efficacy of the serious game or VR inserted within a system of learning. The latter is an example of *simulation as the subject* of research.

Simulation as the Subject

The intended goal of studying a game or VR is often to determine the educational effectiveness of the game. A number of factors should be considered when designing a study for this purpose. Though, in general, there is good evidence to support the educational effectiveness of simulation as a training method, the educational value of a particular game or VR tool cannot be assumed, particularly for higher order outcomes such as behavior change or patient outcome changes. Producing a high-fidelity VR simulation is challenging and often affected by factors such as budgetary constraints and technological limitations. These factors may reduce the educational impact of the game or VR simulation.

Cook and Beckman highlight the strengths of the randomized posttest only design which is well-suited for this type of game research, assuming sufficient numbers and randomization [9]. With a smaller or single-center cohort, a randomized pre-post study may be more appropriate. Taking advantage of portability, distribution, and asynchrony allows for larger sample sizes and multi-center participation to fulfill this requirement. Comparative studies should be conducted to assess how learning from the game compares to more traditional methods of training. Such studies may also help to inform how to best incorporate the game into existing training curricula.

Simulation as the Environment

In this type of analysis, the simulated environment is used as an experimental model to study factors affecting human and systems performance [7]. Perhaps more than any other type of simulation, games and VR allow for standardized and reproducible scenarios and could thus be beneficial for studying a wide range of performance shaping factors including individual and team performance, environmental effects, as well as technological, systems, and patient factors. Examples from the world of manikin-based simulation include comparisons of intubation devices tested on standardized airways [10], or documentation of variations in care between different facilities [11]. Because of selective fidelity, validation is critical for these platforms to make clinically relevant conclusions applicable to the real world. Studies that use serious games and VR as an environment to examine professional behavior or safety threats are rare in the literature. For example, validity evidence is emerging for a serious game on disaster triage management [12] and for pediatric resuscitation management [13].

Because simulation as the environment requires a high level of fidelity to generalize findings to the real world, serious games and VR development can be particularly costly and time-consuming to manufacture the perfect clinical environment. Although this is a limitation of designing high-fidelity, multi-player games to simulate clinical environments, there are growing resources to prevent starting from scratch. There are open source and purchasable resources for available human anatomy, hospital architecture, and even assets such as equipment, healthcare staff models, and programmed behavior, vocabulary, or movements. Examples include Applied Research Associates' BioGears (www.biogearsengine.com/) and Kitware pulse physiology engine (physiology.kitware.com/).

Unique Variables for Games/VR

Serious games can capture precise data including actions performed within the game and time to actions. Web-based games can also provide researchers with system-wide data including information such as Internet Protocol (IP) addresses, user IDs, and login and logout times. For multi-player games, interactions and the timing of interactions between players can be tracked. By recording actions and paths taken by players during game play, investigators can better appraise gamers' decision-making process and reaction times; this process can serve as the basis for assessment of learning in serious games. Researchers may be able to better understand what goes on in the minds of the learners through the players' actions and choices. A theoretical construct used outside of healthcare to describe this type of data collection is termed the *information trail* [14]. Loh et al. describe a deliberate data tracking framework that reveals not just the completion of objectives, but the *process* and the movements learners used to get there. It tends to answer questions about *what, when, where, and how,* but not necessarily *why*, which would require debriefing [14].

Data Collection Methods for Games/VR

The collection methods for data used for game/VR analytics can be separated into two categories: in-situ and ex-situ. In-situ collection occurs in the game itself (e.g., logging game-play events), whereas ex-situ is data collected outside of the game play (e.g., post-play surveys).

In-situ Data Collection: Most game engines have or can be adapted to interact with a Data Collection Engine (DCE) that allows for easy acquisition of in-game events (e.g., assessments and treatments clicked on by the player, doses of medications and fluids selected by the player through user interfaces, etc.). DCEs can provide both detailed and summary data that can be utilized by players, educators, and researchers after game play.

In some circumstances, game researchers may wish to view the actual game play either remotely or post-game play. Several options exist for screen recording. A number of software programs designed for recording computer screen videos exist. Researchers should be aware, however, that the simultaneous use of a screen recorder may slow down game speed unless run on a computer with high processing speed and graphics capacity. An alternative method for recording game play is the use of a High-Definition Multimedia Interface (HDMI)-cloner box. These devices can capture screen video and audio and transmit them to a remote monitor or storage device without slowing game speed.

Ex-situ data collection: During initial development and beta testing of a new game/VR, developers will often want to assess players' satisfaction with the game. A number of survey-based tools have been developed for usability testing. These include the System Usability Scale (SUS) [15], the Software Usability Measurement Inventory (SUMI) [16], and the Questionnaire for User Interaction Satisfaction (QUIS) [17].

Practical Aspects: Development Phase

Subject matter experts (SMEs) must work very closely with the development team. Whether the serious game or VR is used as an educational intervention or as an investigative methodology [7], development must focus on the educational or assessment objectives. Developers often emphasize physical fidelity, rather than the functional fidelity, despite evidence within educational simulations emphasizing the latter (particularly in situations with significant budgetary constraints) [18]. With the advent of widely available physiology engines (ARA BioGears, Kitware PULSE, and HumMOD, etc., described above), the functional fidelity of physiological reactions can be maximized at lowered programming costs. However, SMEs must pay particular attention in anticipating the wide range of actions that high-performing and low-performing subjects may do in the simulated setting. These also include aligning the timing of physiological and treatment changes both within the game time and to real time in a thoughtful manner. Development teams benefit from aligning their work with frequent inputs from SMEs, as each team are likely to make assumptions about user behavior.

We emphasize that the research data plan must be ready prior to the development work. This also includes plans for how data will be transmitted to the researchers and must take into account whether performance data in the game requires secure data transfers (e.g., where will the data reside and where can they be copied?). Institutional Ethics and Review Boards may have additional restrictions on how data are stored, particularly if storage is cloud-based and protected healthcare information (PHI) is included in the dataset. Data that may inform how a user may perform in their workplace are also subject to additional privacy and confidentiality concerns.

Data collection and filtering must be integrated into all games and VR from the beginning, as substantial memory and processing is used to save and process granular data. The manner in which data are curated must be agreed upon. For example, when measuring time durations (e.g., time to chest compression) that are common in simulated medical activities, developers will need clear guidelines on when timing begins and ends, particularly if the scenario uses a strong branch-chain logic and conditional events. Most developers will be able to provide raw data using a *.csv file format common to spreadsheet-type data, but often the data will need to be summarized or cleaned prior to analysis or even displayed to the user. A plan that clearly specifies the research outcome variables and how they will be analyzed will assist the developers in appropriate data acquisition. As another example, capturing gaze data during VR is possible, but requires additional programming and substantial processor power to record during gameplay.

Data collection within VR or games depends on the interactivity and the hardware involved. Standalone VR devices (Oculus Go, Samsung Gear VR, etc.) can record positioning in 3 degrees of freedom but no other positional data. Full VR devices at the time of this publication (Oculus Rift S, HTC Vive Pro) can record the subject's position in all 6 degrees of freedom and potentially gaze pattern. VR or serious games that use their own controller can record the timing and pattern of actions, including hesitancy, inaction, or even urgency if a key or button was pressed repeatedly very quickly. Developers typically use these types of in real time data to further the simulation or game, but recording these data for later use is memory- and processor-intensive, and should be planned in advance. It is not possible for the developer – without knowledge of the research question nor outcome variables desired – to prioritize which data to keep and export without SME and research expert input.

Implementation Phase

Conducting research using games or VR is different than simply asking participants to use the software, and several implementation considerations are recommended. Because games or VR requires participants to learn new skills immediately, which includes game mechanics that they may not be familiar with (e.g., commands, buttons, rules), there is an inherent concern for construct-irrelevant variance [19]. That is, their performance within the game or VR (and even their frequency of use) may be influenced in part by their facility and skill in the platform. Construct-irrelevant variance is a known entity in K-12 games [20], but is infrequently addressed in medical simulation. Sources of construct-irrelevant variance include typing speed and skill, equipment quality (e.g., poor quality speakers vs. headphones), familiarity with control pads, familiarity with common game mechanics, or even vertigo with fast-moving VR.

To account for construct-irrelevant variance in serious games and VR research, we strongly recommend the construction of a tutorial that immerses the user with the specific controls and game mechanics necessary for optimal performance, preferably with no hint of the content that is introduced in the proper game or VR. We also recommend collecting tutorial performance data, both to quantify the level of familiarity in the environment as a covariate in data analysis, and to document improvements in successive tutorials as evidence that construct-irrelevant variance is actively minimized in the research. To that end, if the research study requires multiple playthroughs of the game or VR it is possible that simply playing the game will improve performance, as their facility with the controls and environment will gradually improve as a form of maturation bias known as the carryover effect. There are statistical ways to measure and account for carryover effect [21]; however, if the order of gameplay content can be randomized among a larger sample as in a crossover study, that can also attenuate carryover bias.

Because serious games and VR allow research activities to be done in remote areas, including participants' own homes, the physical environment in which the research activities occur may be varied, adding another source of construct-irrelevant variance. The physical environment includes phenomena like floor space (particularly for VR), distractors (additional people, other electronics, pets), screen size, and processing speed for their own machines. Internet speed connections may also influence game performance. It may be necessary to standardize the physical environment by requiring the study to be completed in a more controlled and consistent setting.

Analysis and Dissemination Phase

Outcome variables common to many healthcare game and VR research studies can include all levels of evaluation, such as satisfaction, knowledge, behavior, and even patient-related outcomes. The allure of data collection using games resides in the large amount of behavioral and performance data available, including time-to-action, choices or selections made, and even pauses or inactive time, which could denote inaction, hesitation, or indecision. Just like any simulation-based research, the research question(s) and methodology must be declared well before the development and implementation of the final product.

Careful attention must be made to the interpretation of game performance. Depending on the game mechanics, navigation of a long menu screen may compound a time-to-critical action variable, for example. Alternatively, a particular branch chain logic that 'ends' a game early may not allow a participant to demonstrate all of their accrued knowledge or performance if the Game Over screen appears early. Establishing some correlation of game performance with clinical performance provides validity evidence for the use of the game or VR, and is often the sentinel research plan with a developed game or VR.

Validity evidence for the content and use of a healthcare game or VR is of interest to a variety of parties. Game developers and hardware developers often lack data on non-entertainment products, and any validity evidence within the healthcare organizations can distinguish their products from their competitors. Healthcare educators would also be interested in validity evidence before implementing games or VR into an already busy curriculum. Healthcare networks and patient safety advocates would value validity evidence of games and VR similarly to the way simulation can be used to uncover latent safety threats. Finally, funders and organizations sponsoring the monetary investment in the development of these systems should also recognize the value proposition for these games, as healthcare games and VR do not have the same profitability potential as as games intended for entertainment.

Closing

Serious games and VR are powerful tools that have distinct advantages and disadvantages when used to conduct simulation-based healthcare research. Researchers are advised to select the modality of the simulation (e.g., serious game vs. VR vs. manikin) appropriate to the fidelity of the simulation and the outcomes being investigated, either for simulation as intervention or simulation as environment. Unique elements of performance data capture include developing an information trail for in situ data capture, asynchronous debriefing, and specific user interface surveys already validated in the non-healthcare literature. Because both serious games and VR requires significant upfront development, SMEs and researchers should work very closely with developers to facilitate successful data capture and analyses.

References

1. Susi T, Johannesson M, Backlund P. Serious games – an overview. New York: Springer; 2015.
2. Bedwell WL, Pavlas D, Heyne K, Lazzara EH, Salas E. Toward a taxonomy linking game attributes to learning: an empirical study. Simul Gaming. 2012;43(6):729–60.
3. Chang T, Pusic MV, Gerard JL. Screen-based simulation and virtual reality. In: Cheng A, Grant VJ, editors. Comprehensive healthcare simulation – pediatrics. 1st ed. Cham: Springer International Publishing; 2016. p. 686.
4. Chang TP, Weiner D. Screen-based simulation and virtual reality for pediatric emergency medicine. Clin Pediatr Emerg Med. 2016;17(3):224–30.
5. Ellaway R. Reflecting on multimedia design principles in medical education. Med Educ. 2011;45:766–7.
6. Curtis MT, DiazGranados D, Feldman M. Judicious use of simulation technology in continuing medical education. J Contin Educ Health Prof. 2012;32(4):255–60.
7. Cheng A, Auerbach M, Hunt EA, Chang TP, Pusic M, Nadkarni V, Kessler D. Designing and conducting simulation-based research. Pediatrics. 2014;133(6):1091–101.
8. Cook D, Beckman TJ. Current concepts in validity and reliability for psychometric instruments: theory and application. Amer J Med. 2006;119(166):e7–e16.
9. Cook DA, Beckman TJ. Reflections on experimental research in medical education. Adv Health Sci Educ Theory Pract. 2010;15(3):455–64.
10. Fonte M, Oulego-Erroz I, Nadkarni L, Sanchez-Santos L, Iglesias-Vasquez A, Rodriguez-Nunez A. A randomized comparison of the GlideScope videolaryngoscope to the standard laryngoscopy for intubation by pediatric residents in simulated easy and difficult infant airway scenarios. Pediatr Emerg Care. 2011;27(5):398–402.
11. Kessler DO, Walsh B, Whitfill T, et al. Disparities in adherence to pediatric sepsis guidelines across a spectrum of emergency depart-

ments: a multicenter, cross-sectional observational in situ simulation study. J Emerg Med. 2016;50(3):403–415.e401-403.
12. Cicero MX, Whitfill T, Munjal K, et al. 60 seconds to survival: a pilot study of a disaster triage video game for prehospital providers. Am J Disaster Med. 2017;12(2):75–83.
13. Gerard JM, Scalzo AJ, Borgman MA, et al. Validity evidence for a serious game to assess performance on critical pediatric emergency medicine scenarios. Simul Healthc. 2018;13: 168–80.
14. Loh CS, Anantachai A, Byun J, Lenox J. Assessing what players learned in serious games: in situ data collection, information trails, and quantitative analysis. Paper Presented at: 10th International Conference on Computer Games: AI, Animation, Mobile, Educational & Serious Games (CGAMES 2007); 2007.
15. Brooke JSUS. a 'quick and dirty' usability scale. In: Jordan P, Thomas B, McClelland IL, Weerdmeester B, editors. Usability evaluation in industry. Boca Raton: CRC Press; 1996. p. 189–94.
16. Kirakowski J, Corbett M. SUMI: The software usability measurement inventory. Br J Educ Technol. 1993;24(3):210–2.
17. Chin JP, Diehl VA, Norman KL. Development of an instrument measuring user satisfaction of the human-computer interface. In: Proceedings of the SIGCHI Conference on Human Factors in Computing Systems; 1988; Washington, DC, USA.
18. Norman G, Dore K, Grierson L. The minimal relationship between simulation fidelity and transfer of learning. Med Educ. 2012;46(7):636–47.
19. Downing SM. Threats to the validity of locally developed multiple-choice tests in medical education: construct-irrelevant variance and construct underrepresentation. Adv Health Sci Educ Theory Pract. 2002;7(3):235–41.
20. Kerr D. Using data mining results to improve educational video game design. J Educ Data Min. 2015;7(3):1–17.
21. Cleophas TJ. Interaction in crossover studies: a modified analysis with more sensitivity. J Clin Pharmacol. 1994;34(3):236–41.

Overview of Computational Modeling and Simulation

6

Roland R. Mielke, James F. Leathrum, Andrew J. Collins, and Michel Albert Audette

Overview
Scientific research involves the formulation of theory to explain observed phenomena and using experimentation to test and evolve these theories. Over the past two decades, computational modeling and simulation (M&S) has become accepted as the third leg of scientific research because it provides additional insights that often are impractical or impossible to acquire using theoretical and experimental analysis alone. The purpose of this chapter is to explore how M&S is used in system-level healthcare research and to present some practical guidelines for its use. Two modeling approaches commonly used in healthcare research, system dynamics models and agent-based models, are presented and their applications in healthcare research are described. The three simulation paradigms, Monte Carlo simulation, continuous simulation, and discrete event simulation, are defined and the conditions for their use are stated. An epidemiology case study is presented to illustrate the use of M&S in the research process.

Practice Points

- There are three main simulation paradigms: Monte Carlo simulation, continuous simulation, and discrete event simulation, however a hybrid simulation combining any two paradigms is also possible.
- The Monte Carlo simulation paradigm refers to the methodology used to simulate static, stochastic system models in which system behavior is represented using probability.
- The continuous simulation paradigm refers to the methodology used to simulate dynamic, continuous-state, time-driven system models.
- The discrete event simulation paradigm refers to the methodology used to simulate dynamic, discrete-state, event-driven system models, such as a queuing model.
- Modeling methods include system dynamics models and agent-based models; both methods frequently are used for complex healthcare and medical systems, including epidemiological applications surveyed here.

Introduction

Modeling and simulation (M&S) long has been used for education and training in the healthcare domain. Most medical practitioners are familiar with the use of visual models and simulations and simulation-based instructional applications to enhance the transfer and acquisition of knowledge. They also are familiar with the use of task trainers, medical mannequins, and immersive interactive virtual reality for training where the objective is to control performance variability (i.e., minimize error) by improving trainee reliability. However, the use of M&S as a computational approach to support and enhance healthcare research is a more recent and perhaps less familiar topic for medical practitioners. The focus of this chapter is to explain how M&S is used in system-level healthcare research and to present some practical guidelines for its use.

Computational modeling and simulation (M&S) refers to the use of models and simulations, along with the associated

R. R. Mielke
Electrical Engineering, University of Wisconsin-Madison, Madison, WI, USA

Department of Modeling, Simulation and Visualization Engineering, Old Dominion University, Norfolk, VA, USA
e-mail: rmielke@odu.edu

J. F. Leathrum · M. A. Audette (✉)
Department of Modeling, Simulation and Visualization Engineering, Old Dominion University, Norfolk, VA, USA
e-mail: jleathru@odu.edu; maudette@odu.edu

A. J. Collins
Department of Engineering Management and Systems Engineering, Old Dominion University, Norfolk, VA, USA

D. Nestel et al. (eds.), *Healthcare Simulation Research*, https://doi.org/10.1007/978-3-030-26837-4_6

analysis, visualization, and verification/validation techniques, to conduct a simulation study. The subject of a simulation study is usually described as a system. A *system* is a combination of components that act together to perform a function not possible with any of the individual components. A system that is the subject of a simulation study is called the *simuland*. A *model* is a mathematical or logical representation the simuland. Selection of a model must consider both the relevant features of the simuland and the questions about the simuland that are to be addressed. A *simulation* is a process for executing a model. Selection of a simulation methodology depends on the mathematical characteristics of the model.

Historically, M&S has been viewed as an important research tool in numerous disciplines or application domains. Research in most domains often proceeds through a sequence of phases that include understanding, prediction, and control [1]. The initial phase is used to gain an understanding of how events or objects are related. An understanding of relationships among objects or events then allows the modeler to begin making predictions and ultimately to identify causal mechanisms. Finally, knowledge of causality enables the user to exert control over events and objects. Research moves from basic to more applied levels as progression is made through these phases. For example, the Human Genome project was undertaken to understand the complete sequencing of chromosomal DNA in human beings. Knowledge of the human genome helps to make predictions regarding genetic variation and can lead to more reliable diagnostic tests and medical treatments applied at the genetic or molecular levels.

M&S is closely linked to all phases of research. At the more basic levels, research is guided heavily by theory. Models are often used to represent specific instances of theories, to differentiate between competing theories, or to exhibit underlying assumptions. Likewise, simulations are used to test predictions under a variety of conditions or to validate theories against actual conditions. At the applied levels, simulations also are used to control events and objects. Simulations in the form of mock-ups or prototypes are used in the creation of products and systems to validate predictions regarding operational requirements, specifications, and user/customer satisfaction.

Although this description of the research process admittedly is simplistic, it does underscore three important points regarding M&S. First, M&S is intimately related to all phases of the research process. M&S is used to generate and refine the theories that help us understand our world as well as the technology we use to interact with the world. Second, the description is generic and highlights where M&S can be applied in any domain where individuals are engaged in research. Thus, biologists, chemists, sociologists, economists, and historians all can use M&S to help formulate research questions, conduct experiments, evaluate theories, and add to their respective bodies of knowledge. Third, the description also shows the different aspects of M&S emphasized along the basic/applied research continuum. Thus, at the basic end, M&S is used more as a research tool whereas at the applied end, it is used either to create products or even may be a product in and of itself.

The remainder of this chapter is organized in four sections. In the first section, Simulation Methodologies, we focus on simulation paradigms. The three simulation paradigms are defined in terms of the system classifications associated with the simulation model. In the second section, Selected Modeling Methods, we describe two modeling approaches often used in healthcare research, system dynamics models and agent-based models. An example of applying M&S to healthcare research is presented in the third section, Example Healthcare Applications. An epidemiology problem is investigated using different modeling approaches and simulation methods to illustrate some of the practical issues that must be considered. In the fourth section, Conclusion, several challenges associated with applying M&S in healthcare research are identified and briefly discussed.

Simulation Methodologies

In this section, we identify the three simulation paradigms, Monte Carlo simulation, continuous simulation, and discrete event simulation, and discuss the process for selecting an appropriate paradigm. Selection of a simulation paradigm depends primarily on the characteristics of the model that is to be simulated. Model characteristics are defined in terms of the mathematical properties of the functional representation for the model. Each simulation paradigm is designed for use with models having a specific combination of these system characteristics. A fourth simulation methodology, hybrid simulation, refers to simulation methodologies that consist of utilizing two or more simulation paradigms to simulate a single simuland model.

System Characteristics

A model often is represented mathematically using the definition of a *function* [2]. A function is a mathematical construct consisting of three components, the domain set X, the codomain set Y, and the rule of correspondence Γ. The domain set consists of the set of system inputs $x(t) \in X$, the codomain set consists of the set of system outputs $y(t) \in Y$, and the rule of correspondence consists of the mapping of inputs to outputs denoted as $\Gamma : X \rightarrow Y$ or $\Gamma\{x(t)\} = y(t)$. The *system state* at time t_0, $q(t_0)$, is the (minimal) information about the system at t_0 such that the output of the system for

$t \geq t_0$ is uniquely determined from this information and the system input for $t \geq t_0$. The *state space* Q of a system is the set of all possible values that the state may take.

System characteristics are defined as all inclusive, mutually exclusive descriptor pairs that are based on the mathematical properties of the model functional representation. Definitions for these descriptor pairs are presented in the following.

- Static or Dynamic – A system is said to be *static* if the system output at time t_i is dependent only on the system input at time t_i. A system is said to be *dynamic* if the system output at time t_i depends on the system input for $t \leq t_i$. Dynamic systems are called systems with memory while static systems are called systems without memory. The output of a static system at time t depends only on the input to the system at time t. The output of a dynamic system at time t depends on both the input to system and the state of the system at time t.
- Deterministic or Stochastic – A deterministic system is a system in which all system outputs are deterministic. A *stochastic* system is a system in which one or more system outputs have uncertainty or variability. In this case, the system output is characterized as a random process and a probabilistic framework is required to describe system behavior.
- Continuous-State or Discrete-State – A *continuous-state* system is a system in which the state space Q consists of elements q(t) that assume a continuum of real values; that is, $q(t) \in R$ (real numbers). Examples of continuous-state systems include many physics-based systems where system variables (position, velocity, magnitude) have real number values. A *discrete-state* system is a system in which the state space Q consists of elements q(t) that assume only discrete values; that is, $q(t) \in I$ (integer numbers). Examples of discrete-state systems include many service systems where system variables (people counts, resource counts, part counts) have integer number values.
- Event-Driven or Time-Driven – In discrete-state systems, state changes occur only at distinct instants of time as variable values change instantaneously from one discrete value to another discrete value. With each state transition, we associate an event. Further, we attribute the state transition to the occurrence of the event. Thus, an *event* is a specific instantaneous action that causes a state transition and we say that systems that exhibit such behavior are *event-driven* systems. In continuous-state systems, the system state generally is obtained by solving differential equation representations of the system. In such systems, state changes can occur simply because time advances, even when there is no input to the system. We say that systems that exhibit such behavior are *time-driven* systems.

Simulation Paradigm Definitions

There are three simulation methodologies, called *simulation paradigms*, for simulating a model: Monte Carlo simulation paradigm; continuous simulation paradigm; and discrete event simulation paradigm. Selection of a simulation paradigm is based upon the system characteristics associated with the model utilized to represent the simuland. The need for three simulation paradigms is due to the differences in the mathematical properties of the model functional representations. The simulation paradigms are defined in the following.

- Monte Carlo Simulation Paradigm – The *Monte Carlo simulation paradigm* refers to the methodology used to simulate static, stochastic system models in which system behavior is represented using probability. The underlying model usually is a random experiment and associated probability space.
- Continuous Simulation Paradigm – The *continuous simulation paradigm* refers to the methodology used to simulate dynamic, continuous-state, time-driven system models. The underlying model usually is a set of differential equations that describe simuland behavior. Simulation output often is a time-trajectory of some simuland state variable. The simulation methodology consists of starting from some initial system state and repeatedly solving numerically the differential equations for very small increments of the time variable. This paradigm usually is used for natural systems where it is possible to associate differential equations with system behavior.
- Discrete Event Simulation Paradigm – The *discrete event simulation paradigm* refers to the methodology used to simulate dynamic, discrete-state, event-driven system models. The underlying model is usually a representation of a discrete event system [2] such as a queuing model, a state automata model, a Petri net model, or an event graph model. Simulation output often is a sequence of state variable values evaluated at event times. The simulation methodology consists of starting from some initial system state and repeatedly updating the system state at the occurrence of each event. Event management is conducted using an event scheduling strategy in which a future event list is updated at each event time. This paradigm usually is used for service systems where it is possible to associate event descriptions with system behavior.

The process for selecting a simulation paradigm is illustrated in Fig. 6.1. It is clear in this figure that once a model of the simuland is developed, the resulting simulation paradigm that is required to simulate the model is also determined. Often however, there is some flexibility in deciding how to

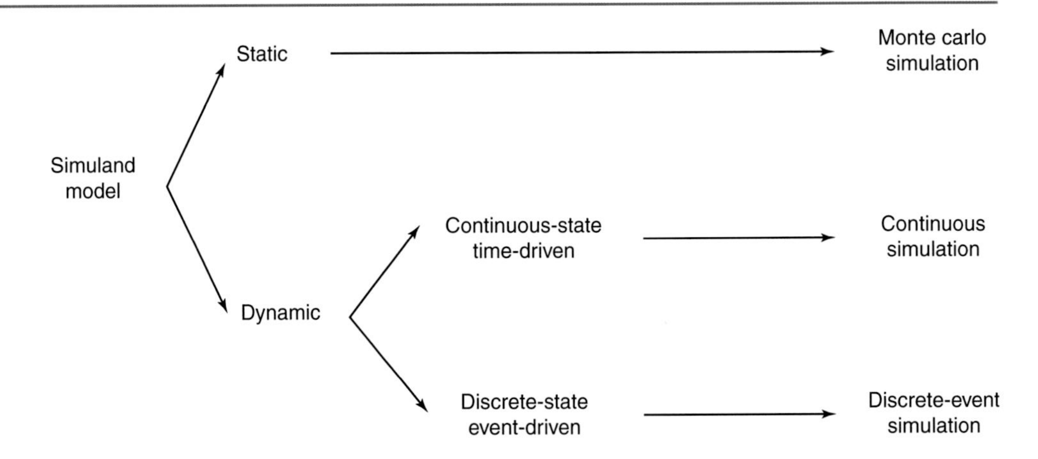

Fig. 6.1 Simulation paradigm selection process

develop the simuland model, thus providing some flexibility in choice of simulation paradigm.

A fourth simulation paradigm, the hybrid simulation paradigm, is sometimes defined; however, this term refers to a simulation methodology that employs concurrently two or more of the simulation paradigms, as defined above, to simulate a single model. For example, the hybrid methodology might be useful when simulating a continuous-state, time-driven model that operates in two different modes. A discrete event system model might be used to change operating modes, while different continuous simulation models might be used to represent system operation in each of the two modes. For example, this situation easily could occur when simulating a model of human physiology.

Selected Modeling Methods

There are numerous methods for developing models and one of the early challenges in any M&S project is the selection of an appropriate modeling method. While each project is unique, there are several guiding principles that apply in all situations. The starting point always is a detailed investigation of the simuland and enumeration of the objectives for the study; that is, identification of the questions about the simuland that the simulation study is to address. The simuland must be modeled so that relevant simuland features are included in the model at a resolution (level of detail) sufficient to address study questions. It is convenient if the model can be developed so that it fits within one of the three simulation paradigms. If that can be done, then there are well-defined procedures for simulating the model and a host of available M&S tools or environments that may be applicable. If the model characteristics do not fit into one of the three simulation paradigms, then a unique simulation methodology must be crafted for that model.

In this section, we introduce two modeling methods, system dynamics models and agent-based models. Both methods frequently are used to describe complex healthcare and medical systems, but each provides a very different perspective of system operation. Both modeling methods address dynamic systems, but can be formulated as either continuous simulation models or discrete event simulation models. In the third section of this chapter, Example Healthcare Applications, both modeling approaches are used to address disease epidemiology. The systems dynamics model is developed as a continuous simulation model while the agent-based model is developed as a discrete event simulation model.

System Dynamics Models

System dynamics models consist of the combination of two components, a stock and flow diagram and a causal loop diagram. A stock is some quantity that is accumulated over time by inflows and depleted by outflows. Stock can only be changed by flows. Thus, stock can be viewed as an integration of flows over time, with inflows adding to the accumulated stock and outflows subtracting from the accumulated stock. Variables representing stock levels usually comprise the state variables for a system dynamics model. A causal loop diagram is a diagram that shows how different system variables and parameters are interrelated. The diagram consists of nodes representing variables or parameters and edges representing relationships between nodes. A positive labelled edge denotes a reinforcing relationship while a negative labelled edge denotes an inhibiting relationship. In system dynamics, the causal loop diagram is used to show how the system state variables and parameters influence the stock inflow rates and outflow rates. The system dynamics model results in the definition of a set of state variable equations describing the dynamical behavior of the modeled system. Ideally, the model state variables and parameters are selected to correspond to specific characteristics of simuland. An example system dynamics model is shown in Fig. 6.3 in the next section.

Numerous applications of system dynamics can be found in healthcare and medical simulation research. In healthcare, the areas of application span disease and substance abuse epidemiology, health care capacity analysis and optimization, and patient flow studies in clinics and emergency care facilities. Examples of disease epidemiology research include heart disease and diabetes studies centering on the impact of prevention and rehabilitation on public health costs [3]. In addition, there have been HIV/AIDS simulation efforts emphasizing virological and behavioral features of the epidemic while portraying the consequences in a simple graphical form [4] as well as the impact of antiretroviral therapy [5]. There also are simulation models for evaluating the possible effects of a screening and vaccination campaign against the human papilloma virus and the impact on cervical cancer [6]. Recent substance abuse epidemiology research centers in particular on cocaine and heroin abuse. For example, a system dynamics model that reproduces a variety of national indicator data on cocaine use and supply over a 15-year period and provides detailed estimates of actual underlying prevalence [7] has been reported. Clinical capacity and flow studies include an optimization study of an Emergency Room [8] in which a system dynamics model is used to investigate the interaction between demand patterns, resource deployment, hospital processes, and bed numbers. One of the findings is that while some delays to patient care are unavoidable, delay reductions often can be achieved by selective augmentation of resources within the unit.

Agent-Based Models

Agent-based models [9] are composed of three components, agents, an environment, and a set of agent relationships or interactions. Agents are self-contained, autonomous objects or actors that represent components of the simuland. An agent has inputs, representing communications from other agents or perceptions from the environment, and produces outputs representing communications to other agents or interactions with the environment. An agent often has a purpose, trying to achieve some goal or to accomplish some task, and the capability to modify behavior over time to improve performance in accomplishing objectives. An environment may be as simple as a grid or lattice structure that provides information on the spatial location of an agent relative to other agents, or may consist of complex dynamic models capable of supplying environmental data that may influence agent behavior. It is the rules of agent interactions, both with other agents and the environment, that are at the heart of any agent-based model. These interactions are usually conducted at the local spatial level with the agents interacting myopically with their immediate neighbors, but can also occur through other environmental projections such as a social network. These interactions might be direct with agents exchanging information, or indirect with an agent deciding to move because it is surrounded by too many neighbors.

It is the combination of many agents interacting simultaneously with each other and with the environment that can lead to emergent behavior within the simulation of agent-based models. Agent-based models are developed at the micro-level through defined agent interactions, but are used to provide insight at the macro-level by observation of the collective behavior of agents. A key property of agent-based models is that even relatively simple rules of agent interaction can result in highly complex collective agent behaviors. Another advantage of agent-based models is their capability to accommodate agent heterogeneity. Agent heterogeneity refers to agents that have different characteristics; they may start with different resources, they may have different tolerances, and they may react differently. The facility for incorporating heterogeneous agents in an agent-based model allows modelers to more closely represent the great diversity that is present in almost all natural systems.

Agent-based modeling primarily is a decision-support modeling methodology. It often is used to develop and test theories and to provide insight into complex system behavior. In the biological sciences, agent-based models have been used to model cell behavior and interaction [10], the working of the human immune system [11], and the spread of disease [12]. Agent-based epidemic and pandemic models can incorporate spatial and social network topologies to model people's activities and interactions. The focus is on understanding conditions that might lead to an epidemic and identifying mitigation measures. Agent-based modeling is one means to utilize the vast healthcare data pool to analyze the impacts of health-related policy decisions on the general public, especially when it would be impracticable, costly, or potentially unethical to use live experiments to evaluate these policies. Agent-based models and simulations allow researchers to experiment with large simulated autonomous and heterogeneous populations to see what phenomena emerge and to evolve theories about these phenomena.

Example Healthcare Applications

The study of the spread of diseases provides a rich domain for selecting examples to illustrate the significance of choosing a modeling methodology. In this section, we develop epidemiological models using a systems dynamics modeling approach and an agent-based modeling approach. The purpose of these examples is to demonstrate that the selection of a modeling methodology has a direct impact on the level of resolution and the uses that can be made of the information that result from simulating the model.

Heath et al. [13] have proposed three different levels for characterizing models based upon the level of understanding concerning the simuland. The levels are called Generator models, Mediator models, and Predictor models. A Generator is a model developed with limited understanding of the simuland and its use is limited primarily to determine if a given conceptual model/theory is capable of generating observed behavior of the simuland. A Mediator is a model developed with a moderate level of understanding of the simuland and it is used primarily to establish the capability of the model to represent the simuland and to gain insight into the characteristics and behaviors of the simuland. A Predictor is a model developed with full understanding of the simuland and it is used primarily to estimate or predict the behavior of the simuland under various operating conditions and environments. A first step in the development of a conceptual model for a simuland is to select a model methodology. This decision often is based on the (strike) developer's level of understanding concerning the simuland. It is important to recognize that this decision has a direct impact on how we can use the simulation results.

System Dynamics Approach to Epidemiological Modeling

A basic system dynamics approach to modeling the spread of an infectious disease within a population is known as *compartmental modeling*. In this approach, the population is partitioned into compartments or subgroups and the model is designed to show how the population of each subgroup changes as the disease progresses. Five different compartmental models are shown in Fig. 6.2. In this figure, each box represents a population compartment and the compartment variable indicates the population of that compartment. The selection of a model is made to best represent the specific disease being studied. For some diseases such as mumps, members of the susceptible population move to the infectious population when they come in contact with another member of the infectious population. Members of the infectious population eventually move to the recovered population and as a result cannot be re-infected. This model is called the SIR model. Other diseases such as strep throat do not grant immunity to those that recover and thus route those recovering back to the susceptible population. This model is called the SIS model. Diseases such as measles provide maternally derived immunity to young infants who do not move to the susceptible population until growing out of the maternal immunity stage. This model is called the MSIR model. Still another model subdivides those in the infectious population into an exposed population where members have been exposed to an infectious person but are not yet contagious. Eventually, members of the exposed population move to the infectious population. This model is called the SEIR model. Other partitions separate the infectious population into a subgroup that is infectious but displays no symptoms and a subgroup that is infectious and displays normal symptoms of the disease. This model is called the SI_CIR model; Typhoid Mary is a classic example of a member of the infectious carrier population. It is interesting to note that the same compartment definitions can be applied to characterizing the states of an individual modeled as an agent in an agent-based epidemiological model.

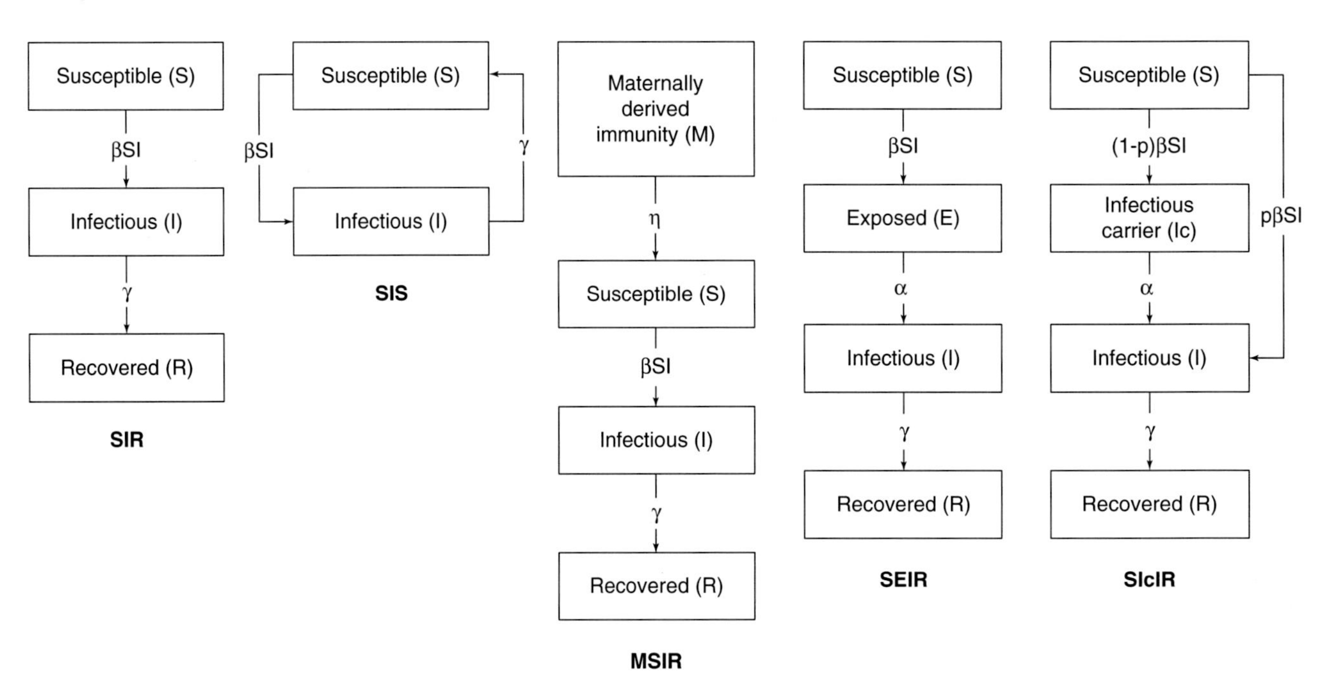

Fig. 6.2 Compartmentalized populations for common epidemiological models

We have chosen to use the SIR system dynamics model as an example. The compartmentalized population diagram shown in Fig. 6.2 is used as the stock and flow diagram. Causal relationships, that relate population flow rates between population subgroups to the compartmentalized populations and the flow parameters for infection rate constant b and recovery rate constant k, are added to the stock and flow diagram to complete the systems dynamics model. State variable equations are developed from the model and result in three first-order differential equations that express the time rate of change for the subgroup populations. The complete system dynamics model, including the resulting model differential equations, is shown in Fig. 6.3.

The SIR model is simulated using the continuous simulation paradigm. We set the population N = S + I + R at 7,900,000 people and it is assumed that N remains constant over the duration of the simulation. It also is assumed that initially ten people are in the infectious population, no people are in the recovered population, and the remaining people are in the susceptible population. The infectious rate constant b is set to 0.50 infectious contacts per day per infected person and the recovery rate constant k is set to 0.33 indicating the fraction of infectious people recovering per day. The simulation is run for a period of 150 days. The simulation results are shown in Fig. 6.4.

The simulation output for the SIR system dynamics model clearly show how the compartmentalized populations change as a function of time as the infectious disease runs its course. The model facilitates investigating how changes to the initial population distribution, the infection rate constant b, and the recovery rate constant k affect the spread of the disease over time and the portion of the population impacted during the disease lifecycle. However, this model provides no information about how physical interactions between infectious people and susceptible people impact disease spread and the eventual severity of the outbreak. However, such information might be essential if an objective of the study were to identify methods to mitigate the spread of disease.

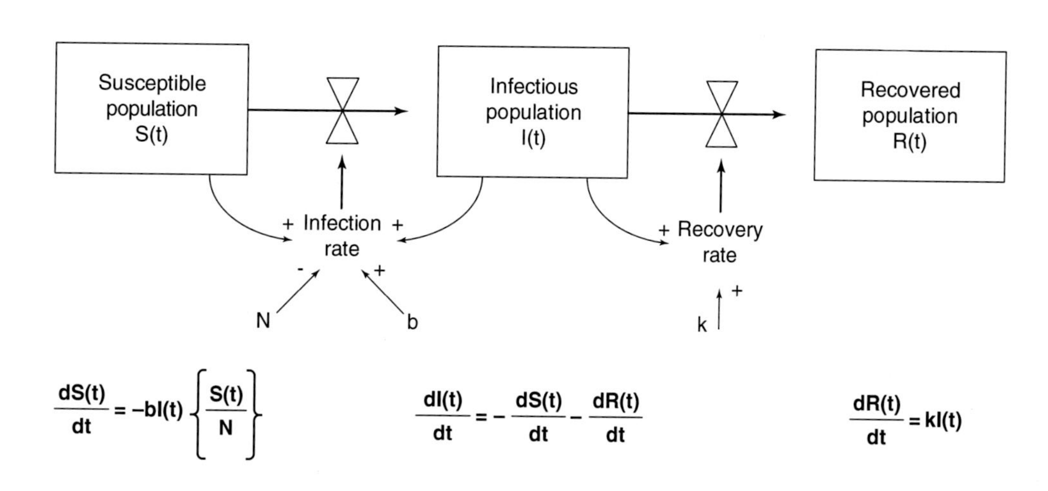

Fig. 6.3 Complete systems dynamics model for SIR example

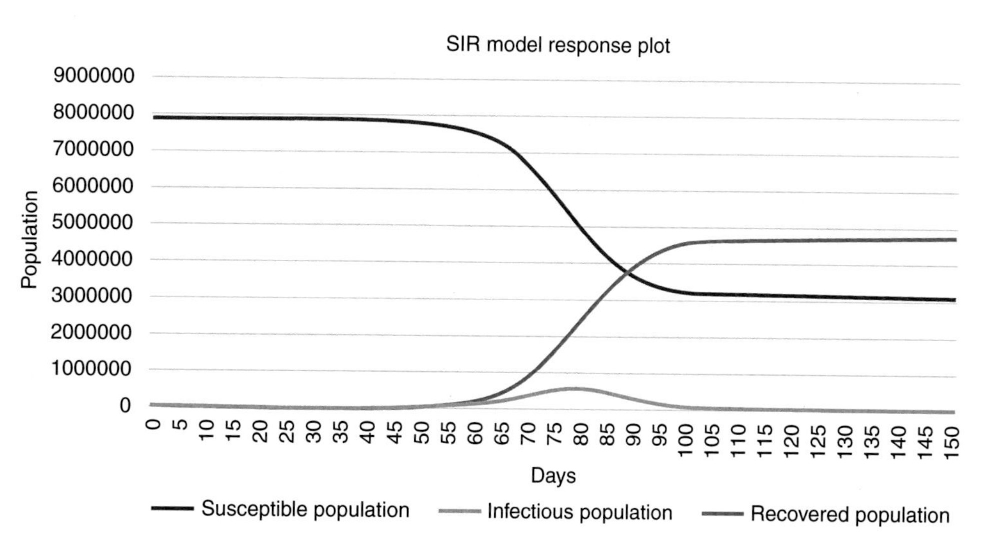

Fig. 6.4 Simulation results for SIR system dynamics model

Agent-Based Approach to Epidemiological Modeling

An agent-based modeling approach presents the opportunity to investigate at greater resolution the causes for the spread of an infectious disease. Our investigations using the SIR system dynamics model showed that disease spread is not due to population subgroup sizes, but rather is due to interactions between infectious individuals and susceptible individuals. Since agent-based models are developed at the individual level, this modeling method facilitates adding much greater detail about how individuals interact.

An agent-based model for the spread of mumps in a small urban environment is presented in [14]. In this model, the agent environment is augmented using geographical information system (GIS) data that identify where individuals live, where they are likely to travel during daily activities, and how they are likely to travel. Individuals are represented as agents. Agent state information includes an activity state, with values representing work/study, leisure, commuting, and a disease state that takes its value from the SEIR states of susceptible, exposed, infectious, and recovered. This state information, when combined with the GIS information, adds considerable detail as to how susceptible and infectious individuals make contact. The flow diagram describing the corresponding agent logic is shown in Fig. 6.5. The flow diagram determines when a susceptible individual comes in contact with an infectious individual and then adjusts the infection rate constant according to the population density at that location.

The model is initialized by distributing the population (1000 individuals) to their home locations in the urban area. In this example, it is assumed that 999 individuals start in the susceptible state and one individual starts in the infectious state. The model is simulated using the discrete event simulation paradigm and the size of the four population subgroups is reported as output. The simulation output is shown in Fig. 6.6. The model allows investigation of how daily behaviors of individuals impact the spread of disease.

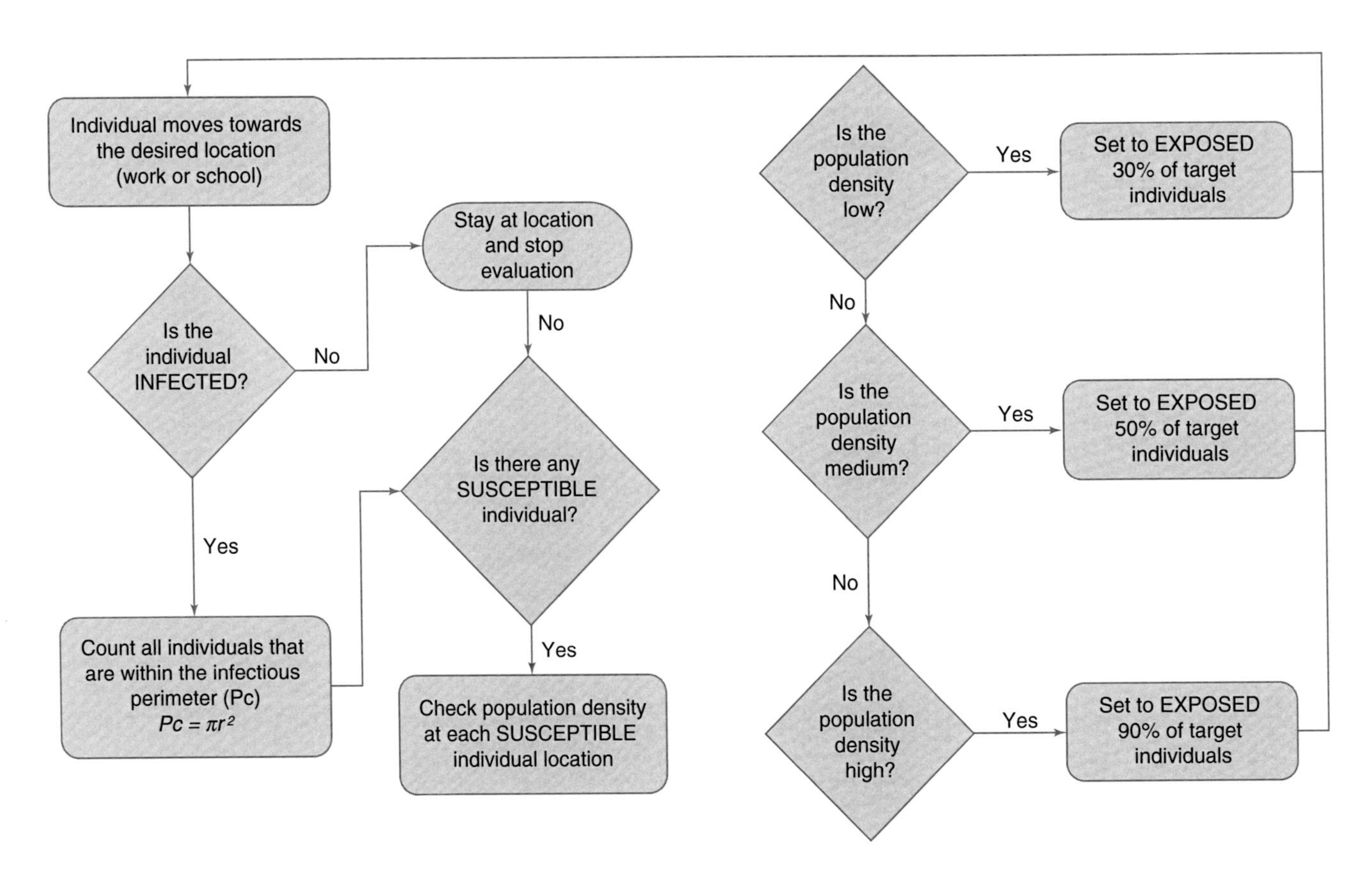

Fig. 6.5 Flow diagram for infection rules describing disease spread. (From [14])

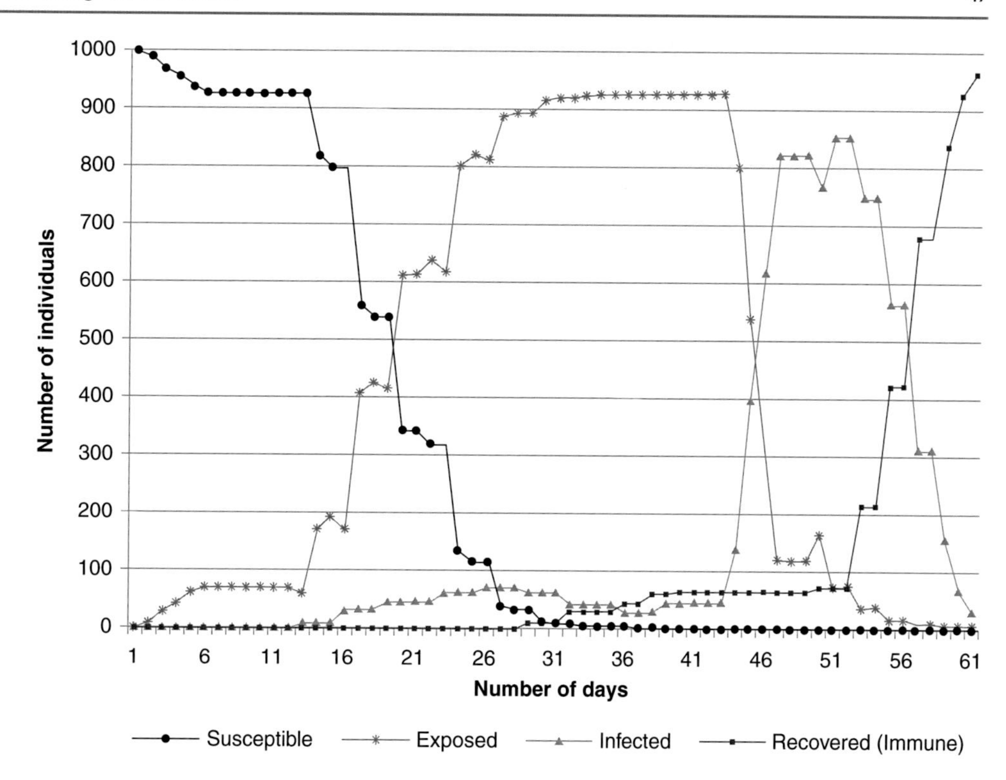

Fig. 6.6 Simulation results for SEIR agent-based model. (From [14])

Conclusion

In this chapter, we have presented a brief overview of how computational modeling and simulation can be used to support healthcare research. In particular, we have described two modeling approaches, system dynamics models and agent-based models, commonly used in healthcare research. Examples showing the use of these models in the epidemiology domain are used to demonstrate the importance of selecting an appropriate model; that is, a model having sufficient resolution to address the questions being asked about the simuland.

References

1. Mielke RR, Scerbo MW, Gaubatz KT, Watson GS. A multidisciplinary model for M&S graduate education. Int J Simul Process Model. 2009;5:3–13.
2. Cassandras CG, Lafortune S. Introduction to discrete event systems. 2nd ed. New York: Springer; 2008.
3. Luginbuhl W. Prevention and rehabilitation as a means of cost-containment: the example of myocardial infarction. J Public Health Policy. 1981;1(2):1103–15.
4. Roberts C. Modelling the epidemiological consequences of HIV infection and AIDS: a contribution from operations research. J Oper Res Soc. 1990;41:273–89.
5. Dangerfield B. Model based scenarios for the epidemiology of HIV/AIDS; the consequences of highly active antiretroviral therapy. Syst Dyn Rev. 2001;17:119–50.
6. Kivuti-Bitok L. System dynamics model of cervical cancer vaccination and screening interventions in Kenya. Cost Eff Resour Alloc. 2014;12:26.
7. Homer J. System dynamics modeling of national cocaine prevalence. Syst Dyn Rev. 1993;9:49–78.
8. Lane D. Looking in the wrong place for healthcare improvements; a system dynamics study of an accident and emergency department. J Oper Res Soc. 2000;51:518–31.
9. Macal C, North M. Tutorial on agent-based modelling and simulation. J Simul. 2010;4:151–62.
10. Thorne B, Bailey A, Peirce S. Combining experiments with multi-cell agent-based modeling to study biological tissue patterning. Brief Bioinform. 2007;8(4):245–57.
11. Tong X, Chen J, Miao H, Li T, Zhang L. Development of an agent-based model (ABM) to simulate the immune system and integration of a regression method to estimate the key ABM parameters by fitting the experimental data. PLoS One. 2015;10(11): e0141295.
12. El-Sayed A, Scarborough P, Seemann L, Galea S. Social network analysis and agent-based modeling in social epidemiology. Epidemiol Perspect Innov. 2012;9(1):1.
13. Heath B, Hill R, Ciarallo F. A survey of agent-based modeling practices (January 1998 to July 2008). J Artif Soc Soc Simul. 2009;12(4):9.
14. Perez L, Dragicevic S. An agent-based approach for modeling dynamics of contagious disease spread. Int J Health Geogr. 2009;8:50. https://doi.org/10.1186/1476-072X-8-50.

Part II

Finding and Making Use of Existing Literature

Seeking, Reviewing and Reporting on Healthcare Simulation Research

7

David O. Kessler, Marc Auerbach, and Todd P. Chang

Overview
On commencing any research project, a vital step is to conduct a thorough review of the literature. Over the last decade, healthcare simulation has matured as a distinct research field. However, the conceptual underpinnings for many areas of research within the field can be found in journals of other disciplines. This underscores the importance of adopting a broad search strategy when investigating literature that might aid with a current research question. In this chapter, we discuss how to seek and critically appraise the literature in healthcare simulation.

Practice Points

- Conducting a literature search for a healthcare simulation topic is an essential first step in research.
- A literature search in healthcare simulation involves querying within the clinical subject matter, simulation process, and simulation modality domains.
- Critically appraising healthcare simulation research requires consideration of the *validity, results, relevance and generalizability* of individual studies.
- Reporting guidelines can aid in the critical appraisal of other studies and the development and dissemination of new research.

D. O. Kessler
Vagelos College of Physicians & Surgeons, Department of Emergency Medicine, Columbia University, New York, NY, USA
e-mail: dk2592@cumc.columbia.edu

M. Auerbach
Pediatrics and Emergency Medicine, Yale University School of Medicine, New Haven, CT, USA
e-mail: marc.auerbach@yale.edu

T. P. Chang (✉)
Department of Pediatric Emergency Medicine, Children's Hospital Los Angeles/University of Southern California, Los Angeles, CA, USA
e-mail: tochang@chla.usc.edu

Introduction

There are many reasons why somebody might embark in a search for literature on a topic within simulation. One might be trying to initiate a new educational program, study a specific clinical or safety problem, develop a new technology, or design a study to test out a hypothesis on any of the above. Learning and exploring what's already been investigated is an important step before embarking on a new program or research project. Finding all of the relevant literature can be a challenge in a relatively young research field such as healthcare simulation, as often the vocabulary for certain concepts may not be well articulated or as standardized as in more established fields. Many of the theoretical concepts underlying healthcare simulation have their origins in other fields such as psychology, human factors, or education. Certain methodologic concepts in simulation, such as debriefing, have deeper roots in the literature of non-healthcare fields such as aviation, military, or even the engineering industry. To make the best use of the literature, simulation specialists and researchers should first identify their goals in conducting the search. This in turn will guide specific strategies that will be most helpful to answer the question or questions being asked. Even if the goal is simply to "keep up", becoming familiar with how to find and critically appraise the literature germane to your field is a useful skill. This chapter outlines the initial steps of forming that inquiry and developing expertise: *Finding & Making Use of Existing Literature.* We define *literature* as the cumulative works – in any medium, writing or otherwise – of experts within the field that has some level of review or peer-review.

Conducting a Literature Review

A *literature review* locates and curates a focused set of evidence designed to characterize a phenomenon regarding simulation or answer a specific question. It can be exhaustive

D. Nestel et al. (eds.), *Healthcare Simulation Research*, https://doi.org/10.1007/978-3-030-26837-4_7

or narrowly focused. Although analysts and those with expertise in library science can be very helpful in conducting an exhaustive review of the literature, most queries can easily be started alone using relevant search terms on an electronic or online database.

When conducting the first literature search in a new field, the hierarchy of evidence should be considered. For example, larger, sweeping reviews such as meta-analyses, systematic reviews, or scoping reviews would provide a pre-curated landscape of the literature. Each of these review articles will have its own citations that can be specifically found for more detailed study. When probing deeper into a field for simulation research, searching for specific peer-reviewed manuscripts with randomized-control trials, observational trials, and pilot studies will be useful. Case reports or descriptions of simulation scenarios and letters to the editor are lesser in impact but may still be valuable. The strength of evidence follows an hierarchy that is often used in clinical studies, but is also applicable to simulation studies [1]. Figure 7.1 provides a graphical representation of the levels of evidence in scholarly work for simulation, with examples in Table 7.1. Although these types of studies may not be reflected exactly in simulation research, the framework provides guidance on the strength of the findings reported.

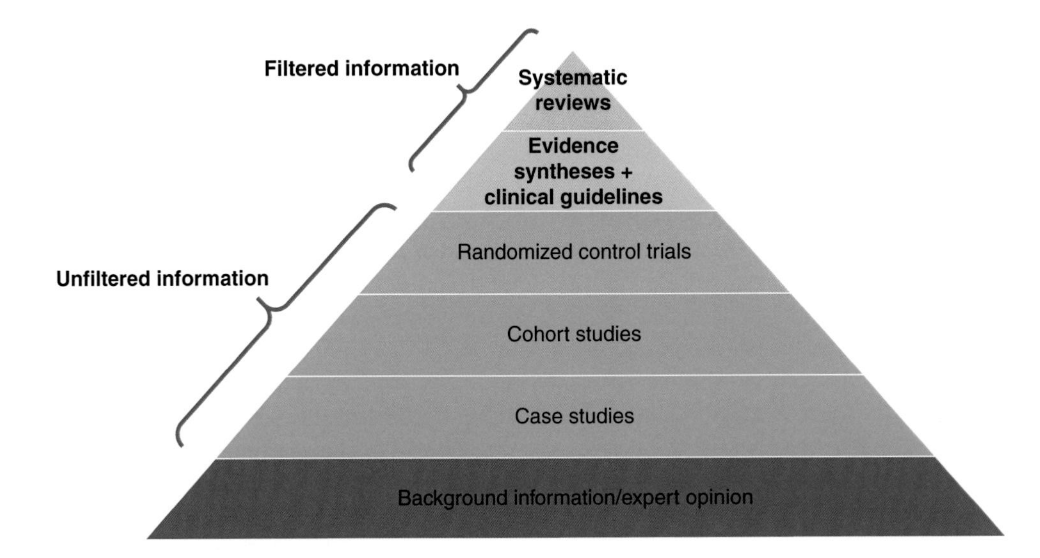

Fig. 7.1 A graphical representation of the levels of evidence in scholarly work for simulation

Table 7.1 Pyramid of evidence examples

Level of evidence	Reference	Synopsis
Meta-analysis or systematic review	Khan et al. Virtual reality simulation training for health professions trainees in gastrointestinal endoscopy. *Cochrane Database Syst Rev* 2018;8:CD008237	The number of studies was small, but virtual reality appears to complement standard simulation practices. Insufficient evidence exists to recommend using VR as a replacement for other simulation modalities
Evidence synthesis/clinical guideline/scoping review	Williams et al. Simulation and mental health outcomes: a scoping review. *Adv Simul (Lond)* 2017;2:2	A 5-stage scoping methodology found a variety of simulation-based education methods revealed in 48 articles; no patient outcomes were reported in them
Randomized Control Trial (RCT)	Cheng et al. Optimizing CPR performance with CPR coaching for pediatric cardiac arrest: A randomized simulation-based clinical trial. *Resuscitation* 2018;132:33–40	Resuscitation teams were randomized to having a trained CPR coach or no CPR coach during simulated resuscitations to determine CPR quality. CPR Coach improved overall CPR quality
Cohort study/observational study/quasi-experimental study	Auerbach et al. Differences in the Quality of Pediatric Resuscitative Care Across a Spectrum of Emergency Departments. *JAMA Pediatr* 2016;170(10):987–94	A prospective chort study evaluated multiple types of emergency department (ED) teams in pediatric simulations. Differences in performances were found among pediatric vs. general EDs
Case study/single-arm study	McLaughlin et al. Impact of simulation-based training on perceived provider confidence in acute multidisciplinary pediatric trauma resuscitation. *Pediatr Surg Int* 2018;34(12):1353–62	Self-reported confidence improved over time following a series of trauma simulations
Case report/technical report	Tjoflåt et al. Implementing simulation in a nursing education programme: a case report from Tanzania. *Adv Simul* 2017;2:17	A thorough description of a simulation implementation, including context, content, and evaluation

There are two principal styles of literature review. A *horizontal* literature review is shallow but wide. This style seeks to get a quick, birds-eye-view of the literature landscape, in which scholarly articles or chapters are gathered from a series of purposeful search phrases and specific databases. Because search databases are updated frequently, all searches should be documented with date/time, the database chosen, and the exact search phrase. This eliminates redundancy when having to update the literature in the future. A *vertical* literature review builds on the horizontal literature review and seeks to deepen understanding. Once several promising literature are found, the vertical review will find further generations of literature from the citations of the initial set; this allows delving into older but potentially strong, sentinel or foundational articles.

Searching by Category

Although Reporting Guidelines for simulation-based research recommend the term 'simulation' or 'simulated' to be searchable in the Title or Abstract [2], simply searching by these two terms will be insufficient to narrow the retrieval. Searches should start with keywords or phrases that represent one of three phenomena of study.

The first possible search is *subject matter*. Examples include an actual clinical condition (e.g., 'cardiac arrest'), clinical therapy (e.g., 'knee arthroscopy'), or phenomenon (e.g., 'mass casualty'). Most simulation activity strives to replicate – to varying degrees – a real healthcare situation, and the subject of inquiry should begin with that which is being simulated. Retrieval results using this method often have non-simulation based literature, which can be helpful to characterize the subject itself.

The second search inquiry is that of *a process*. Finding literature on a process that is agnostic to the clinical scenario is often used for skills or behaviors, such as 'teamwork,' 'response time,' 'decision-making.' Searching initially for process can provide a large number of non-simulation literature, but also manuscripts that cross into different specialties or disciplines. This may be useful when studying a process from a truly interdisciplinary manner. For example, when conducting a literature review prior to studying teamwork performance during a cardiac arrest scenario, searching initially for teamwork may yield literature from operating rooms, labor & delivery, critical care, and from pre-hospital care. There will likely be a collection from a human factors standpoint, descriptions on patient safety initiatives, or actual simulation scenarios.

The third inquiry is for a *specific modality*. These include specifying the search term to 'manikin,' 'serious game,' or 'simulated patient.' This search strategy should yield useful literature on experiences within simulation confined to the modality and perhaps show a breadth of possibilities and uses that are not often considered. It is important to note spelling conventions internationally differ (e.g., manikin vs. mannequin) and to note changes in the terms used over time (e.g., a shift from standardized patient to simulated patient). Technical articles outlining specific applications of these modalities tend to be found well using this method.

Sources

In the literature review process, the database through which the search queries are conducted are vital. Different databases have a particular theme within their scholarly work. For example, the Cochrane database (https://www.cochranelibrary.com/) hosts many large-scale systematic reviews on various medical content. PubMed is a searchable database (https://www.ncbi.nlm.nih.gov/pubmed/) hosted by the US National Library of Medicine and National Institutes of Health that serves as a repository for indexing clinical and basic science research pertinent to the field of healthcare. Although research in the fields of psychology, engineering, and simulation are represented within this database, PubMed is not an exhaustive source for all of the publications in those fields. Additional examples of databases that are useful to the simulation research community include ERIC (https://eric.ed.gov/), a database with scholarly publications within education, and PsycInfo, a database on psychological studies and findings (https://www.apa.org/pubs/databases/psycinfo/index.aspx). Both of these can be useful when amassing literature on simulation as an educational endeavor or when simulation is used to study human behavior. Additional databases may be available in the technical realms, such as engineering (IEEE) or biomedical engineering. Meta-databases that are agonistic to discipline include Web of Knowledge and Google Scholar, which can be both comprehensive but potentially overwhelming.

It is possible, given the relative youth of simulation as a scholarly field, that the phenomenon of interest is not well represented in a literature review. This represents an opportunity for a simulation researcher to fill in the gap through conducting a scoping review or a systematic review and publishing it as a foundational piece of literature. Further details on how to do this are outlined across this book.

Synthesizing and Appraising the Evidence

Before embarking on a literature review it is important to determine the intended "output" from the review. If the aim is to publish the output as a manuscript (e.g., systematic review) the level of rigor is much higher than if the aim is to provide the foundation for a research project and/or a summary of the current evidence in an unpublished format. Critical appraisal of research is *the process of assessing and interpreting evidence by systematically considering its validity, results, relevance and generalizability*. There are diverse resources that can be used to support the process of appraising research including established guidelines related to the research methodology (randomized trials – CONSORT [http://www.consort-statement.org/], observational studies – STROBE [https://www.strobe-statement.org/index.php?id=strobe-home], qualitative studies – COREQ [https://academic.oup.com/intqhc/article/19/6/349/1791966] and guidelines related to the topic of research (health-professions education – MERSQI [3], quality improvement – SQUIRE [http://www.squire-statement.org/index.cfm?fuseaction=Page.ViewPage&PageID=471].

When authors intend to publish the findings in a systematic review or meta-analysis they should reference the best practices from those fields (Cochrane [http://training.cochrane.org/handbook], PRISMA [http://www.prisma-statement.org], BEME [https://www.bemecollaboration.org/]). There are a growing number of guidelines and these have been populated into a single repository on EQUATOR (https://www.equator-network.org/).

Guidelines can be used to help to describe best practices in conducting and reporting research and in a relatively "young" field of research such as simulation it is likely that articles will vary in their quality and compliance with these guidelines. There are a variety of established methodologies for rating the quality of research that should be used evidence to inform how research findings inform practice (GRADE – Grading of Recommendations Assessment, Development and Evaluation – [http://www.gradeworkinggroup.org]).

Simulation-Based Research Reporting Guidelines

The unique features of simulation-based research highlight the importance of clear and concise reporting of research. In 2016 the first simulation-based reporting guidelines were published by four simulation journals in an effort to describe simulation specific elements that are recommended for the reporting of randomized trials and observational studies. These include *Simulation in Healthcare*, *Advances in Simulation*, *Clinical Simulation in Nursing*, and *The British Medical Journal Simulation & Technology-Enhanced Learning* journals [2]. The authors stress the need to be transparent about the role of simulation in the research and clarity if simulation is the subject of research or as an investigational methodology for research.

Simulation as the subject of research often is grounded in theory that has been described in psychology, education and other disciplines and it important for authors to describe this theory. This measurement in this type of simulation research often relies on detailed descriptions of the context and training of the trainers as well as raters involved in the study. A major benefit of simulation as a research methodology is the reproducibility and standardization of a clinical event. This requires authors to describe the simulation and assessment methods in sufficient detail including the simulators, scenarios, assessment tools, and the methods of data collection. In describing the assessment methodology authors should provide evidence to support the validity argument in the context of their work [4]. The context of the simulation must also be described in detail- for example if the simulation was conducted in the clinical environment or a simulation center. Participants' prior experiences with simulation may impact their performance in the study, therefore it is important to describe this when reporting on the population of interest. This should also include details about how participants were oriented to the simulator, environment and equipment. Due to the word constraints providing these details often requires the use of appendices or online supplements to the work. Lastly, the limitations of simulation-based outcomes and how they relate to patient outcomes/generalizability should be provided.

Future efforts should iteratively update these guidelines and work is needed to create similar extensions to other existing guidelines for simulation-based research.

Closing

In closing, finding and making use of the existing literature is important to optimally conduct simulation activities and/or simulation research in the most evidence-based fashion. Knowledge of how and where to search for scholarly publications, practice in critically appraising the literature, and disseminating one's own scholarly activity through standardized methods of reporting, will help to move the field forward and upward in the climb towards better evidence and best practice.

References

1. Burns PB, Rohrich RJ, Chung KC. The levels of evidence and their role in evidence-based medicine. Plast Reconstr Surg. 2011;128(1):305–10.
2. Cheng A, et al. Reporting guidelines for health care simulation research: extensions to the CONSORT and STROBE statements. Adv Simul. 2016;1:25.
3. Reed DA, et al. Association between funding and quality of published medical education research. JAMA. 2007;298(9):1002–9.
4. Cook D, Hatala R. Validation of educational assessments: a primer for simulation and beyond. Adv Simul. 2016;1:31.

Systematic and Nonsystematic Reviews: Choosing an Approach

8

David A. Cook

Overview

Systematic reviews and purposive (nonsystematic) reviews serve valuable and complementary roles in synthesizing the results of original research studies. Systematic reviews use rigorous methods of article selection and data extraction to shed focused, deep light on a relatively narrow body of research, yet of necessity may exclude potentially insightful works that fall outside the predefined scope. Purposive reviews offer flexibility to address more far-reaching questions and pursue novel insights, yet offer little assurance of a balanced perspective on the issue. This chapter reviews the strengths and weaknesses of each approach, and suggests specific questions to help researchers select among these approaches. Different approaches to quantitative and narrative research synthesis, including meta-analysis, are also described.

Practice Points

- Systematic and purposive (nonsystematic) reviews serve valuable and complementary roles.
- Thoughtful evidence synthesis is arguably the most important part of any review; both quantitative and narrative approaches are effective.
- In choosing a review approach, reviewers might ask: What is the purpose of the review? What is the current state of the literature? and, Which set of limitations matter more?

Health professions education research has shown tremendous growth in recent years, and with this comes an increased need for articles that synthesize the findings from individual original research studies. Research syntheses (often called "review articles") serve at least two distinct yet complementary purposes: they provide a succinct summary of what is known about a given topic, and they highlight gaps in our understanding that may warrant increased attention in future research.

Various labels are applied to reviews of different "types," including systematic reviews, narrative reviews, critical reviews, scoping reviews, realist reviews, rapid reviews, and state-of-the-art reviews. Indeed, one group described 14 distinct review types [1]. However, I find such categories difficult to reliability discriminate in practice, not only because there are no universally-accepted definitions but because the boundaries overlap. For example, nine of these 14 review types previously mentioned were variations of a "systematic" review (e.g., qualitative systematic review, rapid review, and systematized review).

I prefer a simpler approach that classifies review articles as systematic or non-systematic (or, to use a less judgmental term, "purposive"). As will be elaborated later on, "systematic" reviews use a defined and reproducible approach in selecting articles and extracting data. Purposive reviews follow a more strategic and adaptable approach to selection and extraction. While some researchers disparage purposive (non-systematic) reviews, others criticize systematic reviews. Yet I believe that both systematic and purposive reviews have strengths and weaknesses that make them more or less appropriate depending on the researcher's purpose or question. Another distinguishing feature of review types is whether they synthesize the original research findings using quantitative or qualitative methods; these distinctions are even more blurred than those for article selection and data extraction.

The goal of this chapter is to provide guidance to readers (i.e., would-be writers of reviews) on how to align their choice of review type and methods with their purpose. I will highlight a fundamental conceptual distinction between systematic and purposive reviews, present three questions to guide the selection among review types, describe approaches

D. A. Cook (✉)
Mayo Clinic Multidisciplinary Simulation Center, Office of Applied Scholarship and Education Science, and Division of General Internal Medicine, Mayo Clinic College of Medicine and Science, Rochester, MN, USA
e-mail: cook.david33@mayo.edu

D. Nestel et al. (eds.), *Healthcare Simulation Research*, https://doi.org/10.1007/978-3-030-26837-4_8

to data synthesis, and conclude with a seven-step approach to planning a review that focuses on principles relevant to all review types. I will also touch briefly on three types of review that may contain elements of both systematic and purposive reviews, namely realist [2], scoping [3, 4], and state-of-the-art reviews.

Strengths and Limitations of Systematic and Purposive Reviews

Systematic reviews use predefined criteria for study inclusion and seek to extract the same information from each study, which usually includes a formal appraisal of study methodological quality. They often (but not always) use quantitative approaches to synthesis, which may include meta-analysis. The systematic approach identifies a comprehensive list of studies relevant to the research question, and distills presumably important information about each study. If done well, it defines the current state of research as regards chosen topic. It also helps to identify research gaps (e.g., populations or interventions notably absent among the studies found), characterize methodological strengths and deficiencies across studies, and avoid the bias that might arise from selecting only studies that support the author's preconceived position. The systematic approach would (in theory) allow another investigator to replicate the results and arrive at similar conclusions. However, reliance on a specific search strategy and distinct inclusion criteria prevents the systematic review from pursuing findings and ideas that are broadly relevant and potentially insightful but strictly fall outside the predefined scope. They are often perceived as narrow, sterile, and detached from the practical complexities of daily life. Systematic review are like lighthouses – they cast a powerful beam that illuminates the intended area of study, but leave the rest of the ocean in the dark. In addition, systematic reviews are not free of bias; every review involves countless decisions including those regarding the scope, search strategy, inclusion criteria, data selected for extraction, extraction process, and presentation of results.

Purposive reviews allow the researcher to reflect broadly upon a theme, drawing upon research, frameworks, and philosophy both within their field and from other fields (e.g., outside of health professions), and thereby yield insights that a systematic review could never achieve. Strategic selection of articles, unencumbered by the rules of the systematic review, further allows researchers to pursue ideas and findings that emerge unexpectedly during the process of the review, and to include a diverse spectrum of research methods. Discordant findings can be used to identify novel insights. Rather than comprehensively define the current state of evidence (what works), purposive reviews tend to address more far-reaching questions and generate novel insights about why and how. However, there is no guarantee that the articles cited represent a balanced perspective on the issue; relevant work could have been inadvertently missed or even deliberately ignored. Purposive reviews act as a floodlight, illuminating a large area immediately near the source, but missing possibly important regions that lie farther away.

The strengths and limitations of systematic and purposive reviews parallel those of quantitative and qualitative research [5]. Both quantitative research and systematic reviews prefer large samples (of human participants or research studies) and emphasize systematic sampling. To minimize error, researchers seek that all subjects/studies be as similar as possible, and differences are viewed as error to be averaged out if possible. By contrast, both qualitative research and purposive reviews emphasize purposive, iterative sampling that shapes and is shaped by emerging insights. Rather than large samples, these approaches emphasize integrating information from multiple sources (triangulation). Differences between subjects/studies are viewed as opportunities to identify novel insights, often through extended data collection and new subjects/studies. Quantitative and qualitative research approaches are generally accepted as complementary, and the same should be true of systematic and purposive reviews.

Specific Review Subtypes

Realist Reviews and Scoping Reviews

Realist [2] and scoping [3, 4] reviews have received increased attention and use in recent years. Each employs a systematic albeit nonlinear approach to article selection and data synthesis, and can be considered a type of systematic review. However, they merit special attention because they have distinct purposes and defined methods, and some methods share features of purposive reviews.

The realist review was introduced as "a new method of systematic review designed for complex policy interventions" [2]. The realist approach is also well-suited for educational activities, which are typically complex. Realist reviews seek to elucidate the theoretical foundations of a given intervention or phenomenon, with particular emphasis on contextual influences (i.e., what works, for whom, in what context, and why). They are systematic in the sense that they use rigorous, transparent, and reproducible methods to search and synthesize the literature. However, a realist review explicitly involves a search for relevant theories and uses these theories to interpret the evidence found. Additionally, the search strategy and selection criteria typically evolve during the review and may include purposive elements [2].

Scoping reviews seek to provide a comprehensive snapshot of the literature in the field. They too are systematic in their use of rigorous, reproducible methods. However, in contrast to the traditional systematic review in which the search strategy, inclusion criteria, data extraction items, and data analysis are largely pre-planned, each of these components typically evolves during the course of a scoping review. The authors identify up-front the scope of the review and define preliminary criteria for each component, but then add to and adjust each component as their understanding of the field grows. Scoping reviews may or may not report the actual results of any study, and sometimes include non-original-research literature such as editorials and other reviews. A well-done scoping review will identify and catalog the key terms, concepts, interventions, outcomes, and study designs extant in the field, thereby creating a map to guide future researchers and reviewers.

State-of-the-Art Reviews

State-of-the-art reviews provide an analysis of current work in the field, typically using a specific (recent) date as the criterion for inclusion (e.g., the last calendar year or past 5 years). They can otherwise adopt the methods of any other review type, with the corresponding strengths and weaknesses. The chief advantage is the emphasis on recent work, which is particularly important in a fast-moving field.

Options for Synthesizing the Evidence

All literature reviews extract evidence of some kind from the publications identified, such as numeric data, statistical test results, or themes. Synthesizing this evidence effectively is arguably the most important part of any review. I consider the synthesis approach separately from the review "type," since both systematic and purposive reviews can appropriately use a broad spectrum of methods to synthesize and report their findings. Indeed, since text reporting and data visualization are inextricably linked with the synthesis process, there are an essentially infinite number of possible approaches to synthesis. Broadly speaking, however, synthesis approaches can be viewed as quantitative – presenting results as numbers; and qualitative – presenting results as a narrative (words). Whether quantitative or qualitative, data synthesis is an art that requires reviewers to put themselves in the shoes of the reader to anticipate and answer their questions, and provide relevant, succinct, and self-explanatory summaries and visualizations of supporting data.

Quantitative synthesis includes meta-analysis and a variety of other methods for reporting and integrating numeric data. Meta-analysis is simply a statistical technique that averages ("pools") the results of several research studies, and estimates the magnitude of between-study differences (heterogeneity or inconsistency) that could signal important differences in interventions, participants, settings, outcome measures, or study designs. Meta-analysis can also be used to examine, and hopefully explain, such inconsistencies. Although meta-analysis and systematic review are often colloquially viewed as interchangeable, in fact they are distinct. Many systematic reviews (likely a majority) use non-meta-analytic methods to synthesize results. Conversely, meta-analysis could, in principle, be applied to any review type; however, it is rarely employed in purposive reviews since most researchers would consider the results misleading in the absence of a systematic identification of studies (i.e., the pooled "best estimate of effect" would be inaccurate if any relevant studies were omitted). The question often arises: Is it appropriate to pool these results using meta-analysis? The answer always depends on the question asked; pooling across different populations (e.g., medical students and residents), interventions, outcomes, and study designs may or may not be appropriate depending on whether the resulting number makes sense and helps to answer the question. As I stated previously, "The most challenging aspect of conducting a meta-analysis … is determining whether the original studies address a common question or framework. Analytic measures of inconsistency can help with this determination, but ultimately this is a conceptual – not a numeric – decision" [6]. Performing a meta-analysis does require skill with the statistical technique, but more important is to know what analyses are needed to support a meaningful and practical message.

Non-meta-analytic numeric synthesis can use a variety of tables, figures, and text to effectively report numeric data without pooling, for both systematic and purposive reviews. However, such reporting should emphasize the magnitude of effect (effect size) rather than the results of statistical tests. Effect sizes for different study designs include raw or standardized differences in scores, correlation or regression coefficients, and odds ratios. Reporting only the results of statistical tests (e.g., "Three studies found a statistically significant benefit.") – so-called "vote counting" – is flawed for at least two reasons. First, vote-counting ignores the magnitude of effect: a large difference may be non-significant if the sample size is small, whereas even tiny differences will reach statistical significance with a large sample size. Second, it relies on a fixed notion of statistical significance (the P threshold of 0.05, while commonly used, is in fact arbitrary).

Most reviews – including systematic reviews and even meta-analyses – employ at least some features of qualitative (narrative) synthesis. Narrative synthesis is hard work! In addition to avoiding vote-counting, reviewers must not present a "litany of the literature" in which the results of each study are described in turn with only minimal integration. Rather, a good synthesis will first interpret and integrate the findings to reach a "bottom line" message that incorporates the strengths, weaknesses, inconsistencies, and gaps in the evidence, as well as potential moderators such as populations, study designs, and contextual factors; and will then report this message together with a succinct summary of the evidence that supports the message. Narrative synthesis works with both numeric data and qualitative data.

Which to Use?

Deciding which type of review to employ depends on the answers to at least three questions.

First, and usually most important, *what is the purpose of the review*? Traditional systematic reviews address focused questions within a defined field. They seek to provide a comprehensive snapshot of current evidence within that field, including a bottom-line appraisal of "Does it work?" They typically identify areas in which evidence is lacking either from a paucity of studies, or from shortcomings in the available studies. Purposive reviews tend to address broader, far-reaching, and less defined questions. They seek to integrate findings across fields, often focusing on "Why or how it works?" in addition to the simpler "Does it work?" They likewise identify areas of needed research, but typically frame these as thematic deficiencies rather than limitations in the number or quality of studies. Some purposive reviews even redefine the question itself, refocusing or reframing our understanding of and research priorities for the field. Scoping reviews seek to present a snapshot of the published literature in a specified field. Realist reviews seek to understand the theoretical foundations for the selected intervention, with emphasis on contextual interactions (what works, for whom, in what context).

Second, *what is the current state of the literature*? Of course, answering this question is one of the reasons to do a review; but the researcher should have some sense of the answer. If there are lots of studies, and especially if the studies are of high quality and/or address very similar questions (e.g., the same type of intervention or the same population), then it might be reasonable to pursue a comprehensive listing and quantitative synthesis of these studies using a systematic review. Conversely, if there are few relevant studies, or the available studies reflect a variety of approaches, participants, interventions, or questions, then a purposive review might be more appropriate. In this case, the purposive review would allow the researchers to look beyond these few studies to identify work done in other fields, or work that addresses other questions and illuminates the topic even if not directly relevant. Of course, one could do a systematic review of very few studies, or a purposive review of a large body of evidence. A scoping review is helpful if the state of the field – the vocabulary, theories, interventions, outcomes, and overall volume of evidence – are truly unknown. A realist review requires a modest number of studies to explore the possible contextual interactions, and compose meaningful evaluation of the underlying theories.

Third, *which set of limitations matter more*? Systematic reviews are limited by reviewers' preconceived notions (biases) that may manifest in the inclusion criteria, the data selected for extraction, the processes of inclusion and extraction, the presentation of results, and the final conclusions. Adherence to the protocol prevents reviewers from pursuing interesting findings that fall outside the scope of the question and inclusion criteria. Most systematic reviews fail to accommodate the complexity of social interventions and interactions. Finally, the implicit trust that many readers naively render to systematic reviews could be viewed as a limitation. By contrast, purposive reviews suffer from clearly subjective inclusion criteria and data extraction, but at least they avoid any pretense of objectivity [7]. They also avoid the constraint of adhering to a protocol, and can more readily accommodate complexity. Scoping reviews are limited by incomplete appraisal of study quality, and by limited synthesis of evidence. Realist reviews are limited by the absence of quantitative synthesis and by the subjectivity encountered in identifying relevant theories and original research studies. Finally, both purposive and realist reviews require that the reviewers possess a fairly advanced understanding of the topic (i.e., to purposively identify relevant studies, theories, and conceptual frameworks), whereas in systematic and scoping reviews this understanding can develop over the course of the review.

Choosing the synthesis approach is dictated primarily by the needs of the emerging message. The message, and thus these needs, can be anticipated up front (e.g., if the purpose is to quantitatively summarize current evidence, then a meta-analysis may be required). However, in many cases the ideal synthesis approach evolves as reviewers examine the accumulating data and contemplate how best to share the insights they are discovering. For example, the data might simply not support a planned meta-analysis, or a graphical representation of numeric data may be added to complement a planned narrative synthesis. As I conceive the tentative synthesis approach during the planning stage, I typically write out a rough draft of the Results section, including sketching key tables and figures.

Two often-cited considerations should *not* be part of these decisions: time and team size. All of these review types require a substantial investment of time to do well; none of them should be viewed as a "fast track to a publication." Time will be determined more by the volume of literature reviewed than the review type per se. All of these review types also require a team approach; a minimum of two reviewers, and often substantially more, is required in conducting a high-quality review to avoid systematic bias, minimize random error, deepen insights, enhance interpretation, and distribute the workload.

A Seven-Step Approach to Planning a Review

In closing I will share seven tips for planning a review of any type, based on points I outlined previously [8].

Clarify the Question

All research projects begin with a clear question or purpose, and literature reviews are no exception. As acknowledged above, questions can differ widely, variously focusing on identifying problems, clarifying theory, testing theory, quantifying impact, or mapping the current state of the field. Borrowing from a framework first proposed for original research studies [9], a review's purpose "might be classified as description (historical or descriptive overview), justification (synthesis of evidence to identify the current state with weak reference to a conceptual framework), or clarification (synthesis of evidence to understand mechanisms, identify gaps, and build a conceptual framework)" [8]. Although all these purposes have merit, clarification studies tend to advance our understanding more than descriptions or justifications [9].

Pick an Approach That Matches the Question

Once the question has been identified, reviewers must select the review type most appropriate to answer that question. As outlined above, these decisions revolve around systematic vs. purposive approaches to study identification and data extraction, and quantitative vs. qualitative approaches to data synthesis.

Plan Defensible Methods

Ideally, reviewers will develop and follow a written plan for conducting the review. The planned methods will depend upon the question and the selected review type. Methods for systematic reviews have been described in books [10, 11], journals [12, 13], and online resources [14, 15], and *reporting* guidelines like the Preferred Reporting Items for Systematic Reviews and Meta-Analyses [16] (PRISMA) highlight important methodological considerations. Guidelines for scoping [3, 4] and realist [2, 17] reviews have also been outlined. By contrast, purposive reviews are much more flexible, and as such do not have universal standards. However, principles of high-quality *original* qualitative research can provide guidance in conducting a high-quality qualitative *literature synthesis* as well; these principles include clarification of purpose, recognition of researcher assumptions and perspectives (reflexivity), working in research teams, purposeful sampling, thoughtful analysis that makes a conscious effort to consider alternate perspectives, and detailed presentation ("rich description") of evidence that both supports and counters the bottom line message.

Set the Stage

Just as with original research, the Introduction should set the stage for the review by summarizing relevant literature to justify the need for a review on this topic, and to clarify relevant theories and frameworks. In justifying the need, reviewers should highlight the strengths and shortcomings of relevant previous reviews rather than citing original research, since relevant original research studies will typically be identified during the review and then cited in the results. Shortcomings in previous reviews do not necessarily arise from methodological weaknesses; they can also arise from differences in their scope, age, type, and synthesis approach. The Introduction should clarify how these shortcomings leave an important gap in our understanding, and how the proposed review will fill this gap.

Organize and Interpret to Share a Clear Message

Reviewers often focus their efforts on identifying and selecting studies and extracting information from them. However, it is equally important – and often more challenging – to effectively synthesize the results of these studies into a meaningful and well-supported message. A review is only as good as its bottom line message; the method of synthesis is a vitally important means to that end.

Appraise Study Quality and Explore the Impact on Review Conclusions

Depending on the review question and scope, the evidence collected and reviewed might take many forms – randomized trials, non-randomized experiments, correlational studies, surveys, assessment validation studies, and various forms of qualitative research. Each of these study designs has "best practice" features that, if followed, strengthen our confidence in the results. These features should be taken in account when drawing conclusions from the data synthesis. The appraisal of study quality is a formal part of most traditional systematic reviews; in all other review types, the quality appraisal should be equally thorough albeit perhaps less formal. Importantly, it is not enough to just describe or enumerate the various design features (as I have often seen done). Rather, the strengths and weaknesses of a given study should determine the degree to which those results influence the bottom-line conclusions of the review. Finally, although method quality checklists have been developed for many study designs, it is the specific design feature (e.g., randomization, loss to follow-up, blinded assessment) not a total quality score that should be emphasized in conducting this integration. The relative importance of one design feature over another may vary for different reviews; a rote one-size-fits-all approach is discouraged [18].

Report Completely

Reviewers must present a complete and transparent report of what they did and what they found. In reporting their methods, reviewers should describe in detail what they did and the key decisions they faced. There are no "standard procedures" for any review type, and nothing can or should be taken for granted. In reporting their findings, they should describe in detail both the processes of the review (such as inter-rater agreement on inclusion or extraction, the source and number of studies considered and included, or the conceptual frameworks considered while interpreting results) and the methods and results of the included studies. Reporting guidelines (not to be confused with method quality appraisal tools) such as those for systematic [16], realist [17], and scoping [3, 4] reviews can, when available, remind reviewers what information to report. Limitations both in the review methods and in the number, quality, and relevance of the original research studies will influence the findings of the review. These limitations should be acknowledged and, as noted above, accounted for in formulating the synthesis.

References

1. Grant MJ, Booth A. A typology of reviews: an analysis of 14 review types and associated methodologies. Health Inf Libr J. 2009;26:91–108.
2. Pawson R, Greenhalgh T, Harvey G, Walshe K. Realist review – a new method of systematic review designed for complex policy interventions. J Health Serv Res Policy. 2005;10(Suppl 1):21–34.
3. Arksey H, O'Malley L. Scoping studies: towards a methodological framework. Int J Soc Res Methodol. 2005;8:19–32.
4. Levac D, Colquhoun H, O'Brien KK. Scoping studies: advancing the methodology. Implement Sci. 2010;5:69.
5. Cook DA. Narrowing the focus and broadening horizons: complementary roles for nonsystematic and systematic reviews. Adv Health Sci Educ Theory Pract. 2008;13:391–5.
6. Cook DA. Randomized controlled trials and meta-analysis in medical education: what role do they play? Med Teach. 2012;34:468–73.
7. Eva KW. On the limits of systematicity. Med Educ. 2008;42:852–3.
8. Cook DA. Tips for a great review article: crossing methodological boundaries. Med Educ. 2016;50:384–7.
9. Cook DA, Bordage G, Schmidt HG. Description, justification, and clarification: a framework for classifying the purposes of research in medical education. Med Educ. 2008;42:128–33.
10. Cooper H, Hedges LV, Valentine JC. The handbook of research synthesis. 2nd ed. New York: Russell Sage Foundation; 2009.
11. Higgins JPT, Green S. Cochrane handbook for systematic reviews of interventions. Chichester: Wiley-Blackwell; 2008.
12. Cook DA, West CP. Conducting systematic reviews in medical education: a stepwise approach. Med Educ. 2012;46:943–52.
13. Cook DJ, Mulrow CD, Haynes RB. Systematic reviews: synthesis of best evidence for clinical decisions. Ann Intern Med. 1997;126:376–80.
14. Higgins JPT, Green S. Cochrane handbook for systematic reviews of interventions 5.1.0 [updated March 2011]. Available at: http://handbook-5-1.cochrane.org/. Accessed 31 May 2018.
15. Campbell Collaboration. Campbell Collaboration Resource Center. Available at: https://campbellcollaboration.org/research-resources/research-for-resources.html. Accessed 31 May 2018.
16. Moher D, Liberati A, Tetzlaff J, Altman DG. Preferred reporting items for systematic reviews and meta-analyses: The PRISMA statement. Ann Intern Med. 2009;151:264–9.
17. Wong G, Greenhalgh T, Westhorp G, Buckingham J, Pawson R. RAMESES publication standards: realist syntheses. BMC Med. 2013;11:21.
18. Montori VM, Swiontkowski MF, Cook DJ. Methodologic issues in systematic reviews and meta-analyses. Clin Orthop Relat Res. 2003;413:43–54.

Part III

Qualitative Approaches in Healthcare Simulation Research

Introduction to Qualitative Research in Healthcare Simulation

9

Debra Nestel and Aaron W. Calhoun

Overview

In this chapter, we introduce and illustrate features of qualitative research through published healthcare simulation literature. We also outline key concepts in qualitative research and offer definitions of commonly used terms. Qualitative research is largely concerned with social phenomena, and hence studies people either directly or indirectly. This means that qualitative studies are subject to specific ethical considerations. We also acknowledge the many challenges that researchers with a primarily clinical background may have with qualitative research given that they have been educated almost exclusively in a biomedical/quantitative science model. Finally, we orient readers to the structure of the section and share valuable resources.

Practice Points

- Qualitative research is primarily person-focused, asking questions of *how and why*.
- Qualitative researchers accept that there are multiple realities and truths that are *constructed* by participants and researchers.
- Data in qualitative research is often textual – transcribed talk from interviews, focus groups or naturally occurring conversation (e.g. debriefings post-simulation). However, data can also be visual (e.g. simulation scenarios).
- Qualitative researchers practice *reflexivity* relative to the research focus, study participants and research process.

Introduction

Qualitative methodology, defined as a research approach that focuses primarily on the collection and analysis of person-focused data, is a relative newcomer to the field of simulation. In the past, much of the research produced by the simulation community has been quantitative in nature. This valuable work has been used to address such diverse issues as the impact of simulation on medication errors [1] and the use of simulation to assess performance in critical situations [2]. By the same token, however, there are questions that quantitative approaches are ill suited to address; in particular, questions of "how" or "why" social phenomena occur. Qualitative research is designed to address questions of this nature.

In this chapter, we introduce the key concepts for qualitative research and define key terms. These are illustrated through examples from the literature. We then discuss a number of practical considerations including ethics and common challenges. The chapter concludes with a brief outline of the rest of the section.

Key Concepts in Qualitative Research

Ontology and Epistemology

At the core of the difference between qualitative and quantitative research are issues of *ontology* and *epistemology*, and it is unwise to proceed without a firm grasp of these concepts. A simple distinction is that ontology addresses questions about the nature of reality, while epistemology addresses questions about the nature and meaning of knowledge and how we come

D. Nestel (✉)
Monash Institute for Health and Clinical Education, Monash University, Clayton, VIC, Australia

Austin Hospital, Department of Surgery, Melbourne Medical School, Faculty of Medicine, Dentistry & Health Sciences, University of Melbourne, Heidelberg, VIC, Australia
e-mail: debra.nestel@monash.edu; dnestel@unimelb.edu.au

A. W. Calhoun
Department of Pediatrics, University of Louisville School of Medicine, Louisville, KY, USA
e-mail: aaron.calhoun@louisville.edu

D. Nestel et al. (eds.), *Healthcare Simulation Research*, https://doi.org/10.1007/978-3-030-26837-4_9

to know. Most healthcare practitioners that have studied in a biomedical paradigm will, by default, hold to a positivist or post-positivist position in terms of these questions. By way of definition, *positivism* refers to a theoretical stance that views reality and knowledge in objective terms (i.e. they are things that are discovered "out there") and further understands them to be best investigated via the scientific method. In contrast, much qualitative research is performed from a *constructivist* perspective, which sees reality and knowledge as mental constructs that are not so much uncovered as "built" through human inquiry. Given what may be a 'default' position of positivism for many researchers with a clinical background, the material that follows on qualitative research may seem somewhat strange. Nevertheless, we encourage you to stay with the material, as qualitative approaches have much to offer in terms of better understanding the rich social processes that occur in educational and other simulation-oriented activities. It is also worth noting that these understandings are not mutually exclusive, and may readily complement each other depending on the area of investigation or the questions being asked. Bunniss and Kelly (2010) offer a helpful description of these concepts relative to medical education research [3].

Features of Qualitative Research

Flick (2014) identifies four "preliminary" features of qualitative research – appropriateness of methods and theories; perspectives of the participants and their diversity; reflexivity of the researcher and the research; and use of a variety of approaches and methods [4]. Here we briefly consider each of these. As we proceed, remember that qualitative research is most interested in seeing phenomena in context with all their complexity intact, while quantitative research tends to adopt a more reductionist approach that seeks to control variables and manage environmental conditions whenever possible.

Appropriateness of Methods and Theories

Qualitative research usually seeks to answer questions that start with "how", "why" or "what" rather than questions with binary responses starting with "does". Further, it seeks to deepen understanding of complex social phenomena. In qualitative research, there is an acknowledgement that variables cannot be controlled and that the phenomenon under investigation can best and perhaps only be understood in its context. This usually leads to data sources such as *talk* – as it occurs naturally (e.g. briefing prior to a team-based immersive simulation) or through interviews or focus groups (e.g. with simulation practitioners and/or learners about their practice and their learning). This *talk* is then transcribed and the resulting *text* used as data for analysis. Other sources include *visual data* obtained through direct observations and/or digital recordings of practice (e.g. team-based immersive simulations) and *documents* and other artefacts (e.g. reflective writing, policies, scenarios etc.). The methods used need to be carefully selected to fit the research purpose. Care must also be taken to acquire data that represents what the study participants actually think or feel and are not only driven by the conceptual schema of the investigator. Therefore, questionnaires that ask open-ended questions may yield valuable data for qualitative research while those with closed-ended or limited option answers (which typically depend to some extent on the preconceived notions of the investigator) would usually be unsuitable. It is important to note that simply asking open-ended questions in a survey format does not guarantee quality data, as responses cannot easily be probed or contextualized.

Perspectives of the Participants and Their Diversity

In qualitative research, researchers study participants' thoughts, feelings, values and experiences. Researchers acknowledge that participants will likely vary in these dimensions and that these variations are sought and valued. Subjectivity is celebrated rather than minimized. This approach reflects the ontologies of qualitative research and respects the existence of multiple experienced realities compared with the positivist and post-positivist positions where a single universally appreciated reality is assumed. Practically speaking, this means that sampling takes different forms. The values associated with sampling in quantitative studies such as randomisation, have little importance in qualitative research. We return to sampling below.

Reflexivity of the Researcher and the Research

"Reflexivity is an attitude of attending systematically to the context of knowledge construction, especially to the effect of the researcher, at every step of the research process" [5]. In qualitative research, researchers *position* themselves with their data. This positioning acknowledges the *subjective* nature of qualitative research. While this is unavoidable (and in many ways advantageous), it is also important that this be explicitly acknowledged by the researcher for the sake of those considering their work. Thus, as a process, reflexivity helps the researcher and others to remain aware of their thoughts, feelings, values and reactions to the research process. These actions of *reflexivity* then become part of the *data* itself. In practice, this is enacted by researchers making notes on their reactions during idea conception, when searching for and reading literature, while making decisions about study design, during data collection (e.g. interviewing), at transcription, at analysis (individually and as a team) and in the selection and presentation of findings. It is valuable to keep a reflexive diary or journal for each project with entries dated, which has the additional benefit of setting up an audit trail of actions, thoughts and values. Reflexivity should also be explicitly addressed in the final published work. We return to this idea later.

Variety of Approaches and Methods

As you read introductory resources on approaches to qualitative research you will find different types of classifications. These can be confusing to new researchers in this field. We offer examples of approaches described by eminent qualitative researchers and methodologists – Creswell and Poth (2019) [19] and Flick (2014) [4]. Table 9.1 represents a taxonomy of qualitative approaches based on their work.

Key Concepts in Qualitative Research

It is beyond the scope of this chapter to offer detailed descriptions of all the key concepts in qualitative research. We have selected four concepts for discussion here – code/category/theme, sampling, triangulation, and saturation/sufficiency, while other concepts that you are likely to encounter are listed and described in Box 9.1. We also direct readers to the recommended readings for further information. Finally, we address the notions of quality in qualitative research, eloquently described as *trustworthiness*.

Code/Category/Theme

As stated above, qualitative research commonly addresses narrative data as a means of gaining insight into the thoughts and experienced realities of the participants. A typical approach to analysis thus begins with a single (or a set of) written transcript/s derived from the participants. After familiarization with the text, the analyst then begins to annotate or highlight aspects of the text that appear especially meaningful, further giving them brief labels for ease of reference. These labels are referred to as *codes*. The analyst's goal is to generate as many codes as needed to encapsulate the content of the transcript. As the process continues, the analyst will typically find that these codes will cluster into *categories*. These, in turn, will guide the researcher toward a deeper sense of the underlying factors that shape the data. These factors, when articulated as coherent subjects, are referred to as *themes* [6]. While the above description seems linear, it is important to note that this process is iterative, and new categories or themes may well prompt the re-coding of certain aspects of the transcript [6]. This iterative process forms a vital part of qualitative methodology.

Sampling

Although *sampling* in qualitative research may use randomization approaches similar to those used in quantitative research, most commonly *purposeful* or *theoretical* sampling is used. That is, participants are selected for the study because they are known to have relevant experience. If this approach is adopted, it is important for researchers to be able to clearly explain why and how participants are considered as having

Table 9.1 Taxonomy of qualitative research approaches from Creswell and Poth [6] and Flick [4]

	Definition
Narrative research	"'Narrative' might be the phenomenon being studied, such as a narrative of illness, or it might be the method used in a study, such as the procedures of analysing stories told" [6]. The context in which the story is told is also important. Narratives may be drawn from existing text or collected for the specific purpose of research
Grounded theory	"The basic idea behind this approach is that theories should be developed from empirical material and its analysis" [4]. "A grounded theory study is to move beyond description ad to generate or discover a theory, a 'unified theoretical explanation' [7] for a process or an action" [6]. The theory that is developed is *grounded* or *derived* from the data, i.e. from participants' experiences. See Chap. 18
Ethnomethodology	"Ethnomethodology is interested in analysing the methods people use for organising their everyday lives in a meaningful and orderly way – how they make everyday life work. The basic approach is to observe parts of mundane or institutional routines…" [4]
Ethnography	An ethnography focuses on an entire culture-sharing group. "…typically it is large, involving many people who interact over time…". "…the researcher describes and interprets the shared and learned patterns of values, behaviours, beliefs and language of a culture-sharing group" [6, 8]
Phenomenology	Attempts to describe the common meaning of the phenomenon being studied to all participants. It seeks "to reduce individual experiences to a description of a universal essence." [6]
Case study	"… case study research involves the study of a case (or cases) within a real-life, contemporary context or setting [9]. This case may be a community, a relationship, a decision process, or a specific project." [6]
Hermeneutic approaches	"The study of interpretations of texts in the humanities. Hermeneutic interpretation seeks to arrive at valid interpretations of the meaning of a text. There is an emphasis on the multiplicity of meanings in a text, and on the interpreter's foreknowledge of the subject matter of a text." [4]

relevant experience. For example, Krogh et al. (2016) described how their participants were identified by purposive sampling after a process of nomination of expert debriefers associated with a national faculty development program (Box 9.1) [11]. Another common approach is *convenience* sampling, in which researchers work with participants simply because they happen to have access to that group. An additional technique is called *snowball* sampling. Here an initial participant is identified who then refers the researchers to additional participants whom they think may have valuable insights, and so on. Finally, sampling may also be undertaken to seek confirming or disconfirming data or 'cases' which are identified after analysis has been commenced.

In all these approaches, it is important to recognize that the sampling process is deliberate rather than random or probabilistic. Representativeness in sampling is not as important a consideration in qualitative research as it is in quantitative research since it is acknowledged that there are multiple realities and therefore each person is their own representative [10]. It is usually not possible to identify the exact number of participants for a qualitative study in advance, as there is no analogue to the power and sample size calculations common in quantitative approaches. In the end, the quality of sampling is judged by the fit between the question and the chosen methodology and not on the overall number of participants or the statistical diversity of the final study participants.

Triangulation

Triangulation is an important concept and refers to the use of different elements and methods (i.e. data sources, analysts, etc.) to bolster and augment each other's findings. Triangulation recognizes that no single data source can reveal all realities. One way this can be accomplished is by combining different types of data sources (i.e. interviews, observations, documents etc.). It may also be approached by the use of different analysts or coders, which assures that multiple researchers bring their experiences to bear on a question or problem. Decisions about triangulation are often influenced by resources since multiple data sources, methods and analysts all require additional time, finances, or personnel.

Saturation

An additional concept worth exploring is that of *data saturation*. Simply put, saturation refers to the theoretical point at which no new observations or insights are identified in data during analysis. One common question that is frequently asked is the minimum number of data points (e.g. interviews, focus groups, documents, etc.) needed to achieve saturation. This, however, stems from a misunderstanding of the term as saturation is never completely reached even with large samples but is instead (using more quantitative terminology) approached asymptotically. More important than the overall number of data points is the ability to effectively argue decisions made about saturation in the final published work. An alternative term – *sufficiency* is increasingly used to reflect researchers' confidence that they have collected and analysed enough data to make meaningful interpretations. The topic of saturation is contested in literature. Saunders et al. (2018) offer valuable considerations in conceptualising and operationalising *saturation* in qualitative research [12].

Quality in Qualitative Research

Researchers familiar with quantitative research may at first find it challenging to consider quality in qualitative research. This is because qualitative research has a different purpose than quantitative research. While it is often an important goal in quantitative research, the 'control' of variables is not part of the repertoire of qualitative researchers. Guba (1981) has helped to make meaning of quality in qualitative research by defining four constructs directly comparable to the quantitative concepts of validity, reliability, generalizability, and objectivity that can be used to judge overall study quality [9]. However, there are tensions in this process for some qualitative researchers, who feel that these efforts seemingly *default* to a more quantitative approach, thereby stripping the qualitative process of its inherently subjective nature. Still, given that many researchers in healthcare simulation are likely to come to qualitative research from a quantitative paradigm, we adopt this process to ease the transition.

The four constructs offered by Guba and reported in Shenton (2004) are reported in Table 9.2 [10]. By way of summary, *credibility* is used in preference to internal validity; *transferability* in preference to external validity/generalizability; *dependability* in preference to reliability; and, *confirmability* in preference to objectivity [15]. Together, these constructs can inform the quality of qualitative research, that is, its *trustworthiness*. In Table 9.2, we draw on Shenton's work to share a contemporary take on strategies researchers may consider to achieve this goal. In addition to these criteria, which primarily relate to a study's methodological rigor, Patton (2002) adds *researcher credibility* (which depends on training, experience and quality of previous work) as well as the overall *philosophical orientation* of the study [14].

Alternative approaches to this schema also exist. One influential approach was proposed by Flick (2014), who posed eight questions for consideration when assessing qualitative research. We have listed these in the first column in Table 9.3 [4]. Each question is answered using published

Table 9.2 Approaches to assuring quality and trustworthiness in qualitative research – Adapted from Patton [14]. Some items are repeated against the different quality types

Quality	Strategy
Credibility	Ensure the research methods are well suited for your study purpose
	Develop familiarity with the culture of the setting in which your study will occur
	Use sampling approaches that are fit for your purpose
	Use triangulation of methods, participants, study sites and researchers
	Adopt techniques to promote study participants' honesty and/or naturalistic behaviours
	Use iterative questioning in data collection
	Seek negative or disconfirming examples in your data
	Practice individual and team reflexivity throughout your study
	Invite peer review of your study processes (where ethics permits)
	Report your qualifications and experience to conduct the study
	Offer member checks of data collected – "participant validation" (e.g. transcriptions of talk) and analysis (e.g. in conversation or in writing)
	Share examples of *thick descriptions* of the phenomenon being studied (these are detailed personal descriptions)
	Locate your findings in previous research as part of extending the *conversation*
Transferability	Provide background data to establish the context of your study and detailed description of the phenomenon in question to allow comparisons to be made
Dependability	Use "overlapping methods" (e.g. visual data – observations, Textual data – naturalistic conversation; interviews etc.)
	Provide details of study description to allow your work to be repeated
Confirmability	Use triangulation of methods, participants, study sites and researchers
	Provide statements of your beliefs and assumptions (stance) about the phenomenon you are studying
	Declare shortcomings or limitations in your study's methods and their potential effects
	Provide details of study description to allow the integrity of research results to be examined
	Use diagrams to demonstrate an "audit trail" through the research process
	Invite peer review of your study processes (where ethics permits)

Table 9.3 Orientation to qualitative research illustrated through two studies

Authors	Krogh, Bearman, Nestel [11]	McBride, Schinasi, Moga, Tripathy, Calhoun [16]
Article title	"Thinking on your feet" – a qualitative study of debriefing practice	Death of a simulated pediatric patient: Toward a more robust theoretical framework
Journal	Advances in simulation	Simulation in healthcare
1. What is the issue and what is the research question of a specific study?	This study sought to explore the self-reported practices of expert debriefers. After summarising current research on debriefing practices, the authors argue that although there is much theory-based research on ideal debriefing, we do not have insight into what debriefers actually do when they debrief Q: "What are the debriefing practices of expert debriefers after immersive manikin-based simulations?"	This study sought to explore and further develop a framework (itself generated via qualitative means) for understanding the reaction of learners to unexpected mannequin death Q: "How is unexpected mannequin death perceived by individuals practicing at an institution where mannequin death is deliberately excluded?"
2. How is the study planned; which design is applied or constructed?	The study was an interview-based study of 24 peer nominated "expert" debriefers. The authors report: "With the researcher as an active interpreter, data was analysed inductively, with a continued awareness of researchers' own preconceptions and backgrounds." Although sampling of participants was complex, the authors reported clear criteria for inclusion	This study consisted of six focus groups composed of physicians and nurses. Focus groups were conducted in semi-scripted format. Clear criteria were used for group participation
3. How adequate is qualitative research as an approach for this study?	A qualitative approach is appropriate since it was an exploratory study of practices. Following this study, a quantitative study design could be used to measure the prevalence of the practices	A qualitative approach is appropriate given the researchers' desire to explore and better model a complex social phenomenon. Individual relationships within the final model could then be examined quantitatively
4. Is the way the study is done ethically sound?	The recruitment and selection of participants for the study was ethically sound. The ethical issues may have included variations in perceived power differences between the research team and the study participants. Some study participants may have felt obliged to participate due to their involvement in a national government funded project. The interviews were conducted by the first author who was largely unknown (at that time) to the participants The nomination process also may have led some participants to feel unhelpfully judged. However, only the research team had access to the nominees	The recruitment and selection of participants for the study was ethically sound. Ethical issues included potential triggering of stressful reactions. Participation was entirely voluntary and thus could be rescinded

(continued)

Table 9.3 (continued)

Authors	Krogh, Bearman, Nestel [11]	McBride, Schinasi, Moga, Tripathy, Calhoun [16]
5. What is the theoretical perspective of the study?	The study was interpretivist in approach and used thematic analysis.	This study used a constructivist grounded theory approach. The study was designed as the next phase of a prior study by Tripathy et al. (2016) that created a theoretical framework for conceptualizing unexplained mannequin death. This model was deliberately not referenced, however, during the initial phases of the analysis
6. Does the presentation of results and of the ways they were produced make the way in which the researchers proceeed and the results came about transparent for the reader?	The topic guide for interviews is included as an appendix. One researcher checked professional transcriptions against the audio-recordings Four analytic steps were described: (1) development of a coding framework (inductively); (2) higher order themes of "practice" and "development of expertise" were analysed in more depth; (3) all transcripts were reanalysed with the 36 codes identified in step 2; and, (4) a further critical review of the data was undertaken. The authors used NVivo. The results were presented as four main themes. Examples of transcript content were offered to illustrate themes and subthemes	The study clearly presents the analytic process used. Analytic steps included: (1) development of initial codes by independent analysis without reference to the initial framework, (2) subsequent triangulation and integration of these codes, again independent of the initial framework, (3) comparison of final codes with the original Tripathy framework, and (4) integration of the new data into the framework. Microsoft Word was uses for coding. The results were presented as seven themes arranged into a large-scale model. Examples of transcript content were offered to illustrate themes
7. How appropriate are the design and methods to the issue under study?	The study design and methods are suitable for the issue under study. However, alternative approaches could include the observation of expert debriefers to see what they actually do rather than what they say they do. It would also be helpful to digitally record debriefers debriefing and then ask them to explain what they are doing as they watch the recording with the researcher	The study design and methods are suitable for the issue under study. However, additional information was not gathered from further focus groups based on the results, limiting the number of analytical iterations. It is thus possible that the framework may bear further modification
8. Are there any claims of 'generalisation'* made and how are they fulfilled?	The authors report consistencies in the debriefing practices of experts. The authors use a theory – 'practice development triangle' to make meaning of their findings. The authors write: "Qualitative studies like this one are dependent on the researchers' approach and preconceptions when data is extracted and interpreted. As with all other interpretive qualitative research, the results are not reproducible or generalizable in a quantitative sense, but the commonalities across our broad sample suggest the findings may have relevance in other simulation-based education contexts"	The authors report large-scale consistency between the themes they obtained and those present within the initial Tripathy framework with the exception of themes that could only be derived from their specific population. The authors state that the model may well generalize to experienced pediatric clinicians (which formed the study population), but appropriately refrain from generalizing beyond this

Flick uses the term 'generalizability while Patton in Shenton proposes 'transferability'

research of interest to healthcare simulation practitioners. The first article by Krogh et al. (2016) explores the debriefing approaches of expert debriefers [11] while the second explores the applicability of a framework for understanding learners' reactions to unexpected death of a mannequin [16]. By way of disclosure, each of the authors of this chapter were senior authors on one of these articles. We recommend you read the articles and Table 9.3 as a means of further orientation to qualitative research.

Practical Considerations

As this introductory chapter draws to a close we address several practical issues that arise when conducting qualitative research. First, we enumerate several ethical considerations that apply, perhaps uniquely, to qualitative studies. We then consider potentially valuable software packages that can assist the researcher, followed by a discussion of issues associated with manuscript preparation and dissemination. Finally, we discuss several practical challenges that the new researcher can face.

Ethical Considerations in Qualitative Research

We have established that qualitative research is largely concerned with social phenomena and as such people are usually being studied. This means that there are important ethical considerations for every qualitative study. While this topic is developed in Chap. 34, we draw attention to several specific considerations particularly applicable to qualitative research.

First, as a method devoted to social phenomenon, qualitative research can be used to address emotionally intense situations. In our field, these could include research studies using data from simulation scenarios in which learners have made errors that would have been potentially *fatal* if performed in real settings, from scenarios triggering negative emotions in participants or from scenarios in which clinicians perform below expected standards. Given the potential for strong emotional responses or threats to identity, researchers must recognized that the process of gathering this data may require learners (study participants) to reflect on their experiences, which in turn may generate stress, anxiety, sadness, or other strong emotions. In these situations, it is incumbent on the researcher to provide emotional assistance or counselling if needed or requested.

A second issue concerns the personal nature of such deep reflections. As stated above, qualitative research is concerned with exploring social phenomenon in all their richness, which in some cases will entail participants revealing thoughts or concerns that they may not wish to be widely shared and may even request to withdraw from the study. Researchers need to respect participant confidentiality. Further, because qualitative research often uses small data sets from specific contexts, it may be possible for participants to be identified in research reports. Researchers must take great care to de-identify participants and to be cautious about guaranteeing anonymity. The relationship of the researcher/s to the study participants may also include power differentials that could influence recruitment, consent to participate, analysis and reporting. There are ethical issues to be considered at all phases of the qualitative research process, and we encourage researchers new to qualitative research to pay particular attention to the human research ethics of their projects.

Software for Qualitative Research

It is worth making reference to software that can assist in the management of data. Unlike quantitative data where the software actually performs the mathematics of the statistical analysis, qualitative analysis software mainly assists with data management. While software may assist with coding data, linking concepts, creating concept maps, memo writing and content searching, the actual analysis is performed by the researcher. The use of software is mainly a matter of personal preference, and opinions as to the relative benefit vary. It may be worth considering the use of software when datasets are very large (which is rare), for situations in which multiple people are analysing, for longitudinal studies (benefits of storing data), for specific types of analyses(template, a prior coding, content analysis) where the analytic framework is clearly defined from the outset of the study, and on the rare occasions when counting is undertaken. It is important to note that the use of a software package should not be viewed as lending additional credibility to the research, and for more limited datasets the use of common word processors containing highlight functions, or even post-it notes, may be a more economical alternative. Ultimately, the strength of the analysis lies with the researcher and not with electronic adjuncts. Box 9.2 lists commonly used commercial software with links to further information. Review the websites to decide which might best suit your needs, and consider asking other researchers or even opting for a free trial if you are interested in a particular product.

Writing Up Qualitative Research

As with quantitative research, qualitative research must also have checks and balances to ensure standards. Increasingly scientific journals are requesting qualitative studies to adhere to published standards. One such standard is the *Consolidated criteria for reporting qualitative research* (COREQ) checklist, which was designed for studies using interviews and/or focus groups [17]. Similarly, the *Standards for Reporting Qualitative Research* (SRQR) offers valuable guidance on reporting research of any qualitative research design [18]. Guidelines can also aid researchers in determining the best ways to report important aspects of study methodology, analytic processes and findings. However, given the diverse ontological and epistemological positions of qualitative researchers – some researchers consider these types of guidelines as restrictive and reductionist. As a result, how researchers interpret 'quality' must be considered within the theoretical stance taken by the researchers. Journals will vary in the extent to which they require research presented in certain ways. It is worth looking at published studies in your target journal to gauge expectations.

One helpful metaphor to consider when preparing the manuscript is that of a conversation regarding the phenomenon of interest. That is, each new study is an attempt to join or further the existing conversation (or, perhaps on occasion to start a new thread) [19]. It is therefore vital that qualitative researchers have a solid, clear grasp of what has been said before (i.e. what is known), and what aspects of the conversation require expansion prior to writing the manuscript. Here, the "problem, gap, hook" heuristic articulated by Lingard (2015) may have value [19]. In this approach, the initial impetus to perform a study (i.e. the problem) is clearly stated, followed by a rigorous exploration of what is and is not known (i.e. the gap). This is followed by a simple statement of what will be examined or done within a study and, importantly, why others should care about these results (i.e. the hook). This approach can also be helpful when initially framing the study as well.

Common Challenges Researchers May Have with Qualitative Research

Many researchers trained in the biomedical paradigm, where a randomised controlled trial in which extraneous variables are meticulously managed and controlled is seen as the gold standard for research, may have trouble with a qualitative process in which such variability is not only tolerated but actively encouraged. As stated at the beginning of this chapter, there are many complementary ways to understand the nature of reality and human knowledge. In Chap. 1, we drew on work summarised by Bunniss and Kelly that describes these concepts, which reflect the overall worldview of the researcher. Such a worldview is not monolithic, however, and can vary significantly depending on the nature of the phenomenon under consideration. Roger Kneebone (2002), who first trained as a surgeon before embarking on career focused on education in a range of settings, offers a concept from physics – total internal reflection – as an analogy to illustrate the challenges he faced when first reading social sciences literature and hence qualitative research [20]. He writes:

> Total internal reflection is that phenomenon whereby light is reflected from the surface of a liquid without penetrating it. A goldfish in a goldfish tank therefore can only see clearly within the water he swims in. He is physically unable to see what is outside except by jumping out of the water and looking around him, a process both uncomfortable and hazardous. … For me, jumping out of the positivistic goldfish tank led to a sudden barrage of new impressions. My main problem lay in having no road map of the territory beyond my tank. I therefore started my reconnaissance by simply wandering about and trying to spy out the land. I did this by exploring a variety of books and articles. Immediately I ran into a difficulty – the language. [20]

Likewise, standardization is usually not as valued within qualitative research as it is more quantitative paradigms. This may seem strange and surprising to researchers who have been trained to *control* as many variables as possible as part of good research design. Depending on what is being studied, however, such control may have little relevance to the ultimate outcome of the study. A related difference is the limited (and contested) use of pilot studies in qualitative research [21]. One of the authors of this chapter (DN) often considers their use in her work, particularly when she anticipates complexity with specific processes associated with the proposed research such as informed consent, data capture, transcription and analysis [22]. In this instance, performing a pilot can offer invaluable insight into the best way to conduct the primary study. Pilot data may even be used in the main analysis. Further, in qualitative research it is also possible to alter topic guides and observation schedules as the research progresses based on initial results. This possibility may seem quite strange if you are more familiar with quantitative research, but is in keeping with the directed, iterative nature of the qualitative approach.

Another difference lies in how qualitative research is reported. This is perhaps most obvious in the methods section of the manuscript. First, full reporting of qualitative study methods can be quite lengthy. Many healthcare journals, however, have somewhat restrictive word counts that are more appropriate for quantitative studies, which in turn can lead to compromised quality as researchers attempt to pare down their descriptions to fit. Increasingly, journals are enabling the publication of researchers' actions against COREQ and SRSQ standards as appendices. Second, it is often important to describe the professional experience of researchers. The purpose of this information is to offer insight (albeit briefly) into the credibility of the research team and their experience of the phenomenon being studied. An example of this can be seen in the reference article by Krogh et al. [11]:

> The three authors have extensive experience with simulation-based education and different debriefing approaches in a variety of contexts and simulation modalities. KK has a medical background while MB and DN are both experienced health professional and simulation educators with extensive experience conducting qualitative research.

Orientation to Chapters in the Section

The chapters that follow represent a deeper dive into the world of qualitative research methods as applied to healthcare simulation. In Chap. 10, Bearman outlines key concepts in qualitative research design using a metaphor of landmark features – purpose, worldview and approach to research design. She describes fundamental elements of qualitative research by linking these landmarks to the practicalities of writing the research question, describing the relationship between the researcher and the research (reflexivity) and selecting an appropriate *methodology* (and method) or approach. In Chap. 11, Smitten and Myrick offers further insights to conceptual and theoretical frameworks in qualitative research.

Our focus then shifts to specific methods: In-depth Interviews (Chap. 12), Focus Groups (Chap. 13), Observational Methods (Chap. 14), Visual Methods (Chap. 15), and Survey and Other Textual Data (Chap. 16). Given the use of audio-visual technology in healthcare simulation, we expect that visual methods will be increasingly important in our field. Although on the surface these methods may seem straightforward, you will see that there are many considerations and nuances in the way the methods are employed that can influence the quality of the data collected. Collecting data is of course only part of the research process and we further focus on one specific method, the Transcription of Audio-recorded Data, in Chap. 17. Finally, we examine various specific analytic methods, including Grounded Theory (Chap. 18); Thematic analysis and Qualitative Content Analysis (Chap. 19); and the Analysis of Naturally Occurring Data including conversations and recorded events (Chap. 20).

Concluding Thoughts

We hope that you enjoy your engagement with qualitative research. The chapters in this section explore focused topics and are illustrated with examples from healthcare simulation research. We believe that the content will help to equip you with the knowledge and skills needed to appreciate and undertake qualitative research for the advancement of field.

Box 9.1 Key terms used in qualitative research

Category – Codes from preliminary analysis of data are later arranged into categories that encompass key features.

Code – "Words that act as labels for important concepts identified in transcripts of speech or written materials" [10]. Creation of codes is usually the first level of analysis in inductive analysis. In deductive or a priori analysis, a code book is established before analysis and is based on a theory/concept being investigated.

Convenience sampling – This approach to sampling is often reported in qualitative research and reflects a non-probabilistic (i.e. non-random) technique in which participants are selected because of their accessibility and proximity to the researcher.

Data saturation – Refers to the, "Point at which no further new observations or insights are made" [10].

Saturation, its importance, and its determination is a contested topic among qualitative researchers. The phrase *data sufficiency* is being increasingly used to report researchers' satisfaction with the amount of data collected for meaningful interpretation.

Grounded Theory (GT) – This is a systematic approach to social sciences research that seeks to develop a theory that is *grounded* or *inductively derived* from the original data. It has a long and contested history. Charmaz (2008) offers a contemporary approach to Constructivist GT (CGT) that highlights the researchers' relationship to the data. In her words, CGT "… begins with the empirical world and builds an inductive understanding of it as events unfold and knowledge accrues" [23].

Member checking – A process whereby transcripts (for example) are sent to participants for them to check the accuracy and meaning of content (respondent/participant validation). Sharing the final analysis with study participants addresses another important aspect of member checking. This may be done in conversation or in writing.

Memo – "A document written in the research process to note ideas, questions, relations, results etc. In grounded theory research, memos are building blocks for developing a theory" [4].

Methodology – This refers to the framework in which the researchers locate their work and influences the phrasing of the research question, the selection of participants, techniques for data collection, analytic moves and team membership, and more. Researchers seek to alignment of all of these elements within their research project to attain consistency or methodological coherence.

Methods – Refers to the techniques used to collect data (e.g. semi-structured focus groups, observations etc.)

Purposive sampling – "Choosing a sample of participants in a deliberate manner as opposed to a random approach. This can include sampling of typical cases, sampling of outlier cases, or sampling participants with representative characteristics" [10].

Qualitative content analysis – Hsieh and Shannon (2005) describe three distinct approaches to qualitative content analysis – Conventional, Directed and Summative. The first two approaches resemble Thematic Analysis (see below), while the third is an important point of difference in which enumeration of themes/codes occurs [24].

Reflexivity – "A concept of research which refers to acknowledging the input of the researchers in actively co-constructing the situation which they want to study. It also alludes to the use which such insights can be put in making sense of or interpreting data" [4].

Sensitizing concepts – These are the ideas, concepts or theories that influence a researcher as they design, implement, analyse and report their research. "Concepts that suggest directions along which to look and rest on a general sense of what is relevant" [4].

Stance – Similar to reflexivity, stance is the declaration by the researcher of their position (i.e. their stance) relative to the topic under investigation.

Thematic analysis – This is a common form of analysis of textual data. "A method for identifying themes and patterns of meaning across a dataset in relation to a research question; possibly the most widely used qualitative method of data analysis, but not 'branded' as a specific method until recently" [25].

Theme – "Coherent subjects emerging from the data" [10].

Theoretical sampling – "The sampling procedure in grounded theory research, where cases, groups, or materials are sampled according to their relevance for the theory that is developed and on the background of what is already the state of knowledge after collecting and analysing a certain number of cases" [4].

Theoretical framework – The theoretical underpinning of a study (e.g. cognitive load theory, expertise theories etc.). It may also refer to the methodology (see above).

Triangulation – "The combination of different methods, theories, data, and/or researchers in the study of one issue" [4].

Trustworthiness – "Corresponds to the validity or credibility of the data. Multiple methods can be used to establish trustworthiness, including triangulation, participant checking, and rigor of sampling procedure" [10].

Box 9.2 Examples of qualitative software

Nvivo https://www.qsrinternational.com/nvivo/home

ATLAS.ti https://atlasti.com/

MAXQDA https://www.maxqda.com/

QDA Miner https://provalisresearch.com/qualitative-research-software/

Quirkos https://www.quirkos.com/index.html

webQDA https://www.webqda.net/o-webqda/?lang=en

Dedoose https://www.dedoose.com/

HyperRESEARCH http://www.researchware.com/

Aquad http://www.aquad.de/en/

Transana https://www.transana.com/

References

1. Hebbar KB, et al. A quality initiative: a system-wide reduction in serious medication events through targeted simulation training. Simul Healthc. 2018;13:324–30.
2. Gerard JM, et al. Validity evidence for a serious game to assess performance on critical pediatric emergency medicine scenarios. Simul Healthc. 2018;13(3):168–80.
3. Bunniss S, Kelly D. Research paradigms in medical education research. Med Educ. 2010;44:358–66.
4. Flick U. An introduction to qualitative research. 5th ed. London: SAGE Publications Ltd; 2014.
5. Cohen, D. and B. Crabtree. Qualitative research guidelines project. 2006. [cited 3 Sept 2018]; Available from: http://www.qualres.org/HomeRefl-3703.html.
6. Creswell J, Poth C. Qualitative inquiry and research design choosing among five approaches. 4th ed. Thousand Oaks: SAGE Publications, Inc; 2018.
7. Corbin J, Strauss A. Basics of qualitative research: Techniques and procedures for developing grounded theory. 3rd ed. Thousand Oaks: Sage; 2007.
8. Harris M. The rise of anthropological theory: a history of theories of culture. New York: Crowell; 1968.
9. Yin RK. Case study research design and methods. 5th ed. Thousand Oaks: Sage; 2014.
10. Sullivan GM, Sargeant J. Qualities of qualitative research: part I. J Grad Med Educ. 2011;3(4):449–52.
11. Krogh K, Bearman M, Nestel D. "Thinking on your feet"—a qualitative study of debriefing practice. Adv Simul. 2016;1:12.
12. Saunders B, et al. Saturation in qualitative research: exploring its conceptualization and operationalization. Qual Quant. 2018;52(4):1893–907.
13. Guba E. Criteria for assessing the trustworthiness of naturalistic inquiries. Educ Comm Technol J. 1981;29:75–91.
14. Patton M. Qualitative research and evaluation methods. 3rd ed. Thousand Oaks: Sage Publications Inc; 2002.
15. Shenton A. Strategies for ensuring trustworthiness in qualitative research projects. Educ Inf. 2004;22:63–75.
16. McBride ME, et al. Death of a simulated pediatric patient: toward a more Robust theoretical framework. Simul Healthc. 2017;12(6):393–401.
17. Tong A, Sainsbury P, Craig J. Consolidated criteria for reporting qualitative research (COREQ): a 32-item checklist for interviews and focus groups. Int J Qual Health Care. 2014;19(6):349–57.
18. O'Brien BC, et al. Standards for reporting qualitative research: a synthesis of recommendations. Acad Med. 2014;89:1245–51.
19. Lingard L. Joining a conversation: the problem/gap/hook heuristic. Perspect Med Educ. 2015;4(5):252–3.
20. Kneebone R. Total internal reflection: an essay on paradigms. Med Educ. 2002;36(6):514–8.
21. Morse J. The pertinence of pilot studies. Qual Health Res. 1997;7:323.
22. Morrison J, et al. "Underdiscussed, underused and underreported": pilot work in team-based qualitative research. Qual Res J. 2016;16(4):314–30.
23. Charmaz K. The legacy of Anselm Strauss for constructivist grounded theory. In: Denzin N, editor. Studies in Symbolic Interaction. Bingley: Emerald Group; 2008. p. 127–41.
24. Hsieh H, Shannon S. Three approaches to qualitative content analysis. Qual Health Res. 2005;15(9):1277–88.
25. Braun V, Clarke V. Using thematic analysis in psychology. Qual Res Psychol. 2006;3:77–101.

Recommended Resources

Creswell JW. Qualitative inquiry and research design: choosing among five approaches. 3rd ed. California: Sage; 2013.

Flick U. An introduction to qualitative research. 5th ed. London: SAGE Publications Ltd; 2014.

Kneebone R. Total internal reflection: an essay on paradigms. Med Educ. 2002;36(6):514–8.

Schwandt T. The SAGE dictionary of qualitative inquiry. 4th ed. Thousand Oaks: SAGE Publications Inc; 2014.

Silverman D, editor. Qualitative research: issues of theory, method and practice. 3rd ed. London: SAGE Publications Ltd; 2011.

Silverman D. Interpreting qualitative data. 5th ed. London: SAGE Publications Ltd; 2014.

Key Concepts in Qualitative Research Design

10

Margaret Bearman

Overview
This chapter provides an outline of key concepts in qualitative research design for healthcare simulation. It explores three landmarks that provide orientation to researchers: (1) a defined purpose for the research; (2) an articulation of the researcher's worldview; and (3) an overarching approach to research design. In practical terms, these translate to: writing the qualitative research question; articulating the relationship between the researcher and the research (reflexivity); and selecting an appropriate methodology. Three methodologies – grounded theory, phenomenology and qualitative description – are outlined and contrasted with a particular emphasis on different analysis traditions. The de facto use of methods as methodology is briefly explored and the importance of coherence across the three landmarks (as opposed to simple adherence to a particular tradition) will be emphasized. Finally, research credibility is introduced as a holistic, dynamic and tacit concept that is highly dependent on researchers and research context.

Practice Points

- Purpose, stance and approach act as three landmarks to orient your qualitative research study.
- Writing qualitative research questions is assisted by focussing on a particular phenomenon.
- The researcher's 'stance' should be articulated early – in particular their own research worldview and its relationship to the specific question at hand.
- Qualitative research has many diverse methodological traditions such as grounded theory, phenomenology and qualitative description.
- Establishing credibility in qualitative research is a dynamic, holistic process, and is not mechanistic.

- **Qualitative** "Of or relating to quality or qualities; measuring, or measured by, the quality of something. In later use often contrasted with *quantitative*." [1]
- **Research** "Systematic investigation or inquiry aimed at contributing to knowledge of a theory, topic, etc., by careful consideration, observation, or study of a subject." [1]

Introduction

Defining qualitative research is not an easy task. It is a term which comes with many weighty traditions, approaches and uses. However, in broad terms, qualitative research is the systematic study of social phenomena, expressed in ways that *qualify* – describe, illuminate, explain, explore – the object of study. 'Qualification' is firmly entwined with subjectivity. Therefore, it is inevitable that qualitative research means very different things to different people and one of its features is the rich diversity of its many approaches. This chapter presents my own views and experience in undertaking qualitative research in healthcare simulation, presented as key concepts that I believe underpin these various traditions.

I recognise that the language of qualitative research can be dense and difficult, and will therefore try to keep it simple. In doing so, I am walking a fine line between making the ideas accessible and over simplifying. I highly recommend that interested readers explore the literature provided for further reference.

Within this chapter, I present three 'landmark' concepts within qualitative research. I call them landmarks because I think of them as tall buildings; no matter which way you turn, you can always see them and use them to orient yourself. These are: *purpose* (e.g. as represented by a qualitative research question); *stance* (relationship of the researcher to the research); and *approach* (often encapsulated within a particular methodology). Finally, I discuss issues of coherence and credibility, which span these landmarks.

M. Bearman (✉)
Centre for Research in Assessment and Digital Learning (CRADLE), Deakin University, Docklands, VIC, Australia
e-mail: margaret.bearman@deakin.edu.au

D. Nestel et al. (eds.), *Healthcare Simulation Research*, https://doi.org/10.1007/978-3-030-26837-4_10

Landmark 1: Purpose (How to Write a Qualitative Research Question)

Research is conducted for many reasons. Some people divide the world between scholarship of teaching – applied research that informs what the decisions teachers make in their local contexts – and scholarship of research, which is closer to 'discovery of new knowledge' [2]. In my view, it is most useful to think about what you wish to achieve. If you want to improve what your colleagues and students do right now, then you will have different drivers than if you want to write a paper that influences practices across universities and countries. In either case, if you are also seeking publication then it is important that your paper presents some addition to what is generally known. This may be a description of a novel program (healthcare simulation journals do publish these) or it may be a more theoretically oriented piece of research. In some instances, qualitative research may fit your intended aims.

Qualitative research achieves particular purposes. In general, it is intended to illuminate a particular object of study in-depth. This type of research provides an understanding of the experiences, insights and actions of those 'on the ground', in ways that are challenging (or impossible) to quantify. Qualitative researchers value holistic, tacit and complex data. To contrast the two traditions, if a quantitative study seeks to demonstrate if simulation training can improve particular skills, a qualitative study may look at why this is (or is not) so. A qualitative approach may provide insight into the complexities of the experience, such as the enabling relationship of peer feedback or the emotional impact of failure.

So how do you know if qualitative research is the right fit for your particular needs? One of the best ways to identify what it is you are wishing to achieve, is to write a research question.

Research questions present a real opportunity to focus your research and to ensure that you are researching something that is meaningful (and publishable). Qualitative research questions are somewhat different than typical intervention-oriented quantitative questions. Qualitative research questions are concerned, at their core, with understanding a *phenomenon*. Part of the challenge in writing qualitative research questions is that identifying the central phenomenon of interest is not always straightforward. For example, consider a series of simulated patient activities designed to 'teach empathy' to health professional students. If this is our object of study, there are a range of possible qualitative research questions. One might be: what influences the development of empathic behaviours in a simulated patient activity for health professional students? Another might be: how do health professional students experience a simulated experience designed to teach empathy? Or a third: how can debriefing and feedback after a simulated patient activity support development of empathic behaviours in health professional students? All these different research questions are about simulation and empathy but they give rise to very different research designs and very different outputs.

Writing research questions is hard; not so much because the form is inherently difficult but because they require researchers to clearly articulate some key issues about the research itself. Qualitative research questions are more emergent than quantitative ones, they can shift and change over time. However, it is very helpful to be as clear as possible when starting out. See Table 10.1 for three heuristics for writing qualitative research questions.

Table 10.1 Three heuristics for writing a qualitative research question

Heuristic 1 Frame with how or what (or similar)

Starting off with the right pronoun can really help in thinking about the research. Consider the pronouns within the research question examples given in the text:

What influences the development of empathic behaviours in a simulated patient activity for health professional students?
How do health professional students experience a simulated experience designed to teach empathy?
How can debriefing and feedback after a simulated patient activity support development of empathic behaviours in health professional students?

'Did' or 'is' questions lead to a yes or no answer (e.g. did empathic behaviours develop?) and are therefore less suited to qualitative research. Other possibilities are: 'in what ways?', 'why?', 'to what extent?' or 'for what reasons?'

Heuristic 2 Focus on the particular phenomenon you are interested in

Focussing on the phenomenon can be the most challenging part of identifying the research purpose. In our example questions the phenomena always include the elements of the simulated patient activity, however there are three different foci:

What influences the *development of empathic behaviours in a simulated patient activity* for health professional students?
How do health professional students *experience a simulated experience designed to teach empathy*?
How can *debriefing and feedback after a simulated patient activity support development of empathic behaviours* in health professional students?

Heuristic 3 Indicate the scope of the study by providing context details

It can be very useful to bound the study by putting the context within the question; this helps both focus the object of study and avoids over generalising from the research. In our examples:

What influences the development of empathy skills in a simulated patient activity for *health professional students*?
How do *health professional students* experience a simulated experience designed to teach empathy?
How can debriefing and feedback after a simulated patient activity support development of empathic behaviours in *health professional students*?

Landmark 2: Stance (Relationship of the Researcher to the Research)

Qualitative researchers often start by declaring their *stance* or worldview. A 'worldview' (sometimes called 'research paradigm') is how researchers view reality and knowledge (sometimes respectively called ontic or epistemic beliefs) [3, 4], which in turn strongly influences how they think about and conduct qualitative research. For example, if you think that objective research is the 'gold standard' and that the most important knowledge is measurable, then this will necessarily shift how you will value interview data and therefore design your qualitative study. On the other hand, if you feel that peoples' experiences are never measurable, then you will approach the task in a different way.

Based on your worldview, you may find yourself seeking out different theories that are relevant to the phenomenon at hand. For example, if you are interested in understanding how health professionals learn empathy in simulated environments, you may find yourself attracted to theories of mind. Qualitative research doesn't always require theoretical foundations, however, like all research, it does require that the study will be situated within the broader literature. In some forums, a theory-informed approach is a sign of credibility of the research.

Articulating and documenting your worldview, theories and beliefs is the start of a process called 'reflexivity'. 'Reflexivity' is a means whereby researchers consider their relationship with the research and research participants throughout the course of the study [5]. Reflexivity articulates how researchers are influencing the research, and how the research is influencing the researchers. I can give an immediate example of this type of reflexive articulation. As I am writing this chapter, I am doing so as a person who has published many qualitative research papers, many of which concern simulation, and therefore I have a strong stake in seeing my views of what qualitative research 'is' being represented in a text on research methods. Additionally, in the process of writing this chapter I also revisited Pillow's excellent paper [5] which extended my own notion of reflexivity. This explicit declaration allows you, the reader, to understand some of what informs my writing of this chapter.

For those who have quantitative background, discussions of 'reflexivity' may draw comparisons with discussions of 'bias'. Importantly, reflexivity is not about 'reducing bias' but rather the recognition and declaration of the nuanced complexity the researchers and their contexts bring to the research. Reflexivity is sometimes addressed near the end of a study; I suggest it is useful to think about from the beginning. In our example of our simulated patient activities that 'teach empathy', different researchers may articulate very different thoughts and experiences at the outset. For example, if team members are also responsible for developing the program then they will have different frames of reference than if they have been independently contracted to study this program as part of an overall study of best practices. Moreover, some in the team may hold the view that empathy cannot be taught and others than it should always be taught. This will strongly influence the whole of their approach. By articulating these ideas, consideration can be given as to how to shape the research design to account for these necessary stances. It may also happen that, while conducting the research, views may shift. For example, the person who once thought that empathy can always (or never) be taught may come to see differently.

Landmark 3: Approach

The final landmark identifies the overarching frame of the research design. This is often called "methodology" although sometimes this notion is conflated with "methods". It is worth distinguishing these two terms. Methodology literally means the study of methods while methods are specific techniques for conducting the research. Methods in qualitative research usually refer to: deciding who the participants or contexts are (sampling); data collection processes (such as observation, interviews, focus groups or visual methods); analysis processes, where data is interpreted and articulated into its most salient elements; and output processes, where the analysis is presented for others, usually as text or diagrams. Some qualitative methodologies align with very particular methods, others do not.

In general, qualitative methodologies refer to those qualitative approaches which have been theorised and documented. They are very diverse, to the point of being confusing, and to add to the challenge they are constantly developing and changing. The qualitative methodology that is most frequently mentioned in the healthcare simulation literature is "grounded theory", but other methodologies are also mentioned, I'll briefly describe three of these contrasting traditions: grounded theory, phenomenology and qualitative description. As with all qualitative traditions, there are variations and subforms within each approach; what follows is an overview sketch with an example.

Grounded Theory

Grounded theory methodologies are designed to generate an interpretation of a particular context, which explains the phenomenon being studied (the "theory") [6]. Sampling, analysis and data collection occur together, so that the processes of participant selection and data collection depend on the current analysis. Analysis techniques are well defined and have strongly informed the whole field of qualitative analysis. In particular, what's called 'open coding' – where small units of meaning within the text are identified and labelled – is a key technique [7], commonly used in many qualitative approaches. In grounded theory, these units are clustered into higher order categories, as is also found in other approaches [7]. Another noteworthy process is "constant comparison", where the researcher constantly compares their analysis with the data, to look for divergences and convergences. To give an illustration of a grounded theory study, Walton, Chute and Ball [8] sought to understand "how do [nursing] students learn using simulation?" The authors describe a series of iterative data collection and analysis leading to a conceptual model, which illustrates how simulation allows students to "negotiate the role of the professional nurse".

Phenomenology

Phenomenology contains a range of methodologies which seeks to understand the "lived experience" of participants [9]. In this way, the research uncovers what people experience, which cannot be understood except through their perspective. This means that the participants are always sampled from those who have experienced the phenomenon themselves, and participant numbers tend to be low. The emphasis is on the holistic understanding, including what a person does and feels as well as thinks. There are many different forms of phenomenology. Some phenomenological analyses result in lists of themes, others in narrative descriptions that represent the distilled experience of the learner. For example, my doctoral study was a psychological phenomenological investigation of a 'virtual patient' encounter that outlined a representative narrative of how medical students experienced the interaction. This indicated a strong affective response to the virtual patient, which entangled feelings about the fictive 'patient' with the limitations of the educational technology [10].

Qualitative Description

The purpose of qualitative description is to describe the phenomenon as the participants themselves see it as closely as possible. This approach "does not require a conceptual or otherwise highly abstract rendering of data" [11]. It tends to draw on thematic analysis methods [12, 13], with findings most often represented through a series of themes and subthemes. Participant selection tends to be very practical in nature, and is often based on which potential participants would be best placed to provide views on this particular issue. To present another example from my own research experience, Krogh, Bearman and Nestel investigated video-assisted-debriefing (VAD) through interviewing 24 expert debriefers who worked with manikin-based immersive simulation [14]. The paper sought to describe how participants used VAD. We labelled (coded) the transcripts against a conceptual model of simulation education, using these labels to extract out those portions relevant to debriefing, which were then 'open coded' using the grounded theory technique described above. The findings outline very practical themes: the impact of the audio visual program, various educational approaches, and how the debriefers balanced benefits and limitations.

Different Forms of Analysis

As is illustrated by the above examples, different traditions have different ways of working with data. One of the key distinguishing factors is how much interpretation the researchers bring to the data. In other words, how much do they read 'between the lines', seeking to understand the fundamental concepts underlying the data? The grounded theory example above is highly *interpretive*. The model does not necessarily align with anything that the students said directly but conceptual categories are built iteratively through constantly comparing the data through an integrated process where interviews and conducting the analysis inform each other. The same is true for the phenomenological study. While the findings reflect the participants' experience, the distillation process required the researcher to identify what is essential about the collective experiences, which will therefore look distinctively different to the participants' own words. On the other hand, the qualitative description is highly *descriptive* (as is suggested by the name). The themes in this example are thus very close to what the participants themselves said and they describe the activities the participants undertake. This difference between description and interpretation strongly shape what findings will look like and is important to think about before embarking on the research.

The above three studies also illustrate how different traditions treat coding. Coding can be *inductive*, where the researcher starts to categorise and cluster the text without nominating any particular codes or themes beforehand. On the other hand, coding can be *deductive*, where the researcher applies a pre-existing framework and then extracts section of transcripts that relate to these codes. Inductive and deductive coding are often done sequentially. In some types of phenomenology and other narrative methodologies, the singular narrative is never broken down; it always remains a whole. In this way, individual sections of transcripts are never extracted and the whole experience is always represented together.

This is called *holistic* coding. Coding that leads to themes, categories or sub-categories is often called *thematic* coding (or sometimes thematic analysis).

Coherence: Keeping Everything in Alignment

These three landmarks – the qualitative research question (purpose), researcher orientation and reflexivity (stance) and methodology (approach) – inform each other. Indeed, one of the core concepts underpinning qualitative research is the maintenance of alignment between these landmarks. The research question should match with the methodology and the worldview underpinning the research. This notion is generally referred to as *coherence*. Coherence is a challenging concept and, in my view, it is not reducible to the sum of its parts. Often, in qualitative research, it is a dynamic endeavour whereby the researchers are constantly making adjustments to ensure that purpose, stance and approach align appropriately. This is illustrated in Table 10.2, where our three empathy questions are matched to example contexts, researcher orientation and methodologies. I have also added a row with possible sampling approaches, data collection methods and analysis techniques to illustrate the relationship between the landmark concepts and possible qualitative methods.

When Methods Become Methodologies...

While purists often hold to the importance of situating work within a particular tradition, this is less critical in healthcare simulation qualitative research, and in many cases specific methodological traditions are not invoked. Instead, researchers report what they have done, drawing heavily on particular analysis methods. In this instance, aspects of the methods become a de facto methodology (e.g. 'thematic analysis' or 'mixed methods'). This is perfectly acceptable and in my view, does not *necessarily* diminish the contribution of the research provided the first two landmarks are clearly defined, the methods are well documented and reported, and the de facto methodology (approach) aligns with the research purpose and researcher stance. In other words, coherence is more important than slavish adherence to a particular way of doing things. However, coherence is not the only marker of credibility for qualitative research studies.

Credibility in Qualitative Research

How do we know if a qualitative study is credible – rigorously conducted with trustworthy findings? The key markers of rigour in a quantitative study – such as generalisability and bias reduction– do not apply to qualitative research. While there are quality indicators they tend to vary from tra-

Table 10.2 Comparing three different qualitative studies around the same simulation experience

	Example 1: Educators wanting to understand how they can optimise the use of simulation to teach empathic behaviours	Example 2: A simulation expert undertaking doctoral studies is studying a range of different programs. Her topic is developing empathy through simulation	Example 3: Experienced researchers are building a program of research in debriefing and feedback
Landmark 1: qualitative research question	What influences the development of empathic behaviours in a simulated patient activity for health professional students?	How do health professional students experience a simulated experience designed to teach empathy?	How can debriefing and feedback after a simulated patient activity support development of empathic behaviours in health professional students?
Landmark 2: example of orientation to research	Researchers are teachers who want to understand the influence of their simulation on their students. They declare and note their multiple role as clinicians, researchers and teachers and their strong belief that empathy can and should be taught. They will adjust the program based on the research	Researcher is a doctoral student who wants to understand what it is like to experience simulated empathy. She declares and notes her view that simulation is a profoundly emotional experience. She draws widely from the simulation literature	Researchers are a team who have previously studied the role of debriefing and feedback in developing 'non-technical skills'. They are highly expert debriefers and declare and record their own view on what constitutes 'good' debriefing
Landmark 3: methodology	Qualitative description, as the researchers seek to articulate what teachers and students consider influences on learning empathy	Phenomenology, as the researcher wishes to describe the 'lived experience' of empathy holistically	Grounded theory, as the research team wish to build a conceptual understanding of how simulation might promote empathy
Methods (aligned with landmarks)	Participants includes all students and teachers involved in a particular simulation activity. Data collection methods might be interviews and focus groups, followed by thematic analysis	Participant numbers will be small (under 10 participants) and sampled purposefully. Data collection will be long form interviews, designed to elicit the 'story' of experience. Analysis will be holistic	Participant numbers are emergent depending on findings. Data collection methods are a mix of observation and interview. Analysis and data collection are undertaken together

dition to tradition. For example, iteration of analysis and data collection is important to a grounded theory approach but not essential for, say, qualitative description. Moreover concepts like 'saturation', 'member-checking' and 'triangulation' are often invoked as quality indicators but often used uncritically [15]. I would regard some kind of reflexivity as a sign of credibility, but reflexivity can be very tokenistic and can actually reduce rather than enhance credibility in some instances. In essence, defining and articulating credibility in qualitative research is not straightforward.

One of the most challenging issues with credibility in qualitative research is that it is highly contextual – both with respect to the context of the study and the context of the resulting publication. Indeed, the markers of credibility in healthcare simulation research are not the same as for the higher education literature. Moreover, in my experience what is considered credible can vary from journal to journal. The stance of journal editors – their view of what reality and knowledge is – can strongly influence what they deem is good research. In a nutshell, credibility is a dynamic and tacit notion, and bound up in professional judgements.

These shifting sands can make establishing credibility a rather tricky proposition. Rather than enumerate various features that might constitute credibility across a range of qualitative research study, I think it is more important to recognise the complexity of the landscape. In my view, the key features of credibility are that: (1) it is dependent on the three landmarks of purpose, stance and approach, (2) it cannot be achieved by following a series of rules or a checklist and, (3) its presentation within the final publication is highly dependent on the audience. This means that for each study it is incumbent on the researchers to make a case for the rigour and trustworthiness of their research.

Closing

This brief overview has introduced key concepts underpinning qualitative research such as purpose, stance, methodology, methods and credibility. I have chosen to highlight those that I think are particularly important, however, any introduction to qualitative research cannot capture the immense complexity of this field. The remaining chapters in this section will help expand many of the ideas that I have so briefly mentioned. Many of the methodologies and methods in the general education literature are infrequently seen in healthcare simulation, so there is enormous opportunity for future research to reveal new understandings of how we learn, teach and practice using simulation methodologies.

References

1. Oxford English Dictionary 2018.
2. Boyer EL. Scholarship reconsidered: priorities of the professoriate. Lawrenceville: Princeton University Press; 1990.
3. Nestel D, Bearman M. Theory and simulation-based education: definitions, worldviews and applications. Clin Simul Nurs. 2015;11(8):349–54.
4. Bunniss S, Kelly DR. Research paradigms in medical education research. Med Educ. 2010;44(4):358–66.
5. Pillow W. Confession, catharsis, or cure? Rethinking the uses of reflexivity as methodological power in qualitative research. Int J Qual Stud Educ. 2003;16(2):175–96.
6. Lingard L, Albert M, Levinson W. Grounded theory, mixed methods, and action research. BMJ. 2008;337(aug07_3):a567-a.
7. Watling CJ, Lingard L. Grounded theory in medical education research: AMEE guide no. 70. Med Teach. 2012;34(10):850–61.
8. Walton J, Chute E, Ball L. Negotiating the role of the professional nurse: the pedagogy of simulation: a grounded theory study. J Prof Nurs. 2011;27(5):299–310.
9. Tavakol M, Sandars J. Quantitative and qualitative methods in medical education research: AMEE guide no 90: part II. Med Teach. 2014;36(10):838–48.
10. Bearman M. Is virtual the same as real? Medical students' experiences of a virtual patient. Acad Med. 2003;78(5):538–45.
11. Sandelowski M. Focus on research methods-whatever happened to qualitative description? Res Nurs Health. 2000;23(4):334–40.
12. Miles MB, Huberman AM, Huberman MA, Huberman M. Qualitative data analysis: an expanded sourcebook. Thousand Oaks: Sage; 1994.
13. Clarke V, Braun V. Thematic analysis. In: Teo T, editor. Encyclopedia of critical psychology. New York: Springer; 2014. p. 1947–52.
14. Krogh K, Bearman M, Nestel D. Expert practice of video-assisted debriefing: an Australian qualitative study. Clin Simul Nurs. 2015;11(3):180–7.
15. Varpio L, Ajjawi R, Monrouxe LV, O'brien BC, Rees CE. Shedding the cobra effect: problematising thematic emergence, triangulation, saturation and member checking. Med Educ. 2017;51(1):40–50.

Additional Resources

Braun V, Clarke V. Successful qualitative research: a practical guide for beginners. London: Sage; 2013.
Cresswell JW. Qualitative inquiry and research design: choosing among five traditions. Singapore: Sage; 1998.
Savin-Baden M, Major CH. Qualitative Research: The essential guide to theory and practice. London: Routledge; 2013.

Refining Your Qualitative Approach in Healthcare Simulation Research

11

Jayne Smitten

Overview
A knowledge of potential research approaches is foundational to the execution of acceptable scholarship within the qualitative domain. The presence or absence of an adequately defined theory based approach within the research can lead to divergent outcomes (i.e. rejection versus acceptance) in publishable academic writing. Therefore, the selection of a quality approach and its subsequent integration throughout the fabric of a research endeavor is critical to the success of any qualitative research process. This process naturally applies to simulation research in its many dimensions. In this chapter, we introduce various theoretical and conceptual perspectives to generate further understanding and preparation for qualitative research. Exploration of a few qualitative research approaches will be presented. The process used to select the qualitative approach will be discussed with a view toward how such an approach can be integrated into a qualitative research study.

Practice Points

- Theoretical and conceptual approaches differ, with theoretical approaches functioning at a higher level of abstraction than conceptual approaches.
- The theoretical and/or conceptual approach will have a profound influence on how the study is structured and conducted.
- Four core constructs, the *problem, purpose, significance* and *research question(s)*, should be considered when choosing an approach.
- The final theoretical and/or conceptual approach must be congruent with the researcher's own worldview.

J. Smitten (✉)
College of Health and Society, Experiential Simulation Learning Center, Hawai'i Pacific University, Honolulu, HI, USA
e-mail: jsmitten@hpu.edu

The Essence of Theoretical and Conceptual Approaches

Understanding the essence of the theoretical or conceptual approaches is essential when researchers venture into the world of qualitative research. Many novice researchers, however, are uncertain as to how a ***conceptual*** and ***theoretical*** approach is selected. What are the differences between approaches and, more importantly, how are they used? These terms are often used interchangeably and have been debated in the research literature [1–4]. However, from our view (and for the purposes of this chapter) these terms will ***not*** be treated as identical. The essence of a theoretical approach, or what on occasion is described as a 'framework' in the research literature, is that it is based on a *pre-existing* conjectural foundation that has been determined and validated in the scholarly realm. Conceptual approaches, on the other hand, are more particular, and are established with regard to how the researcher actually frames the exploration of the research question [5, 6]. Thus, they have direct bearing on how the research problem is determined and grounded within the phenomenon that is being explored [5, 6]. Table 11.1 further distinguishes these two approaches. As can be seen from the examples within the table, theoretical approaches are typically more general, and address the more fundamental ideas within which a study belongs. Conversely, a conceptual approach is based on more specific concepts or variables within the research study. The theoretical approach is often considered a more 'formal and higher' level of abstraction than the 'lower' level conceptual approach.

The term 'framing' (as used above within the word 'framework') refers to how the researcher interprets the research findings and connects them to other knowledge. It is important to remember, however, that qualitative research generally does not use pre-defined theoretical 'frameworks' as does quantitative research. Instead, qualitative research is about ***uncovering*** the approach by which the data can best be understood through the research process itself. Thus, theo-

D. Nestel et al. (eds.), *Healthcare Simulation Research*, https://doi.org/10.1007/978-3-030-26837-4_11

Table 11.1 Distinguishing theoretical & conceptual approaches in simulation research

Theoretical approach	Conceptual approach
Symbolic Interactionism [7] • A down to earth scientific approach to studying human group life and human conduct • Example: Study of research participants in a simulation teaching and learning environment, including but are not limited to, the use of symbols, words, gestures or interpretations to convey meaning [7, 8]	• Exploration of the theoretical teaching/learning and/or leadership concepts/principles for a research study [15] • Conceptualizing by observing and scientific study on how people learn based on their own personal understanding and knowledge of the world • Examples: Study of educational leadership using a simulation approach [16]; Exploration of inquiry-based science teaching, using a learning cycle approach based on constructivist principles with emphasis on the investigation of phenomena [15–18]
Grounded Theory Method [9–13] • Paradigm of inquiry providing a scientific approach that legitimizes acquisition of research data from social psychological processes [9–12] • Method especially useful for researching unanswered questions in the social psychological realm that require the development of more robust theoretical underpinnings to support future research [8–14]	*Principles of Good Practice in Undergraduate Education* [8, 19] • Seven Good Practice principles that may provide context in a simulation teaching and learning research study may include [19]: (1) encouragement of contact between students and faculty (2) development of reciprocity and cooperation among students (3) use of active learning techniques (4) provision of prompt feedback (5) emphasis of time on task (6) communication of high expectations (7) respect of diverse talents and ways of learning

retical or conceptual approaches ***do not actually guide*** qualitative research as they do in quantitative research. That being said, the researcher is still required to be well versed or familiar with the literature and the research findings in the area or focus of the study. Without such an understanding of what is known both empirically and theoretically about the topic being researched, the researcher will be unable to derive a useful qualitative approach.

Notwithstanding the differences between the conceptual or theoretical approaches, in this chapter we will concentrate specifically on ***preparation for the research study*** by exploring how the theoretical approach functions as the foundational structure, vision and focus for the research process. Selection of an appropriate and fitting theoretical approach early in the design process is critical and will provide the organizational foundation for the literature and resource review (including consideration of timing for certain qualitative approaches, such as grounded theory, that can be sensitive to this), design, methodology and analysis (i.e. examination, evaluation, consideration) processes. The theoretical foundation is thus imperceptibly threaded into the fabric of the entire qualitative research study.

Over the past few decades the number and diversity of possible qualitative research methods has expanded significantly [20, 21]. The key consideration in any research is that the *research question is the driving force* behind selection of the appropriate method to address that question. With the many approaches currently available, the general theoretical context chosen is usually an expression of how the researcher wishes to conceptually approach the research question and ultimately portray the results [20, 21].

In many ways creating a work of qualitative research is analogous to building a sculpture, musical composition, or other work of art. Using this metaphor qualitative research may thus be conceptualized as a creative theoretical process (and not simply a method or technique) that centers on two important questions: ***what is being explored,*** and ***how will the data be understood?*** Research endeavors include 'pushing the boundaries' throughout the entire complex process, ensuring emphasis on the emerging qualitative theoretical approach as the fundamental and foundational support of the research story.

Grant and Osanloo [22] emphasize that a theoretical framework is the "blueprint for the entire qualitative research inquiry. It serves as the guide on which to build and support the theoretical approach and further defines how it will philosophically, epistemologically, methodologically, and analytically approach the dissertation as a whole" [22]. While the aforementioned quotation is focused on dissertation work, the theoretical approach may be applied to qualitative research endeavors. This theoretical blueprint, as described by Grant and Osanloo [22] correlates to the construction of a home, which involves an "exterior view" (i.e. elevation drawing) that provides a structure and global perspective to the research problem, as well as "interior view" (i.e. floor plan) that uses the framework notion to organize the concepts and goals of the study.

It must also be recognized that there may not always be an explicitly pre-determined theoretical approach. A theoretical approach may, in fact, not be described until after gathering adequate data to account for the theoretical underpinnings of a research study. A posteriori theoretical approaches may be developed as one's study is designed with the actual emergence of the data, as in a grounded theory methodological approach [9–13]. An example of this process is a simulation research study where the researcher is seeking to explore how healthcare educators are prepared to facilitate and influence the educational process in the human patient simulation environment [8]. The qualitative research method in this case would guide the study with the emphasis on development and subsequent emergence of a theoretical structure. The emerging data would thus establish/create a posteriori the chosen theoretical approach to further understand the phenomenon. Moreover, the actual formulation of the theoretical approach evolves from the rigorous simulation research data gathering process. This one example, which showcases the iterative sequence of events within a single study that leads to the selection and application of a theoretical approach, highlights the essential power and complexity of qualitative research [8].

Constructs of Theoretical Approaches to the Research Process

According to Grant and Osanloo [22], four constructs apply to each potential research approach: the ***problem***, ***purpose***, ***significance*** and ***research question(s)***. These constructs are critical for guiding the choice of research design and data analysis and should be used to define the overall research process and evidence gathering techniques that will be used. The ***problem statement*** is essential and defines the root issue of the research. The ***purpose*** justifies the study, answering specific queries on what one hopes to gain or learn from the study. The ***significance*** links the importance and value of the research study. The aforementioned three constructs describe how the theoretical approach connects to the problem, relates to the purpose, and links to the importance of the research study. The final construct, the ***research question(s)***, is complementary to the base or theoretical approach and transforms the above elements into specific areas of investigation on which concrete studies can be built. The final questions posed in the research study will serve to exemplify the relationship between what is known and what problem or subject is being explored. The theoretical approach chosen thus provides the solid base on which the overall shape of the research design is constructed. See Table 11.2 for examples of how these four constructs can be applied to a specific qualitative research study [8].

Table 11.2 Constructs for guiding qualitative research process

Constructs	Examples
Problem	Apparent need for research related to the preparation and application of high-fidelity human patient simulation (HPS) in healthcare education
Purpose	To *explore the process* in preparing healthcare educators in the use of high-fidelity HPS as a teaching/learning approach in undergraduate healthcare education
Significance	Paucity of research available addressing the preparation of healthcare educators in the use of HPS for the reality of their teaching and learning practice
Research question(s)	What is the social/psychological process used to prepare healthcare educators in the use of high-fidelity HPS as a teaching/learning approach for undergraduate healthcare education? How are healthcare educators prepared to facilitate, guide and influence the teaching/learning process in the high-fidelity HPS environment?

Note: The examples above are adapted from the first author's (J.S.) dissertation [8]

Examining Epistemological Beliefs Towards Research Design

Our fundamental beliefs are known to be influenced by our assumptions, values and ethics, and thus will influence our choice of theoretical approaches in research [23–25]. There is no right or wrong answer to this question, no 'one size fits all' theory that works with every research query. Instead, it is the researcher's responsibility to identify their ***own belief systems*** and give due consideration to their own epistemological values when determining an appropriate theoretical endeavor. By way of definition, epistemology refers to the study of knowledge itself and how it is discovered, created, and/or interpreted. Following are examples of epistemic standpoints that may provide foundational qualitative research approaches to study human and social behavior within the field of simulation-based or health profession. These include but are not limited to: positivism and post-positivism, interpretivism, symbolic interaction, feminism, phenomenology, and post-modernism [21].

Qualitative researchers can choose from a multitude of approaches that may have commonalities but may also exhibit great diversity. Curiously, Willis [21] claims there is much more 'paradigm diversity in the qualitative genre than in the quantitative approach' (p.147). The underpinning characteristics of qualitative research are known to include the 'search for contextual understanding' and the 'emergent approach' to guide the researcher in their quest [21]. Contemporary qualitative research continues to evolve and expand from a diversity of paradigms into approaches that contribute to a further understanding of our human and social behaviors within the fields of simulation-based and/or health professional education [21–24].

The epistemological foundation of the qualitative research approach generally encompasses the interpretivist or constructivist paradigms. Constructivism addresses reality as socially constructed; findings are literally created as the exploration proceeds within the research process. This perspective views the meaning of research data as a construct that is established by the research team. In contrast, the quantitative approach typically delineates its findings within the positivistic paradigm, which focuses on objective knowledge that exists "out there" and is discovered via the use of established, valid tools and patterns of statistical inference.

Constructivist and interpretivist perspectives are often used interchangeably in the qualitative research community. In our view this is not strictly accurate, however, as the interpretivist paradigm not only perceives phenomena as socially constructed, but also recognizes the collection and interpretation

Table 11.3 Theoretical and conceptual approach checklist: selection and integration for your qualitative research

What is the professional discipline? (Why does it matter?)
Does the theory, if applicable, match with the methodological plan?
Does the research study methodology draw from the principles, concepts and tenants of the theoretical or conceptual approach?
Do the three (3) constructs (problem/purpose/significance) align well with the qualitative theoretical or conceptual research approach?
Do the research questions require modification for a priori or posteriori theoretical approach?
How does the research approach inform the literature review?
Does the theoretical or conceptual approach undergird the conclusions, implications and recommendations of the data analysis?

Adapted from [22]

of socially constructed data as inherently subjective as well. Those adhering to this view interpret text (and other data) based on "socially constructed realities, local generalizations, interpretive resources, stocks of knowledge, intersubjectivity, practical reasoning and ordinary talk" [25].

When choosing an overall theoretical or conceptual approach, it is thus critical that the researcher connect their larger worldview (constructivism, interpretivism, etc.) to the four constructs (***problem, purpose, significance*** and ***research question[s]***) discussed above in the most congruent manner possible. Only this approach can provide a foundation reliable enough to serve as the base for a qualitative study. This stance is further exemplified by Maxwell's words [25]: "The function of this theory is to inform the rest of your design-to help you to assess and refine your goals, develop realistic and relevant research questions, select appropriate methods, and identify potential validity threats to your conclusions. It also helps you justify your research." [p.]. These points are critical toward creating and building a focused qualitative research design. Table 11.3 provides a checklist that embodies these principles.

A Worthwhile Struggle

The qualitative research paradigm embraces a vast array of theoretical and conceptual approaches. The chapter focuses on elucidation of these approaches and answers the query of 'Why do I need a theoretical approach at all?' Continued improvement of one's research skills, overall understanding, and working knowledge of the theoretical approaches available are important phases in the research journey. Ongoing consultations with advisors, colleagues, editors, mentors and peers are vital to this process, and can provide key insights into how theoretical approaches can be better integrated into your research. Learning more regarding the application of the diverse theoretical approaches will inevitably enhance your perspective and potential repertoire. Ultimately, the theoretical thread, subtly woven throughout the fabric of the research study, provides vital clarity and enhances the usefulness of the research findings. The effort is well worth it!

References

1. Bell J. Doing your research project: a guide for first-time researchers in education, health and social science. Berkshire: Open University Press; 2005.
2. Green HE. Use of theoretical and conceptual frameworks in qualitative research. Nurse Res. 2014;21(6):34–8. https://doi.org/10.7748/nr.21.6.34.e1252.
3. Merriam SB. Qualitative research: a guide to design and implementation. San Francisco: Jossey-Bass; 2009.
4. Schultz JG. Developing theoretical models/conceptual frameworks in vocational education research. J Vocat Educ. 1988;13: 29–43.
5. Astalin AK. Qualitative research designs: a conceptual framework. Int J Soc Sci Interdiscip Res. 2013;2(1):118–24.
6. Camp WG. Formulating and evaluating theoretical frameworks for career and technical education research. J Vocat Educ Res. 2001;26(1):27–39.
7. Blumer H, Denzin N. Symbolic interactionism. In: Smith J, Harre R, Langenhove L, editors. Rethinking psychology. Newbury Park: Sage; 1995. p. 43–58.
8. Smitten J. Nurse educators preparing for the use of high-fidelity human patient simulation: a process of finding their way [PhD dissertation] [Edmonton (AB)] University of Alberta 2013. Retrieved from: http://www.era.library.ualberta.ca.efbb92eb.
9. Glaser BG. Conceptualization: on theory and theorizing using grounded theory. Int J Qual Methods. 2002;1(2):23–38.
10. Glaser BG. Grounded theory review. Mill Valley: Sociology Press; 2005.
11. Glaser BG, Strauss AL. The discovery of grounded theory: strategies for qualitative research. Chicago: Aldine; 1967.
12. Strauss A, Corbin J. Basic of qualitative research: technique and techniques and procedures for developing grounded theory. Newbury Park: Sage; 1998.
13. Chenitz WC, Swanson JM. From practice to grounded theory: qualitative research in nursing. Melon Park: Addison-Wesley; 1986.
14. Parker B, Myrick F. The grounded theory method: deconstruction and reconstruction in a human patient simulation context. Int J Qual Methods. 2011;10(1):73–85. https://doi.org/10.1177/160940691101000106.
15. Kolb AY, Kolb DA. Learning styles and learning spaces: enhancing experiential learning in higher education. Acad Manag Learn Edu. 2005;4(2):193–212.
16. Duran LB, Duran E. The 5E instructional model: a learning cycle approach for inquiry-based science teaching. Sci Educ Rev. 2004;3(2):49–58.
17. Shapira-Lishchinsky O. Simulation-based constructivist approach for education leaders. Educ Manag Adm Leadersh. 2014:1–17. https://doi.org/10.1177/1741143214543203
18. McHaney R, Reiter L, Reychav I. Immersive simulation in constructivist-based classroom E-learning. Int J E-Learn. 2018;17(1):39–64. Waynesville, NC USA: Association for the Advancement of Computing in Education (AACE). Retrieved November 28, 2018 from https://www.learntechlib.org/primary/p/149930/
19. Chickering AW, Gamson ZF. Seven principles for good practice in undergraduate education. Racine: The Johnson Foundation, Inc., Wingspread; 1987.

20. Molasso WR. Theoretical frameworks in qualitative research. J Coll Charact. 7:7. https://doi.org/10.2202/1940-1639.1246.
21. Willis JW. Foundations of qualitative research: interpretive and critical approaches. Thousand Oaks/London: Sage; 2007. https://doi.org/10.4135/9781452230108.n5.
22. Grant C, Osanloo A. Understanding, selecting, and integrating a theoretical framework in dissertation research: creating the blueprint for your 'house. Adm Issues J: Connect Educ Pract Res. 2014;4(2):1–16. https://doi.org/10.5929/2014.4.2.9.
23. Munhall P, Chenail R. Qualitative research proposals and reports: a guide. Sudbury: Jones and Bartlett; 2008.
24. Denzin N, Lincoln Y. The SAFE handbook of qualitative research. Newbury Park: Sage; 2005.
25. Maxwell J. Qualitative research design: an interactive approach. Thousand Oaks: Sage; 2004.

In-Depth Interviews

12

Walter J. Eppich, Gerard J. Gormley, and Pim W. Teunissen

Overview
In-depth interviewing has become a popular data collection method in qualitative research in health professions education. Interviews can be unstructured, highly structured or semi-structured, the latter being most common. A well-crafted semi-structured interview guide includes predetermined questions while allowing flexibility to explore emergent topics based on the research question. In order to collect rich interview data, researchers must attend to key elements before, during, and after the interview. The qualitative methodology used impacts key aspects, including: who performs the interview, who participates in the interview, what is included in the interview guide, where the interview takes place, and how data will be captured, transcribed, and analyzed.

Practice Points

- Be clear about the research question and how the interviews will help to answer this question.
- Prepare for each interview carefully to ensure you capture rich data.
- Remain flexible during interviews; pose predetermined questions and explore emergent topics that are relevant to your study.
- After each interview reflect on how you can improve for the next interview.
- Use additional techniques to elicit the interview such as Pictor, Rich Pictures, or Point-of-view Filming.

Short Case Study
Susan, a health professions educator, wants to study how healthcare debriefings contribute to peer learning among medical students. She has previously engaged in survey research and statistical analysis of survey results. Given the exploratory nature of the research question, she plans to interview both medical students and simulation educators. She is not sure if she is on the right track or where to start and seeks guidance.

Introduction

While quantitative research paradigms dominate healthcare simulation, qualitative research complements existing approaches by exploring how and why simulation promotes learning. Healthcare simulation researchers skilled in quantitative research methods, however, may find themselves challenged with the diversity of qualitative data collection methods. Among these, interviews represent a common and seemingly straightforward approach, although potential pitfalls may prevent the collection of rich data for analysis.

This chapter aims to:

- Explain the role of interviews as a data collection method and their relationship to qualitative methodology
- Differentiate between structured, semi-structured, and unstructured interviews

W. J. Eppich (✉)
Feinberg School of Medicine, Department of Pediatrics, Division of Emergency Medicine, Ann & Robert H. Lurie Children's Hospital of Chicago, Northwestern University, Chicago, IL, USA
e-mail: w-eppich@northwestern.edu

G. J. Gormley
Centre for Medical Education, Queen's University Belfast, Belfast, UK
e-mail: g.gormley@qub.ac.uk

P. W. Teunissen
Faculty of Health Medicine and Life Sciences (FHML), Department of Educational Development and Research, School of Health Professions Education (SHE), Maastricht University, Maastricht, The Netherlands

Department of Obstetrics and Gynaecology, VU University Medical Center, Amsterdam, The Netherlands

D. Nestel et al. (eds.), *Healthcare Simulation Research*, https://doi.org/10.1007/978-3-030-26837-4_12

- Provide a roadmap for designing an interview guide
- Explore supplemental elicitation strategies
- Offer guidance on preparing for and conducting the interview
- Review how to capture and transform interview data for later analysis

Why Interview?

For certain research questions and methodologies, interviews can enable the collection of rich data. As Kvale (2007) notes, "interviews allow the subjects to convey to others their situation from their own perspective and in their own words" [1].We can view interviews as a social dialogue between participant and interviewer, which highlights the critical role of the interviewer in co-constructing knowledge with the interview participants [1]. In-depth interviews lend themselves to exploration of social phenomenon as participants share their life-worlds, or their "lived everyday world" [1]. Many factors contribute to the quality of interview research, including interviewer characteristics, the qualitative methodology used, the sampling strategy, rapport management, and interviewing technique just to name a few. Most of these factors should be detailed in published research reports (see Tong et al. for a detailed reporting criteria) [2].

Why Choose In-Depth Interviewing?

Various forms of interviews exist, including structured, semi-structured and unstructured interviews. Structured interviews apply highly standardized questions to solicit specific data points that lend themselves to quantitative analysis. In contrast, semi-structured and unstructured interviews as used in qualitative research can yield much richer data. Unstructured interviews often supplement field observations used in ethnography, a specific qualitative research methodology [3]. In these, researchers use a more conversational approach to explore behaviors and other phenomenon observed in the field. In-depth semi-structured interviews are the sole or primary mode of data collection in most qualitative research projects [4], making them the most widely used form of interview. Semi-structured interviews combine a series of pre-planned questions in an interview guide with emergent questions or probes depending on the dialogue between interviewer and participant [4].

How Does Interviewing Fit with Various Qualitative Methodologies?

Interviews represent a data collection *method*, much like observations, document analysis, and the review of audiovisual materials [5]. Qualitative researchers employ these methods as part of a larger overarching qualitative research *methodology*. These methodologies not only guide data collection, but also reflect a comprehensive approach and interpretive framework that influences formulation of research questions, sampling strategy, approach to data collection and analysis, and goals of the project. Examples include narrative research, phenomenology, ethnography, grounded theory, or case studies [5]. For example, researchers use a grounded theory methodology to build a theory *grounded* in the data [6, 7]. This approach dictates that data collection and analysis proceed iteratively using the principles of constant comparison and theoretical sampling. Theoretical sampling involves selecting participants for their unique perspectives that inform theory development. While general principles of interviewing apply in most cases, the specific methodology has significant impact on the interview approach. See Cresswell and Poth (2018) for an in-depth discussion of the relevant issues [5].

How to Create an Interview Guide?

The creation of an interview guide represents an essential element of the interview process [1, 7]. From a very practical point of view, obtaining ethics approval will likely require you to outline what questions you will pose and which topics will remain off-limits. For this purpose, the predetermined questions informed by your research questions are usually sufficient. You may also choose a theoretical or conceptual framework to help you design your interview guide.

After the purpose of the study, ground rules for the interview, and any demographic data such as age or professional background has been covered, a broad open-ended question can serve as a point of departure for the interview. Subsequent questions should explore topics related to the research question(s), and, if possible, solicit specific examples that allow study participants to describe their experiences. For example, we propose the following broad opening question for our exemplar study that addresses how debriefings contribute to peer learning: *"Please tell about your prior experiences participating in healthcare debriefings"*. Potential follow-up questions are as follows:

- *"Please describe your most recent debriefing (as participant or facilitator)"*
- *"Please tell me about a recent debriefing that stands out for you and why?"*
- *"Please tell me about the interactions you have with the other participants during a debriefing and how those interactions shape your experience"*
- *"What aspects of debriefings shape what you take away from them?"*

As participants describe their experiences, the astute interviewer will identify relevant issues that emerge from the conversation and explore these in greater details as they relate to the study's research questions. It can be helpful for interviewers to have follow-up questions or 'probes' prepared (see below in "conducting interview").

Enhancing Data Collection During Interviews

While research interviews elicit participants' views, they often fall short of providing a full picture of their experiences. For example, some individuals may find it challenging to articulate their experiences fully. Drawing upon social sciences, a range of techniques can help to enrich interview data collection. These strategies should be planned in advance and reflected in your interview guide. Such techniques can allow interviewers to make a greater connection with the participant's experiences and permit a deeper shared understanding about the subject matter under investigation. A few examples of techniques that could be used in simulation-based research include:

- **Rich Pictures:** In this technique, participants are invited to draw their experiences. This drawing then serves as a prompt for the participant to describe the picture during the course of the interview. See Cheng et al. for an example [8].
- **Point-of-view Filming**: Participants record their first-person perspectives of an activity while wearing digital video glasses *(e.g. learners wear digital video glasses or a body-cam during a simulation-based learning activity).* This film footage is subsequently used to elicit their experience during the interview. Interviews augmented by point of view filming rely less on memory than traditional interviews as both the interviewer and interviewee can pause and replay parts of the recording with particular relevance, elaborating on them as needed. This technique also allows the interviewer to observe and empathize with the participant in light of what transpired [9, 10]. See Figs. 12.1, 12.2, and 12.3.
- **Pictor technique**: Participants use this visual technique to construct a representation of their experiences using arrow-shaped adhesive cards on a chart. Interviewers can then use this representation to elicit the interview and help participants share their experiences in greater detail. See King et al. for an example [11].

How to Recruit and Select Participants?

Multiple sampling strategies exist [5]. Recruiting participants based purely on convenience should be avoided. A purposive sampling strategy seeks to collect data intentionally

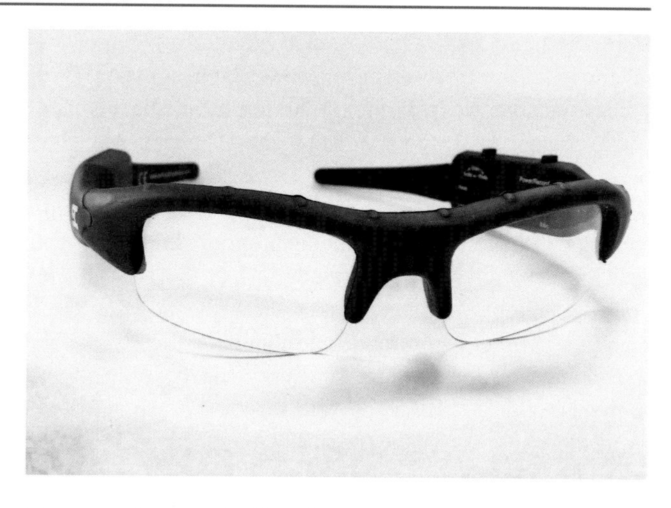

Fig. 12.1 Example of video glass to capture a research participants point of view (PoV) in an activity

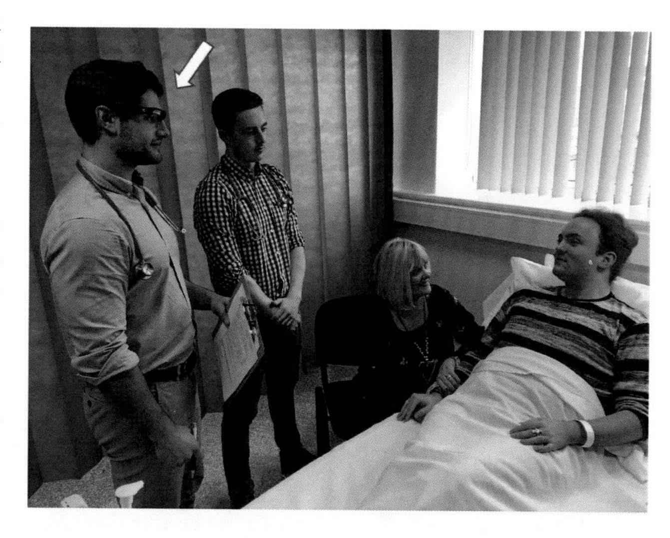

Fig. 12.2 Illustration of a research participant wearing video glasses in a simulated scenario. (Reproduced with permission of Queens University School of Medicine, Dentistry and Biomedical Sciences)

Fig. 12.3 A research participants PoV via footage from video glasses. (Reproduced with permission of Queens University School of Medicine, Dentistry and Biomedical Sciences)

from a range of data-rich informants who possess both similar and disparate views on the phenomenon in question. Both of these perspectives have great value and should be included.

A number of important steps must be accomplished before participants are recruited. First, ethics board approval must be obtained. You may also need to obtain permission from key stakeholders (e.g. training program directors if you seek to recruit physicians-in-training). Once this is complete, consider announcing your study at departmental lectures or meetings in order to inform people about the research and let them know you are recruiting subjects. Send electronic mails to the target group. Based on your responses, you will need to decide who to interview and in which sequence. Here your chosen methodology may provide some guidance. For example, if you are using a grounded theory approach to your analysis, your sampling strategy will be primarily based on participants' potential to shed light on an evolving theoretical model you have identified through constant comparison and iterative analysis. It can also be helpful to deliberately interview participants with differing backgrounds as this sheds light on a range of perspectives. Rather than interviewing medical students in sequence based on year of study (i.e. first years, then second years, etc.), you might instead intersperse early year, later year, and middle year students. Depending on the stage of your research you might also select participants for their potential to share alternate perspectives rather than those who will affirm what you have already found.

Conducting a Research Interview

Now that you have obtained ethics approval and designed your semi-structured interview guide, you are ready to recruit participants, schedule interviews, and prepare for the interviews themselves. Each of these elements include key steps that occur before, during, and after the interview.

Before the Interview

After deciding on the type of interview, interview guide, and recruitment strategy, now you must prepare for the interviews themselves. You must both (a) plan your approach to the interviews in general, and (b) prepare before each and every interview. This section addresses both topics.

General Approach to Interviewing

The researcher must address several important issues well in advance of the initial interview: (a) who will conduct the interviews, (b) what materials and devices will be used to capture data, and (c) what supplies might you need bring for the interview participants.

'Who' Will Perform the Interview?

This is a vital question since the interviewer represents the data collection instrument. In many instances you as the primary researcher will be the person doing the interviews, but in certain cases you may not be the most suitable person. Study participants should feel free to share information without fear of repercussions. Whoever conducts the interview will need to be "reflexive" about their role and clearly consider how their past experiences shape interpretations [12]. Prior relationships between interviewer and study participants may prevent the collection of rich, high-quality data (depending on the nature of those relationships and the research questions). Therefore, the research team should proactively discuss whether or not someone else should do an interview if you have a prior relationship with participant(s). Naturally, whoever performs the interviews should have some interviewing experience to ensure that they use effective questioning and rapport building techniques. Therefore, we advise gaining some experience before your first research interview. One strategy involves recording and transcribing any pilot interviews. By reviewing the audio-recordings and interview transcripts, you will identify areas of improvement. Even better, an experienced research interviewer can listen to portions of the interview and provide feedback on your general interviewing approach as well as specific questioning techniques. You should also consider whether more than one person will conduct interviews for your study and in what circumstances. Discuss expectations for the interviews and use of the interview guides beforehand. Finally, joint review of completed interviews helps get multiple interviewers on the same page. All of these considerations require deliberate forethought and planning.

What Supplies Will You Need for Data Capture?

Data capture usually requires a device with audio-recording capability (e.g. Dictaphone, tablet, or smartphone), although in some instances an observer can take notes if recording is impractical or impossible. Most experienced researchers also bring a second device to as a back-up in case of technical issues. Print your interview guide (preferably in a large, easy to read font) as well as paper and pencil(s) to take notes during or after the interview. Consider jotting some field notes immediately after the interview ends; such reflections may relate to the interview participant and their responses, or to the interviewer and their immediate impressions about that interview.

What Materials Should You Bring for the Interviewees?

Beyond the consent form, you may wish to collect written responses from interviewees regarding demographic characteristics (e.g. contact details, age, occupation, prior training, to name a few); a specially designed form will be required for this step. In addition, special elicitation techniques may require additional supplies, such as paper and pencils for interviewees to draw rich pictures or arrow shaped sticky notes for the Pictor technique. Finally, you will need to plan ahead if you wish to provide participants with a beverage or a snack.

How Will You Respond If a Study Participant Becomes Distressed?

Although a rare event, some interviewees may become distressed or disclose something concerning that may require further attention. You may even need to shift the focus of the interview from 'research' to the 'well-being' of the participant. Ethics committees may wish to know your approach to such situations, which could involve offering the interviewer the choice to continue or end the interview, turning off the recording, or discussing the issue further with either study staff or someone trained in psychosocial support.

How to Prepare Before Each Interview

You will need to arrive at the site well before the interview is scheduled to make sure everything is ready. Ensure that you are in a safe and quiet place where you will not be disturbed. For in-person onsite interviews, think about possible sources of noise that may impact the quality of your audio recording, such background noise from doors opening and closing, people talking, or vehicles passing by open windows. Such noise at inopportune times may make key words unintelligible for transcription. For example, one author realized after the first interview that he had to change from ceramic coffee mugs to paper cups because the sound of the coffee mug being placed on the table interfered with the audio-recording. Consider the structure of the environment as well. Think carefully about the positioning, light and temperature of the space to make sure both you and your interviewee will be comfortable. Explicitly check with your interviewee that (s)he is comfortable before you start the interview. If the interview will take place in an unfamiliar location, know where the bathrooms are and offer participants a bathroom break before the interview begins. You will have less control over these aspects of the environment during interviews via telephone or video-conferencing technology. You will also need to confirm explicitly with participants that the time and location remains suitable for the interview. Invite participants to silence their mobile phone before the interview starts unless pressing matters exist. Review the study's purpose, interview procedure and time frame, and ask them to sign the consent form before you start recording. See Table 12.1 for a pre-interview checklist.

In summary, the axiom, 'fortune favors the prepared', certainly holds true when planning for research interviews. By taking extra supplies, checking and double-checking your equipment and materials, you will be less likely to be surprised by missing items or experience technology failures that prevent you from capturing valuable data.

Table 12.1 Key considerations before an in-depth interview

Pre-interview checklist
Offer restroom break
Offer drinks/snack
Have back-up recording devise available, with sufficient battery power and memory
Silence or turn off cell phones and pagers
Place sign on door ("Please do not disturb—Interview in Progress")

Gather supplies
  Interview guide
  Notes, pens/pencils

Obtain explicit consent to record
Turn recording devices on

The Interview

You are now ready to conduct the interview. Switch on your recording device (and back-up device) and ensure that they are actually recording. Although you will already have obtained consent, re-orient participants to the purpose of the study and provide reassurance about confidentiality. Let participants know how the interview will proceed, and inform them that they should make no assumptions about what the interviewer knows and does not know about the phenomenon in question. Invite participants to be explicit in their responses so that a full picture of their valuable perspectives can be collected. Specifically inform participants of the need to explore their thinking and the role of follow-up questions. Let participants know there are no right or wrong answers to the questions posed; rather, it is their perspectives and experiences that we (i.e. both interviewer and participant) will explore together. At the outset of the interview, invite participants to introduce themselves and perhaps their role *(e.g. a simulation-based educator who conducts debriefings, or a medical student who participates in debriefings).* Such an opening question gives participants an opportunity to settle into the interview by talking about low-risk topics.

Now that the interview has commenced, how you proceed depends on the type of interview you wish to conduct. For example, if you pursue a semi-structured approach, you will have a list of key topics/questions as part of a predetermined interview guide. Informal conversational interviews will be more exploratory, and interviewers will need to allow themselves to be guided by participant responses. Regardless of interview type, interviewer must develop and maintain trust with the interviewees to elicit the richest possible data. At all times interviewers should remain open and interested in the interviewees' perspectives. They should also support interviewees as they share their experiences, especially if the subject matter is sensitive *(e.g. when a participant shares an experience about a simulation in which their performance was sub-standard compared to their peers).*

Open questions are more invitational and provide interviewees with greater agency (for example – instead of asking *'have you ever facilitated a debrief that went well?* – you could reframe this question by asking *'would you like to share a time when a debrief went well for you?'*). Probing questions allow interviewers to gain deeper insights to interviewees' viewpoints and experiences. Such probes might include, *"Can tell me more about that?" "What do you mean by that?" "What makes you say that?" "Can you provide an example?"* This approach to questioning allows both interviewers and interviewees to co-create knowledge about the subject matter under investigation.

As you your interview draws to a close it is important to allow interviewees the opportunity to share any further insights or clarifications. If participants have no further contributions you can formally end the interview, thank the study participant, and turn off the recording device.

Immediately after the Interview

Once the interview formally ends, make sure that participants are comfortable allowing the content of the interview to be included in the study. Depending on the nature of the interview you may want to provide a short debrief for participants. This gives them the opportunity to share their experiences about the interview. You may also, based on your chosen methodology, wish to contact the participants again (e.g. for a follow up interview in a grounded theory study, member checking of your results in a phenomenology study, or provision of a study summary to participants once it is complete). It is important to notify participants about these contingencies and how you will contact them.

Once the participant has left you may which to audio record a short reflection about the interview or jot down some field notes. Doing this allows you to capture key concepts that may be beneficial for analysis and allows you to improve the process for subsequent interviews. Finally, consent forms and paperwork from the interview need to be securely stored. Most interview recordings are now in digital format, and you can upload the recorded interview data files from the device to a password protected and encrypted location approved by your ethics committee. This data can then be transcribed for further analysis.

Post-interview

When the interview ends and you have thanked your participant and parted ways, your top priority is to ensure that you have completely captured the data by checking your audio-recording device. This data should be uploaded immediately if possible to a secure digital location. If your devices didn't record, use a paper and pen to write down as much information from the interview as you can recall (or record a voice memo). Next, go through the notes you made during the interview. Take some time to elaborate on them and add factual observations you remember from the interview (e.g. perhaps how a participant said or didn't say something, or expressed themselves nonverbally, during the discussion of a specific topic). Add your initial interpretations about what you heard from the participant as well. These initial impressions may cover a range of topics, from theoretical concepts that came to mind during the interview to connections to previous interviews. Capture these thoughts as memos in your interview notes.

Most researchers listen to their recording again once the transcript becomes available to ensure its accuracy and familiarize themselves with the data. Consider briefly summarizing what the interview was about and what you learned about the phenomenon you're studying. If your research design includes providing a summary of the actual interview to participants, this can serve that purpose as well. Important questions to ask yourself when doing this include the following:

- What did I learn from this interview?
- What might I do differently next time?
- Should any questions be modified to reflect new lines of inquiry? (This may also require discussion with your research team and even an amendment to your ethics approval.)

In most cases you will wish to transcribe the audio files into textual data for further analysis. This depends, however, on your chosen qualitative methodology. Given its time intensive nature, it can be helpful to use either a commercial transcription service or one based in your institution. These typically incur costs, so it is important to budget for this service when designing your research project. If you submit your data to a third party for transcription, you are responsible for ensuring that all necessary precautions are in place to keep your data safe. In most countries this entails a non-disclosure agreement and/or a user agreement signed by both parties. When you receive transcripts you should de-identify them by removing personal information as per your research protocol. This essential step must be completed before sharing the data with your research team if they are only allowed to work with de-identified data.

Commercial transcription services tend to charge differently for one-on-one interviews versus group interviews or focus groups. You will also need to decide how much detail in transcription your research requires. Depending on your research domain, methodology and aim, transcribing the audio file to an 'intelligent verbatim transcript' may work for you. True "verbatim' transcription means that every utterance (i.e. every stutter, stammer, 'um', cough and laugh) appears in the transcription. This approach increases the expense given the level of detail required. In an 'intelligent verbatim transcript' the transcriptionist will omit fillers like 'um', laughter and pauses from the transcript while preserving the participants' meaning. Some light editing may also be done to correct sentences and delete irrelevant words. If you use intelligent verbatim transcription you should review the transcript for accuracy since some utterances may have significant meaning for your study. Also, some technical/medical terms or jargon may be transcribed as "[unintelligible]" but will be easily recognizable (and potentially significant) to you. This makes your review for accuracy all the more important.

Qualitative analysis software can be used to facilitate the analysis of different forms of interview data, including audio, video (in the case of point-of-view filming) or text files. These software packages themselves perform no analysis, but provide a platform for coding data that allows you to search for specific codes and link them to relevant analytic memos. Such a platform is not essential, but can be helpful depending on the amount of data you will analyze.

Conclusion

As qualitative approaches to research in healthcare simulation expand, in-depth interviewing as a data collection method has become more popular. A well-crafted interview guide incorporates predetermined interview questions while providing flexibility to explore emergent topics that study participants raise. Successful collection of rich interview data demands attention to key elements before, during, and post interview. Researchers must consider the effect of their chosen qualitative methodology on key elements of data to be collected: who performs the interview, who participates in the interview, what to include in the interview guide, where the interview takes place, and how data will be captured, transcribed, and analyzed.

References

1. Kvale S. Doing interviews. London: Sage; 2007.
2. Tong A, Sainsbury P, Craig J. Consolidated criteria for reporting qualitative research (COREQ): a 32-item checklist for interviews and focus groups. Int J Qual Health Care. 2007 Dec;19(6):349–57.
3. Reeves S, Peller J, Goldman J, Kitto S. Ethnography in qualitative educational research: AMEE guide no. 80. Med Teach. 2013;35(8):e1365–79.
4. Dicicco-Bloom B, Crabtree BF. The qualitative research interview. Med Educ. 2006;40(4):314–21.
5. Creswell JW, Poth CN. Qualitative inquiry & research design: choosing among five approaches. 4th ed. Thousand Oaks: Sage; 2018.
6. Watling CJ, Lingard L. Grounded theory in medical education research: AMEE guide no. 70. Med Teach. 2012;34(10):850–61.
7. Charmaz K. Constructing grounded theory. 2nd ed. London: Sage; 2014.
8. Cheng A, LaDonna K, Cristancho S, Ng S. Navigating difficult conversations: the role of self-monitoring and reflection-in-action. Med Educ. 2017;51(12):1220–31.
9. Skinner J, Gormley GJ. Point of view filming and the elicitation interview. Perspect Med Educ Bohn Stafleu van Loghum. 2016;5(4):235–9.
10. Lewis G, McCullough M, Maxwell AP, Gormley GJ. Ethical reasoning through simulation: a phenomenological analysis of student experience. Adv Simul. 2016;1(1):26.
11. King N, Bravington A, Brooks J, Hardy B, Melvin J, Wilde D. The Pictor technique: a method for exploring the experience of collaborative working. Qual Health Res. 2013;23(8):1138–52.
12. Creswell JW. Research design: quantitative, qualitative, and mixed methods approaches. 5th ed. Thousand Oaks: Sage; 2018.

13 Focus Groups in Healthcare Simulation Research

Nancy McNaughton and Lou Clark

Overview
This chapter outlines focus group method as an accessible approach to educational inquiry for live simulation based research. We define focus group method and its conceptual underpinnings, and describe the manner in which it fits into the lexicon of qualitative approaches either on its own or in combination with other techniques and tools. The chapter also includes a conversation between two researchers about various concerns and questions regarding how to run a focus group.

Focus group research is a useful qualitative method for simulation educators seeking to turn their daily work into scholarship that reflects the voices of many participating stakeholders. While this is a seemingly straight-forward method, our goal here is to illustrate to colleagues the complexities and nuances of best practices for implementing focus groups.

Practice Points

- Introduce group agreements at the beginning of a focus group by gathering ideas from everyone about what they need from each other and the facilitator in order to speak freely.
- Create a focus group guide with open ended questions specific to your research topic.
- Focus groups are social engagements and should provide a comfortable environment in which participants can share their thoughts.
- Power dynamics need to be taken into consideration during all phases, from design and deliberation of group composition to facilitation of a group and in the analysis.
- Have a plan in place prior to running a focus group in the event that a member becomes distressed. Not all topics are suitable for a focus group format and this too needs to be taken into consideration during the planning phase.

Introduction

Focus group research method represents a narrative approach to gathering information in the form of a group conversation that has broad appeal across professions, including an array of marketing, political, business, and organizational development groups. In health professional education—and for simulation educators in particular—focus groups are a valued method, either on their own or in conjunction with others, for exploring issues, explaining social phenomena, and deepening our understanding about how people make meaning from their experiences. Groups who may benefit from this method include (but are not limited to) learners, simulated patients (SPs), faculty members, and subject matter experts. Focus groups are an increasingly popular method within Simulation Based Education (SBE) for exploring a range of topics including quality assurance and safety within simulation design, clinical skills learning, SP recruitment, training, scenario development and briefing and debriefing. Due in part to its adaptability to multiple purposes, formats and groups the complexity and challenge of focus group method tends to be underestimated by new researchers.

This chapter will explore focus group method as one of many approaches within qualitative research with the goal of helping new researchers understand and integrate best practices.

N. McNaughton (✉)
Institute of Health Policy, Management and Evaluation, University of Toronto, Toronto, ON, Canada
e-mail: n.mcnaughton@utoronto.ca

L. Clark
Hugh Downs School of Human Communication, Arizona State University, Tempe, AZ, USA
e-mail: leclark2@asu.edu

D. Nestel et al. (eds.), *Healthcare Simulation Research*, https://doi.org/10.1007/978-3-030-26837-4_13

The main body of the chapter will take the form of a conversation between two researchers in order to illustrate some of the most frequently asked questions and common concerns about focus groups as a qualitative approach.

Conceptual Considerations

Focus groups as a method fit within a social constructivist paradigm that views reality (ontology) as socially negotiated or constructed and knowledge (epistemology) as a product of the social and co-constructed interaction between individuals and society. More importantly, focus groups as a method of data gathering fit under a methodological umbrella known as phenomenology, which is concerned with how people make meaning from their experiences in the world. The researcher engaging in focus groups is interested in participants' ideas, interpretations, feelings, actions and circumstances [1].

Background

Focus groups were originally described as "focused interviews"or "group depth interviews". The technique was developed after World War II to evaluate audience response to radio programs [2]. The method was later adopted by broadcasting, business, marketing, and organizational development professionals and further developed within the sociology discipline by Robert K. Merton and colleagues as an ideal way of collecting data on a wide range of social and professional phenomena [2, 3].

Focus groups came into the education realm in the 1970s during a time of growing interest in participatory approaches for carrying out research [4]. As mentioned above, today they fit conceptually within a social constructionist paradigm and can be used as a valuable data collection method that is sensitive to people whose voices are not traditionally included in research. However, focus groups are not only used for exploratory and descriptive research, but also for more practical purposes such as conducting needs assessments, developing consensus guidelines as well as a way to follow up on quality assurance initiatives. Researchers in the simulation field can benefit from a knowledge of focus group method as it is an approach well-suited to gathering the perspectives and experiences of the many stakeholders involved in simulation.

For the purposes of our chapter focus groups are defined as:

> (…) group discussions organized to explore a specific set of issues… The group is focused in the sense that it involves some kind of collective activity… crucially, focus groups are distinguished from the broader category of group interview by the explicit use of the group interaction as research data. [5]

A focus group as defined above by Kitzinger suggests an interactive format in which topics may be addressed and explored broadly by participants. The focus group leader is essentially a facilitator, guiding the discussion and making sure participants stay on topic while allowing unplanned for revelations from within the group. The role is key to collecting relevant and meaningful data from interactions within your groups.

One of the most important elements of focus group method is the dynamic that is created between participants. This dynamic will affect the quality of the information that is collected and requires deft management by the focus group leader. Power dynamics need to be taken into consideration during all phases, from design and deliberation of group composition to facilitation of a group and in the analysis. When thinking about who to put together into a focus group there are many considerations. You are fundamentally creating a social and conversational environment in order to hear about ideas and stories that may require trust and sharing. Therefore it is important to consider the homogeneity or heterogeneity of a group especially as it relates to the question or issue that is being explored.

Remember a key aim of focus groups is to be able to record and explain the meanings and beliefs that influence participants' feelings, attitudes and behaviours [6].

Rationale

The advantages of focus group method are many. First, nuances between points of view can be attended to by the focus group leader, allowing for clarification, follow-up, and expansion on ideas. Next, non-verbal responses to a topic can be captured which supplement (or contradict) verbal responses. The leader acting as facilitator is following and probing ideas that are presented by participants. At the same time participants can develop their own ideas by listening to the opinions of others (the group effect). On a more practical level, there is the convenience of collecting many perspectives on a topic in one place and at one time, potentially at a lower cost than if the individuals were interviewed separately. In the end focus group transcripts should capture the words of participants allowing for potentially greater depth of meaning and nuance to be revealed and perhaps new insights to be gathered through the use of language itself. As has been pointed out, focus groups allow flexibility for researchers with respect to design and format, group makeup, tools used, and topics that can be covered.

In the end data produced by focus group method comes from people for whom the topics have relevance and may have greater face validity than other means of collecting information. Resources to help new researchers learn about, plan, conduct, and analyze focus groups are plentiful. A number of these resources are highlighted in the reference list at the end of the chapter.

Conversation between Simulation Educators on Focus Group Method

This section of the chapter features a conversation between two simulation educators, (SE1 and SE2). The first educator (SE1) is new to qualitative research and focus group method and is trying to decide if it is the appropriate method to use for an upcoming project in which SPs' experiences portraying emotionally challenging cases will be explored. The second educator (SE2) shares experiences of using focus groups with SPs and offers practical tips.

SE1: *I want to know about how my SPs are feeling when they portray emotionally challenging cases. Should I do an interview with each of them or a focus group? What data would a focus group provide that an interview won't? Could I do both? Which one would I do first, and why?*

SE2: Having done both individual interviews and focus groups with SPs during my research, I think the choice depends on your research goals and also on whether or not you will be asking sensitive questions. Since you are exploring emotionally challenging portrayals it seems that sensitive material may come up. For example, I am conducting a study for which I asked SPs about their experiences on this very topic, including portraying patient's experiencing domestic violence and who recently had a spouse die (e.g. breaking bad news). During individual interviews some of the SPs portraying the domestic violence patients surprised me, because I thought we did a careful job of screening them before we cast them to make sure no SP portraying a domestic violence patient experienced it in real life. It turns out some had experienced domestic violence in their own lives, and felt that participating made them feel positive and proactive. This is important information that I'm not sure would have come out in a focus group. Alternately, when I conducted a focus group for the same study and asked similar questions about motivation to participate, they had a robust discussion about the power of feedback following SP encounters. Several noted that their ability to provide feedback to learners motivated them to undertake portraying emotionally challenging cases. While feedback came up during individual interviews, the focus group provided a rich discussion which then informed individual interviews. This experience demonstrates the power of focus groups both in their own right and also as a qualitative method that may be used successfully in conjunction with individual participant interviews. It also shows how sensitive material may emerge in any qualitative study, so it is always important to keep in mind ethical considerations especially if you are including vulnerable populations [7].

SE1: *We were talking about what I would consider formal research just now. Related to this, I am getting a sense that my SPs are not happy about the way that they are being used for breaking bad news roles. I have heard that some of them don't want to do them anymore. Should I do a focus group on this, or is it more of a program evaluation?*

SE2: Often, a focus group may feel like a program evaluation and vice versa. To distinguish between the two, it is useful to consider whether the goal in collecting information is to improve the educational experience of learners, or to describe and explain the SPs experience in addition to program improvement. If the goal is basically to improve the educational experience, a program evaluation is the best choice. If you seek to study the experience and explain it—in addition to improve the education experience—I recommend using focus groups.

SE1: *What type of protocol is needed if I pursue projects as research in addition to program evaluations?*

SE2: No formal protocol is needed to do program evaluations. However, often educators conducting program evaluations realize research questions and interests manifest in the program evaluation process—I recommend SP Educators file a broad Institutional Review Board (IRB/REB) application with their institution about researching/exploring work protocols with SPs. For example, SP educators including myself have often become interested in studying work conditions of SPs during the training process that occurs during event preparation, such as how SPs are impacted when portraying emotionally challenging patient cases [8, 9]. This way, when issues such as emotionally challenging cases come up, the research protocol will be there for SP Educators to capture and explore routine debriefing with SPs as research. If an IRB protocol is in place and you make debriefing with SPs a routine practice, then it will not feel strange to SPs when you discuss challenging cases or events.

In terms of how to conduct a focus group, it is a good idea to begin by developing a focus group guide. This is essentially a list of questions for your participants based on research goals which facilitators use to ensure research questions are standardized and topics are being addressed in each focus group that is part of the same study. The IRB, (Institutional Review Board) office at your institution may offer a template on their website for both focus group guides as well as individual interview guides. (If you have not already, it will help you to become familiar with the IRB office at your institution). I usually include several open-ended questions, each followed by a few probing or follow-up questions. Follow-up questions enable you as researcher to explore interesting responses that you want to hear more about. When you have probing or follow-up questions ready to go, this will prevent you from fumbling for words or missing additional information related to a valuable topic you might not have anticipated.

SE1: *Is a focus group guide different from an interview guide?*

SE2: There are perhaps more similarities than differences. Because a focus group is a social event, however, there may be a certain amount of inhibition for participants to share their thoughts and feelings at the top of the session. There are number of methods a focus group leader can use to create a feeling of safety and we have included some ideas about these in the section below entitled *Tips for running a focus group*.

SE1: *Once you've developed a focus group guide, how do you address key logistical issues such as number of participants and the length of the focus group? Also, how do you as a facilitator balance the contribution of very vocal participants with quieter ones?*

SE2: There are several important logistical factors to consider when facilitating a focus group. First, I recommend including between 6 and 10 participants—enough so varying perspectives will be heard but not too many so that each participant has ample opportunity to contribute. While you may have multiple stakeholders you wish to include in a study, (e.g. SPs, physicians, students) I recommend careful consideration of who to include in which groups. For example, physicians and students included in a group of SPs would likely influence the responses. Be especially careful to consider any power differentials in terms of role.

During your introduction, you should inform participants as to the nature and goals of the research study using an IRB approved informed consent document or research preamble. A preamble or description of the research without obtaining consent is sufficient if there is no significant risk to you participants as determined by the IRB and the study is placed in the exempt category. If the IRB classifies your research with the exempt category, you can simply use this document or preamble to inform the participants about the study and give them the option to continue or decline. Those electing to participate should keep the informed consent document for their reference. When research studies pose a potential significant risk to participants, they may be classified by the IRB as necessitating a full IRB board review. This may occur if your study is assessed to pose a risk to vulnerable populations such as actual patients; please note educational studies with SPs are often classified exempt as SPs are not considered actual patients.

Following the informed consent process, make sure each participant and facilitator put on name tags to alleviate awkwardness in terms of addressing one another. The facilitator should begin by establishing a comfortable environment in which participants may share their thoughts and opinions. Establishing a comfortable environment includes ensuring—to the best of their ability—confidentiality. At this early point in the focus group it is also helpful to share with the group how long they can expect to be with you, where the facilities are located, information about reimbursement (if any). Most importantly remind the participants that the conversation is being recorded and to try not to talk over one another or interrupt or carry on side conversations as you want to capture everyone's ideas.

Next, the facilitator should draw upon their focus group guide to ask questions of the group. Ideally, the group will begin a dialogue within itself, so that the facilitator is guiding the conversation when needed but simultaneously stepping back so participants may interact freely with one another. As a facilitator, I feel most successful when participants are engaged with one another purposefully and on target with the research study aims while I am offering an observant and supportive nonverbal presence. Should you have one or a few participants who are dominating the conversation, you may choose to facilitate this situation in a variety of ways, but it is important to encourage other members to speak up without alienating the dominant voices. You could also make a broad claim at the beginning and throughout that it is important to hear from each member of the group. To encourage participants who are not speaking out, you may call them by name to ask their opinions. If these facilitation strategies do not work and you still have one or a few dominant voices, be direct in acknowledging their contributions but asking the dominant group members to temper their participation so that other group members may contribute.

For timing the focus group should run—at the longest—between 60 and 90 min—and this may be influenced by the number of participants with fewer (e.g. 6) taking less time and the maximum recommended (e.g. 10) taking more time. Use at least two audio or video devices to record the focus group, as it is always important to have a backup recording device so you do not lose valuable data due to a technical mishap. A cell phone with voice memo capability is always another option for a backup device, but we urge you to have a primary recording device independent of a cell phone. You will also want to consider how the recorded audio will be downloaded for later transcription. To learn more about transcription and data management—please see Chap. 17 by Nicholas, Clark, and Szauter [11].

You should be mindful of the time as the conversation continues. Once you have asked all the questions you've planned on, or if time runs short—consider guiding the conversation to a conclusion. I recommend doing this directly by asking the group a broad question to signal that time is running short such as "Since our time together is ending in a few minutes is there anything else anyone wants to add that hasn't been discussed yet?" Once participants have offered any last remarks, thank them for their time, reassure them again regarding confidentiality, and offer to be available in case questions arise following this session.

Transcribing the resulting narrative data is another important piece of assuring the quality of your research. This topic is covered in detail in Chap. 17 of this volume [11]. Given the iterative nature of qualitative research listening to the recordings of the various focus groups as you proceed is helpful with respect to shaping subsequent groups you may want to hold on your topic. Tiberius has published a helpful one page guide that I took into my first focus groups as a reminder for myself.

SE1: *How do I know how many focus groups I should run?*

SE2: Most researchers agree that there is no magic number of focus groups for the successful completion of your data collection. There are a number of considerations to think about here. The principle of saturation or data sufficiency is the most relevant, however this will be affected by your sampling strategy which will have a direct effect on your decisions about group composition.

Saturation

There are different kinds of "saturation" (theoretical, data, member). To answer your question, however, the number of focus groups you plan for depends on when you feel that you have reached a point where no new information is being collected. Saturation point determines the sample size in qualitative research as it indicates that adequate or sufficient data has been collected for a detailed analysis. This may mean that even if you plan for a particular number of groups it my change depending on your decision about the amount of data you feel is necessary to adequately answer your question. Along with these considerations is the understanding that running more groups is not necessarily better. However, Crabtree and Miller suggest that when focus groups are to be the sole source of data collection a minimum of four to five focus groups is recommended. Barbour suggests that nominally three or four focus groups are advisable if you want to conduct across group analysis looking for patterns and themes [3].

Sampling

Sampling for focus groups involves a researcher's strategic choices about how different group configurations may impart a range of ideas and insights into a research question [1]. This will have an impact on how many groups you plan to run. There are different kinds of sampling, such as "purposeful" sampling in which participants are chosen based on pre-determined set criteria that best suits your research topic. For example, consider a group made up of recent retirees talking about pensions. This group may involve people from different educational and economic backgrounds or not. It is up to the researcher to delimit the group according to what they want to hear about from the group. Other common kinds of sampling in qualitative research include "convenience" sampling which as the name implies is more logistically informed by who you as a researcher realistically have access to, and "snowball" sampling which asks participants to share names of others who may be helpful for the researcher to contact for future groups. As with other decisions in qualitative approaches sampling and saturation are iterative and may change as your research is underway. An explicit rationale for your sampling strategy is important for reporting your findings.

Group Composition

Depending on your research topic you may want to plan homogeneous groups such as all nurses or all SPs. A homogeneous sample involves people who may have a number of shared criteria, (i.e., age, socio economic status, profession) or share a common relationship to the issue being explored. For example, SPs who have all portrayed roles in a psychiatry OSCE [8]. Heterogeneous groups on the other hand are made up of people from disparate backgrounds (social, economic, ethnic, gender, educational, and professional) and with diverse experiences with the topic being explored in the group. As a caution because of the potential for uneven power relations such group can be tricky to run. Imagine having a group of patients all from different walks of life sharing their thoughts about fair access to free healthcare. You will get a rich variety of perspectives and will need to make sure the most privileged participants are not dominating while the less advantaged voices get lost or silenced. One of the advantages of heterogeneous group compositions is that participants do not know each other and everyone comes to the meeting without pre-set assumptions about the other people in the group. When heterogeneous groups are well run the information can be very rich. It all depends on what you are looking for out of your data collection.

Tips for running a focus group

Starting the Focus Group

Once you have finished the informed consent or preamble process and introductions, it is important to begin the focus group with a question that puts your participants at ease. Ideally, the first question should be relatable for participants so that they may connect it with their own experiences. A relatable first question will support participants who may feel awkward sharing their experience with strangers or people they may not know. Most participants however, will overcome this once they get to know each other. Other techniques that may be helpful to encourage participants to share their experiences include the use of a visual stimuli or trigger such as a video or a paper case that presents the participants with a dilemma relevant to the subject matter. This can help alleviate initial discomfort by focusing participants on the issue or topic in a more external way, providing a bridge to their stories through discussion about a common item. In this way you can facilitate a comfortable environment for participants to relate their experiences.

Negotiating Power Dynamics in the Group

Power needs to be taken into consideration at all stages of your focus group design and delivery. There can be a tendency for dominant individuals to want to lead a group towards consensus. One way to counter this is to be explicit during your introduction about your interest in everyone's ideas across a range of perspectives and your lack of interest in agreement on a topic. Introducing group agreements at the beginning of the focus group by gathering ideas from everyone about what they need from each other and the facilitator in order to speak freely is also helpful, For example, turn cell phones off, do not interrupt each other, etc. If someone is dominating the group by taking up too much time than you can bring everyone back to the group agreement as a reminder to share the space. This can be done gently by taking what has been shared by the one person and asking for other's opinions or views about the statement. A questioning approach by the moderator is important to the process of making participants feel valued. As mentioned earlier, periodically go around the group to make sure everyone has an opportunity to answer questions and share their thoughts.

Redirecting Participant Eye Contact

Often participants will look at the focus group leader when responding to a question, (especially at the beginning of a session) rather than engaging with each other. Ideally, as the session progresses, participants should make eye contact with one another which is a nonverbal signal that they are engaged with the conversation. One way to begin this process is to cast your eyes around the group when the person who is answering the question is responding. The speaker's eyes will often follow that of the moderator's around the group and in this way both the speaker and the moderator invite individuals from the rest of group to get involved in responding.

What If No One Is Saying Anything?

This is the most common anxiety experienced by first time focus group leaders. One of the hardest skills to master is comfort with silence, which leaves space for the others in the group to jump into the conversation. Above all, care must be taken to preserve the social space of the group. Leaders must always be mindful of the dynamic that is occurring between people and the impact on participants' ability to contribute their thoughts. Some people just take longer to feel comfortable in a group than others and as leaders our responsibility is to provide the opportunity for them to be heard.

What if you notice someone is reacting strongly to the discussion or getting upset with something that is said?

Establishing emotional and physical safety for focus group participants is crucial. If you notice that a participant is reacting strongly to the discussion they may be experiencing the topic or the focus group as a reminder of a previous experience or simply be deeply affected by what is being revealed within the group. We refer to this as being "triggered". If you notice this happening it is important to act immediately to support the participant. Triggering can involve both emotional and physical discomfort when a topic resonates profoundly and could manifest in a number of ways such as recalling a painful memory, the desire to immediately leave the session, or even to burst into tears. So, it is important to have a plan in place prior to running the focus group in the event that a member becomes distressed. Not all topics are suitable for a focus group format and this too needs to be taken into consideration during the planning phase. More sensitive topics or those in which sharing confidential information is important for the research will require a one on one focused relationship between the researcher and the subject.

One cannot not always know ahead of time what may trigger a participant. In a recent study with patient instructors who are HIV positive (PHA-PI's) we conducted focus groups following their exchange with second year medical students who were providing them with a positive diagnosis of HIV [10]. The study itself was rich and very positively received by the students, preceptors, and HIV positive Patient Instructors. In the focus groups following the sessions, however, some of the PI's were triggered by the discussion - not the study itself but the focus group discussion afterward triggered unwanted memories and emotions that came vividly back to some of them. Discussions about experiences of loss, discrimination, rejection stuck with after few of the participants after leaving the group and going home. We had foreseen this possibility and planned for health professionals to be available on site and on call. In this way we were able to speak with and connect those who requested help to immediate healthcare support ad follow up counselling services as needed. While this is an extreme example, such triggering can occur even for seemingly benign topics. Ultimately, we cannot know if someone in the group has had an experience with the topic or with another individual in the group that the discussion may reopen. It is our ethical responsibility as researchers to be ever aware of the possibility for unintended harm to participants though our research processes, and we must do our best to mitigate this [11].

Conclusion

Focus group method is a useful qualitative approach for simulation educators seeking to turn their daily work into scholarship while reflecting the voices of many participating stakeholders. While this is a seemingly straight-forward method, it is, in reality, a quite complex and nuanced technique. This method offers busy professionals the opportunity to gather a variety of perspectives on relevant educational issues in a brief period of time. Additionally, focus group method may be combined with individual participant interviews to strengthen data by triangulating it—in order to build common themes and findings from multiple voices that offer a variety of perspectives on a common topic. As with all research methodology involving human subjects, care should be taken to ensure confidentiality and respect for participants. Potential ethical issues must also be considered from the inception of the design to the final analysis and reporting.

References

1. Stalmeijer R, McNaughton N, van Mook WNKA. Using focus groups in medical education research: AMEE guide no. 91. Med Teach. 2014;36(11):923–44.
2. Stewart DW, Shamdasani PN. Focus groups: theory and practice, Applied social research methods series, vol. 20. Newbury Park: Sage Publications; 1990.
3. Barbour RS. Doing focus groups. London: Sage Publications Ltd; 2007.
4. Freire P. The pedagogy of the oppressed. Harmondsworth: Penguin; 1970.
5. Kitzinger J. The methodology of Focus Groups: the importance of interaction between research participants. Sociol Health Illn. 1994;16(1):103–21.
6. Rabiee F. Focus-group interview and data analysis. Proc Nutr Soc. 2004;63:655–60.
7. Clark L. Ethics of working with vulnerable populations. In: Matthes JP, Davis CS, Potter RF, editors. The international encyclopedia of communication research methods. Hoboken: Wiley-Blackwell; 2017.
8. McNaughton N, Tiberius R. The effects of portraying psychologically and emotionally complex standardized patient roles. Teach Learn Med. 1999;11:135–41.
9. Bokken L, Van Dalen J, Rethans JJ. The impact of simulation on people who act as simulated patients: a focus group study. Med Educ. 2006;40:781–6. https://doi.org/10.1111/j.1365-2929.2006.02529.x.
10. Jaworsky D, Chew D, Thorne J, Morin C, McNaughton N, Downar G. et al. From patient to instructor: honouring the lived experience. Med Teach. 2012:34(1):1–2
11. Szauter K, Nicholas C, Clark L. Transcription and data management. In: Nestel D, Hui J, Kunkler K, Scerbo MW, editors. Healthcare simulation research: a practical guide. Cham: Springer; 2019. Chapter 17.

Additional Resources

Tiberius R. The Focus Group guide: University of Miami: Miller School of Medicine; 2006.
Crabtreet BF, Miller WL. Doing qualitative research. Thousand Oaks: Sage Publications; 1999. p. 1999.
Creswell JW. Qualitative inquiry and research design. Choosing among five approaches. 3rd ed. Los Angeles: Sage Publications; 2013.
Krueger RA. Focus groups: a practical guide for applied research. Newbury Park: Sage Publications; 1988.
Shankar PR, Dwivedi NR. Standardized patient's views about their role in the teaching-learning process of undergraduate basic science medical students. J Clin Diagn Res. 2016; https://doi.org/10.7860/JCDR/2016/18827.7944p.1–5.
Smeltzer SC, Mariani B, Gunberg Ross J, de Mange EP, Meakim CH, Bruderle E, Nthenge S. Persons with disability: their experiences as standardized patients in an undergraduate nursing program. Nurs Educ Persp. 2015:398–400. https://doi.org/10.5480/15-1592.
Wagenschutz H, Ross PT, Bernat CK, Lypson M. Impact of repeated health behavior counseling on women portraying an overweight standardized patient. J Nutr Educ Behav. 2013;45:466–70.

Observational Methods in Simulation Research

14

Birgitte Bruun and Peter Dieckmann

Overview

Observational data collection, analysis, and interpretation for research in and with simulation research entails choices regarding strategy, techniques and tools that should be made in the same process as the formulation of the research question and the development of concepts that will guide the study. Depending on the research question and purpose, observational data collection may be combined with other research methods. Analysis of observational data may draw from different research traditions with implications for interpretation. Ethical considerations should be on-going from conceptualising a research project to writing up the results.

Practice Points

- Observational data collection can be useful in studies applying both deductive and inductive reasoning.
- The units of analysis in research are never "just there" to be observed, but must be conceptually delineated from the endless, dynamic stream of events in the world.
- It may be useful to separately consider observation strategy, observation technique and tools for observation to strengthen the study design.
- Observation strategies, techniques and tools influence how units of analysis appear, so choosing them should be considered carefully in relation to how they will work together with a given research problem, research tradition, analytical concept, and observation ethics.
- The quality of tools to assist observational data collection is not inherent, but depends on how the tool "cuts the cake" and how this "cutting" works together with other elements in a research project.

B. Bruun (✉)
Copenhagen Academy of Medical Education and Simulation (CAMES), Capital Region of Denmark,
Center for Human Resources, Copenhagen, Denmark
e-mail: Birgitte.bruun@regionh.dk

P. Dieckmann
Copenhagen Academy of Medical Education and Simulation (CAMES), Capital Region of Denmark, Center for Human Resources, Copenhagen, Denmark

Department of Clinical Medicine, University of Copenhagen, Copenhagen, Denmark

Department of Quality and Health Technology, University of Stavanger, Stavanger, Norway
e-mail: mail@peter-dieckmann.de

Introduction

The literature based on observational data collection in and with simulation research offers many examples of inspiring observation techniques and tools. This chapter presents examples, but the main purpose of the chapter is to take a step back to contemplate the matching of method and technique with the purpose and problem of a given research project that uses observations. Even the most sophisticated observation technique loses its allure and helpfulness if it is badly matched with the research question, the conceptualization of the units of analysis, the overall purpose of the research, or the theoretical orientation of the study.

This chapter is divided into five sections. First, we present some of the many aspects that might be illuminated in and with simulation research through observation. In the second section we present two research traditions, the post-positivist and the constructionist tradition, as two overarching frames for posing (observation-based) research questions that have profound implications for choice of method. Next follows a

D. Nestel et al. (eds.), *Healthcare Simulation Research*, https://doi.org/10.1007/978-3-030-26837-4_14

section on useful considerations regarding choice of observation strategy, techniques and tools, and a section on how observations are turned into data. The last section of the chapter discusses ethics of observation.

Our emphasis is on observational data collection for research in and with simulation, although many of our considerations may also be relevant for readers who apply observation in simulation for training and learning purposes, or observations in other settings.

What can be observed in and with Simulation Research and what cannot?

Observation is useful as a way of exploring what people actually do, which may not always be the same as what they think or say they do. Observation can offer data on embodied practices, acts, tasks, processes, flows, events, incidents, interactions, verbal exchanges, repetitions, succession, duration, pace, and movements in physical space. Examples of what observation is used for include the analysis of case-irrelevant-communication [1], investigating the psychometric properties of the use of an observational tool to rate teamwork [2], or comparing actions of anaesthesiologists in a simulated setting with their actions in a clinical setting [3].

Such units of observation are never "just there" and ready to be observed. They all need to be turned into discrete entities through definition or other conceptual work before it is possible to identify and "see" them in any way useful for research. Besides, any interaction within a healthcare team, their interaction with the simulator, and with the environment is so complex that it is impossible to observe "everything". Therefore, it is necessary to define the scope of attention carefully: is it where a team-member looks that is of interest, or the micro-expressions in her face, while she interprets the monitor? If you are studying tasks, how are tasks broken down and coded into actions? In a study of anaesthesia professionals, a moment of physical inactivity can be coded as "idle", whereas the very same moment can also be broken down and coded into "gathering information", or "decision making" or indeed "idle" if anaesthetists themselves are involved in coding [4]. In this way, a task group can be broken down into actions in many ways with profound implications for the data that is produced. Here it is important to establish the "difference that make a difference" [5] which happens in a dialogic process between observation, conceptual framework and research problem – either before the study begins as is often the ambition in the post-positivist research tradition, or as the study unfolds, as it may happen in the constructionist tradition.

Meanings, intentions, interpretations, values, emotions, stress, cognitive processes, etc. cannot be inferred from observation alone. If these aspects are included in a research question, observational data collection will need to be combined with other methods, such as interviews, questionnaires, measurements or physiological measures like EEG or skin resistance. Each method offers a partial and positioned picture of the problem. Method, concepts and analytical object stand in a mutual relationship to each other, so what you see depends on what concepts and methods you "see" with, knowingly or unknowingly.

Combining methods does not offer a more whole or more complete picture, but a facetted snapshot combined of more than one source of data. Possible tensions between what is "seen" through one and the other method offer a chance to add to the nuance of analysis and to generate new research questions. Combinations of observations and other visual methods with verbalised data are described in more detail in Chap. 15.

When studying complex phenomena, e.g. mental models, decision making, team work, communication, safety, or transfer of learning, extra care is needed in delineating exactly what insights can be produced in relation to particular concepts through observation or through observation in combination with other methods. Simulation educators will know this challenge from debriefing practice, where questions are needed to understand the frames behind actions [6].

Once one or more units of analysis have been identified a next step could be to consider the overall analytical orientation to observation and observational studies. The overall analytical orientation may be given already by the domain that your study is conducted in, for example human factors, safety studies, simulation studies, or improvement of healthcare. Each domain has its own traditions and has come to prioritize certain theoretical, practical, and empirical approaches over others. Your task is then to make explicit what aspects of your problem that this analytical orientation makes visible (how it "cuts the cake") and what aspects that remain invisible. It has implications for the analysis, for example, whether a study focuses observations on people practicing a particular profession or on interactions between professionals in a team; whether observations focus on the diagnostic steps taken or on the communication about results, etc.

Observational studies are sometimes defined as directed at something or someone in their 'natural setting'. In simulation research, what is observed is not (completely) naturally occurring, but set up with a particular purpose in mind. Researchers should make explicit how and to which extent this purpose – and possible differences between simulation practice compared to clinical practice – play a role in the way they frame their objects of analysis and their research questions. A study comparing clinical settings with simulation settings using observations, showed, for example, a great match of actions, when looking at the core task (providing anesthesia in this case), but differences in actions around the core task. For example, study participants clean the anesthesia equipment during an operation, while such an "additional" behavior was not observed in the simulation setting.

Post-positivism or Constructionism?

How observational data are collected, interpreted, and reported is dependent on the overall research tradition that the study is embedded in.

In the post-positivist medical research tradition observational studies are often understood in contrast to randomised controlled trials, as focusing on the naturally occurring. Observations are treated as measurements. Observational studies do not entail intervention or control group as such, but register specified events or developments over time. Various observation techniques may be applied to answer often pre-defined and closed research questions aimed at counting occurrences and at quantifying. This approach to knowledge generation is often deductive, built on a general theory looking for confirmation or variation in specific instances. A challenge is to present the general theory explicitly as a theory, and not as a fact: as stated above, a first step is to single something out as an observable event out of the stream of ever-changing events [7]. This is not a neutral act, but a matter of interpretation. Still, partiality, or observer bias, is actively sought minimized in various ways.

In the constructionist research tradition observation is often applied as a method to produce qualitative insights into individual and group actions, interactions, practices and "doings" over time and in context. Such studies aim to describe the difference that makes a difference. Observation of social processes is often based on open-ended and explorative questions aimed at seeing and describing something (a) new. This approach to knowledge production is often inductive, moving from observations of the specific to generating more general concepts or theory. A challenge in this approach is to make explicit exactly how to generalise findings from one context to another. Observers recognize predispositions as a condition for generating knowledge and may make a point out of explicating their position in the field, for readers to take this positionality into account.

Summing up the difference between these approaches, a deductive study might, for example investigate, which interactions can be sorted into an already given behavioural taxonomy, whereas an inductive study could try to establish a new taxonomy based on a "fresh" view of what is observed. Both research traditions are useful for observational data collection in and with simulation. The crucial consideration is to explicitly reflect on the match between purpose, problem, method, and tradition of research. When the links between these elements are strong, research results will be powerful. Explicit reflection about theoretical orientation makes the difference between casual, everyday observations and purposeful, systematic observation for research, whether it draws from the post-positivist or constructionist tradition.

It can be tempting to combine methods drawing from different research traditions, e.g. counting instances of interruption (however defined) in combination with interviews about the effects of those interruptions. Such combinations can be quite demanding in terms of analysis and theoretical underpinnings. Consider, for example, the possible outcomes of such an interruption study, where there are many interruptions, but most of them are not described as disturbing. Is the number of disturbances then more important than the subjective feeling of being disturbed? And how should the study handle the uncertainty that the experience of being disturbed might not actually reflect the consequences of interruptions for safety, for example, that might be unknown to the people interviewed? In interdisciplinary research, it is important to continuously make explicit how observation strategies, techniques and tools from different disciplines configure objects of analysis differently, and where connections and differences between the different objects and views onto them may emerge with implications for what can be "seen". Again, possible tensions between what is "seen" through one or the other method offer a chance to add to the nuance of analysis and to generate new research questions.

Observation Strategies, Techniques and Tools

The development of your research design can start from any element in your project and will typically involve some adjustments before the choice of elements becomes well justified. In this process, it may be useful to separately consider observation strategy, observation technique and possible tools for observation.

- **Observation strategy (the plan for producing your data):** Will you zoom in on particular phenomena from the start of the research or will you begin by attempting to observe and record "everything"? Or rather, is the intention to observe and look for nothing in particular at the beginning of research? Are you looking for paradoxes, patterns, break-up of patterns, or changes over time? Or are you trying to identify the key problem confronting a group? [8] Over the course of your research will your research be divided into phases and will you move back and forth between "wide-angle" and "narrower" observation? What is the timing, duration, and sequence of your observations? Will you apply repetitive observations to explore change and development? How long time do you have for a single observation and how long is the time frame in which you can observe? Consider also, how much time you will have for the analysis of your data. How many people are involved in the observations and in which roles? How might their perspectives differ and how might differences influence your study?
- **Observation techniques (the way you execute your plan):** Will you observe through direct visual contact yourself? Where will you be physically located for what

purpose? Will you employ assistants and how will you instruct them? Might filming provide more accurate data for your research purpose? Many simulation centres offer the possibility of live-view projections from a simulation area to a research area (often a control room). This possibility allows for several perspectives onto the object under observation, by including more than one camera. It is also possible to record the events for later analysis where play back in slow-motion or fast forward can be useful (see Chap. 15), but a cost of recording, if you are not in the room at the same time, may be loss of insight into what happens outside the lens of the camera(s) that may influence what the camera records.

- **Observation tools (media of observation):** Observations are visual impressions of a person. As such they are very volatile, transformed into memories virtually at the same moment, they are happening. To use such observations in research they typically need to be transformed into a more permanent trace than the memory of a person, ranging from an informal oral report for a colleague to a detailed transcript or completed checklists, timetables, or behavioural marker systems (ANTS [9], NOTSS [10], etc.). Here it is important to note that the quality of a tool is not inherent to it, but a matter of how it is applied in the production of data, in the analysis of data, and in the way conclusions are drawn from the data. This includes, for example, whether those, who are using the tool are familiar with it, use it in the same spirit, understand the underlying definitions in a similar way, and also whether they use a similar observation activity (e.g. where do different observers look in the beginning of a scenario, whom do they concentrate on, how narrow is their focus?). Electronic and analog logs, lists or tables can be used to create a durable trace of the observations more or less in real time, while observing [4, 11]. These can be elaborated upon from memory, film or other sources after observing depending on the need for detail and being aware of the way the medium shapes the observation data. If adding to observations after the end of the session time makes a great difference: the sooner, absolutely the better.
- **The role of the observer:** How will you act in relation to the people and events that you observe? This role can range from being completely detached (no interaction with those observed) to completely involved (participating observation), with degrees of involvement in between. Each mode of involvement has implications for the data you can produce [12]. If you choose to be detached what might be the implications for your observations if you are also invisible to the observed by observing, for example, from a control room or afterwards from film?
- **The effect of being observed:** In a research study, participants will typically be aware of and informed about being observed and the so-called "Hawthorne effect" will be relevant. The knowledge about being observed will change behavior to an unknown degree. Whether this is considered a source of bias or a condition for producing knowledge, it may be useful to consider how many observations of the same phenomenon that might be needed to get a good grasp of both norm and variation.

Selected Examples of Studies that Involve Observations in Simulation Related Research

Manser and colleagues used an observation scheme to compare actions in simulated and clinical settings and concluded that the core activity in both settings as largely comparable, but also described characteristic differences between the settings. The core tasks were seen as comparable, task elements not belonging to the core were lost in simulation.

- Manser T, Dieckmann P, Wehner T, Rallf M. Comparison of anaesthetists' activity patterns in the operating room and during simulation. Ergonomics. 2007;50 (2):246–60.

Kolbe and colleagues observed communication patterns during anaesthesia in high- and low performing teams and could describe some of the features of communication in high performing teams, by investigating patterns of interaction.

- Kolbe M, Grote G, Waller MJ, Wacker J, Grande B, Burtscher MJ, et al. Monitoring and talking to the room: autochthonous coordination patterns in team interaction and performance. J Appl Psychol. 2014;99 (6):1254–67.

Escher and colleagues analysed video recordings of simulation scenarios and described the impact that the different communication channels between the simulation educators and the simulation participants have on the flow of the scenario.

- Escher C, Rystedt H, Creutzfeldt J, Meurling L, Nystrom S, Dahlberg J, et al. Method matters: impact of in-scenario instruction on simulation-based teamwork training. Adv Simul (Lond). 2017;2:25.

Nyström and colleagues unfolded the benefits of conceptualizing debriefings from a sociomaterial point of view.

- Nystrom S, Dahlberg J, Edelbring S, Hult H, Dahlgren MA. Debriefing practices in interprofessional simulation with students: a sociomaterial perspective. BMC Med Educ. 2016;16:148.

Transforming Observations into Data

Once observations have been recorded as sound and image on film, digitally, in charts, as drawings, numbers, keywords or more detailed text they only transform from "raw" into data when related to a research question and suitable analytical concepts. The transformation from observation to data may happen through coding and classifying observations, for example into groups of tasks or actions, or sorting observations into themes and sub-themes for further analysis. During the coding process the recorded trace of a visual, auditive, or other impression is acknowledged or filtered as a relevant piece of data. Particularly in qualitative studies, but also sometimes in quantitative studies, not all observations will end up being relevant for analysis, and the distinction between relevant and irrelevant may develop gradually with the process of analyzing data. Here, inductive studies may build their "filter", or conceptual framework, as they go, whereas deductive studies apply filters that were created previously and perhaps in other contexts.

An essential aspect of transforming observations into data, which may seem trivial but is not, is the art of coding. The delineation and sub-division of categories and codes is as much an analytical task as delineating your object of analysis. Small variations in how an observation is coded and turned into data can make a big difference in the analysis. When developing codes an aim may be to eliminate ambiguity and variation, but you can also aim to be as explicit as possible about the variation that a category or code embraces, and about possible implications of ambiguities for analysis and interpretation. Making variations and ambiguities explicit, for example in a logbook of codes, is useful already when designing a study, and may be on-going until well into the analysis process. This explication is particularly relevant when more people are involved in coding or rating observations.

Observation Ethics

Studying other people entails some basic considerations about ethics. Chapter 15 discusses special ethical considerations, when working with visual material. Permission to carry out the study must, of course, be sought in the appropriate formal as well as informal fora. Apart from properly informing all the observed, and all others in the setting, in good time in writing and orally you can encourage questions and ensure that all the observed have opportunities to ask their questions. When asking for consent, anonymity and confidentiality should be ensured in all records and reports from the study, unless other is agreed upon with the observed. If justified in relation to the research problem it is possible to observe without informing those observed or to "hide" the actual object of the observations. Study participants might be informed about being observed, but might not be informed about what exactly the researchers are interested in, or might even be informed in a "misleading" way. For ethical reasons, such study conditions should always be disclosed and "debriefed" after the end of the study.

The ethical dictum to "do no harm" is not limited to protecting study participants while observing them, but may also be relevant on a more overall level when forming research questions (is there a bias or unfortunate preconception in the way the research question is stated?), over the conduct of observation (are the observed given a chance to learn about and respond to findings?), to writing it up (are the observed represented in a nuanced way?).

Finally, the edict to "do no harm" also applies for the researcher her- or himself. Consider ahead how your topic, your choice of methods, and your interaction with the observed might put you in awkward or unwanted situations, where you might see things, that you would prefer not to see. Consider your own limitation as much as possible beforehand and how you may prevent getting into situations you would rather avoid. You might still be surprised, but anticipation might at least help you in dealing with the surprise. If you engage assistants for observing or filming it is even more important to discuss ethical aspects of their position with them in advance. Follow up with them during and after the study.

If ethical issues come up for observers during a study they should be handled as such, but depending on the topic and approach of the study they may also be important clues to what is at stake for the observed. In this way ethical issues are not obstacles, but possible keys to a deeper understanding of a topic.

Closing

Research in and with simulation quite often applies observation as method. For this reason, it is important to reflect on the possibilities and limitations of observation in simulation and to carefully and separately consider observation strategy, technique and tools in relation to a given research problem, as well as the concepts that frame it, and the ethics of observation.

References

1. Widmer LW, Keller S, Tschan F, et al. More than talking about the weekend: content of case-irrelevant communication within the OR team. World J Surg. 2018; https://doi.org/10.1007/s00268-017-4442-4.
2. Zhang C, Miller C, Volkman K, Meza J, Jones K. Evaluation of the team performance observation tool with targeted behavioral markers in simulation-based interprofessional education. J Interprof Care. 2015;29(3):202–8. https://doi.org/10.3109/13561820.2014.982789.

3. Manser T, Dieckmann P, Wehner T, Rall M. Comparison of anaesthetists' activity patterns in the operating room and during simulation. Ergonomics. 2007;50(2):246–60. https://doi.org/10.1080/00140130601032655.
4. Rall M, Gaba DM, Howard S, Dieckmann P. Human performance and patient safety. In: Miller RD, editor. Miller's anesthesia. Philadelphia: Elsevier Saunders; 2015. p. 106–66.
5. Bateson G. Form, substance and difference. In: Steps to an ecology of mind. Chicago: University of Chicago; 1972. p. 455–71.
6. Rudolph JW, Simon R, Dufresne RL, Raemer DB. There's no such thing as "nonjudgmental" debriefing: a theory and method for debriefing with good judgment. Simul Healthc. 2006;1(1):49–55. https://doi.org/01253104-200600110-00006 [pii]
7. Ringsted C, Hodges B, Scherpbier A. "The research compass": an introduction to research in medical education: AMEE guide no. 56. Med Teach. 2011. https://doi.org/10.3109/0142159X.2011.595436.
8. Wolcott HF. Confessions of a "trained" observer. In: Transforming qualitative data. Description, analysis, and interpretation. Thousand Oaks: Sage Publications; 1994. p. 149–72.
9. Fletcher G, Flin R, McGeorge P, Glavin R, Maran N, Anaesthetists PR. Non-Technical Skills (ANTS): evaluation of a behavioural marker system. Br J Anaesth. 2003;90(5):580–8. http://www.ncbi.nlm.nih.gov/pubmed/12697584. Accessed May 2, 2018.
10. Yule S, Flin R, Maran N, Rowley D, Youngson G, Paterson-Brown S. Surgeons' non-technical skills in the operating room: reliability testing of the NOTSS behavior rating system. World J Surg. 2008;32(4):548–56. https://doi.org/10.1007/s00268-007-9320-z.
11. Manser T, Wehner T. Analysing action sequences: variations in action density in the administration of anaesthesia. https://link.springer.com/content/pdf/10.1007%2Fs101110200006.pdf. Accessed 24 Apr 2018.
12. Spradley JP. Participant observation. New York: Holt, Rinehart and Winston; 1980.

Visual Methods in Simulation-Based Research

15

Peter Dieckmann and Saadi Lahlou

Overview
In this chapter we discuss theoretical and practical considerations when using visual methods for research. We outline the nature and origin of visuals, describe the research process around visuals, and the purpose and possible output formats of visual methods. Visuals can be produced for a research project, or a project can be built around existing visuals. The visual itself can be the data of interest, or it can be used to elicit verbal or other responses in study participants. With ever more powerful technology it is important to consider the ethical aspects in using visuals for research: during planning of the study, during its conduct, and dissemination. Considerations need to include the context in which the visual was produced and in which it is shown, as framing of the visual can substantially change its meaning.

Practice Points

- Visuals can be used to describe how simulation scenarios unfold in practice; technical possibilities enrich human perception.
- Participants can become active co-researchers when involved in the production of the visual, or when their comments on the visual are elicited.
- By their nature, visuals can 'paint a thousand words' and therefore usually need some verbal interpretation.
- Ethical considerations include the production, analysis, and dissemination of research using visual methods.

Introduction

Interactions in clinical and simulated practices are dynamic and involve various entities: humans, machines, devices. Social and organizational rules scaffold this interplay [1]. Simulations render this complexity and add a layer of complexity, where certain entities "stand for" other entities (e.g. a simulation manikin representing a patient) [2]. Visual methods provide unique research possibilities for capturing data on this dynamic interplay, for data analysis and data presentation [3, 4]. Especially, when used in combination with verbal methods (like interviews), visual methods have the potential to capture both the observable (behavior) and internal (experience) aspects of activity. Study participants can be active during the production of the visual (e.g. drawing how they see the team dynamic) and can thus engage in a new way in research [5].

Visual methods comprise many media [6]: photographs, drawings [5], paintings, maps, videos, etc. We use the word "visual" to refer to this diversity. Visual methods are becoming more widespread in medical education [7, 8] and simulation [9]. Within simulation-based research the most widely used visuals are audio/video recordings of scenarios, films in teaching, and pictures or diagrams in presentations.

This chapter discusses theoretical assumptions and practical, technical, and ethical considerations on the use of visual methods in research. We consider visual methods as modes of investigation, not as the object of study. Because of their

P. Dieckmann (✉)
Copenhagen Academy of Medical Education and Simulation (CAMES), Capital Region of Denmark, Center for Human Resources, Copenhagen, Denmark

Department of Clinical Medicine, University of Copenhagen, Copenhagen, Denmark

Department of Quality and Health Technology, University of Stavanger, Stavanger, Norway
e-mail: mail@peter-dieckmann.de

S. Lahlou
Department of Psychological and Behavioural Science, London School of Economics and Political Science, London, UK
e-mail: s.lahlou@lse.ac.uk

D. Nestel et al. (eds.), *Healthcare Simulation Research*, https://doi.org/10.1007/978-3-030-26837-4_15

prominent role in healthcare (e.g. in simulation), we focus on of audio-video recordings. We hope to inspire other researchers to engage in the use of visual methods. To get started we recommend reading Bezemer [7], Pauwels [10], Jewitt [11] and Derry [12].

Theoretical Standpoints

We build on a framework distinguishing three aspects: the nature and origin of the visual, the research focus and design, and the format and purpose of the outcome [10].

Nature and Origin of the Visual

Visuals can be produced for the purpose of the research (e.g. asking a team to draw the hierarchy in their team) or for another reason (e.g. photographs from the team development day). Visuals can show the study participants (e.g. scenario videos) or be produced by them (e.g. visual time lines of scenario events). Visuals can be created by the research team as data, or in order to use them as stimuli in a study of participants' reactions to those stimuli [13], or to validate rating instruments for performance [14] etc. While visuals can be reused, their origin and purpose must always be considered and specified, because they impact contents.

Visuals are an abstracted point of view. They can try to preserve as many features of the "original" as possible, or they can aim for emphasizing what is relevant for a given research interest. While tradeoffs are inevitable, their rationale must be explicitly considered: a detailed material used without reflection on its relation to what it actually pictures can be less useful than material of lower quality and scope of which the relation to the phenomenon is well discussed.

Research Process Around the Visual

The "same" visual may be used for very different purposes depending upon the process in which it is enshrined. Different aspects can be addressed: the visual itself (e.g. who is depicted where in the team drawing), its production process (e.g. atmosphere while drawing the team picture), reactions it elicits (e.g. the reflections about a drawing), or the interactive practices it triggers (e.g. whether hierarchy depicted in the drawing replicates itself while talking about the drawing). What is seen as relevant in the visual itself or what it elicits will depend on the theoretical framework of the research team. This framework must be made explicit.

Purpose and Format of the Visual Describing the Output

The purpose of the visual is related to the aim of the study in which it is used. In principle, visuals can help, for example, to get a detailed understanding of some aspect of the world [15], to compare aspects of the world or people with each other [9], or to explore the meaning that human beings assign to the world [8]. Whether a film is presented as "standard practice" in an instruction session or as "occasion for reflexivity" to the team whose action was filmed induces very different cognitive (including emotional) dynamics among participants.

In short, visuals should be considered as part of a larger project and explicit reflection on the conditions of production, purpose and use is good practice. This is not just methodological hygiene: these aspects are an integral part of *using* visuals to produce some effect in the viewers; integrating them in the design of research will considerably improve and magnify the impact of the visuals for the purpose at hand. Just saying "I'll show a video" is a first but incomplete stage of designing a good process.

Output of research about visuals is usually in text form or might include snippets from the visual material (e.g. a paper with supplementary videos online). Currently, visuals are not widely accepted as a means of scientific communication [12]. We advise careful storing and documenting current visuals as they are made (including keeping them in maximal resolution in original format) for the days such material will be publishable in full visual format or as reference after publication.

By their nature, visual methods depict what is visible and are blind to inner states of humans, machines and organisations. Visuals can indeed replace a thousand words, but without setting words to them, interpretations might be wide-ranging (or worse). Therefore, in most cases, we use the visual *in combination with* verbal descriptions providing data on what is not visible in the visual (context, comments of participants about the activity). Consider videos: they illustrate *behavior*, but what we are usually interested in is *activity*: "behavior is what subjects do, as described from the outside by an external observer. It is an external description of objective phenomena. In contrast, *activity* is what subjects do, experienced from their own perspective." [1, p.425–426].

Ethical Considerations

Ethical aspects of collection, processing, storing and uses of visuals are especially sensitive and must be addressed [12, 16]. Institutional Review Boards may not be very familiar with the features of this type of research and thus reluctant to approve studies [17]: consider explicitly these aspects *before* production.

During the production of visuals, care should be taken about decency (e.g. which body parts are shown in what way) and, if wished for, anonymization throughout the whole process. Anonymization can be done through blurring image parts (e.g. faces) and muting voices (replaced by subtitles, if needed). Legal advice might be necessary, if data are relevant for patient safety and quality of care, to clarify whether or not visuals might be used as "evidence" in a legal sense. Clear agreements are needed about who might see which parts of the material in which context and form, described typically in informed consent forms [18]. For sensitive material, consider destroying the video immediately after analysis, and keeping, for example, only the recording of the discussion about the video.

Trust needs to be built between the researchers and study participants. In our practice with video-recordings, we offer participants after the recording to take the only copy of the material home *before* we researchers have access to it. We typically ask them to decide, within three days, whether they are happy with us going further with the analysis of their material. They can delete the recording, or we help in deleting parts of it. Only then do we get the copy back and start analyzing [19]. The fear of a visual showing an error can at times be addressed by reminding participants that the visual will show how they did their best in difficult circumstances [16]. Another way around this issue is to explain that we are not interested in recording errors, because we can get just as useful data by asking how participants avoid such errors. In our experience, cases of participants finding *after the fact* that their records contain something embarrassing are extremely rare. Nevertheless, concern about recording such sensitive events is widely prevalent *before* participants agree to record. So, this concern must be addressed frankly and explicitly because the concern is real even though unfounded. Making explicit how such cases would be addressed is essential both for participants and for ethics boards.

Besides the visual itself, its framing in terms of captions and descriptions is important to consider from an ethical perspective. The more powerful digital tools get, the higher is the potential for manipulations.

Beware that, once published or shared, visual material can be used in contexts that potentially changes the message it sends by using different captions or even by manipulating images. The material shared should be considerate of the rights of those being depicted. Ethical principles and norms on how to deal with these issues are still work in progress (see Mok et al. [20] or Wiles et al. [21]). Ethical considerations should be documented both at the stage of initial production (informed consents, ethics boards) and in their subsequent use (publications, presentations).

Finally, when sharing visuals (e.g. for analysis or in presentations) copyright and issues of "fair use" can become relevant if material is used that was produced by others [10].

Practical Examples of Capturing and Producing Visual Data

Visuals allow for a new perspective upon what is supposedly known [16, 22]. Seeing oneself from the outside on a video [16], investigating an action in slow motion [3], using colour to express social relationships are but a few examples of providing viewers with a reflexive distance with a phenomenon. Putting an action camera onto a (simulated) patient's forehead while being maneuvered through the hospital will provide interesting footage to stimulate discussion. Video recordings of a scenario from the bird's eye perspective, or from several different angles will literally provide a new perspective and visual feedback. Subject-centered perspectives make the limitation of one's perspective visible and can be used in an effective way to elicit commentaries later on [23]. Seeing another person's recording from a subject-centered perspective can provide unique new insights. Eye-tracking systems allow fine-grained analysis of gaze. In combination with other physiological data they can provide extensive data to investigate perception and cognition [24]. Other possibilities include asking patients to take a digital camera and record what patient safety challenges they see in a department.

Electronically produced visuals can go beyond the mere capturing of an event in real time. High speed cameras allow seeing details invisible to the human eye (e.g. micro-expressions during decision making). Photographs of many instances of the same event can be combined to literally show the "corridor of normal performance" [22]. Within healthcare work (and simulation), a study identified eight types of leadership behavior from video recordings of simulations [15]. Time lapse photography makes it possible to capture developments and movements that are too slow for the human eye or take too long for one study to investigate (e.g. movements of people in a department over the course of 24 hours) [21].

Where the participants are producing the visual, the instructions they receive are important (e.g. expressing the key points of their opinion by drawing the most important three points they see). The person producing the visual will – to a large extent – influence what is recorded, for example, through perspectives, zooming, start and end of the recording. With multiple study participants, one can appreciate the different understandings of the "same" concept (e.g. leader and team members). Participants can use photographs [25], drawings, or staged video clips etc. Participants likely also need support, because of a (perceived) lack of drawing skills – a widespread phenomenon [26].

Eliciting Data from Visuals

Elicitations aim to supplement visuals with elements that are not seen, like cognitive, emotional and social influences. They can: (1) focus on reconstructing or re-enacting past thoughts, feelings, and impressions; (2) trigger narratives around the depicted material to reach a thicker description; and, (3) stimulate reflections and individual development of the person producing the elicitation [11]. Elicitation is often a more mutual exchange between the researcher and the study participant and their interaction can itself be recorded and become relevant research data [6, 27]. This potential to involve was used to include patients within patient safety research [28]. Standardized visuals can serve as triggers in research, where the actual data collected comprise the reactions to the visuals [29, 30].

Eliciting data from visuals can be supported throughout phases of research [10, 12]. During the production of a visual it is possible to take note of interesting pieces or parts (e.g. setting a mark on a video recording). Macro-level coding can be used to get familiar with the material and to identify and manage points of interest in the material (e.g. identifying the beginning and ending of significant subparts on a video). This step makes it easier to compare visuals from different persons or conditions. Software products are available to support this step electronically and continue to develop [12]. Narrative summaries about the whole visual or its parts can help in recording a certain way of seeing and interpreting the visual at a point in time or by a specific person. At times further visuals (e.g. diagrams or time lines) can help to condense the information. Transcriptions can then be used in micro-level analysis. There are several guides that can help in deciding about the level of detail and the best form in a transcript [31].

In our own practice, we use replay interviews in relation to first-person perspective video recordings mostly. Participants see their own videos and comment about the goals they try to achieve or to avoid [4, 20]. They then describe what was important for them with respect to the overall aim of the study in which they take part. They have a very active role in this process, often controlling the replay of the video and engaging more with the conversation than just answering questions.

Besides such more qualitative elicitations, it is also possible to elicit quantitative data – automated or manual. Timings can be measured or patterns described, measured, or compared; over time, between conditions or persons.

Analyzing Visuals

There are two ways of handling large amounts of data that visual methods offer: The detailed analysis of short typical segments (often in combination with transcripts), or the overview of a subset of criteria to be investigated across the body of material available (typically based on codes assigned to the material and the quantitative analysis of such codes) [11]. There are several software products that facilitate the analysis of visuals, for example, by allowing for tagging of certain elements of a visual and assigning them to categories [10, 11]. The analysis of visuals can hardly expect to capture "all" the contents of the visuals, as they are too rich. It is essential to know what to look for in the material: coding thousands of video frames would not be realistic. But when the researcher knows what to look for, usually the events are not so frequent in the material, and hours of video turn out to contain only a few minutes, or a few dozen occurrences, of relevant data. Therefore, clarifying what exactly the research question is becomes crucial for the analysis of visuals. When the items to code are theoretically driven, we advise the following: scan some of the material to figure out what occurrences of data seem "interesting" and relevant for the research question. Once such "interesting occurrences" are defined, comb the whole data set to extract only such similar occurrences for coding. Systematic detailed coding can then be performed on the (much smaller) extracted material. This strategy is called "retrospective-sampling" [19].

A study using video in clinical practice about the use of alarms estimated the cost of capturing, managing, and analyzing on hour of video to about 300 US dollars [32]. It might therefore be interesting to explore ways of sharing visual data across research groups [12, 18].

Limitations

There are limitations to visual methods. Some participants might find a recording device disturbing; participants asked to produce the visual might feel performance pressure. When using visuals from already existing sources, this problem might be less pronounced and human beings might get more familiar as recording device become present in more and more areas.

Despite currently being a little cumbersome, the handling of visuals is likely to become more efficient. A study using an automated analysis of the moving pixels and a video recording of a simulated patient/doctor interaction could distinguish movement patterns between two conditions, where in one condition the simulated doctor listened actively and mainly typed in the other condition [33]. Such systems can provide valuable help in identifying areas of interest in large corpuses of visual data, which then could also be analysed in more detail. As one uses more and more visuals, the proper management of such data (documentation of conditions of production and analysis etc., see above) becomes indispensable to avoid mess, ethical issues, and mistakes: start now!

In summary, visuals have great potential for research and training. They can paint a thousand words, they make things visible that are invisible to the human eyes otherwise. They provide excellent cues to help participants explain their internal mental states. What they show, however, is a specific perspective that needs to be made explicit: the context in which a visual is produced and used has a great impact about the conclusions that are drawn from visual research. Therefore, ethical and methodological awareness during the planning, conduct, analysis and interpretation of visual research is very important. Reflection must include what happens after the study is finished as well. Explicitly addressing such issues is not only methodological hygiene and ethical good practice: this reflexivity and the resulting documentation will considerably augment the quality and value of the visuals, as well as their future usability and impact.

References

1. Lahlou S. Installation theory. The societal construction and regulation of behaviour. Cambridge: Cambridge University Press; 2017.
2. Dieckmann P, Gaba D, Rall M. Deepening the theoretical foundations of patient simulation as social practice. Simul Healthc. 2007;2(3):183–93.
3. Bezemer J, Cope A, Korkiakangas T, Kress G, Murtagh G, Weldon SM, et al. Microanalysis of video from the operating room: an underused approach to patient safety research. BMJ Qual Saf. 2017;26(7):583–7.
4. Lahlou S, Le Bellu S, Boesen-Mariani S. Subjective evidence based ethnography: method and applications. Integr Psychol Behav Sci. 2015;49(2):216–38.
5. Cristancho S. Eye opener: exploring complexity using rich pictures. Perspect Med Educ. 2015;4(3):138–41.
6. Banks M, Zeitlyn D. Visual methods in social research. 2nd ed. London/Thousand Oaks, California: SAGE; 2015.
7. Bezemer J. Visual research in clinical education. Med Educ. 2017;51(1):105–13.
8. Rees CE. Identities as performances: encouraging visual methodologies in medical education research. Med Educ. 2010;44(1): 5–7.
9. Doberne JW, He Z, Mohan V, Gold JA, Marquard J, Chiang MF. Using high-fidelity simulation and eye tracking to characterize EHR workflow patterns among hospital physicians. AMIA Annu Symp Proc. 2015;2015:1881–9.
10. Pauwels L. The SAGE handbook of visual research methods. London: SAGE Publications Ltd; 2011. Available from: http://methods.sagepub.com/book/sage-hdbk-visual-research-methods. 2018/03/06
11. Jewitt, C. An Introduction to Using Video for Research. 2012. NCRM Working Paper. NCRM. (Unpublished). Available online: http://eprints.ncrm.ac.uk/2259/; Accessed: 02. August 2019.
12. Derry SJ, editor. Guidelines for video research in education. Recommendations from an expert panel. 2007. Available online: http://drdc.uchicago.edu/what/video-research-guidelines.pdf, Accessed 06 Mar 2018.
13. Preston C, Carter B, Jack B, Bray L. Creating authentic video scenarios for use in prehospital research. Int Emerg Nurs. 2017;32:56–61.
14. Jepsen RM, Dieckmann P, Spanager L, Lyk-Jensen HT, Konge L, Ringsted C, et al. Evaluating structured assessment of anaesthesiologists' non-technical skills. Acta Anaesthesiol Scand. 2016;60(6):756–66.
15. Sadideen H, Weldon SM, Saadeddin M, Loon M, Kneebone R. A video analysis of intra- and interprofessional leadership behaviors within "The Burns Suite": identifying key leadership models. J Surg Educ. 2016;73(1):31–9.
16. Iedema R, Mesman J, Carroll K. Visualising health care practice improvement: Innovation from within. London: Radcliffe Publishing; 2013.
17. Harte JD, Homer CS, Sheehan A, Leap N, Foureur M. Using video in childbirth research. Nurs Ethics. 2017;24(2):177–89.
18. MacWhinney B. Talkbank. Available from: https://talkbank.org.
19. Saadi L. How can we capture the subject's perspective? An evidence-based approach for the social scientist. Soc Sci Inf. 2011;50(3–4):607–55.
20. Mok TM, Cornish F, Tarr J. Too much information: visual research ethics in the age of wearable cameras. Integr Psychol Behav Sci. 2015;49(2):309–22.
21. Wiles R, Prosser J, Bagnoli A, Clark A, Davies K, Holland S, et al. Visual ethics: ethical issues in visual research. Review paper. London: ESRC National Centre for Research Methods; 2008.
22. Dieckmann P, Patterson M, Lahlou S, Mesman J, Nyström P, Krage R. Variation and adaptation: comment on learning from good performance in simulation training. Adv Simul. 2017;2(21). Open Access: https://advancesinsimulation.biomedcentral.com/articles/10.1186/s41077-017-0054-1
23. Lahlou S. Observing cognitive work in offices. In: Streitz NA, Siegel J, Hartkopf V, Konomi S, editors. Cooperative buildings integrating information, organizations, and architecture CoBuild 1999, Lecture notes in computer science. Berlin/Heidelberg: Springer; 1999.
24. Holmqvist K. Eye tracking: a comprehensive guide to methods and measures. Oxford/New York: Oxford University Press; 2011.
25. Wang C, Burris MA. Photovoice: concept, methodology, and use for participatory needs assessment. Health Educ Behav. 1997;24(3):369–87.
26. Edwards B, Edwards B. Drawing on the right side of the brain. Definitive. 4th ed. New York: Tarcher/Penguin; 2012.
27. Lahlou S, Le Bellu S, Boesen-Mariani S. Subjective evidence based ethnography: method and applications. Integr Psychol Behav Sci. 2015;49(2):216–38.
28. Collier A, Wyer M. Researching reflexively with patients and families: two studies using video-reflexive ethnography to collaborate with patients and families in patient safety research. Qual Health Res. 2016;26(7):979–93.
29. Hillen MA, van Vliet LM, de Haes HC, Smets EM. Developing and administering scripted video vignettes for experimental research of patient-provider communication. Patient Educ Couns. 2013;91(3):295–309.
30. Evans SC, Roberts MC, Keeley JW, Blossom JB, Amaro CM, Garcia AM, et al. Vignette methodologies for studying clinicians' decision-making: validity, utility, and application in ICD-11 field studies. Int J Clin Health Psychol. 2015;15(2):160–70.
31. Clark L, Birkhead AS, Fernandez C, Egger MJ. A transcription and translation protocol for sensitive cross-cultural team research. Qual Health Res. 2017;27(12):1751–64.
32. MacMurchy M, Stemler S, Zander M, Bonafide CP. Research: acceptability, feasibility, and cost of using video to evaluate alarm fatigue. Biomed Instrum Technol. 2017;51(1):25–33.
33. Hart Y, Czerniak E, Karnieli-Miller O, Mayo AE, Ziv A, Biegon A, et al. Automated video analysis of non-verbal communication in a medical setting. Front Psychol. 2016;7:1130.

Survey and Other Textual Data

16

Michelle A. Kelly and Jo Tai

Overview

Surveys are commonly used in healthcare simulation research and evaluation. Many surveys have been created to fit specific simulation requirements or areas of interest. However, it is important to ensure that surveys have appropriate rigour for reliable representation and meaningful interpretation of data. This chapter covers concepts important to basic survey design for qualitative research in healthcare simulation education. However, surveys can be used simultaneously to also source quantitative data. The benefits and limitations of using surveys, as well as practical and ethical considerations in the collection and use of textual data are discussed. This chapter also presents the options of exploring alternate textual data sources, such as educational materials, course guides, reflective assessments, as well as data from online/social media platforms. More comprehensive guides on how to develop, administer and analyse survey and other textual data can be sourced from the resources and references included in this and other related chapters of the text.

Practice Points

- Developing surveys requires adequate time and resources to ensure a rigorous 'tool' and meaningful data.
- Where possible, consider using (or adapting) an existing survey before creating a new one.
- The choice of online or paper survey should be made in relation to the context of your research and population as well as the convenience for participants and researchers.
- Other useful textual data sources can be sourced from educational materials, online and social media platforms.

Introduction

When to Use Surveys

Surveys, also commonly referred to as questionnaires, can provide insights about concepts, ideas or opinions which can otherwise be difficult to quantify [1]. There are both theoretical and practical reasons for using surveys. While this chapter sits in the qualitative section of this book, surveys are frequently used in quantitative designs, however the types of questions asked and data collected are likely to look different (see Table 16.1 for examples of different question formats). Mixed-methods designs also commonly include a survey component to either inform subsequent steps in the research, or to triangulate with or corroborate other data. Surveys are a relatively static instrument for data collection, so the area of interest and related questions must be well defined. Practically, surveys are useful for gaining a wide (and possibly representative) sample easily, can be managed in person (paper-based) or online, and provide data that can be readily analysed, compared with, for example, data from interviews or focus groups.

Purpose

Depending on the research question and design, 'one-off' surveys offer opinions at a point in time, a 'snap shot' of the

M. A. Kelly (✉)
Simulation and Practice, School of Nursing, Midwifery and Paramedicine, Curtin University, Perth, WA, Australia
e-mail: Michelle.Kelly@curtin.edu.au

J. Tai
Centre for Research in Assessment and Digital Learning (CRADLE), Deakin University, Melbourne, VIC, Australia
e-mail: Joanna.tai@deakin.edu.au

D. Nestel et al. (eds.), *Healthcare Simulation Research*, https://doi.org/10.1007/978-3-030-26837-4_16

Table 16.1 Examples of qualitative & quantitative survey questions and formats

Qualitative survey question	Quantitative survey question–Visual analogue scale	Likert scale
What was authentic about the simulation?	Please place an X on the line to indicate how authentic you found the following components of the simulation. (Where 0 = not at all, 10 = completely authentic) Mannequin appearance 0 ______ 10 Scenario 0 ______ 10 Resuscitation equipment 0 ______ 10 Anaesthetist (as played by simulation educator) 0 ______ 10	Please circle the number to indicate your agreement with the following statements on a scale of 1 = strongly disagree to 5 = strongly agree The mannequin had a realistic appearance 1 2 3 4 5 The scenario felt comparable to real life situations 1 2 3 4 5 The resuscitation equipment was similar to what I normally use 1 2 3 4 5 The anaesthetist role was well represented 1 2 3 4 5
Categorical responses	**Qualitative survey question**	
Do you recall any specific element/s from the preparation for fieldwork (at university) that influenced how you interacted with the older people during clinical time? Yes ☐ No ☐ Do not recall ☐	Which specific element/s from the preparation for fieldwork (at university) influenced how you interacted with the older people during clinical time? Can you give some examples?	
Do you feel your perceptions of older people have changed since the (intervention) session? In what ways? Yes ☐ No ☐ Do not recall ☐ (Then some free text response)	How have your perceptions of older people changed since the (intervention) session?	

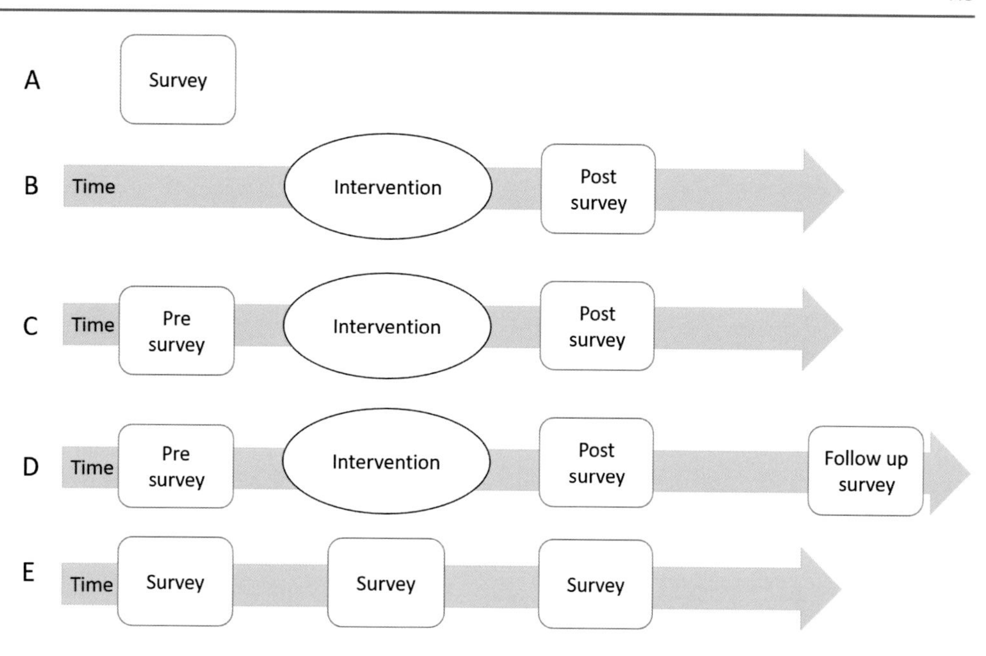

Fig. 16.1 Survey timing. A – single ('one off') cross-sectional survey. B – post-intervention only (to evaluate intervention). C – pre & post intervention (can be used to assess change in attitudes or perceptions). D – pre & post with follow-up (to determine longer term effects). E – Time series which may capture change over time without any particular intervention

area of interest. Time series or longitudinal, repeated surveys offer more of an ongoing sense of the information (concept or construct) of interest. Pre/post surveys, often associated with an intervention, can capture the immediate or downstream impact of an activity depending on the time points chosen. Figure 16.1 reflects different survey administration options. In addition to data for research purposes, surveys can be a vehicle for learner feedback or for program evaluation. Ensuring the chosen survey is the right tool for the job is key to collecting appropriate and useful data to address the research question(s), elicit meaningful feedback for learners or contribute to program development. Table 16.1 provides some examples of how the question format may influence the type and depth of responses.

Benefits

A survey can be an efficient and low-cost method of data collection within educational settings. Non-observable behaviours, attitudes, experiences, and beliefs can be solicited from the participant through purposeful questions. Determining or seeking agreement on the 'construct' of the topic, that is, what are you trying to measure, is an important preparatory step when choosing or developing surveys. Irrespective of the type of survey (scores, free text responses) the questions need to accurately represent the construct so appropriate inferences can be drawn [2]. At times, survey questions or wording may need to be adapted to better match the population of interest, type of practice or country context. Scales, a series of questions which focus on a particular construct, can capture different facets of the same area of interest and hence yield richer data [1].

While most often associated with quantitative survey design, the ideas which inform scale design can equally inform the design of free text questions for research or evaluation. Gehlbach, Artino and Durning [3], offer a useful seven-step guide to assist with scale design, which in summary includes:

- sourcing information from literature and interviewing the population of interest;
- synthesising these data to develop items;
- seeking expert validation using cognitive interviewing techniques; and
- pilot testing.

Alignment Between Research Questions, Methodology and Methods

The decision to use a survey needs to be considered in relation to the research questions, which influence but are also influenced by the chosen methodology. The underlying theory supporting a particular research method impacts on if, and how, a survey is used. Methodologies used in healthcare education research are likely to support the use of a survey as one means of data collection (see [4] for an overview of common methodologies used in healthcare education research). The survey needs to be able to provide data which can aid in answering the research question. For instance, a survey should be able to answer the question "Is the simulation an authentic experience for nurses?", as the question deals with perceptions and experiences which need to be captured directly from the participant. However, the question "do junior doctors perform the resuscitation protocol better after simulations?" would require more than just participant self-report on the survey, as it points to a skill. Here, a measure of performance such as an Objective Structured Clinical Examination (OSCE) where an external observer uses established criteria to assess the participant would be more appropriate. Similarly, a study based in a phenomenology tradition might use a survey to provide simple data on participants' experiences, while one based in a narrative inquiry tradition (where and how the story is told is also important) might require interviews or audio diaries to fully capture participants' stories: Chaps. 9 and 10 deal with alignment within the research in more depth.

Challenges

As with all forms of data collection, getting an appropriate number of participants to respond may be a challenge. For surveys, this can be expressed as the 'response rate': the number of responses received, compared to the number of invitations issued (i.e. the sample population of interest). A higher response rate will ensure the data is more representative of the sample, and more trustworthy conclusions can be reached. Where a survey is advertised freely (e.g. via a newsletter or website) then the response rate will be an estimate based on the total potential population reached. Survey design is important: Shorter surveys may lead to higher response rates if completion is not perceived as too onerous. However, the breadth of data may be limited and result in responses that are self-report. Response rates have been found to be higher for paper, compared to online surveys, though the content and target population may also have influenced completion [5]. The challenge is to ensure questions are clear, focussed and exhibit validity [6]. Detailed explanations of validity (and reliability) are covered in Chap. 26, but here we acknowledge the common terms of face validity, construct validity and scale or question validity. In essence validity refers to the accuracy of, for example, survey questions to accurately reflect the concept being investigated. Face validity may be reached when a panel of experts confer agreement that a range of survey items 'measure' let's say authenticity or visual fidelity. Further and more rigorous testing of survey questions or items is required to achieve other levels of validity [7, 8]. Artino et al. [9] highlight pitfalls to be aware of when writing survey items and approaches to minimize confusion; for instance ensuring an item is a question, not a statement; and the delicate territory of creating negatively framed items.

Free text response questions are often interspersed within or added at the conclusion of surveys, providing opportunity to gather richer data from participants. LaDonna, Taylor and Lingard [10] recommend judicious use of questions which seek open responses, in that such questions should be closely aligned with the intent of the research aims and objectives. If the intent of free text responses is to gain deeper insights into the *why, how and what* of participants' answers, questions need to be focussed and appropriate and analysis should be conceptualized beforehand to ensure robust insights about the phenomena being explored [10]. The instructions or prompts for such questions should be perceived as an open invitation to provide more context or reasons for chosen responses. However, the text length should be limited, and stated as such in the instructions, to enable meaningful contributions and manageable analysis.

The language or discourse and unconscious bias of the researcher can influence participants' responses to the survey questions [1]. Hence development or testing the survey with a sample of the target population is an important step in your research methods. Table 16.2 provides examples of how survey questions were refined following research team input (drawn from the chapter authors' own, unpublished works). If translating a survey into another language, additional care must be taken as there may not be exact equivalence of words, phrases or terms across languages. Sousa and Rojjanasrirat [11] offer multi-step processes to ensure rigor and accurate meaning of surveys for different languages. The simplest approach is a forward or one-way translation from the source language to the target language, but blind or double blind back translation with qualified translators, followed by psychometric testing confers the most comprehensive and reliable outcome. In relation to different cultures, further considerations may be necessary to how survey questions are framed, interpreted and received by participants to avoid offence or embarrassment [11, 12]. Again, it is recommended that a researcher competent in the source language work with a translator who is equally competent with the target language to ensure accurate meaning of survey questions.

Limitations

Depending on the rigour of the survey and method of distribution, data is most often unidimensional, that is, a response is provided from pre-determined options (categories or scales) but there is little insight about *why* participants responded in the way they did. In contrast to interactions during interviews or focus groups, additional nuances of meaning for the researcher (e.g. intonation, emphasis, and non-verbal expressions) are not available within the survey method unless participant responses are provided in audio (e.g. telephone) format. Free text responses can also be open to subjective interpretation by the researcher [1], hence the requirement for clear yet open-ended questions or directions.

Maximizing Survey Success

Once you have considered the aforementioned points, here are some tips for maximizing your return on investment with survey data. Making contact with the intended population before administering the survey is a form of marketing, of eliciting awareness and possibly interest in the area of inquiry. Follow-up

Table 16.2 Refinements to survey questions following review by research team members

Question type	Original format	Next iteration	Reason for change/s
5-point ranking scale (strongly disagree to strongly agree)	This simulation is suitable for a face-to-face lecture	This simulation is more suitable for a face-to-face session at university	Improved discernment of the question – representing a variety of face-to-face formats
5-point ranking scale (strongly disagree to strongly agree)	This simulation is suitable for an on-line learning activity	This simulation is more suitable for an on-line learning activity	Improved discernment of the question – rather than answering similarly for this question and the one above
5-point ranking scale (strongly disagree to strongly agree)	I have seen these types of patients in my clinical placements	A further question added: I have cared for patients like this during my clinical placement/s	
Likert scale	Please rate to what extent you feel the following qualities are a feature of the peer assisted learning you've experienced. 1. Takes the pressure off me to know everything (less threatening learning situation)	Please rate to what extent you agree with the following statements. Reported advantages – compared to traditional teacher-led learning, PAL... Is less threatening	
Likert scale	Reassures me that I am at an appropriate stage of learning (allows me to measure my progress against my peers?)"	Reassures me that I am at an appropriate stage of learning (on the right track) Allows me to measure my progress against my peers	As there are two components to the first question, they should be asked separately
Free text response	Do you think self and peer assessment and feedback is of value? What is the value to the giver? Receiver? Teaching staff?	What value does self or peer assessment have and to whom? (e.g. student giver/receiver, teaching staff)	Changed to an open-ended questions to avoid a "yes/no" response with no further detail

communications with participants provides an update on progress of the data collection process, and reminds non-respondents their opinion is still welcome. There is a fine balance, however, with the type, format and frequency of reminders or updates, which on one hand may improve response rates and on the other may be interpreted as a form of harassment.

Ensure surveys can be completed anonymously; requesting people add any identifying factors can negatively impact on response rates. Online surveys often do not have specific questions which lead to identification, and are prefaced with information and directions. Depending on jurisdictional requirements, recorded informed consent or formal ethical board review may not be necessary. Usually, proceeding to the survey implies participant consent. Coded paper surveys may be associated with a participant consent form, which can then be separated and managed to ensure data is de-identified. If conducting a pre/post or multiple time points paper surveys, using coloured paper for the 2nd survey and another colour for a 3rd survey can help to decrease confusion and assist with achieving complete data sets. Table 16.3 summarizes key considerations when selecting, modifying and administering surveys.

Table 16.3 Key considerations when selecting, modifying and administering surveys

Focus	Considerations
Connectedness	The survey questions should align with the relevant research question/s, context and theory
Robustness	Where possible, use established surveys/ items which are reliable and valid
	Determine and report the validity and reliability of the survey with your population
	If adapting, modifying or translating questions, seek permission from the author/s
	Avoid leading or value laden questions; test questions on your target population prior to commencing the research
Clarity	Readability, literacy level, translation accuracy, visual design (see Artino and Gehlbach [13] for pitfalls e.g. labelling, spacing, and solutions)
Engagement	Structure, length, opening paragraphs and instructions, personable, ensuring confidentiality.

Ethical Issues and Intellectual Property

Compared to other forms of data collection, surveys may seem to have relatively few ethical issues. Though specific legislation will vary depending on country and region, at a most fundamental level, principles such as the Charter of Fundamental Rights of the European Union (http://www.europarl.europa.eu/charter/pdf/text_en.pdf) give baseline operating requirements. National bodies such as the National Institute of Health also have their own guiding principles for ethical research (https://www.nih.gov/health-information/nih-clinical-research-trials-you/guiding-principles-ethical-research), so it is important to refer to codes which apply to your context (e.g. Australia has a National Statement on Ethical Conduct in Human Research – https://www.nhmrc.gov.au/book/national-statement-ethical-conduct-human-research).

Regardless of jurisdiction, the standard concepts of autonomy, beneficence, non-maleficence, and justice apply, as discussed below. In addition, as with other forms of research data, data storage and data transfer should be considered. If using a commercial online survey platform, there may be others who are able to access the data, and the terms of service may indicate that data is stored overseas, invoking international data protection and privacy laws.

Autonomy

Potential participants should be able to freely choose to participate in the survey. Unlike other forms of data collection, consent to participate in research by completing a survey can be indicated by the completion and return of the survey. This may be more complicated in online settings, where partial data is recorded if participants exit prior to completion. The information provided prior to commencement therefore needs to be clear about what will happen to data submitted when the participant withdraws from or does not complete the survey. Depending on the research and jurisdictional requirements, partial data may be excluded from analysis.

Beneficence

Participants should be made aware of the benefits of completing the survey, which may not be direct benefits to themselves. These benefits should usually outweigh any inconvenience or potential harm arising from the research. The survey should be kept to a reasonable length, collecting targeted data related to the research question, rather than including a range of items in the event that something interesting might arise. The time commitment for completing the survey should be clearly noted in the information for participants, and be reasonable.

Non-maleficence

Harm to participants or others is a serious concern in research. While harms are likely to be minimised in survey-based data collection, some questions may still have emotional and/or psychological consequences, especially around sensitive topics. Careful consideration of wording may mitigate these risks, but safeguards should always be in place (e.g. access to counselling or other support services).

Justice

It could be argued that using a survey is more "just" as it is easier to collect and represent data from a wider range of participants. For example, if collecting a short survey following a conference, an online survey could be used rather than handing out paper surveys or doing interviews. In this case, justice is done as it is less onerous, and 'more accessible' to those who were not present at a particular session, as long as delegates have agreed to being contacted. It is also likely to allow for a wider range of opinions than, for instance, selected interviews with people who are able to give up a longer amount of time. Justice is also seen as being in terms of balancing the risks and benefits (as outlined above).

Other Textual Data

Other types of textual data may also be rich sources of opinions for analysis. In addition to gathering data directly from people on their experiences, perceptions, and understanding, textual sources from education, hospital and online settings may also be used as forms of data collection. Table 16.4 outlines the potential value which may be gained from exploring these types of data. Many of these sources (e.g. Facebook, Twitter, blogs/websites) may also be avenues for recruitment for surveys.

Here, care must be taken in how the data is handled, as these materials have been created for other purposes. Considerations include:

- Are you able to contact the creator/owner of the material?
- Does the owner of the material know that their work is being used for research purposes? If materials are not in in the public domain, then consent or permission for their use is likely to be required.
- What is being done about identifiable information? (e.g. quotes from social media may be able to be traced back to the creator; are there likely to be harms involved?)
- Will you need ethical/institutional review board approval for using these sources retrospectively?

Analysing Survey Data

Depending on the type of survey data and method of collection, data management and analysis is the next phase. If you have used an online survey through portals such as Qualtrics (https://www.qualtrics.com) or SurveyMonkey (https://www.surveymonkey.com) the data can be immediately accessed. If paper surveys have been used or data is from social media

Table 16.4 Potential data sources and their value for research in healthcare simulation education

Data source	Potential value & Issues for consideration
Education	
Unit guides/course outlines, curriculum maps	These materials may add background depth and context to research in determining participants' prior experiences, or add another dimension to studies of simulation curricula
	Permission may be required from the institution to include them in analysis, especially if extracts are to be published
Written simulation scenarios	Scenarios are likely to contain more fine-grained information which may be of interest, including analysis of objectives, topics, simulated patient roles, format of simulations, and even equipment required
	Permission is likely to be required from the creators of the materials
Reflective writing pieces	Written reflections from students may deliver further insight into the impacts of simulations and educational experiences [14]
	It is likely that a consent process will be necessary to include these in a research project – opt-out consent may be an option if large numbers of de-identified pieces are used
	Researchers should note however that the accuracy of these reflections may be compromised, as it has been demonstrated that students may tailor their reflections to perceived requirements [15]
Evaluation & feedback form	Evaluation and feedback forms are frequently collected at several levels – the activity/session, the program, and the institution. Free text comments (if any) may assist in understanding participants' perspectives
	Though the data has usually already been collected, in stricter jurisdictions, some form of retrospective ethical clearance is required to allow the information to be published
Online discussion forums	Depending on the education design, participants may be asked to commence or continue discussions in an online format. These may be fruitful sources of additional data if they have been set up well
	Here, an opt-out consent process may be sufficient if all contributors are de-identified (both as sources, and in-text if they are referred to by name)
Hospital-based	
Critical incident reports/sentinel event report data	When seeking to demonstrate impact, hospital-based data may provide some clues, both for the initial development of a project, and also in evaluation. While numerical statistics form a large part of this information, incident descriptions are usually free-text entries
	Significant consideration of ethical and legal implications are required for use of this data; and approval is required. It is likely this data would need to be properly de-identified prior to access
Online and Social media	
Twitter – via hashtags or accounts	Tweets found through particular hashtag (labelled topic) or from a particular account (i.e. individuals, or an organisation) may assist in providing perspectives on a phenomenon of interest, usually in the form of discussions. Hashtags are frequently used for labelling tweets connected to conferences, e.g. #SimHealth, #imsh17, #SSHSummit, #SESAM. Some hashtags contain strong discussion around healthcare education – e.g. #FOAMed (Free, open access, medical education). If Twitter is used as part of an educational intervention, tweets from individual accounts may be useful to evidence discussion and engagement [16]
	While public tweets are essentially free for all to access, and many ethical review boards may allow Twitter data to be included without individual permissions. In the context of educational programs, there may be requirements for students to be informed if their contributions created for educational purposes are planned to be used for other (i.e. research) purposes
Facebook	Facebook, like Twitter, may also be a source for data on groups of people interested in specific topics, or could be used in formal education interventions where social interactions are required [17]. It is likely that permissions should be asked, especially if the researcher has to access a closed group for the data
Blog posts, comments, websites	There are a number of blogs and websites (run by individuals, groups, or organisations) devoted to simulation education. These sources may provide evidence of ongoing commentary and opinion on various topics
	Unlike Twitter and Facebook it may be more difficult to determine who is responsible for content, so care must be taken when attributing sources and asking for permissions

sources, information needs to be entered or uploaded manually into a database or file created specifically for the research. Handling and management of all data need to adhere to the ethical principles agreed to within the jurisdictional approval. All data need to be checked for completeness, accuracy and quality prior to analysis [18]. If preferred, data can be exported into other analytical software programs (such as IBM's SPSS or QSR International's NVivo). To ensure appropriate rigor, recommended frameworks and step-by-step processes are expected for results to be reliable, reportable and representative of the research aims and objectives [19]. Chapters 17, 19, 24, 28 and 29 offer more detailed guidance about managing and analysing various forms of data.

Closing

Textual data from surveys and other sources can provide rich insights about non-observable behaviours, and constructs such as beliefs, attitudes experiences and opinions. These qualitative data sources contribute to the broader perspectives of the area/s of interest and can complement or corroborate quantitative data.

To ensure rigorous output, we have highlighted important steps to consider when selecting, modifying or designing surveys for healthcare simulation research and evaluation. Interpreting survey data is an important step in the process of drawing and applying meaning from the data sources. As the healthcare simulation community matures, ensuring best practices in research design from the outset will enable comparisons across populations or larger, multisite research studies to build the evidence related to the effectiveness and impact of simulation.

References

1. Artino AR, La Rochelle JS, Dezee KJ, Gehlbach H. Developing questionnaires for educational research: AMEE guide no. 87. Med Teach. 2014;36(6):463–74. PubMed PMID: 24661014. Pubmed Central PMCID: 4059192.
2. Wu M, Tam HP, Construct JT. Framework and test development—from IRT perspectives. In: Wu M, Tam HP, Jen T, editors. Educational measurement for applied researchers: theory into practice. Singapore: Springer; 2016. p. 19–39.
3. Gehlbach H, Artino AR, Last Page DSJAM. Survey development guidance for medical education researchers. Acad Med. 2010;85(5):925. PubMed PMID: 00001888-201005000-00043.
4. Tai J, Ajjawi R. Undertaking and reporting qualitative research. Clin Teach. 2016;13:175–82.
5. Nulty DD. The adequacy of response rates to online and paper surveys: what can be done? Assess Eval High Educ. 2008 . 2008/06/01;33(3):301–14.
6. LaDonna KA, Taylor T, Lingard L. Why open-ended survey questions are unlikely to support rigorous qualitative insights. Acad Med. 2017;93:347–9. PubMed PMID: 29215376.
7. Cook DA, Zendejas B, Hamstra SJ, Hatala R, Brydges R. What counts as validity evidence? Examples and prevalence in a systematic review of simulation-based assessment. Adv Health Sci Educ Theory Pract. 2013;19:1–18.
8. Adamson K, Gubrud P, Sideras S, Lasater K. Assessing the reliability, validity, and use of the Lasater clinical judgment rubric: three approaches. J Nurs Educ. 2012;51(2):66.
9. Artino AR, Gehlbach H, Durning SJAM. Last page: avoiding five common pitfalls of survey design. Acad Med. 2011;86(10):1327. PubMed PMID: 00001888-201110000-00038.
10. LaDonna KA, Taylor T, Lingard L. Why open-ended survey questions are unlikely to support rigorous qualitative insights. Acad Med. 2018;93(3):347–9. PubMed PMID: 00001888-201803000-00013.
11. Sousa VD, Rojjanasrirat W. Translation, adaptation and validation of instruments or scales for use in cross-cultural health care research: a clear and user-friendly guideline. J Eval Clin Pract. 2011;17(2):268–74.
12. Kelly MA, Ashokka B, Krishnasamy N. Cultural considerations in simulation-based education. Asia Pac Schol. 2018;3:1–4. https://doi.org/10.29060/TAPS.2018-3-3/GP1070.
13. Artino AR, Gehlbach HAM. Last page: avoiding four visual-design pitfalls in survey development. Acad Med. 2012;87(10):1452. PubMed PMID: 00001888-201210000-00035.
14. Kelly MA, Fry M. Masters nursing students' perceptions of an innovative simulation education experience. Clin Simul Nurs. 2013;9(4):e127–e33.
15. Maloney S, Tai J, Lo K, Molloy E, Ilic D. Honesty in critically reflective essays: an analysis of student practice. Adv Health Sci Educ. 2013;18(4):617–26.
16. Zimmer M, Proferes NJ. A topology of Twitter research: disciplines, methods, and ethics. Aslib J Inf Manag. 2014;66(3):250–61.
17. Aydin S. A review of research on Facebook as an educational environment. Educ Tech Res Dev. 2012;60:1093–106.
18. Hagger-Johnson G. Introduction to research methods and data analysis in the health sciences. Hoboken: Taylor and Francis; 2014.
19. Cheng A, Kessler D, Mackinnon R, et al. Reporting guidelines for health care simulation research: extensions to the CONSORT and STROBE statements. BMJ Simul Technol Enhanc Learn. 2016;2:51–60.

Additional Resources

How to pre-test and pilot a survey questionnaire. Tools4Dev. Practical tools for international development. http://www.tools4dev.org/resources/how-to-pretest-and-pilot-a-survey-questionnaire/.

Survey methodology. Statistics Canada http://www5.statcan.gc.ca/olc-cel/olc.action?objId=12-001-&ObjType=2&lang=en&limit=0, https://select-statistics.co.uk/calculators/sample-size-calculator-population-proportion/.

Survey techniques: relative advantages and disadvantages http://www.cdc.gov/hepatitis/partners/Perinatal/PDFs/Guide%20to%20Life%20Appendix%20C.pdf.

Transcription and Data Management

17

Catherine F. Nicholas, Lou Clark, and Karen Szauter

Overview

You may think that transcription is the simple yet tedious process of typing recordings of conversations or interviews before data analysis can begin. Nothing could be further from the truth. As you plan your research project, a series of questions must be asked and answered to assure that the transcripts support the data analysis process. Questions like "What level of detail should be included?" "How should the data be represented?" "Who should do the transcribing? "What recording and transcription equipment is needed? "What ethical considerations need to be taken into account?" "How do I keep track of all the data?" In this chapter we present a process for qualitative researchers to select the transcription approach best suited to their project.

Practice Points

- Transcription approach depends on the research question and methodology.
- Ensuring high quality recording will improve usability.
- Date file management is key to accurate data analysis.
- Thoughtful reflection on one's own stance or position needs to be considered during the transcription process.

Each researcher makes choices about whether to transcribe, what to transcribe and how to represent the data in text [1 p66].

Introduction

The overall goal of transcription is to create an accurate and usable document. Not until the 1990s [1–4] did the impact of transcription on "how participants are understood, the information they share and what conclusions are drawn" surface in the literature [5 p1273]. Rather than transcription being the mundane process of converting the spoken word into the written word, it was recognized as the first step of the analytic process. As a result, no single approach to transcription can be applied as each transcript is unique to the research. While there is no standardized approach, several general principles apply [6, 7]:

- The approach should match the research methodology.
- The choice must be balanced against the cost and time to produce the final transcript.
- Transcription conventions (symbols to represent different aspects of verbal and non-verbal conversation) must be easy to understand so the data can be managed and analyzed.
- A style guide should be created to reduce error, increase the usability and accuracy of the final transcript.
- Identifying information must be removed to reduce bias and assure confidentiality.

C. F. Nicholas (✉)
Clinical Simulation Laboratory, University of Vermont, Burlington, VT, USA
e-mail: cate.nicholas@uvm.edu

L. Clark
Hugh Downs School of Human Communication, Arizona State University, Tempe, AZ, USA
e-mail: leclark2@asu.edu

K. Szauter
Department of Internal Medicine, Department of Educational Affairs, University of Texas Medical Branch, Galveston, TX, USA
e-mail: kszauter@utmb.edu

D. Nestel et al. (eds.), *Healthcare Simulation Research*, https://doi.org/10.1007/978-3-030-26837-4_17

The goal is to find an approach to accurately capture the unique way participants tell their stories and capture the complexities of talk while producing a useable and accurate transcript [8, 9].

Overview of Transcription Approaches

Oliver [5] presents transcription strategies on a continuum. One side is the naturalized approach which results in a verbatim transcript. Every element of speech (overlapping speech, laughter, pauses, tone, nonverbal etc.) is included in as much detail as possible. This approach is best used for conversational, discourse and narrative analysis, which look at speech patterns that occur between people or how ideas are shared. The drawback to the naturalized approach is an increase in transcription errors due to the complexity of the transcription process and additional time and expense to complete. Using this method, every hour of tape requires 6–7 h of transcription. The other side is the denaturalized approach where these elements are not included as the focus is on the meanings and perceptions of the participants. The approach is best suited to ethnography, thematic analysis, grounded theory and mixed methods research. Transcription time is reduced to 4 h for every 1 h of tape but essential data may be lost. Figure 17.1 depicts this continuum.

Variations or hybrid approaches to these transcriptions practices [10] may be better suited to answer your research questions. Remember the goal is to produce a transcript that serves the needs of the research [11]. In order to reach that goal, many questions need to be asked and answered at the beginning of the research design. You may also find that the transcription approach changes over the course of the research and that those questions need to be reconsidered.

Determining the Transcription Approach for Your Research

Who will use the transcript, for what will it be used, and what features will best serve your research? In order to choose the best approach, consider the following questions:

Question 1: Do you plan to audio/video record your participants? Recordings can be saved and reproduced allowing for repeat analysis and can be made available to other researchers. If the focus of the research is sensitive, audio/video recording may be considered intrusive. You may choose direct scribing which allows the participants to create their lived experiences, to discover their narrated selves, and decide what they would like to share [12 p811]. You can also use field notes, journals or other data capture methods. If you plan to record, continue to Step 2.

Question 2: Based on the research questions and chosen methodology, what and how much needs to be recorded and transcribed (selectivity)? Research that seeks to understand how people speak to each other (conversational, discourse, narrative) would include verbal and non-verbal elements of speech. See Table 17.1 for examples.

Question 3: How will talk will represented in the written text? Regardless of who does the transcription (researcher, professional transcriptionist or other), a style guide can guide the process [3, 5, 9]. The style guide should include:

1. Uniform layout of the final transcript: font style and size, spacing between lines, margins, numbered lines, and page numbers.
2. Speaker labeling:
   (a) Remove identifying information (participant names, ethnicity, role, gender, and demographic data) to reduce bias and assure confidentiality.

Table 17.1 Verbal and non-verbal elements of speech

Verbal element	Example
Spoken word elements	Tone, inflection, cadence and pace Dialects, slang, translations
Patterns of conversation	Turn-taking: the manner in which orderly conversation normally takes place- influenced by culture and gender Overlap: speaking at same time Politeness strategies: speech that expresses concern for another and minimizes threat to their self esteem Repair: correction to speech error
Non-verbal element	Voice quality, rate, pitch, loudness etc.
Non-word elements	Reactive or response tokens: (hm, huh, oh, mhm) Discourse marker: (oh, well, you know, I mean) Laughter, crying, sighs Silence, short or long pauses
Fine/gross physical movements	Waving Pointing Nodding Hand gestures

Fig. 17.1 Transcription strategy continuum

(b) You can use letters, numbers, or pseudonyms. Be careful as label choices may influence analysis in an unintended way. For example, if you label the speakers doctor and nurse, you may be introducing bias.
(c) Create a master list of participant names and identifiable variables separate from transcripts.

3. Speaker representation: Does each speaker receive their own paragraph with a line in-between speakers? Would transcribing like a play or in columns work better than standard prose?
4. Glossary of terms and concepts specific to the research: technical, medical, clinical or simulation specific terms.
5. Transcript symbols: How will you represent spoken word elements, patterns of conversation, non-word elements and relevant gestures in the text? Will you use a standardized convention like the Jefferson Transcription System (which is used to look for speech patterns and allows you to annotate speech with movement and interaction between or among participants), or will you adapt or create your own? It is important to create a system that is user friendly and contributes to transcript accuracy. See Table 17.2 for further details.

Question 4: Transcription: Who will do the transcription? You, a colleague, a professional transcription service or another option? This choice is based on methodology, resources and time. Here are some considerations:

1. Researcher as transcriptionist: If your research requires a naturalized approach you may choose to transcribe the recording yourself. By doing the transcription yourself you begin the analysis. Remember to save frequently! This approach provides consistency and allows researchers to develop a close relationship with their data. In addition to being the first step in the data analysis process, researchers may improve their own interviewing practices. The main limiting factors, include time and potential limitations in keyboarding skills. If the latter is an issue consider using voice activated software [13, 14]. See Table 17.3 for transcribing tips.

Table 17.2 Sample of transcript symbols

Element	Symbol
Non-relevant	(…)
Pause	*(pause)*
Tone	{sarcastic)
Loud	IN CAPS
Unclear	*(unclear)*
Different speakers	A,B,C etc.
Unclear speakers	A? B?
Speaking at same time	C/D

2. Professional transcribers or transcription services, volunteers (such as graduate students), or study participants: If your research allows a denaturalized approach you may be able delegate to another. Professional transcribers may be costly but can be much faster than researchers. Volunteers or graduate students may offer free labor, but with free labor comes potential challenges with commitment to task completion within a certain timeframe and general reliability. Study participants have been used in a variety of ways from fact checkers to co-transcriptionists. To increase accuracy, consider the transcriptionist as part of the research team and provide training. Provide information about the research goals, setting and participants as well as the style guide. Clarify any questions and make changes if it improves the guide. Ask them to transcribe a section of the recording. Check for accuracy and answer questions about using the style guide. Review one completed interview before moving to others [3].
3. Computer generated transcripts or online interviews provide another alternative to traditional transcription: Computer generated transcripts or online interviews are already transcribed, and errors are greatly reduced or all together absent. Computer interviews can also provide accessibility to other institutions and for groups who may have limited mobility to attend face to face interviews [15]. Computer generated transcripts are best for one on one interviews or in group settings where there is very little overlapping talk and all participants speak standard dialects of the base language. Regardless of your choice, the transcription process is time intensive and should be considered when developing your project timeline [1, 6, 11–13, 16–20]. See Table 17.4, which outlines both challenges and solutions for focus group transcription.

Table 17.3 Transcription equipment and ergonomic tips

Transcription equipment
Computer with large amount of memory and fast processor
Voice recognition software (e.g. Dragon Naturally Speaking)
Transcription machine with foot pedal
Ergonomic tips
Desk and chair providing maximum support and comfort
Regular rest and stretch breaks

Table 17.4 Transcribing focus group recordings

Challenges	Solutions
Many or overlapping voices	Listen to identify 7–8 speech markers for each #1 – says "like" frequently, has a low, scratchy voice #2 – raises voice at end of sentences and says "you know" frequently
Capturing the hierarchal structure of the group	Identify voices that dominate the conversation

Practical Considerations for Increasing Recording Quality and Usability

The quality of the recording, and therefore the transcript, is related to the equipment, environment and the operator [4, 6, 13, 16].

Equipment: Choose recording equipment that is high quality yet affordable, with a fast speed or high quality setting. It will use more memory or tape but produces a clearer recording. Be sure the recording device runs on both house current and batteries. If possible use an external microphone (such as a wireless lapel microphone) to optimize the sound. Avoid voice activated features as they may not record the first few words. Always have spare batteries.

Consider readily available online recording solutions like SpeakPipe (https://www.speakpipe.com/voice-recorder), NCH Software (http://www.nchsoftware.com/software/recording.html) or Apowersoft (https://www.apowersoft.com/free-audio-recorder-online).

Environment: Chose a quite space with no background noise or interruptions. Avoid public spaces like restaurants. Locate the recording device so all voices can be heard. Set the recording device on a stable surface. If video recording, place the camera to capture all essential elements needed for analysis (i.e. room layout, body orientations, relationships to other and simulation modality).

Preparation: Practice with the equipment to determine best settings. If the environment changes, recheck settings. Check equipment for record and playback prior to each session.

Recording session: Test the equipment again. Identify session (interviewee, date and time) at beginning and ending of recording. If appropriate, ask all participants to speak one at a time, slowly and clearly.

After the session: Listen/watch the recording taking notes as you go. Adjust or add transcription symbols to the style guide as needed. Label the tape, make a copy and keep all recordings secure and safe from extreme temperatures.

Managing Data: Working with the Tapes and the Transcripts

Data file management is essential to optimize the subsequent analysis of the data. Without a data management plan, you may find your data has confusing file names, is lost, or is incomplete.

The method of data collection will be guided by the primary research question. A system for file labelling should be established at the beginning and adhered to throughout the project.

Although the content of the generated transcription data file will be used in the analysis, the primary source of the data should be documented with as much detail as practical. When available, this includes demographics such as age, sex, ethnicity, socio-cultural identifiers, and context of the data collection. This information will serve to identify the data source, and may also serve an important role in the analysis of the data.

The person(s) obtaining the data also must be documented and included in the data file. A description of the process and the number of persons obtaining data is important to note, and the use of tags that link the specific data set with interviewer identifiers should be considered.

Finally, an unedited, "original" copy of the transcription should be kept in storage. Depending on the level of detail adhered to in the transcription, editing or "cleaning" of the file may be needed to format the transcription prior to analysis [3, 9, 16, 18].

Here are four steps to follow:

1. Create a file naming system that includes:
   (a) Date
   (b) Name of study
   (c) Participant ID number or pseudonym
   (d) File format (i.e. rtf or mp4)
   (e) Type of data collection method (interview, video, field notes etc.)
   (f) Site of data collection (name simulation lab, name of clinical setting, geographic location)
   (g) Name of interviewer or focus group facilitator
   (h) Demographics meaningful to your research project
   Include the file name in footer of all study documents (Box 17.1).

Box 17.1 Sample data label

Date: 12.28.2017
Study: Debriefing vs feedback
Interviewee: A
Type: One on one interview
Site: CSL/UVM
Interviewer: Nicholas

2. Create a step by step data tracking system (Table 17.5)
   (a) Digital audio/video and consent:
      i. File name is given to both
      ii. Consent reviewed for completeness and filed
      iii. Original audio file is saved
      iv. Audio file is downloaded and given file name
      v. Word file created, named according to system
      vi. Coversheet prepared
      vii. Downloaded audio file and matching word file ready for transcription
3. Follow transcription process and style guide

Table 17.5 Sample data tracking system

Debriefing vs Feedback study master list						
File name/file format	Interview date	Interviewer	Participant name	Age	Level	Location
A.rtf	12.28.2017	Nicholas	Anne Levine	25	Med student-2	CSL/UVM
B.mp4	12.29.2017	Clark	Burton Calhoun	21	Senior nursing student	ASU
C.rtf	12.30.2017	Szauter	Carlos Santiago	45	Simulation educator	UT Galveston

4. Transcription completed
   (a) Audio file and word file reviewed for completeness and correctness
   (b) Feedback provided to transcriptionist
   (c) Finalized for analysis

In summary, management of volumes of transcribed material requires careful organization, agreed-upon labelling standards at the outset, and meticulous documentation of each step of the process. Decisions made throughout the process of data gathering, the transcription process, and the approach to the transcribed document all impact the validity of the research outcome.

Ethics of Transcription

There is a growing body of literature exploring ethics in relation to transcription and transcribers [21, 22]. The transcriber's personal bias towards the research subject and participants including cultural differences can influence the final transcription [4, 20]. No matter who the transcriber is, it is important that they practice self-reflexivity—thoughtful reflection on the effect of one's own bias in the research process [4, 14, 20]. This is to ensure the most accurate and relevant data is compiled in the most ethical manner considering all stakeholders (including study participants) [5, 14]. Researchers and transcribers may take various approaches designed to empower study participants during and following the transcription process [3, 5, 12, 15, 21].

Researchers can:

- Give participants the opportunity to review, comment, and amend transcripts—a process known as member checking.
- Work with participants to determine how they would like their speech portrayed.
- Enlist participants as transcribers, allowing them to amend and alter previously recorded interviews.
- Use direct scribing so interviewer and participants can look at a computer screen as the interviewer types participant responses in real time.

Transcriptionists can:

- Keep notes regarding any biases they become aware of during the transcription process.
- Refer questions regarding interpretation of data to the interviewer for clarity.
- Share their impressions with participants following data transcription.

Ethical considerations need to be reviewed for the transcribers themselves [22, 23]. Transcribers have reported negative emotional and physical impacts including negative feelings (e.g. sadness, anger) and persistent memory of distressing details that can result in sleeplessness and headaches. Transcribers may suffer secondary trauma when working on particularly sensitive or traumatic material. Researchers can support transcriptionists by discussing the nature of sensitive research with them to prepare them. Additionally, researchers can encourage transcriptionists to take advantage of technology and join on-line social media or chat groups for transcriptionists where they can debrief difficult material in closed group settings. This is crucial as transcribers often report feeling isolated since many telecommute.

Dissemination

Description and reflection about transcription should be viewed as an essential component of a manuscript. Providing a description of steps taken to ensure audiotape quality, transcription guide, and training for transcriptionists will allow others to assess the trustworthiness of the data as well as the interpretations drawn from it [4, 20].

Closing

While no one transcription process exists for all studies, remember:

1. Match your transcription process to your research.
2. If possible, include your transcriptionist as part of the research team to increase content accuracy and study validity.
3. Set up a data management protocol early in the process to increase efficiency, accuracy and transparency.
4. Consider the ethical implications of your research subject on the transcriptionist.

References

1. Lapadat JC, Lindsay AC. Transcription in research and practice: from standardization of technique to interpretive positionings. Qual Inq. 1999;5(1):64–86.
2. Sandelowski M. Focus on qualitative methods. Notes on transcription. Res Nurs Health. 1994;17(4):311–4.
3. Tilley SA. Conducting respectful research: a critique of practice. Can J Educ Revue canadienne de l'education. 1998;23:316–28.
4. Poland B. Transcription quality. In: Gubrium JF, Holstein JA, editors. Handbook of interview research: context and method. London: Sage; 2002. p. 629–50.
5. Oliver DG, Serovich JM, Mason TL. Constraints and opportunities with interview transcription: towards reflection in qualitative research. Soc Forces. 2005;84(2):1273–89.
6. Bailey J. First steps in qualitative data analysis: transcribing. Fam Pract. 2008;25(2):127–31.
7. Bird CM. How I stopped dreading and learned to love transcription. Qual Inq. 2005;11(2):226–48.
8. Chadwick R. Embodied methodologies: challenges, reflections and strategies. Qual Res. 2017;17(1):54–74.
9. Clark L, Birkhead AS, Fernandez C, Egger MJ. A transcription and translation protocol for sensitive cross-cultural team research. Qual Health Res. 2017;27(12):1751–64.
10. Davidson C. Transcription: imperatives for qualitative research. Int J Qual Methods. 2009;8(2):35–52.
11. Hammersley M. Reproducing or constructing? Some questions about transcription in social research. Qual Res. 2010;10(5):553–69.
12. Saldanha K. Promoting and developing direct scribing to capture the narratives of homeless youth in special education. Qual Soc Work. 2015;14(6):794–819.
13. Easton KL, McComish JF, Greenberg R. Avoiding common pitfalls in qualitative data collection and transcription. Qual Health Res. 2000;10(5):703–7.
14. Tracy SJ. Qualitative research methods: Collecting evidence, crafting analysis, communicating impact. Hoboken: Wiley; 2012.
15. Grundy AL, Pollon DE, McGinn MK. The participant as transcriptionist: methodological advantages of a collaborative and inclusive research practice. Int J Qual Methods. 2003;2(2):23–32.
16. Bolden GB. Transcribing as research: "Manual" transcription and conversation analysis. Res Lang Soc Interact. 2015;48(3):276–80.
17. Burke H, Jenkins L, Higham V. Transcribing your own data. ESCRC National Centre for Research Methods; 2010.
18. Ashmore M, Reed D. Innocence and nostalgia in conversation analysis: the dynamic relations of tape and transcript. InForum Qualitative Sozialforschung/Forum: Qual Soc Res. 2000;1:3:73–94.
19. Ayaß R. Doing data: the status of transcripts in conversation analysis. Discourse Stud. 2015;17(5):505–28.
20. Bucholtz M. The politics of transcription. J Pragmat. 2000;32(10):1439–65.
21. Mero-Jaffe I. 'Is that what I said? 'Interview transcript approval by participants: an aspect of ethics in qualitative research. Int J Qual Methods. 2011;10(3):231–47.
22. Kiyimba N, O'Reilly M. The risk of secondary traumatic stress in the qualitative transcription process: a research note. Qual Res. 2016;16(4):468–76.
23. Wilkes L, Cummings J, Haigh C. Transcriptionist saturation: knowing too much about sensitive health and social data. J Adv Nurs. 2015;71(2):295–303.

Additional Resources

e-Source Behavioral and Social Sciences http://www.esourceresearch.org/tabid/36/Default.aspx sponsored by the Office of Behavioral and Social Sciences Research, National Institutes of Health and the Department of Health and Human Services (20 interactive chapters on methodological questions on behavioral and social sciences research).

Development of Standardized Notation Systems

1. Atkinson JM. Heritage J. Jefferson's transcript notation. The discourse reader; 1999. p. 158–66.
2. Du Bois JW. Transcription design principles for spoken discourse research. Pragmatics Quarterly Publication of the International Pragmatics Association (IPrA). 1991;1(1):71–106.

Grounded Theory Methodology: Key Principles

18

Walter J. Eppich, Francisco M. Olmos-Vega, and Christopher J. Watling

Overview

Grounded theory (GT) is a common qualitative methodology in health professions education research used to explore the "how", "what", and "why" of social processes. With GT researchers aim to understand how study participants interpret reality related to the process in question. However, they risk misapplying the term to studies that do not actually use GT methodology. We outline key features that characterize GT research, namely iterative data collection and analysis, constant comparison, and theoretical sampling. Constructivist GT is a particular form of GT that explicitly recognizes the researcher's role in knowledge creation throughout the analytic process. Data may be collected through interviews, field observations, video analysis, document review, or a combination of these methods. The analytic process involves several flexible coding phases that move from concrete initial coding to higher level focused codes and finally to axial coding with the goal of a conceptual understanding that is situated in the study context.

Practice Points

- Since GT requires an iterative approach to data collection and analysis, researchers must plan this approach in advance
- In constructivist GT, researchers must reflect deeply about how their backgrounds, perspectives, and beliefs may impact the research process
- The aim of GT is a situated conceptual understanding of a social process
- While solo researchers can perform GT, research teams with members who have different backgrounds may add value to the analysis and interpretation of findings
- Although GT methodology outlines a stepwise approach, researchers should remain flexible
- Researchers should avoid the pitfall of endless cycles of coding and re-coding in order to achieve an elusive "correct" coding structure; they should remember to move from organizational to conceptual thinking

W. J. Eppich (✉)
Feinberg School of Medicine, Department of Pediatrics, Division of Emergency Medicine, Ann & Robert H. Lurie Children's Hospital of Chicago, Northwestern University, Chicago, IL, USA
e-mail: w-eppich@northwestern.edu

F. M. Olmos-Vega
Department of Anesthesiology, Faculty of Medicine, Pontificia Universidad Javeriana, Bogotá, Colombia

Anesthesiology Department, Hospital Universitario San Ignacio, Bogotá, Colombia
e-mail: folmos@javeriana.edu.co

C. J. Watling
Departments of Clinical Neurological Sciences and Oncology, Office of Postgraduate Medical Education, Schulich School of Medicine and Dentistry, Western University, London, ON, Canada
e-mail: chris.watling@schulich.uwo.ca

Short Case Study

Susan, a simulation educator, wants to study how healthcare debriefings contribute to peer learning among medical students. She plans to interview both medical students and simulation educators. She has heard that grounded theory may be a good qualitative research approach, but she would like to learn more.

Introduction

This chapter has several aims:

- Provide an overview of GT with a focus on constructivist GT
- Highlight the potential applications of GT

D. Nestel et al. (eds.), *Healthcare Simulation Research*, https://doi.org/10.1007/978-3-030-26837-4_18

- Explain the critical role of reflexivity
- Describe how to collect and analyze data using a GT approach
- Offer approaches to evaluating the quality of GT studies
- Identify pitfalls of GT work and how to avoid them

Defining Grounded Theory

GT represents a commonly used qualitative methodology in health professions education [1–3]. GT outlines systematic steps to data collection and analysis with the aim of developing a higher understanding of social processes that is 'grounded' or inductively derived from systematic data analysis [4]. In using GT, researchers aim to make statements about how their study participants interpret reality related to the process under study; hypothesis-testing is *not* the goal of GT.

Classical GT originated in 1967 from work by Glaser and Strauss [5] that outlined concrete steps in deriving theory from empirical qualitative data. Multiple subsequent grounded theorists have modified the initial approach that Glaser and Strauss outlined to address evolving methodological considerations and views about the nature of reality. These various streams led to multiple schools of thought, including the development of a constructivist view of GT. This view was first espoused by Charmaz, and has become popular in health professions education research [2, 6, 7]. Classical GT contends that researchers can and should set aside their perspectives, backgrounds, and beliefs [5]; this view seems increasingly illogical and implausible [2, 6]. Constructivist GT explicitly acknowledges that researchers actively participate in creating knowledge, a stance that sits better with contemporary views of education research [2]. The history of GT is beyond the scope of this chapter; interested readers should refer to published works for a thorough discussion [8, 9] and a critique [10].

Several key features characterize GT across these various traditions [2, 11]:

- An ***iterative approach*** to data collection and analysis, in which researchers complete a small number of interviews and perform an initial analysis, and then use that analysis to inform subsequent data collection, which further guides data collection, and so on. Qualitative studies violate this key feature of GT if analysis occurs only after all data has been collected.
- ***Constant comparison*** during analysis, in which data points are compared with each other and emerging theoretical constructs are continually refined through comparision with fresh examples. As an analytic strategy, constant comparison may find use outside of GT approaches; what characterizes GT is the use of constant comparison in concert with other key features mentioned here.
- ***Theoretical sampling*** is a sampling strategy informed by theoretical considerations. In this approach, the sample is not determined a priori, but rather selected purposefully as the analysis proceeds and theoretical understanding evolves. Theoretical sampling differs from 'purposive sampling', which simply means non-random sampling (i.e. deliberately sampling participants from a population who you anticipate will be informative for the question(s) under study). In contrast, theoretical sampling responds to the evolving data analysis and reflects a way of adjusting the sampling strategy to elaborate developing understandings and concepts more fully [2, 6].

In addition, Charmaz highlights moving beyond the creation of themes and categories to interpretation or an 'interpretative rendering' [6]. Thus, a process of 'theorizing' characterizes constructivist GT, which reflects an effort to lift the analysis from mere categories to conceptual interpretation. Rather than abstract 'grand theory', constructivist GT yields middle-range theory that represents situated conceptual understanding [2, 6]. Given its widespread used in health professions education research, the remainder of the chapter will focus on constructivist grounded theory.

Selecting Grounded Theory for Your Project

Research questions that lend themselves to a GT approach are exploratory in nature. They should be both broad enough to allow exploration but also sufficiently focused to allow researchers to define contexts and potential study participants who can shed light on the process in question [2]. These research questions seek to understand (a) the specific factors that contribute to certain processes (the "what"); (b) the reason behind opinions, behaviors, attitudes and perceptions (the "why); and (c) the reason certain processes occur (the "how").

Researchers have used grounded theory to study a variety of processes. Representative examples include:

- How medical students learn oral presentation skills [12]
- How their clinical supervisors contribute to the development of independent practice in medical trainees [13–15]
- How learning culture impacts feedback practices [16, 17]
- How resident and supervisor interactions impact autonomy and participation in clinical training [18, 19]
- How nursing students learn using simulation and what basic social processes support this learning [20]
- How nursing faculty evaluate student performance in simulation [21]
- How healthcare professions perceive mannequin death in simulation education [22, 23]

As Watling and Lingard (2012) point out, multiple qualitative methodologies are available and researchers should ensure a 'methodologic fit' between the particular research question and the chosen qualitative methodology before proceeding [2]. See Starks et al. 2007 for a comparison between phenomenology, discourse analysis, and grounded theory [24] and Reeves et al. for a discussion about ethnography in medical education [25]. Chapter 11 in this text also addresses this key aspect of qualitative research.

In the case study presented at the outset of the chapter, Susan—our simulation educator—was interested in studying how simulation debriefings contribute to peer learning. A more clearly delineated research question might read: *"How do simulation debriefings contribute to peer learning among medical students?* Since we can view debriefings and peer interactions as a social process that needs more clarification, constructivist GT appears to have a good methodological fit. The research question is broad, yet sufficiently defined to clarify what context (debriefings in simulation-based education) and individuals (medical students and simulation educators) Susan should recruit for her study. Both medical students and simulation educators would be able to shed insights on the issue of peer learning in healthcare debriefings.

Maintaining Reflexivity

Before proceeding, Susan will need to reflect on what experiences she brings to this study, since researchers cannot set aside their own background and perspectives [6]. Through the process of data analysis, researchers make analytical sense of participants' experiences *from their point of view*. That is, they bring to the analysis their motives, interests and preconceptions about the research topic and they engage with participants from the particular perspective formed by these personal experiences [6]. Therefore, reflexivity is a process of 'surfacing' the researcher's perspective, making it visible in order to examine how it shapes the analysis, and potentially alerting the researcher to blind spots this perspective may create.

Reflexivity requires researchers to maintain constant awareness about their role in the construction of knowledge. They should ask themselves: "Who am I? Why do I want to answer this research question? Have I experienced what I am researching? How are these experiences different from my participants? Am I imposing my view on the data? Are these data challenging my view of the problem I am researching?" Researchers must answer these questions and remain mindful about how they impact data analysis to assure accountability, transparency and trustworthiness of the research. To refer again to our example, Susan will need to reflect on her own experiences in simulation and debriefing and on her relationships with the students and fellow simulation educators she intends to interview. In some instances (especially with her simulation educator colleagues) she may determine that she knows them too well to perform the interviews herself. To this end, it can be quite helpful to form a team of researchers who bring different perspectives into the analytic process.

In constructivist GT in particular, memo writing can elucidate the researcher's role in knowledge construction. From the outset and throughout data analysis and collection, researchers should capture their reflections in writing to make them explicit. For example, they might explore the roots of their identity as it relates to their research, their current best understanding of the social process under study and how they may have experienced it, or their relationship with the participants. 'Reflexivity memos' can be quite cathartic and revealing. Becoming aware of preconceptions may be challenging, so we encourage researchers to write about their experiences freely without overthinking them. We always recommend including a section in the final manuscript that captures aspects of reflexivity. This might include a description of the researchers and their backgrounds, their research paradigms, and their relationship with the participants. Reflexivity memos might also serve as an appendix to the manuscript in order to showcase efforts in this regard.

Collecting Data for a GT Study

Multiple data sources and collection methods lend themselves to a constructivist GT approach [2, 3, 6]. While in-depth individual interviews are most common, focus groups, field observations, and document analysis also yield data amenable to GT. Referring again to our example, Susan's aim to collect data using semi-structured interviews, focus groups, or a combination of the two seems an appropriate choice for a constructivist GT study. If conducting such focus groups, we would recommend separating medical students from simulation educators as the learning environment at Susan's medical center might prevent medical students from sharing openly while simulation educators are present. It would also be entirely acceptable—with prior ethics approval—to combine data collection methods (interviews, observations) [6].

Analyzing Data Systematically Using Constructivist GT: The Coding Process

Coding constitutes the backbone of the constructivist GT analytical process. Coding entails labelling segments of data from interview transcripts, field notes, documents, or even videos in order to categorize, summarize and synthesize them. Researchers use these different coding phases to move toward an analytical account of participants' experiences. Coding mediates the construction of theory or conceptual understanding by encouraging researchers to engage deeply with the data. The resulting codes should capture how people make sense of their experiences, how and why they act upon these experiences and how the context influences their responses. To attain these aims, codes should remain close to the data, display action and connect these actions within coherent processes.

Constructivist GT coding includes three phases: (a) open or initial coding; (b) focused coding, and (c) axial coding [6]. During open coding, researchers explore data in order to find patterns that help them advance their analytical ideas and assist new data collection. During focused coding, researchers consolidate open codes into focused codes that accommodate similar pieces of data. These focused codes are compared with subsequent data to assess their fit. Later, axial coding delineates the relationships among codes as well as the areas of commonality, overlap, and divergence. This step may include the creation of a diagram to visually depict the relationships among them. Importantly, this analytical process is not linear. Instead, researchers should remain flexible and move back and forth between each coding phase until reaching theoretical sufficiency [26]. Theoretical sufficiency refers to the timepoint in data analysis where sufficient data have been collected to allow for a sound conceptual understanding of the process under study without inconsistency or discontinuity. Importantly, constructivist GT has moved away from the notion of key findings or themes "emerging" from the data; rather, researchers "identify" codes and themes within the data [26]. Since the researchers themselves drive the analysis, reflexivity as discussed above plays a critical role.

Memo writing serves as a tool to capture analytic considerations regarding patterns in the data as well as evolving relationships between themes and categories [2, 6]. Memos document analytical decisions and their rationales, and in part comprise an audit trail that allows researchers to reconstruct the transparent process of transforming raw data into a theoretical model. Memos also enhance the researcher's ability to engage in reflexivity about their own critical role in analysis. We now describe each step in GT analysis in detail by guiding readers through this process.

Initial or Open Coding

Imagine you are a simulation educator like Susan and you are sitting in front of your first interview transcript. Read carefully, attending to the following questions [6]:

- What is happening in the data?
- What do these data suggest?
- What is explicit or what has been left unsaid?
- From whose point of view?

Begin coding the first interview transcript line-by-line; that is, select a short segment and construct a code that describes and ideally conceptualizes that line. This process of open (or line-by-line) coding takes time, but helps you get to know the data and deconstruct the stories and descriptions in the interview into their component parts. Remember to look for conflict and action in the data, and code accordingly. Gerunds (or the '-ing' form of a verb) fulfil this aim [6, 27]. Examples include: 'perspective taking', 'managing time, 'coordinating with the team', 'contradicting someone', 'speaking up' or 'pushing back'. These codes suggest action and help you focus on *processes*, not *individuals*. Focusing on actions prevents you from making premature conclusions before constructing higher level concepts. Stay close to the data, construct short and sharp codes, and move lightly through data without overthinking this initial phase of coding. You will revisit the data many times, so do not try to 'get it perfect'. Remember that these initial codes are provisional and that you will refine them in later stages. You might also use participants' terms or slang to create initial codes; these are termed in vivo codes. For example, a simulation educator might describe their approach to debriefing as follows: *"Especially during interprofessional debriefings, I try to uncover contrasting perspectives to highlight various viewpoints"*, which yields the code of *'uncovering contrasting perspectives'*. These *in vivo* codes preserve participants' meanings about actions and processes, keeping the analysis grounded in the data. After coding the first 3–4 interviews you should ideally have a general sense of what is happening in the data, what conceptual directions you might take, and what is missing from the data. Researchers can then use these initial impressions to shape the next iteration of data collection and analysis in several ways:

- Possible modification of the interview guide: do some interview items need to be changed or deleted? Should new items be added? Of note, your original request for ethics approval should explicitly state that your interview guide will evolve in response to your developing analysis. However, significant changes to your interview guide may require additional ethics approval.
- Identification of additional areas for exploration and further probing based on issues that emerge during individual interviews.
- Theoretical sampling to solicit specific perspectives about the process under study.

At this early stage of coding it can be helpful—though by no means obligatory—to have more than one member of the research team code one or two full interviews, or even the same segments of one interview. An analytic team meeting provides an excellent context for this process. Such meetings should not focus on 'interrater agreement', but should focus instead on developing analytic momentum through collaborative discussion grounded in the actual data. Although our simulation educator, Susan, might complete her GT study on her own, we would recommend that she strategically invite collaborators for their ability to bring different perspectives to the data analysis.

Focused Coding

After coding several interviews, take a closer look at the list of initial codes. Select those codes that appear frequently or that you believe have a greater significance for your research question, and compare them with each other both within and between transcripts. Do you see overlap? Do some initial codes address similar phenomena which you can combine under a higher order code? Should some initial codes be renamed to elevate their conceptual value, or should they be left unchanged? This process results in **focused codes**.

Focused coding aims to synthesize, analyze and conceptualize larger segments of data through constant comparison. Once developed, researchers should test focused codes throughout the data to determine their conceptual strength. In particular, they should help you understand what is happening in larger segments of data, a characteristic initial codes lack. If your focused codes pass this scrutiny they could be preliminary categories within your evolving theory or conceptual model. Please note that while focused coding gives you analytical direction, you should remain flexible enough to reassess initial codes anew depending on incoming data from fresh interviews through constant comparison. Sometimes you will construct a specific code during initial coding that has high analytical power, and so it will feel natural to elevate it to a focused code (or even to a final category). On the other hand, you may struggle to move back and forth between initial and focused codes since some focused codes may not help you understand new information or are not robust enough to explain all the data you already have. These analytical decisions should be captured in a memo, and could also be the substance of further research team meetings. You might also use 'sensitizing concepts' (concepts gathered from existing literature and theory that provide an overall guiding frame of reference) [28] when constructing these focused codes. This approach should only be used to advance the analytical concepts, however, and should not impose or 'force' [29] theories onto the data.

Axial Coding

After generating robust focused codes, now you can move to **axial coding**. This phase aims to create, develop and specify the main categories within your data, and to further integrate them into a coherent theoretical or conceptual account of the processes involved in participants' experiences. For each category, you should be able to describe the 'when', 'where', 'why', 'who', and 'how' of the category, as well as its respective consequences. Keep in mind that you will need to support each category with representative quotes of text from the interview transcripts. Memo writing will help you keep track these decisions.

The categories themselves, however, only serve as the building blocks of the final theory or conceptual model. You must also address the relationships between them. A robust theory or conceptual framework should have categories that logically relate to each other, should not contain obvious unexplored gaps, and, most importantly, should be able to explain all new data. You may also realize you need additional perspectives to fill gaps in the evolving conceptual model or theory. In particular, you should attend carefully to discrepant examples in your data that appear to contradict each other, as they can contain important explanatory clues. A good developing analysis should account for discrepant examples rather disregarding them. At this phase, the construction of a diagram may be helpful. By creating a visual depiction of your analysis, these diagrams can assist you in: (a) seeing the big picture, (b) moving from coding to conceptual understanding, and (c) simplifying the development of the final theory or conceptual framework. Diagrams may also serve as a great resource to explain your findings in the final manuscript.

Once axial coding is complete, you could claim to have attained theoretical sufficiency [26] and stop data collection. Otherwise, you should go back to early stages of the coding process. We highlight here that coding, at least initially, serves primarily as an organizational strategy that groups similar pieces of data together, thus facilitating analysis. There is a real danger, however, of getting stuck in endless cycles of coding and re-coding without really moving from organizational to conceptual thinking. Researchers should remember that no "correct" coding structure exists. At some point researchers must draw the line on continuous refinements of their coding approach.

Evaluating the Quality of Grounded Theory Studies

The evolution of classic GT into multiple distinct traditions demands a clear delineation of what GT is and it is not. Indeed, researchers often say they have used GT when in fact this is not the case. As we have already highlighted, GT research has key features that characterize it, including iterative data collection and analysis, constant comparison, and theoretical sampling [2, 6]. For example, if researchers design a study in which they perform 15 interviews over three days and start analysing after data collection is complete, this approach does not reflect GT but rather good thematic analysis (see Braun & Clarke, 2006 [30]). The concurrent nature of data collection and data analysis in GT allows researchers to explore the problem deeply and to follow leads they identify in the data, making this iterative approach core to GT methodology.

Suddaby (2006) provides additional useful guidance. GT is _not_: [31]

- an excuse to ignore the literature
- presentation of raw data
- theory testing, content analysis or word counts
- an overly rigid technique to assess data
- perfect or easy
- an excuse for the absence of a methodology

Although the term "theory" features prominently in grounded theory, Timonen and colleagues point out that GT does not necessarily produce fully elaborated theory [11]. In fact, GT may lead to a useful conceptual framework that links concepts without covering all aspects of the process in question. Although theory-building should be an initial goal, practical considerations of the research process may limit theory building due to various constraints, such as lack of access to key informants required for theoretical sampling. Indeed, fully elaborated abstract or "grand theories" require multiple studies in different settings and contexts. Researchers should also keep in mind that a "theory" generated by a single GT study will be very specific to the study context, but may contain concepts that are readily transferrable to other settings.

We must also clarify the role of literature review in GT. Although classical GT [5] recommends that researchers have sufficient grasp of the literature to realize a study is needed, it further states that they should set existing theory aside during data collection and analysis. The notion of the 'blank state' in GT research, however, is a misconception [11, 32]. More recent conceptualizations recognize that delineating areas of focus and research questions first requires an understanding of the literature [11]. In fact, building a research program using GT requires researchers to engage with existing literature and with increasingly robust understandings of a particular problem. Further, constructivist GT recognizes that researchers bring prior background and perspectives to their work [2, 6] making reflexivity all the more important to promote transparency between researchers and those engaging with their work. In order to deepen existing theoretical insights, researchers must remain 'open to portrayals of the world' (p. 4) [11].

A number of criteria exist to assess the quality of grounded theory studies. These often relate to key features of trustworthiness, and overlap with general principles of high impact qualitative research. These criteria include [6]:

- *Credibility:* Do the findings convey a true picture of the phenomenon under study?
- *Transferability:* Do the findings provide sufficient detail about the study's context to allow readers to determine whether the findings apply to other settings?
- *Originality:* Do the findings present new insights or conceptual understandings*?*
- *Resonance:* Do the findings resonate with participants or those in similar situations?
- *Usefulness:* Are the findings useful in the everyday world?

Researchers can demonstrate their rigorous approach to qualitative research by maintaining an audit trail. This audit trail documents the analytic journey and allows researchers to reconstruct how they proceeded from A to B and demonstrate trustworthy and credible findings. In constructivist GT, reflexivity memos, analytic memos, journals, meeting notes, and other records of the research process all contribute to a robust audit trail.

Previous conceptualizations of qualitative research, including grounded theory, seemingly mandated additional steps to ensure objectivity [26]. These steps include triangulation, member checking, and saturation. These notions have been questioned more recently, since they may have been applied to qualitative research without much thought in order to 'tick boxes' that demonstrate a rigorous process to peer reviewers [26]. For example, the idea of 'saturation' has often been heralded as the end point in data collection without clear definitions of what saturation even means. In lieu of saturation, the concept of 'theoretical sufficiency' has gained traction among researchers in health professions education; please see Varpio et al. (2017) for an excellent discussion of the problems with thematic emergence, triangulation, member checking, and saturation [26].

Conclusions

Grounded theory, especially in its constructivist interpretation, has found widespread application in health professions education research. GT's capacity to address the "why", "what", and "how" questions makes it a powerful choice for simulation researchers seeking to understand 'how' and 'why' simulation-based strategies work to promote meaningful learning. Key features of GT include iterative data collection and analysis, constant comparison, and theoretical sampling. Several components of the research process contribute to trustworthy and credible results, namely reflexivity and a comprehensive audit trail that ensures transparent analytic decisions. Attention to these key principles will ensure appropriate designation of the research methodology as 'grounded theory'.

References

1. Harris I. What does "The discovery of grounded theory" have to say to medical education? Adv Health Sci Educ Theory Pract. 2003;8(1):49–61.
2. Watling CJ, Lingard L. Grounded theory in medical education research: AMEE Guide No. 70. Med Teach. 2012;34(10):850–61.
3. Kennedy TJT, Lingard LA. Making sense of grounded theory in medical education. Med Educ. 2006;40(2):101–8.
4. Lingard L. When I say...grounded theory. Med Educ. 2014;48(8):748–9.
5. Glaser BG, Strauss AL. The discovery of grounded theory: strategies for qualitative research. Chicago: Aldine; 1967.
6. Charmaz K. Constructing grounded theory. 2nd ed. London: SAGE Publications Ltd; 2014.
7. Higginbottom G, Lauridsen EI. The roots and development of constructivist grounded theory. Nurse Res. 2014;21(5):8–13.
8. Mills J, Bonner A, Francis K. The development of constructivist grounded theory. Int J Qual Methods. 2006;5(1):25–35.
9. Morse JM, Stern PN, Corbin J, Bowers B. Developing grounded theory: the second generation. New York: Taylor & Francis; 2009.
10. Thomas G, James D. Reinventing grounded theory: some questions about theory, ground and discovery. Br Educ Res J. 2006;32(6):767–95.
11. Timonen V, Foley G, Conlon C. Challenges when using grounded theory: a pragmatic introduction to doing gt research. Int J Qual Methods. 2018;17:1–10.
12. Lingard L, Garwood K, Schryer CF, Spafford MM. A certain art of uncertainty: case presentation and the development of professional identity. Soc Sci Med. 2003;56(3):603–16.
13. Kennedy TJT, Regehr G, Baker GR, Lingard LA. 'It's a cultural expectation...' The pressure on medical trainees to work independently in clinical practice. Med Educ. 2009;43(7):645–53.
14. Kennedy TJT, Regehr G, Baker GR, Lingard L. Point-of-care assessment of medical trainee competence for independent clinical work. Acad Med. 2008;83(10 Suppl):S89–92.
15. Kennedy TJT, Lingard L, Baker GR, Kitchen L, Regehr G. Clinical oversight: conceptualizing the relationship between supervision and safety. J Gen Intern Med Off J Soc Res Educ Prim Care Intern Med. 2007;22(8):1080–5.
16. Watling C, Driessen E, Van Der Vleuten CPM, Vanstone M, Lingard L. Beyond individualism: professional culture and its influence on feedback. Med Educ. 2013;47(6):585–94.
17. Watling C, Driessen E, Van Der Vleuten CPM, Lingard L. Learning culture and feedback: an international study of medical athletes and musicians. Med Educ. 2014;48(7):713–23.
18. Olmos-Vega FM, Dolmans DHJM, Vargas-Castro N, Stalmeijer RE. Dealing with the tension: how residents seek autonomy and participation in the workplace. Med Educ. Wiley/Blackwell (10.1111). 2017;51(7):699–707.
19. Olmos-Vega FM, Dolmans DH, Guzmán-Quintero C, Stalmeijer RE, Teunissen PW. Unravelling residents' and supervisors' workplace interactions: an intersubjectivity study. Med Educ. Wiley/Blackwell (10.1111). 2018;52(7):725–35.
20. Walton J, Chute E, Ball L. Negotiating the role of the professional nurse: the pedagogy of simulation: a grounded theory study. J Prof Nurs. 2011;27(5):299–310.
21. Watts PI, Ivankova N, Moss JA. Faculty evaluation of undergraduate nursing simulation: a grounded theory model. Clin Simul Nurs Elsevier Inc. 2017;13(12):616–23.
22. Tripathy S, Miller KH, Berkenbosch JW, McKinley TF, Boland KA, Brown SA, et al. When the Mannequin Dies, creation and exploration of a theoretical framework using a mixed methods approach. Simul Healthc. 2016;11(3):149–56.
23. McBride ME, Schinasi DA, Moga MA, Tripathy S, Calhoun A. Death of a simulated pediatric patient: toward a more robust theoretical framework. Simul Healthc. 2017;12(6):393–401.
24. Starks H, Trinidad SB. Choose your method: a comparison of phenomenology, discourse analysis, and grounded theory. Qual Health Res. 2007;17(10):1372–80.
25. Reeves S, Peller J, Goldman J, Kitto S. Ethnography in qualitative educational research: AMEE Guide No. 80. Med Teach. 2013;35(8):e1365–79.
26. Varpio L, Ajjawi R, Monrouxe LV, O'Brien BC, Rees CE. Shedding the cobra effect: problematising thematic emergence, triangulation, saturation and member checking. Med Educ. 2017;51(1):40–50.
27. Charmaz K. The power and potential of grounded theory. Med Soc Online. 2012;6(3):1–15.
28. Bowen GA. Grounded theory and sensitizing concepts. Int J Qual Methods. 2nd ed. 2006;5(3):12–23.
29. Kelle U. "Emergence" vs. "Forcing" of empirical data? A crucial problem of "Grounded Theory" reconsidered. Forum Qual Soc Res. 2005;6(2):27. http://nbn-resolving.de/urn:nbn:de:0114-fqs0502275.
30. Braun V, Clarke V. Using thematic analysis in psychology. Qual Res Psychol. 2006;3(2):77–101.
31. Suddaby R. From the editors: what grounded theory is not. Acad Manag J. 2006;49(4):633–42.
32. Urquhart C, Fernandez W. Using grounded theory method in information systems: the researcher as blank slate and other myths. J Inf Technol. Nature Publishing Group. 2013;28(3):224–36.

Analyzing Data: Approaches to Thematic Analysis

19

Gerard J. Gormley, Grainne P. Kearney, Jennifer L. Johnston, Aaron W. Calhoun, and Debra Nestel

Overview

In this chapter, we focus our attention on qualitative research related to healthcare simulation. This chapter explores two related approaches to analysing qualitative data – *thematic analysis* and *qualitative content analysis*. Both of these methods are commonly used in qualitative research, and are considered relatively accessible forms of analysis. We provide a logical approach to their use offering references and a worked example. In addition to introducing each method, we discuss how to ensure rigour and trustworthiness in your research. These issues form an essential part of any qualitative approach.

Practice Points

- Thematic analysis and conventional/directed qualitative content analysis are accessible and relatively straightforward forms of qualitative analysis.
- Careful consideration of the research question is critical in selecting the most appropriate analytic approach. We caution against being 'lured in' by the apparent ease of thematic analysis or qualitative content analysis. All approaches have subtleties and complexities, and neither may be the most appropriate choice for your research question.
- Terms used in thematic analysis and qualitative content analysis are sometimes used interchangeably, but meanings are contextual and can differ in each approach.
- The development of themes is never totally independent of the researchers. Indeed, all researchers using qualitative methodologies will need to grapple with their influence on the data and its interpretation. It is imperative that researchers are aware of and reflect regularly on these influences. Keeping a reflexive diary/notebook and having regular discussions within the research team aids this reflexive awareness, and allows the researchers to be cognisant of their position amongst the data. These steps also provide an important audit trail, which proves the intellectual progression of your work.
- Rigour and trustworthiness in qualitative research are principally reflections of how well the research question, methodology, reported findings and discussion align and cohere.

G. J. Gormley
Centre for Medical Education, Queen's University Belfast, Belfast, UK
e-mail: g.gormley@qub.ac.uk

G. P. Kearney · J. L. Johnston
Centre for Medical Education, Queen's University Belfast, Belfast, UK
e-mail: g.kearney@qub.ac.uk; j.l.johnston@qub.ac.uk

A. W. Calhoun (✉)
Department of Pediatrics, University of Louisville School of Medicine, Louisville, KY, USA
e-mail: aaron.calhoun@louisville.edu

D. Nestel
Monash Institute for Health and Clinical Education, Monash University, Clayton, VIC, Australia

Austin Hospital, Department of Surgery, Melbourne Medical School, Faculty of Medicine, Dentistry & Health Sciences, University of Melbourne, Heidelberg, VIC, Australia
e-mail: debra.nestel@monash.edu; dnestel@unimelb.edu.au

Introduction

Scenario 1

Gabriel is a simulation-based educator in a large medical school. Over time he has remarked that some students persistently fail to engage well with simulation-based teaching activities. This observation is also supported by his colleagues and the literature. However, reasons why this phenomenon occurs remain largely unexplored. Gabriel is keen to carry out qualitative research in this area and is currently considering which analytic approach will

D. Nestel et al. (eds.), *Healthcare Simulation Research*, https://doi.org/10.1007/978-3-030-26837-4_19

best help him understand this phenomenon. After some discussion, it is determined that thematic analysis would be the most appropriate qualitative method for exploring this question.

Scenario 2

In reviewing a debriefing training course with faculty facilitators, a group of researchers observes that the overall understanding of educational concepts and theories seems to directly relate to the quality and ultimate success of the debriefing. The research team are interested in this topic for a number of reasons. First, they want to better understand debriefers' knowledge and articulation of educational concepts and theories, since a core component of the debriefing course focuses on educational concepts and theories that inform this practice. Second, the researchers have also noticed variation in the quality of debriefings and want to better understand what's happening. This interview-based study using qualitative content analysis is a first step in deepening their understanding.

To date, much research in simulation has been *quantitative* in nature. Aimed at clearly delineating the effects of various simulation-based interventions, this approach has served us well in many ways. Quantitative approaches are, however, best suited to 'what is' questions, leaving questions of 'why' or 'how' unaddressed. That is, why do learners behave in different ways during stressful simulations? How do various debriefing approaches effect learners differently? How do hybrid simulations assist learners in developing safer approaches to clinical practice? *Qualitative* research, which usually analyses words (instead of numbers), is common in other fields such as anthropology and sociology, and is gaining popularity in simulation-based education research.

There are many approaches to performing a qualitative analysis, most of which draw from the social sciences. Some of these have been outlined in Chaps. 9–20. In this chapter, we explore *thematic analysis* and *qualitative content analysis* as approaches to conducting data analysis. Specifically, our aim is for readers to:

(1) Develop a working understanding of *thematic analysis* and *qualitative content analysis*
(2) Understand the situations in which these approaches can best be used for analysis

Before You Get Started…

As described earlier in this book, there are several key steps that need to be considered before choosing *thematic analysis or qualitative content analysis*. Newcomers to qualitative research should be cautious of just wanting to *'do a focus group study'* or *'do a thematic analysis study'* without considering the nature of their research. Before you consider methods of data collection *(e.g. focus groups)* or methods of analysing your data *(e.g. thematic analysis)* there are a number of important steps to contemplate.

As with all forms of research, establishing a well-defined research question is of critical importance – the *keystone* of your work [1]. Without this, your study will be at risk of losing its focus and rigour. It is important to realise the difference between a qualitative *research question* and a quantitative *hypothesis*. While the latter predicts a relationship between variables, the former are open-ended, and are not predictive. Research questions are developed in many different ways, and they are often derived from corridor conversations or ideas sparked from practice. Moving from these first ideas towards a useable and creditable research question involves some basic general principles. First, define the topic (or phenomenon) under investigation (why is it important/of interest?). Reading widely around the topic and having conversations with *critical friends* is essential. Second, define the specific problem you wish to address (what is it, and why is it problematic?). Third, conduct a more focused review of the literature with the goal of better delineating the gap in evidence that you would like to address through your research. What is, and is not, known about your topic? Now you are in a position to develop and refine your research question/s!

It is also useful to consider who will be interested in the research, and who will benefit from your research. Only then should you consider which research approach is most suited to your project. Rigour in qualitative research is a reflection of how well the research question, methodology, reported findings and discussion are aligned [2].

What is Thematic Analysis?

Qualitative data is usually composed of words. Making sense of what is said, and sometimes what is meant, is at the core of qualitative research. Qualitative researchers can analyse data in different ways. Many of these qualitative approaches use established educational or social *theory* to help guide the analysis (for example, *activity theory*, *complexity theory*, etc.), and we note that theoretical approaches are becoming more popular in simulation-based research (See examples [3–6]). It is arguably more difficult, however, for new researchers and clinicians to get started from scratch with these methods. Thematic analysis appears more approachable, and also offers a more open-ended approach: it can both be *theory-generating* and *theoretically informed*. Researchers develop *themes* from textual data, and then organise them in a way to enable interpretation to be balanced with the inductive process of discovery they have just undertaken. Thematic analysis can be applied to a wide range of subject areas and types of dataset, such as transcripts from focus groups or interviews, recorded natural conversations (with research permissions, of course) or audio-diaries [7].

One issue which is often confusing to new qualitative researchers is that thematic analysis can be used as an umbrella term, encompassing several approaches, e.g.:

- Template Analysis [8, 9]
- Framework Analysis [10]
- Thematic Analysis (Braun and Clarke) [7, 11]
- Qualitative Content Analysis [12]
  - Conventional content analysis
  - Summative content analysis
  - Content analysis

Some of the approaches lend themselves to a more generic approach whilst others are more bound by ontological (i.e. the nature of reality) and epistemological (i.e. the nature of knowledge) positions. Such positions reflect where the researcher locates their research on the spectrum from realism (i.e. an objectivist stance – where things are as they appear and measuring them thus enables us to arrive at the one "true" version of reality) to relativism (i.e. a subjectivist stance – where no two individuals capture reality in the same way and hence multiple versions are expected). This spectrum lies at the heart of much qualitative research. Although we will not expand on this topic here, it is important for researchers to consider their own ontological and epistemological position and ensure their analysis is consistent with this position.

Qualitative content analysis (QCA) is an alternative approach to traditional thematic analysis, at times indistinguishable from it. The boundaries between approaches are not always clearly specified and are sometimes used interchangeably [13]. They appear to have emerged from different research traditions, and our observations suggest that QCA is more commonly used in nursing research and that it appears more often in research from North America. We describe it here as a further option for qualitative researchers. Hsieh and Shannon describe three distinct approaches to QCA – *conventional, directed* and *summative* [12]. It is in summative QCA that the distinction between approaches is most clearly drawn. We illustrate these three approaches in Boxes 19.1, 19.2, and 19.3 using a fictional transcript of an *interview* from a study related to Scenario 2 above that seeks to explore the educational concepts and theories experienced faculty consider when debriefing team-based simulations.

Box 19.1

Example of a transcript segment in a qualitative research interview between a researcher and interviewee – an experienced debriefer of teamwork simulations. Using QCA (conventional), the researcher has prepared for the analysis by reading the transcript and, using a highlight function, has marked (in bold italic) words and phrases of interest to them relative the research question

Interviewer: I'd like to understand the theories that inform your debriefing approach…
Interviewee: Well, that's interesting. I'm not certain of the names of some theories but I know that I use a few. Should I just tell you what I'm thinking about when I'm approaching a debrief?
Interviewer: That would be helpful.
Interviewee: So, ***I usually have in my mind what I'm going to do*** in the debrief. I appreciate that many of the scenarios that I debrief have ***a strong emotional component so I need to be aware*** of that all the time. I know that there are theories about how some ***really strong emotions can get in the way of learning***. So, yeah, I've really got to be aware of this and acknowledge it to the participants. I often get started in a neutral sort of way though by saying something like – "Thanks for participating." I then usually just ask them how they're feeling. Chances are they'll answer by stating some thoughts rather than emotions but I usually push them a little to state an emotion and ask them about whether this was also how they were feeling in the simulation and if they think it influenced their behaviour. I'm trying to get them to ***make connections between feelings and behaviour***. They're often surprised that others did not even notice that feeling. And even in the debriefing, perhaps that should be especially in the debriefing, they can be pretty emotional too. So, yeah, ***theories about learning and emotions are really important. If you acknowledge the feelings, you can establish rapport*** and ***if they're really upset, well, it might not be the best time to learn***. I may have to come back to them when those emotions are more settled. [Silence]
Interviewer: Is there anything else about emotions and learning?
Interviewee: Maybe, I think emotion is linked with other things too. I also think about ***the time we have available for the debrief: where it will be, if we have capacity for video replay, the number of issues to get through, how well I know the participants, when I'll see them again, where they are in their education*** and other stuff too. So, there are theories behind all of these things. A key thing is simply ***how much information they can take on***. I know there is a theory about ***cognitive load*** – I think that's what it's called. I don't remember the details of it but it helps you think about the ***amount of learning that can happen in any one scenario and debrief***. It makes you think about the ***design of the scenario*** – important things like avoiding distractions if they're taking the participants away from what needs to be learned - unless managing distractions is a ***learning objective*** *[Laughs]*.
Interviewer: It sounds complicated.
Interviewee: The more I talk about it the more I'm realising that. So, I know there's theories about ***learning from peers***, near peers and so on. And theories about ***establishing trust with participants***. I'm always trying to do things that ensure ***respect***. If they're not respecting each other, there won't be much learning. Actually, I set this up in the briefing and just try to keep it going in the debriefing. *[Silence]*
Interviewer: Anything else?
Interviewee: I also do things like ***summarise, I get the participants to summarise*** what we're discussing. ***That gets everyone involved*** and it also means ***we're repeating key ideas***, so that's getting at cognitive load too. ***Not too much information, be clear about it, repeat it, emphasise it, get others to state it in their own words. I usually end the debrief like this***. Oh, and then there's ***learning objectives*** too – that all helps with cognitive load. ***It's being clear about the purpose of the learning, of setting expectations***. *[Pauses]* Look, I guess the whole debriefing is ***a reflection***. Yeah, there's that phrase, ***reflection-on-action***. I am asking them to go back over what has happened, to make sense of it as they did in the moment, in-action, and now with the benefit of hindsight. And, how this experience could be managed in future. This means the things the participants did that were really effective and those that weren't. Of course, ***I try to access their reasons for each of these things, thinking about the conditions that influenced their behaviours***.

Box 19.2

This is an example of the next step in conventional QCA. The researcher has considered the research question and has extracted the relevant text from the transcript – these were highlighted in bold italic in Box 19.2 and now appear as a list in Box 19.3. This data is the basis for the next step. This interim step is not absolutely essential, but can be valuable when you are first doing analysis as it documents carefully and systematically how the codes were derived from the data. Alternative approaches can also be used, including marking up the transcript directly with post-it notes and highlighter pens. It is also important to write notes that document your reactions to the data and decision-making processes. These can then be used as an audit of your analytic moves in order to address issues of reflexivity (defined in Chap. 9)

- ***I usually have in my mind what I'm going to do***
- ***a strong emotional component so I need to be aware***
- ***really strong emotions can get in the way of learning***
- ***to make connections between feelings and behaviour***
- ***theories about learning and emotions are really important***
- ***If you acknowledge the feelings, you can establish rapport***
- ***if they're really upset, well, it might not be the best time to learn.***
- ***the time we have available for the debrief, where it will be, if we have capacity for video replay, the number of issues to get through, how well I know the participants, when I'll see them again, where they are in their education***
- ***how much information they can take on***
- ***cognitive load***
- ***amount of learning that can happen in any one scenario and debrief***
- ***design of the scenario***
- ***learning objective***
- ***learning from peers, near peers***
- ***about establishing trust with participants***
- ***respect***
- ***summarise***
- ***I get the participants to summarise***
- ***That gets everyone involved***
- ***we're repeating key ideas***
- ***Not too much information, be clear about it, repeat it, emphasise it, get others to state it in their own words. I usually end the debrief like this***
- ***learning objectives***
- ***It's being clear about the purpose of the learning, of setting expectations.***
- ***reflection***
- ***reflection-on-action***
- ***I try to access their reasons for each of these things, thinking about the conditions that influenced their behaviours***

Box 19.3

In this example, we begin to create a coding framework based on higher level clustering of key concepts from the initial data extraction. Each text fragment that the researcher identifies as a key concept has been allocated and given a named code. This means that each text fragment can be checked against the coding framework. This process will be repeated for the first few transcripts with the goal of building a coding framework to be applied to all transcripts. Here the highest level code is designated by a number, with subcodes designated by a letter and so on. So, for example, 'Having a Debriefing Plan" should be considered a high level code that incorporates several ideas: "having learning objectives", "considering environmental influences" and "promoting reflection.' Each of the key concepts represented by text fragments can be checked against the code. Although currently arranged as a hierarchy, the coding framework has many points of intersection which will likely be rearranged once the framework is applied to other transcripts. In square brackets, the text fragments illustrate the codes

1. *Intentionality in learning (planning for learning)* [I usually have in my mind what I'm going to do]
2. *Having a debriefing plan* [I usually have in my mind what I'm going to do]
   a. *Having learning objectives* [learning objective]
   b. *Considering environmental influences*
      i. *Including time, location, setting, video review* [the time we have available for the debrief, where it will be, …]
   c. *Promoting reflection* [I try to access their reasons for each of these things…]
3. *Designing for learning*
   a. *Cognitive load* [how much information they can take on]
      i. *Learning objectives* [it's being clear about the purpose of the learning]
      ii. *Repetition including summarisation* [I get the participants to summarise]
      iii. *Emphasis* [… emphasise it…]
   b. *Reflective practice* [reflection-on-action]
4. *Recognising and acknowledging participants' emotions*
   a. *Establishing rapport* [If you acknowledge the feelings you can establish rapport]
   b. *Building trust and respect* [about establishing trust with participants]
5. *Monitoring participants' emotions* [a strong emotional component so I need to be aware]
   a. *The role of emotions for and during learning* [theories about learning and emotions are really important]
      i. *Barriers to learning* [If they're really upset, well, it might not be the best time to learn]
6. *Considering participants' awareness (feelings and behaviour)* [I try to access their reasons for each of these things…]
7. *Relational issues in learning*
   a. *Knowing participants* [… how well I know the participants]
   b. *Peer learning* [learning from peers, near peers]

What Are Codes and Themes, and What's the Difference?

These are common questions from beginning researchers. Research themes can be considered as particularly prominent or repeating features derived directly from data. The process of organising data into themes is at the heart of thematic analysis and qualitative content analysis. Codes are simply the building blocks of themes.

Codes are usually segments of textual data (e.g. from passages in interview transcripts) which researchers identify as relevant to the research question. They can take the form of short phrases constructed by the researcher, or sometimes an actual comment from data ("natural" codes). In interview data, they may aim to capture the essence of individuals' transcribed talk. For example, a code might be *"Feeling anxious performing in front of peers"* as a reason not to fully engage with simulation-based learning activities. Assigning extracts of the text that are relevant to these *codes* is known as *coding*. As the research analysis progresses, *codes* will be developed into *themes* which characterise distinctive features of participants' experiences. To continue with the example, once the study is complete, the "*feeling anxious performing in front of peers*" code may be fused with several other similar codes to form the theme "*negative emotional impacts*." Note that it is not the specific phrasing that distinguishes codes from themes, but rather the iterative, developmental process used to meld and arrange the codes into the final thematic list.

Thematic Analysis

Here, we use *Template Analysis* as an exemplar, since it has been previously used in simulation-based education research [14]. This provides researchers with a largely generic approach to their analysis, regardless of their philosophical stance on research. In the 'additional resources' section at the end of this chapter we have references that describe how to carry out other forms of thematic analysis. Template Analysis is iterative, and follows a number of logical steps (Fig. 19.1).

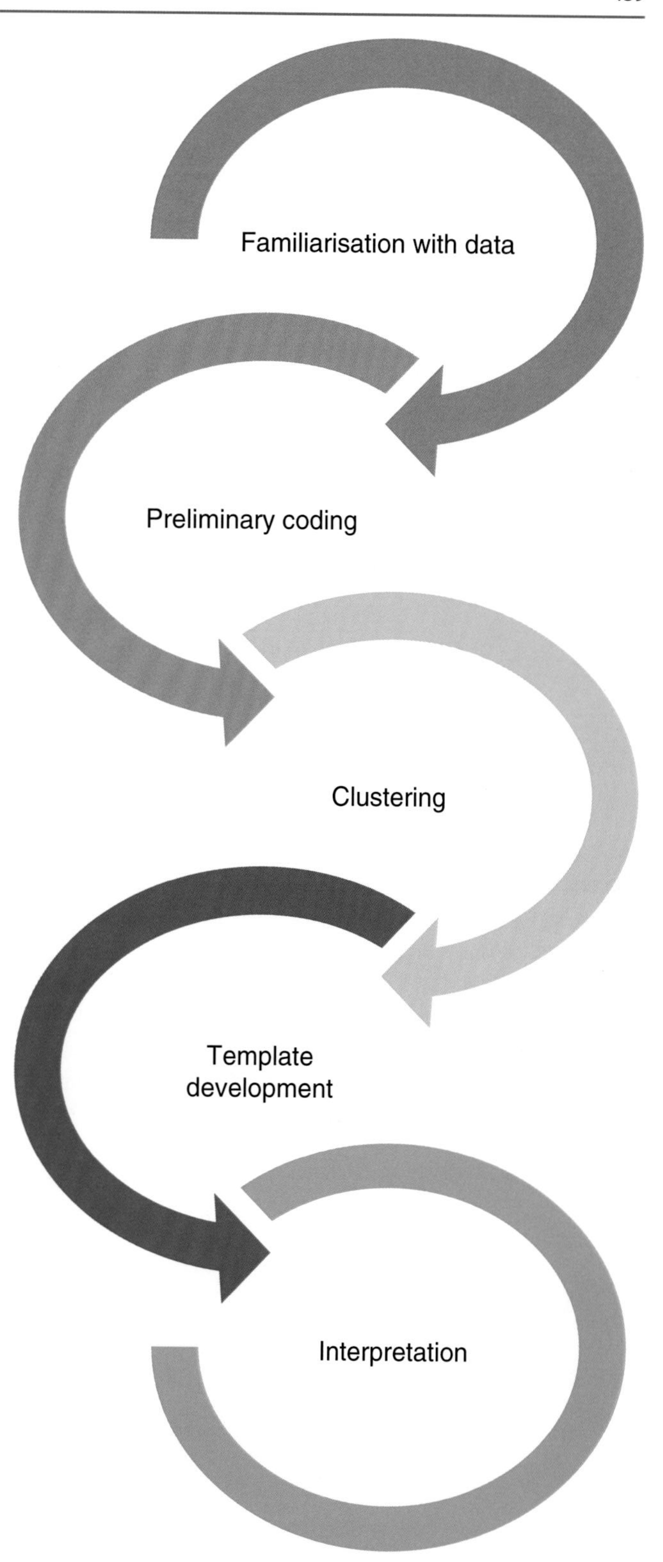

Fig. 19.1 Data analysis schematic for thematic analysis using the template analysis approach

Analytic Approaches to Scenario 1

STEP 1: Familiarisation

Once your dataset has been obtained, the analytical process begins. In the example above, exploring why some students fail to fully engage with simulation-based education, data might consist of focus group transcripts. Researchers involved in the analysis must *immerse* themselves in the data prior to the coding process. Mere *acquaintance* is inadequate. One of the authors describes how she "sits on her hands" (so as not to be able to mark up the transcript) during the first read of the transcript, because it is so easy to get caught up in minutiae of word by word/line by line text before appreciating the whole text. This immersion will ensure that themes will remain firmly rooted in participants' accounts and not simply drift toward the personal perspectives of the researchers in an uncritical way. The presence of multiple analysts can also be of benefit as it explicitly allows for the presence of multiple perspectives. Immersion is achieved simply by reading and re-reading the transcripts. Listening to the original interview recordings is also worthwhile. Some argue that actually transcribing the interview recordings is an excellent method of becoming immersed in the dataset – but pragmatism is also needed as this is not always feasible.

Fig. 19.2 Illustration of researchers assembling preliminary codes into representative groups

STEP 2: Preliminary Coding

Once familiar with the data, the next step is *preliminary coding*. First, highlight any areas that are relevant in addressing the research question. Practically, this can be done in analogue fashion using coloured pens and notes in the margin (several of the authors' favoured approaches). Software packages can be useful but far from essential (See Chap. 9). For the digitally minded, everyday word-processing software will produce similar results. In all cases, note that the researcher, not the programme, does the analytic work: technologies are simply tools to help with the job at hand. Sum up the area that you have highlighted by assigning a preliminary code. Each code should refer to a single simple concept (See below). Once you have finished with this first transcript, do the same with another sample of data (a second focus group, for example), ideally diverse in its content. For example, this time consider views from students who do engage more in simulation-based learning. At this stage (and with this method) you do not need to do this with the entire dataset, particularly if it is large. A subset will often suffice.

STEP 3: Clustering

Now the researcher/s should start to group preliminary codes together (see Fig. 19.2). Clustering means assembling similar preliminary codes into representative groups and assigning them *theme names*. It is from these groupings that the initial themes will be identified. Tentative *a priori* themes (i.e. potential themes that were identified prior to reviewing the dataset, perhaps informed by prior theoretical considerations) can also be introduced at this stage. Once the researcher considers the clusters to have captured the essence of the data and the participants' experiences, they then can progress to developing an *initial template*.

STEP 4: Template Development

At this stage of the analysis, themes identified from *clustering* are assembled into an initial template. Consideration of the relationship between the themes should also take place at this stage, and it may be that some 'minor' themes are subsumed into other themes or discarded. Once this initial template has been developed, the researcher should apply it to another subset of data. In this process, the researcher identifies the text that falls under a theme from the initial template. During this process, themes may be refined and developed. Furthermore, themes from the initial template may be omitted and new themes added. Once this is done the researcher should apply the refined template to another subset of data. Iteratively, the template will be modified until the researcher considers they have a final version of the template. This final template is then applied to the entire dataset to ensure it represents the data, and addresses the research question. Keep in mind that additional codes and themes will likely emerge at this phase. This does not represent a failing of the initial coding, but is an essential part of the process. Qualitative research is by nature iterative.

STEP 5: Interpretation

Once the final template has been achieved and applied to the entire dataset conclusive interpretation can take place. This consists of a deep consideration of how the themes relate to each other, and allows you to make cohesive sense of the findings and review their relevance to the research question. If each member of the research team has been working separately up to now, this step provides an opportunity for the disparate analysts to share their various interpretations of the data. This is a helpful means of *triangulating* interpretations of data (confirming from different perspectives [See Chap. 9]). The final stage of pulling themes together can vary from study to study, depending on epistemological positions. In general, however, consensus should be reached between all the researchers' involved in the study.

Analytic Approaches to Scenario 2

Conventional content analysis (CCA) resembles inductive thematic analysis. Codes are derived from the data (e.g. transcribed talk). These initial codes are arranged into key thoughts or concepts, which are then clustered by their shared meaning. A hierarchy of codes may be produced with descriptions of each code illustrated by data – this can be in the form of a data display table (similar to Template Analysis described above). Alternatively, it can appear as a hierarchical list. In the discussion section of the manuscript, the findings are compared with relevant theories. Boxes 19.1, 19.2, and 19.3 illustrate this process using a transcript segment based on Scenario 2. In Box 19.1, the researcher has highlighted an initial set of 'words and phrases' within the transcript that represent their best sense of the key concepts (codes) within the text. In Box 19.2 they have been extracted, and so they are now out of their context. In Box 19.3, the codes are clustered into key topics or categories. This progression from identifying phrases and creating categories is usually not a linear process, but requires moving back and forth between data, codes, and categories. In our example, latent content (i.e. implicit meaning in the data) is used to create codes. As each transcript is analysed the same process is undertaken. After analysing a few interviews, a single coding framework is created that is derived from the data from each transcript. This framework is then applied to all transcripts seeking confirming and disconfirming data. New codes may be identified during analysis which will mean going back to all transcripts over and over again. As the analytic process continues, these codes will be abstracted to categories and high level themes. If more than one researcher is involved in the analysis, researchers would usually meet after a few of the transcripts have been analysed to share their initial coding and categories. Researchers discuss agreement/convergence and disagreement/divergence in their selections and usually seek to create a single coding framework. However, finalising the coding framework too early can limit the researchers' openness to seeing new ideas in subsequent data. Holding the framework "lightly" is vital at this early stage. Revisit the coding framework after all data has been analysed, make adjustments and go back to the data for refinement, seeking confirming and disconfirming data and also thinking about what is not present in the data. While this can seem a strange process to researchers more familiar with quantitative approaches, it reflects the ontological feature of qualitative research where multiple realities are expected, acknowledged and valued, that naming implicit meaning is accepted, and that sometimes it is important to consider what has not been said in the transcribed talk.

Directed content analysis (DCA) resembles the deductive analytic approach described above *as* a priori coding. In this approach, the coding is established *in advance of* the analysis using existing research and/or theory to construct the coding framework. Rather than theory building, which is often the idea with thematic analysis or CCA, the aim of this approach is usually to 'test' known theories. This approach may also include an inductive component in parallel to the a priori coding in which the researcher simultaneously identifies codes of interest (i.e. that are not included in the a priori coding framework). These additional codes are also clustered and subsequently added to the coding framework.

Once analysis is complete and researchers come to write the discussion section of a manuscript, findings are compared with relevant theories, offering confirmation or disconfirmation and/or extending the theory (this approach can be *theory extending* as well). In our second fictional example (which draws on studies of debriefing), the researchers have established a list of theories that are commonly cited in a literature review of debriefing. Because of the presence of these theories, DCA as a process aligns well both with what the researchers currently know and the nature of their research question. Box 19.4 contains a list of these theories and their applications.

Box 19.4

This is an example of an a priori coding framework of key educational concepts and theories informed by the review of basic texts on debriefing and recent systematic reviews of debriefing practices. This time the coding framework is not derived from the data but developed based on published relevant literature. Although currently arranged as a hierarchy, the coding framework has many points of intersection which will likely be rearranged once the framework is used. For summative content analysis, the number of times the educational concept or theory is described by the interviewee can be counted

1. *Behavioural*
    a. *Experiential learning theory*
    b. *Mastery learning*
        i. *Cognitive load theory*
            1. *Instructional design*
                a. *Learning objectives and learning outcomes*
                b. *Sequencing of learning*
    c. *Deliberate practice*
    d. *Emotion and learning*
2. *Cognitive*
    a. *Cognitive load theory*
        i. *Instructional design*
            1. *Learning objectives and learning outcomes*
            2. *Sequencing of learning*
    b. *Cognitive apprenticeship*
3. *Constructivist*
    a. *Reflective practice*
    b. *Adult learning theory*
    c. *Scaffolding*
    d. *Advocacy inquiry*
    e. *Affective elements of learning*
4. *Socio-cultural*
    a. *Communities of practice*
    b. *Peer assisted learning*
    c. *Co-operative learning*
5. *Socio-materiality and Complexity theories*
    a. *Activity theory*
    b. *Actor-Network theory*

Finally, *summative content analysis* offers completely different insights and is therefore the most distinctive of the QCA approaches. Vaismoradi et al. writes that: *"in spite of many similarities between content and thematic analysis, …. cutting across data, and searching for patterns and themes, their main difference lies in the possibility of quantification of data in content analysis by measuring the frequency of different categories and themes, which cautiously may stand as a proxy for significance."* [13]. This summative content analysis technique is used when researchers are interested in the frequency of particular words. While this approach has not been applied to healthcare simulation in any substantive fashion, the study by Ross et al. is a notable exception [15]. In their study, summative content analysis is used to summarise and make meaning of simulation studies in anaesthesia journals across a decade: *"We found broad acceptance and uptake in anaesthesia with an increase in publications over the time period, mainly attributable to a steady increase in manikin studies. Studies using manikin technology (130/320; 41%) are distinguished as skills/performance studies (76; 58%) and studies focused on the use, testing, and validation of equipment (52; 40%)."* [15].

Other examples of summative content analysis relate to the analysis of participants' responses to simulation-based education. Often free text on evaluation forms, participants' statements may be analysed as positive, neutral or negative. The frequencies of statements in each category are then reported. However, it can be challenging to make meaningful conclusions of these outcomes. This approach is often used as an evaluation rather than a research technique.

Scenario 2 offers a model of a study where this might be useful. In that hypothetical study, the researchers may decide to enumerate the numbers of theories that debriefers in the study identify directly (i.e. the interviewee names the theory) and indirectly (i.e. the interviewee describes the manifestation of the theory without providing the explicit name). So, to continue with the illustration, let us suppose that the interviewees explicitly mention examples of cognitive load theory, reflective practice, the role of emotion in learning, and peer assisted learning. While their level of understanding of these concepts may be unclear from their transcribed talk, the fact that interviewees bring them up by name is an example of direct identification. In contrast, the interviewees do not name advocacy inquiry, but do make statements that fit well within that framework (e.g. "I try to access their reasons for each of these things, thinking about the conditions that influenced their behaviours."). Returning to the researchers' goal of gaining a better understanding of debriefers' knowledge and articulation of educational concepts and theories, it is easy to see the value of this approach as a guide to their ongoing investigations.

Additional Concepts: Reflexivity and Participant Selection

During thematic analysis researchers write reflexive notes about the decisions they make throughout the research process. This enables the researcher to reconstruct in detail the process used to develop the final coding framework and subsequent thematic list, and hence addresses the issue of reflexivity. Reflexivity is defined as the impact the position or perspective of the researcher has upon the results. As we discussed earlier this must be clearly addressed and documented. Reflexivity is an essential part of rigour and allows readers of the final research to clearly understand how the results were obtained.

With regard to study participant selection, participants in qualitative studies are often not selected for *representativeness*. Instead they may be deliberately selected based on their difference from other participants, a process known as purposeful selection. Remember that the goal of qualitative research is to explore and ask 'why' questions, and thus the deliberate inclusion of participants holding widely disparate opinions in a way that does not match their population distribution can be of benefit.

Conclusion

In summary, thematic analysis is a commonly used analytic approach. For those new to qualitative research, thematic analysis is a relatively straightforward form of analysis.

As with all forms of research, however, a critical consideration of the research question is necessary before selecting a method. Qualitative research offers an exciting complement to the quantitative studies most typically found in this field, and the methods we describe here offer a means of accessing this rich group of analytic approaches.

Tips for Carrying Out Thematic Analysis

- It is helpful (although not always practicable) to have a diverse range of professional backgrounds within the research team.
- Hone your research question before deciding which method is the most appropriate to answer it.
- Consider transcribing some or all of the data yourself. Although this may be time consuming there is no better way of immersing yourself in the data.
- While some researchers use commercially available qualitative software packages, others use traditional methods such as highlighter pens and post-it notes! Qualitative software packages are particularly suitable for studies that have large datasets, and allow data to be organised and structured. Always remember, however, that software will not actually carry out the analysis, this is up to you, the researcher! The use of the more traditional methods does not reduce the rigour of the analytic approach.
- Don't rush the analysis – it can take some time. Iterative analysis benefits from team meetings over time. Coming back to the data after a break with 'fresh eyes' will aid the process of refining themes and their interactions.
- Consider member checking. This is where study participants read their transcripts and/or derived themes, and ascertain whether they resonate with their experiences.
- You may have to omit some 'minor' themes (or accept that they will be subsumed into other themes). Be prepared to 'kill your darling' codes!
- The development of themes is never totally independent of the researchers. It is thus imperative that researchers maintain an awareness of, and actively reflect on, these influences. Keeping a reflexive diary/notebook and having regular discussions within the research team aids this reflexive awareness and allows the researchers to be cognisant of their position amongst the data.

References

1. Mattick K, Johnston J, de la Croix A. How to...write a good research question. Clin Teach. 2018;15(2):104–8.
2. Tracey S. Qualitative quality: eight "big tent" criteria for excellent qualitative research. Qual Inq. 2010;16(10):837–51.
3. Battista A. Activity theory and analyzing learning in simulations. Simul Gaming. 2015;46(2):187–96.
4. Battista A. An activity theory perspective of how scenario-based simulations support learning: a descriptive analysis. Adv Simul. 2017;2:23.
5. Fenwick T, Dahlgren MA. Towards socio-material approaches in simulation-based education: lessons from complexity theory. Med Educ. 2015;49(4):359–67.
6. Gormley GJ, Fenwick T. Learning to manage complexity through simulation: students' challenges and possible strategies. Perspect Med Educ. 2016;5(3):138–46.
7. Braun V, Clarke V. Successful qualitative research: a practical guide for beginners. London: SAGE; 2013.
8. King N, Brooks J. Template analysis for business and management students. London: SAGE; 2016.
9. King N. Doing template analysis. In: Symon G, Cassell C, editors. Qualitative organizational research: core methods and current challenges. London: SAGE; 2012.
10. Ritchie J, Spencer L. Qualitative data analysis for applied policy research. In: Bryman A, Burgess R, editors. Analysing qualitative data. Routledge: London; 1994.
11. Braun V, Clarke V. Using thematic analysis in psychology. Qual Res Psychol. 2006;3:77–101.
12. Hsieh H, Shannon S. Three approaches to qualitative content analysis. Qual Health Res. 2005;15(9):1277–88.
13. Vaismoradi M, Turenen H, Bondas T. Content analysis and thematic analysis: implications for conducting a qualitative descriptive study. Nurs Health Sci. 2013;15:398–405.
14. Corr M, et al. Living with 'melanoma' ... for a day: a phenomenological analysis of medical students' simulated experiences. Br J Dermatol. 2017;177(3):771–8.
15. Ross AJ, et al. Review of simulation studies in anaesthesia journals, 2001–2010: mapping and content analysis. Br J Anaesth. 2012;109(1):99–109.

Additional Resources

Tong A, Sainsbury P, Craig J. Consolidated criteria for reporting qualitative research (COREQ)' Consolidated criteria for reporting qualitative research (COREQ): a 32-item checklist for interviews and focus groups. Int J Qual Health Care. 2007;19(6):349–57.

King N. Doing template analysis. In: Symon G, Cassell C, editors. Qualitative organizational research: core methods and current challenges. London: Sage; 2012.

Braun V, Clarke V. Using thematic analysis in psychology. Qual Res Psychol. 2006;3:258–67.

Ritchie J, Spencer L. Qualitative data analysis for applied policy research. In: Bryman A, Burgess RG, editors. Analysing qualitative data. London: Routledge; 1994.

Thematic analysis website: lhttps://www.psych.auckland.ac.nz/en/about/our-research/research-groups/thematic-analysis.html

Naturally Occurring Data: Conversation, Discourse, and Hermeneutic Analysis

20

Lisa McKenna, Jill Stow, and Karen Livesay

Overview

Simulation provides unique and individual learning experiences. Research methods that assist in understanding such experiences can be particularly beneficial for those developing simulation curricula, as well as understanding the student experience. Conversation analysis reveals subtle nuances which govern how people use language to interact. Discourse analysis facilitates understanding of why people act and respond in the ways that they do, particularly focused on how power and knowledge operate. Hermeneutic analysis allows understanding of the lived experiences of individuals in different contexts. These approaches are all underpinned by the view that there are multiple legitimate truths or perspectives in any given situation.

Practice Points

- Conversation, discourse and hermeneutic analysis provide rich approaches to interpreting naturally occurring data.
- Conversation, discourse and hermeneutic analysis recognise the multiplicity of individual experience and existence of multiple *truths*.
- Conversation analysis provides an explicit structure to analyse conversational data.
- Discourse analysis offers approaches for understanding the reasons why people act and respond in the ways they do.
- Hermeneutic analysis facilitates understanding the lived experiences of individuals.

L. McKenna (✉)
School of Nursing and Midwifery, La Trobe University, Melbourne, VIC, Australia
e-mail: l.mckenna@latrobe.edu.au

J. Stow
School of Nursing and Midwifery, Monash University, Melbourne, VIC, Australia
e-mail: jill.stow@monash.edu

K. Livesay
School of Health and Biomedical Sciences, RMIT University, Melbourne, VIC, Australia
e-mail: Karen.Livesay@rmit.edu.au

Introduction

Simulation provides unique and individual learning experiences. Research methods that assist in understanding such experiences can be particularly beneficial for those developing simulation curricula, as well as understanding the student experience. This chapter provides an overview of three particularly useful approaches: conversation analysis, discourse analysis and hermeneutic analysis. The chapter presents an overview of fundamental underpinning theoretical and philosophical perspectives, examples of their previous use in healthcare research, and discussions around potential applications of these approaches to healthcare simulation research.

Conversation Analysis (CA)

Understanding the complex rules governing how people communicate is important if we are to understand how communication impacts work and social interactions. Sociologist Harvey Sacks pioneered the empirical study of naturally occurring conversation when he developed the first rules of conversational sequence from audiotaped calls to a suicide help line. CA as a method developed later as a collaboration between Sacks, Emmanuel Schlegloff and Gail Jefferson [1]. Conversation analysts work with video and audio recordings analysing segments of data, playing and re-playing the interaction. Repeated listening using 'unmotivated looking' allows the researcher to discover what is going on in the data, rather than searching the data for premeditated themes [2].

D. Nestel et al. (eds.), *Healthcare Simulation Research*, https://doi.org/10.1007/978-3-030-26837-4_20

Underpinning Theories/Philosophical Perspectives of Conversation Analysis

CA is grounded in Ethnomethodology, the study of the methods people use to establish and share meaning to produce everyday social order [3]. CA differed from discourse analysis and other sociological approaches of the time because it did not offer an analysis or description of the social setting, gender or other hierarchies, nor did it attribute beliefs and desires to the participants, but focussed on how the mechanism of spoken language produced social order [2, 4]. CA lies at a unique interface in the social sciences, between sociology, linguistics, communication and social psychology [5] revealing how people use language to express subtle differences in social interactions which have social consequences [6]. A fundamental principle of CA is talk-in-interaction or the norms of turn-taking in conversation [5].

The aims of CA research can be summarised by the question "why that now?" CA researchers analyse interactions strictly with reference to the observable behaviour of participants, using highly detailed transcripts to interrogate and describe naturalistic interaction (See Chap. 17). Their primary goal is to identify interaction actions (for example, asking, telling, introducing, announcing) and methodically describe how they are accomplished [7, 8]. Analysis is based on transcribed recording of actual interactions. However, transcripts are still subjective representations of the original data. The transcriber makes decisions about which features of talk, in addition to dialogue, to include. Capturing the minutiae of conversation is achieved by means of Jefferson's unique contribution to CA, the development of methods and extensive notational conventions for transcription analysis (see glossary of transcript signals with introduction) [9] where punctuation symbols show stress, intonation, changes in volume and sound length. To illustrate, a short extract from an original Sacks transcript is compared with a later version fully notated by Jefferson (see Box 20.1) [9].

Box 20.1 Short extract from original sacks transcript compared with a later version by Jefferson

[Sacks GTS trans:1964]

A. I started work at a buck thirty an hour and he said if I work a month you get a buck thirty five an hour and every month there would be a raise-

T. Howd you get the job?

A. I just went down there and asked him for it

[Jefferson, GTS:1;2;3;R;1-5:3-4]

Ken: I started workin etta buck thirty en hour

(0.4)

Ken: en'e sid that if I work fer a month: yih getta buck,h h thi[rty↓fi:ve=

(Dan): [((sniff))

Ken: =>n hour en(.) ev>ry month he uh (·) he rai[ses you] °(·) °]

Dan: [How'dju]g e t th]e jo:b,

Ken: ↑ I js wen' down ther'n ↓a:st eem for it

Analysis of the notated data begins when the researcher notices distinctive social interactions, language and behaviours and then, finding other instances, begins to map out the boundaries of the phenomena. For example, periods of silence and token acknowledgements such as "mm" or "yeah" in a transcript can reflect passive resistance to an idea or instruction. As more examples are collected the analyst is able to define the phenomenon's generic, context independent, properties and apply them to other contexts [1, 10].

How Has Conversation Analysis Been Used in Healthcare Research?

Whilst CA has had a wide ranging impact on social science research it has only relatively recently been adopted by healthcare researchers. Gafaranga and Britten [11] used CA to analyse the opening sequence of General Practitioners' consultations with patients. Their findings challenged the then conventional understandings of the impact of professional language on patient disclosure [11]. Bezemer and colleagues [12, 13] used CA to study specialised forms of talk, the nature of institutions and organisations with a focus on producing detailed analyses of how health professionals interact. Their work provides insights into the development of the surgeon's role, interprofessional teamwork and how order is established and maintained in the operating theatre.

How Can Conversation Analysis Be Applied to Healthcare Simulation Research?

Kendrick [14] asserts that whilst CA is grounded in close observation and inductive generalisations it is time to extend CA research from naturalistic observation to experimental and laboratory studies. The work of Bezemer and colleagues above shows how CA can be used to not only inform the organisation of work practices; but also the design of intra- and interprofessional simulations. Recently Johansson, Lindwall and Rystedt [15] used CA to investigate the use of video in debriefing medical and nursing students following handover communication simulations. Debriefing video data revealed students were able to access a third person perspective of their performance which facilitated reflection on learning.

Discourse Analysis

Where CA allows us to explore the use of conversation and interaction between people through the use of language, discourse analysis allows development of understanding how and why people act in certain ways, and hence may offer unique opportunities for researching simulation in healthcare.

The term 'discourse' is used in different ways, depending on who uses it. For a linguist, discourse examines the sequencing and structure of language, or the way in which speech is used [16]. However, social researchers may look at discourse as relating to social and cultural influences and practices. From that perspective, discourses are seen to shape what we do and the world in which we function. While language is one aspect of that, this view extends to the way in which people interact and present themselves, as well as the clothing they wear, the gestures they use and their attitudes. "Discourses, then, involve the *socially situated identities* that we enact and recognize in the different settings that we interact in." ([17], p.10).

Underpinning Theories/Philosophical Perspectives of Discourse Analysis

Discourse analysis involves examining relationships and how social and cultural factors work to shape and influence them. It contends that there are multiple discourses in action at any one time, hence advocating the existence of multiple views or *truths* operating. While there are many different ways of performing discourse analysis, one approach is Critical Discourse Analysis (CDA). Fairclough [18] outlines that there are three key properties that constitute CDA, that is, it is *relational, dialectical* and *transdisciplinary.* Firstly, relational refers to the analysis focusing on social relations within interactions. Secondly, dialectical explores relations between objects. Finally, transdisciplinary refers the range of different influences, such as economic, political, cultural, and educational, all at play within a context. In another model, Foucauldian discourse analysis (FDA), places emphasis on power and knowledge in relationships, and asking *how* and *why* questions about interactions [19]. Hence, from the discourse analysis research perspective, the researcher is analysing the factors that influence interactions and behaviours and how these operate in situations. Data can be sourced from multiple origins and may be in a range of formats such as interview transcripts, video-recordings, documents, letters or even photographs.

How Has Discourse Analysis Been Used in Healthcare Research?

Discourse analysis has been used across healthcare in a range of ways. Wright et al. [20] conducted a discourse analysis of Canadian newspapers to examine physicians' perspectives on end-of-life care, identifying three predominant discourses: contentions with integrating euthanasia into medicine, whether euthanasia could be distinguished from end-of-life care, and advocacy for palliative care. They concluded it was important for physicians to be aware of how they were portrayed within euthanasia debates and their impact on public views.

In another study, Paz-Lourido and Kuisma [21] interviewed GPs working in primary healthcare to understand existing poor collaboration with physiotherapists. Using discourse analysis, they were able to conclude that poor communication was the result of lack of GP knowledge about physiotherapists and lack of exposure to, and resources to support, interprofessional education. Haddara and Lingard [22] employed discourse analysis to examine in published literature whether there was a shared discourse around interprofessional collaboration (IPC). Their review found at least two different IPC discourses operating, a utilitarian discourse implying that IPC produced better patient outcomes, and an emancipatory discourse implying it was needed to reduce medical dominance. The researchers argued that tensions between these were probably responsible for challenges experienced by educators in successfully operationalising IPC. Being able to manage emerging tensions was viewed as enabling successful IPC implementation.

How Can Discourse Analysis Be Applied to Healthcare Simulation Research?

Discourse analysis offers new and unique perspectives on healthcare simulation research. To date, there have been few studies to employ the approach in such research. In one related study, de la Croix and Skelton [23] examined conversational dominance in consultations between simulated patients (SPs) and third-year medical students. They recorded, transcribed and analysed the conversations, finding that the SPs dominated the interactions which was different to traditional doctor-patient consultations. Despite this, they argued that realism of interactions was the key outcome of such simulations.

Simulation often involves complex, multiprofessional situations. Discourse analysis has the potential to provide important understandings of interactions between different health professionals, such as power relations and how they operate within a range of situations, in both positive and negative ways in simulated settings. Such understandings can enhance preparation of health professionals to work in interprofessional contexts. Discourse analysis also offers opportunities to examine relations between healthcare providers and consumers, particularly in the context of cultural and medical discourses and how they influence actions and interactions. Such learning can directly inform health professionals' skills in effective client communication and empathy development.

Hermeneutic Analysis

Consider a research interview or a narrative derived from an interview. In the narrative, the teller shares their lived experience. This sharing occurs through language. In the narrative example below, Hwei-ru speaks in English but this is not her first language. Her language is transcribed into text and from this, the reader draws meaning through a process of interpretation. Interpretation is the hinge between language and lived experience [24]. To what extent are our interpretations of the text impacted by history, culture, time or linguistic traditions? Text interpretation ordinarily includes inferences as a function of sense making but the hermeneutical position extends to the ontological view that lived experience is an interpretive process. Hermeneutical analysis systematizes the process of interpretation (see Box 20.2) [25].

Box 20.2 Hwei-ru narrative

If young people can open their minds, the world is nice. You know, as a simulated patient that I have told the student that I go to the temple. You know someone believe as a Buddha, someone believe as a, the ones that go to church, but they can't. They have to do it very, very be careful in China back then. They believe in the Buddha only at home. At home and they have to close the door, close the window and everything, don't like the neighbour knows. Well the people, you know the government is very strong. The people only do something under the table. They have to, they kind of angry or something, you know, if somewhere else knows this, maybe tell the government, they having the trouble. I don't believe in the Buddha, so no trouble with me, but I have some family member, they believe in the Buddha. Yeah, they only do something at the night. They do not talk about it you know I am a family member so I know what is happening. You kind of believe and it is as a revolution, yeah, cultural revolution.

Hermeneutic philosophy is the study of interpretive understanding or meaning. Hermeneutic is derived from a Greek word and was originally applied to theological stories to ascertain their "true meaning". Over time, the field has widened to explore understanding of humans in the context of life.

The concept of understanding in hermeneutical analysis embraces duality and subjectivity born of culture, prejudice, tradition and time [26]. The interpretation of text, results from reflection and self-examination by the researcher of their own situatedness and personal perspectives. In an interchange of ideas between researcher and subject (often text), the meaning is derived from understanding the constituent parts while, simultaneously, the parts are understood only by understanding the whole.

The researcher understands that their background, method, purpose and experience will influence their interpretation. The sociocultural and historic influences of interpretation are emphasized, so that another researcher would potentially develop different understandings that focus on alternate aspects and result in somewhat different scenarios. Interpretation is therefore understood to be an understanding reached at a point in time. A different understanding may be achieved at a different time. "Interpretation is always incomplete, perspectival and changing" ([27], p.116).

Underpinning Theories/Philosophical Perspectives of Hermeneutic Analysis

Phenomenology and hermeneutics, while closely associated approaches, can be differentiated through their philosophical bases. Both approaches seek to express knowledge embedded in a context and are sometimes referred to interchangeably. "Phenomenology focuses on a person's lived experience and elicits commonalities and shared meanings, whereas hermeneutics refers to an interpretation of textual language" ([28], p.968). A Phenomenological analysis of Hwei-ru's narrative may focus on the importance of religious observance whereas hermeneutical analysis may expose the power dynamic experienced in China or issues of familial trust in Chinese culture. Hermeneutic analysis is phenomenological to the extent that it does disclose phenomena.

Hermeneutic theory can be divided into four groups according to methodological approaches: Objective Hermeneutics, Hermeneutic Inquiry, Philosophical Hermeneutics and Critical Hermeneutics. It is vital to clarify the approach used due to their differences.

- Objective Hermeneutics, supported by the work of Husserl, is associated with a positivist stance due to the process of "bracketing" [29], a process whereby the researcher attempts to suspend his or her own biases and beliefs (naïve awareness) before data collection [30]. However, many people have contested this notion of objectivity as unachievable.

- Hermeneutic phenomenology was associated with the work of Martin Heidegger, a student of Husserl. Heidegger's approach moved from the primacy of epistemological emphasis (knowledge) to an ontological foundation of understanding or 'verstehen' achieved through 'Being in the world' [30]. The German word 'dasein' (for which there is no exact translation in English) describes a concept of human action and everyday life "Being in the world" and addresses awareness of the co-construction of understanding from within existence, rather than while detached from it.
- Philosophical Hermeneutics is a branch of hermeneutic phenomenology but applies specifically to the work of Gadamer [29]. Building on the work of his predecessors, Heidegger, Dilthey and Schleiermacher, Gadamer refined and championed the hermeneutic circle representing the movement between parts and whole of the text while seeking understanding. Rather than methodological, this cycle represents a manifestation of the worldview we recognise interpreted through particular instances but with steady reference to the worldview that produced it.

Gadamer also offered what he called 'fusion of horizons' to explain the process of looking beyond the immediate to see a larger whole. Gadamer further explained prejudice, history culture and bias as essential components of our personal horizon operating within a present horizon that moves and is reformed through self-reflection and self-awareness. Awareness of this personal horizon enables the researcher to discover unique understanding against his or her own fore-meaning. This negotiation with our prejudice, history and culture enables us to engage with the unfamiliar. The horizon of the researcher and text studied combine or fuse as a merging of perspectives when new understanding is achieved in a dialogical process. Understanding incorporates both researcher and participants (text) without either owning the perspective or having supremacy, as fusion is a function of the individual fore-meaning [26, 29, 30].

Critical hermeneutics regards knowledge as active and influenced by socio-political context. Extending the influence of history, culture and personal stance on an individual's understanding, critical hermeneutics suggests active engagement to question those influences and present alternate meaning that the individual may not perceive themselves.

How Has Hermeneutic Analysis Been Used in Healthcare Research?

Hermeneutic analysis is classified as a constructivist or interpretive paradigm suitable to research, evaluation and policy analysis [31]. The aims include to identify and support understanding of phenomena and to bring them into agreement. Researchers using hermeneutic analysis source most data from transcribed interviews. Participants select the logic and order of language to convey or veil manifestations within their narrative [32]. Researchers need to contemplate the multiple interpretations possible to understand lived experience.

In healthcare, hermeneutic analysis is eminently suitable to focus on experiences of patients and healthcare providers attempting to achieve health or accept illness in the myriad of environments in which modern healthcare occurs. Understanding lived experiences of patients, families or caregivers by answering questions of what and how about human issues is particularly suited to hermeneutic analysis. Health researchers frequently adopt variations in hermeneutical phenomenology. Dowling [33] provides an interesting critique of papers claiming to have used a hermeneutical approach, demonstrating the lack of philosophical basis in the methodology of many studies, despite reference to the influence of hermeneutic phenomenology as well as examples of good work. It is recognised that convergence of philosophical approach and study design are imperative.

How Can Hermeneutic Analysis Be Applied to Healthcare Simulation Research?

Benefits of hermeneutic research in healthcare simulation are multiple. These include the impact of lived experience and it's bearing on learning and understanding. This is particularly important for the practitioner who presents in a simulation scenario with preconceptions, history and culture commensurate with being an adult. In this manner, Pollock and Biles [34] explored what it was to be a learner in a simulation through understanding subjective experience.

From a patient's perspective, hermeneutic inquiry foregrounds their voice, enables consideration of multiple perspectives in a manner that remains patient centric through interplay of the hermeneutic cycle and fusion of horizon. In simulated participant-based research, the characterization is constructed upon understanding of unique and diverse perspectives. Through this style of research authentic understanding of ways of 'being' can be developed. Knowledge generated from practice is useful in simulation-based education to capture the complex and situational aspects of patient or simulated characters.

Nursing research dominates hermeneutical studies in simulation similar to those found in healthcare more generally. An interesting study by Lejonqvist, Eriksson and Meretoja [35] explored expressions of clinical competence in simulation-based learning. This study is noteworthy for adaptation to a scenario as the data source rather than text.

Fore-meaning in this study was informed by earlier research to identify the appearance of clinical competence. Finally, the video-recording of simulations permitted the hermeneutic cycle enactment until all was "unfolded".

Conclusion

Simulation offers unique, but complex individual healthcare learning experiences. Research methods facilitating understanding of these complex and multifaceted experiences can be particularly beneficial for simulation curriculum development, and understanding unique participant and student experiences. Conversation analysis provides a structured methodology for analysing participant interactions including comprehensive, standardised symbols for notating transcripts. Discourse analysis can facilitate understanding why people act and respond in the ways that they do, and how power and knowledge operate within situations. Hermeneutic analysis allows for understanding of lived experiences of individuals in varied simulation contexts. Such approaches acknowledge the existence of multiple truths or perspectives in any given situation, making them optimal in simulation research.

References

1. Sidnell J. Conversation analysis: an introduction. Singapore: Wiley-Blackwell; 2010. p. 1–12.
2. Liddicoate AJ. An introduction to conversation analysis. Tyne & Wear, GB: Athenaeum Press; 2007. p. 1–12.
3. Garfinkle H. Studies in ethnomethodology. New Jersey: Prentice Hall Inc; 1967.
4. Lee JR. Prologue: talking organisation. In: Button G, Lee JR, editors. Talk and social organisation. Philadelphia: Multilingual Matters; 1987. p. 19–53.
5. Hutchby I, Woofitt R. Conversation analysis. 2nd ed. Cambridge: Polity Press; 2008. p. 30–2.
6. Seedhouse P. Conversation analysis as research methodology. In: Richards K, Seedhouse P, editors. Applying conversation analysis. Hampshire: Palgrave Macmillan; 2005. p. 51–266.
7. Shegloff EA, Sacks H. Opening up closings. Semiotica. 1973;8(4):289–37.
8. de Ruiter JP, Albert S. An appeal for a methodological fusion of conversation analysis and experimental psychology. Res Lang Soc Interact 2017;50(1):90–197. Available from: https://doi.org/10.1080/08351813.2017.1262050.
9. Jefferson G. Glossary of transcript symbols with an introduction. In: Lerner G, editor. Conversation analysis: studies from the first generation. Amsterdam: John Benjamin; 2004. p. 13–31.
10. Robson JD, Heritage J. Intervening with conversation analysis: the case of medicine. Res Lang Soc Interact. 2014;47(3):201–18. https://doi.org/10.1080/08351813.2014.925658.
11. Gafaranga J, Britton N. Talking an institution into being: the opening sequence in general practice conversations. In: Richards K, Seedhouse P, editors. Applying conversation analysis. Hampshire: Palgrave Macmillan; 2005. p. 75–90.
12. Bezemer J, Murtagh G, Cope A, Kress K, Kneebone R. The practical accomplishments of surgical work in the operating theatre. Symb Interact. 2011;34(3):398–414.
13. Bezemer J, Korkiakangas T, Weldon S, Kress G, Kneebone R. Unsettled teamwork: communication and learning in the operating theatres of an urban hospital. J Adv Nurs. 2015;72(2):361–72.
14. Kendrick K. Using conversation analysis in the lab. Res Lang Soc Interact. 2017;50(1):1–11.
15. Johansson E, Lindwall O, Rusteldt H. Experiences, appearances and interprofessional training: the instructional use of video in post-simulation debriefings. Int J Comput Support Collab Learn. 2017;12:91–112. https://doi.org/10.1007/s11412-017-9252-z.
16. Gee JP. An introduction to discourse analysis: theory and method. 4th ed. New York: Routledge; 2014.
17. Paltridge B. Discourse analysis: an introduction. 2nd ed. London: Bloomsbury; 2012.
18. Fairclough N. Critical discourse analysis: the critical study of language. 2nd ed. New York: Taylor & Francis; 2013.
19. Springer RA, Clinton ME. Doing Foucault: inquiring into nursing knowledge with Foucauldian discourse analysis. Nurs Philos. 2015;16:87–97.
20. Wright DK, Fishman JR, Hadi Karsoho BA, Sandham MA, Macdonald ME. Physicians and euthanasia: a Canadian print-media discourse analysis of physician perspectives. CMAJ Open. 2015;3(2) https://doi.org/10.9778/cmajo.20140071.
21. Paz-Lourido B, Kuisma RME. General practitioners' perspectives of education and collaboration with physiotherapists in primary healthcare: a discourse analysis. J Interprof Care. 2013;27(3):254–60.
22. Haddara W, Lingard L. Are we all on the same page? A discourse analysis of interprofessional collaboration. Acad Med. 2013;88(10):1509–15.
23. de la Croix A, Skelton J. The simulation game: an analysis of interactions between students and simulated patients. Med Educ. 2013;47:49–58.
24. Ricoeur P. The conflict of interpretations: Essays in Hermeneutics. Ihde D (Ed). A&C Black; London 2004. p. 61–77.
25. Livesay K. Culturally and linguistically diverse simulated patients: otherness and intersectional identity transformations revealed through narrative (Doctoral dissertation, Victoria University, 2016).
26. Wernet A. Hermeneutics and objective hermeneutics. The SAGE handbook of qualitative data analysis. Metzier (Ed). Sage Publications, London 2014:234–246.
27. Geanellos R. Exploring Ricoeur's hermeneutic theory of interpretation as a method of analysing research texts. Nurs Inq. 2000;7(2):112–9.
28. Byrne M. Hermeneutics as a methodology for textual analysis. AORN J. 2001;73(5):968–70.
29. Dowling M. Hermeneutics: an exploration. Nurse Res. 2004;11(4):30–9.
30. Annells M. Hermeneutic phenomenology: philosophical perspectives and current use in nursing research. J Adv Nurs. 1996;23(4):705–13.
31. Lincoln YS, Guba EG. The constructivist credo. Walnut Creek: Left Coast Press; 2013.
32. Ho KH, Chiang VC, Leung D. Hermeneutic phenomenological analysis: the 'possibility' beyond 'actuality'in thematic analysis. J Adv Nurs. 2017;73(7):1757–66.
33. Dowling M. From Husserl to van Manen. A review of different phenomenological approaches. Int J Nurs Stud. 2007;44(1):131–42.
34. Pollock C, Biles J. Discovering the lived experience of students learning in immersive simulation. Clin Simul Nurs. 2016;12(8):313–9.
35. Lejonqvist GB, Eriksson K, Meretoja R. Evidence of clinical competence by simulation, a hermeneutical observational study. Nurse Educ Today. 2016;38:88–92.

Part IV

Quantitative Approaches in Healthcare Simulation Research

Quantitative Research in Healthcare Simulation: An Introduction and Discussion of Common Pitfalls

21

Aaron W. Calhoun, Joshua Hui, and Mark W. Scerbo

Overview

In contrast to qualitative research, quantitative research focuses primarily on the testing of hypotheses using variables that are measured numerically and analyzed using statistical procedures. If appropriately designed, quantitative approaches provide the ability to establish causal relationships between variables. Hypothesis testing is a critical component of quantitative methods, and requires appropriately framed research questions, knowledge of the appropriate literature, and guidance from relevant theoretical frameworks. Within the field of simulation, two broad categories of quantitative research exist: studies that investigate the use of simulation as a variable and studies using simulation to investigate other questions and issues. In this chapter we review common study designs and introduce some key concepts pertaining to measurement and statistical analysis. We conclude the chapter with a survey of common errors in quantitative study design and implementation.

Practice Points

- Quantitative research uses variables that can be measured numerically to test hypotheses.
- Quantitative and qualitative methods form a natural continuum in simulation and educational research.
- True experimental studies, which include a randomized control group, are the most rigorous quantitative designs.
- Hypothesis testing proceeds via a process of statistical inference in which the null hypothesis (i.e. no difference or effect between variables) is rejected and the experimental hypothesis is accepted.
- A number of common errors exist in quantitative research that can be easily avoided if the researcher is aware.

A. W. Calhoun (✉)
Department of Pediatrics, University of Louisville School of Medicine, Louisville, KY, USA
e-mail: aaron.calhoun@louisville.edu

J. Hui
Emergency Medicine, Kaiser Permanente, Los Angeles Medical Center, Los Angeles, CA, USA

M. W. Scerbo
Department of Psychology, Old Dominion University, Norfolk, VA, USA
e-mail: mscerbo@odu.edu

Quantitative research methods focus on variables that can be measured numerically and analyzed using statistical techniques [1–3]. The primary advantage of this approach is its ability to investigate relationships between variables within a sample that can then be inferred to a larger population of interest while more consistently controlling for threats to validity. In most quantitative research involving simulation-based interventions the population consists of the global set of learners and/or practitioners, and the outcome in question is typically a change in the knowledge, skills or attitudes or an alteration in patient outcomes. For example, Auerbach et al compared lumbar puncture (LP) skill among pediatric and emergency medicine interns exposed to a single procedural simulation-based mastery learning course versus the mastery learning course plus an additional just-in-time simulation prior to their first clinical LP [4]. Here the variable of interest is the presence (or absence) of the extra just-in-time simulation, and the primary outcome of interest is the success of the learner at performing their first lumbar puncture (skill) as well as several secondary process measures related to specific content (knowledge) addressed by the intervention. It is important to note that an increasing number of quantitative studies are also using simulation-based modalities as an evaluative mechanism. A recent study examining the effect of a novel cellular phone application on prehospital providers'

D. Nestel et al. (eds.), *Healthcare Simulation Research*, https://doi.org/10.1007/978-3-030-26837-4_21

communication skills showcases this approach. In this study subject communication skills were assessed during a scripted interaction with simulated parents, and evaluations completed during this interaction were used as primary outcome data [5].

The goal of this chapter is to provide a practical foundation for subsequent chapters by linking their contents to issues that commonly arise in quantitative simulation research. The chapter begins with an exploration of the relationship between research questions, theories, and hypotheses, and proceeds from this to a brief review of quantitative study designs, outcomes and assessment, and key components of statistical inference. We close with a review of errors commonly seen in quantitative research. Each of these subsections is linked when appropriate to other chapters that expand upon the ideas raised here.

Defining the Hypothesis: The Role of Frameworks in Quantitative Research

Previous chapters have explored the topic of qualitative research in great depth and we will not reiterate their contents in depth here. As a brief summary, qualitative approaches are usually oriented toward the *synthesis* or *creation* of theories or frameworks and *suggest* hypotheses [6]. In contrast, quantitative approaches typically *begin* with a hypothesis that makes a specific prediction regarding an outcome and then formally tests that hypothesis. Before beginning a quantitative study of any sort it is vital to perform a thorough literature review. Specific attention should be paid to the current state of the research and the presence (or lack thereof) of conceptual or theoretical frameworks that may affect predictions regarding the outcome and/or suggest potential explanatory mechanisms [7, 8]. Without such theoretical work it becomes more difficult to understand what the results of the study may mean, regardless of how well-crafted the study may be. Unfortunately, it is relatively common for researchers in simulation to begin with a question or hypothesis that was formed without the guidance of a thorough literature review or adequate supporting theory [9]. While there may be many reasons behind this, in our experience this often springs from a simple lack of familiarity with the role of theory in framing research questions and/or the most relevant theories to bring to bear on a given hypothesis. It is thus critical that the initial literature review specifically includes other relevant theoretical research or prior qualitative studies of relevance so that potentially applicable theoretical frameworks can be identified early in the research process.

By contrast, what about those simulation-based research questions for which little has been written? It is important to note that relevant theories can often be found outside of an investigator's immediate areas of expertise, and there is a wealth of educational models to which investigators can appeal. The members of the healthcare simulation community come from a myriad of professional backgrounds and fields of inquiry, and can serve as a valuable resource when exploring this broader literature. If no appropriate theory is found, then some degree of descriptive or qualitative theoretical work may be needed to define the context of the overall research question and generate an applicable framework [10–13].

Types of Quantitative Research: Common Study Designs

A number of research designs fall under the quantitative category including descriptive, correlational, quasi-experimental, and experimental studies [14]. Unlike most quantitative designs, descriptive studies do not attempt to infer relationship among variables, but instead are used to create a detailed picture of the current state of the issue or question under study [15]. As an example, consider a study describing communication skills in a series of simulated cardiac arrests. The variables measured are not compared with any other learner characteristics but are presented simply as a distribution with the goal of illustrating the patterns present among the learners. Similar to qualitative approaches, descriptive studies can assist researchers in better defining a research problem or question when the outcome data are collected in a systematic fashion. For example, consider a study in which residency programs are surveyed regarding their use of simulation to train learners in the technical and communication skills associated with successful resuscitation. While the results will not allow researchers to address correlation or causation, they can form the basis of more focused interventions and research questions.

Correlational studies go a step beyond and attempt to elucidate potential relationships between a set of variables. These studies can shape ideas and hypotheses by drawing attention to potentially interesting connections, but *cannot* meaningfully comment on whether the relationship in question is causal in nature. Consider, for example, a study of learners participating in simulation of cardiac arrests during which both technical skills and communication skills were assessed. A correlational study could assess whether technical skill variables such as maintenance of high-quality chest compressions had any relationship with nontechnical skills variables such as assessments of team fixation (i.e., the failure to revise current diagnoses and/or plans despite additional evidence) [16]. Note that in this case it is impossible to determine whether skill evidenced in one of these variables was the source of skill in another. Instead, the most that can be inferred is that they appeared to be related in some way. One could, however, use this information to form a hypothesis regarding the possibility of a causal relationship that would then be amenable to future experimental testing.

Quasi-experimental studies take this process yet another step further by looking at the effect of the presence or absence of one variable on another using a non-randomized comparator group of some sort. To continue with the previous example, imagine that an inverse relationship was detected between number of fixation events and chest compression efficacy and that the research team subsequently hypothesized that team fixation exerts a causative, inhibitory effect on chest compression performance. Here we finally have both independent (number of fixation events) and dependent (effectiveness of chest compressions) variables. An independent variable is defined as a variable that is *actively manipulated or intervened upon* by a researcher to cause a predicted effect while a dependent variable is defined as a measured variable whose value is *predicted* by the independent variable. To test this hypothesis, the researchers design a subsequent study in which an assessment of chest compression effectiveness is conducted before and after an intervention intended to diminish fixation events and compared to chest compression effectiveness measured in a non-random convenience sample of learners who did not receive the intervention. Should post-intervention improvements be found, the researchers can now make stronger arguments about potential causation. As stated above, in quasi-experimental studies subjects are not randomly assigned to intervention and comparison conditions. The overall strength of their conclusions are therefore limited as confounding variables (defined as variables that obscure the effect of the independent variable) often cannot be completely addressed [3]. Many studies of simulation as an educational intervention take this form, and while this approach formed an important part of the growth of our field, more rigorous approaches are now needed and preferred [17].

The final type of quantitative study design is experimental. Such a study retains the temporal flow of the quasi-experimental study while assigning participants at random to the intervention and control groups. This control group, if similar to the interventional (experimental) group in terms of relevant demographic variables, gives the researcher added confidence that differences observed among participants are not confounded with group assignment. This in turn further strengthens the case for causality should a differences of outcomes be found between the two groups of subjects. To continue with the example presented above, if the comparator group is transformed into a true control group that is selected in a randomized fashion and receives the same pre and post chest compression effectiveness assessments, but does not take part in the intervention, it is now a true experimental study. By eliminating potential confounding variables, the randomization process allows the researcher to make far more confident claims regarding causal effects attributable to the intervention. Chapters 23 and 24 provide a thorough discussion of study design and outcome variable selection.

Measuring Outcomes: An Introduction to Assessment

Medical education research is somewhat different from biomedical research as many of the variables that we wish to examine quantitatively are not as well defined. While physical quantities such as temperature and blood pressure are easily represented on a numeric scale, theoretical constructs such as leadership are more difficult to represent (and hence to measure) in this format (see for example, Chap. 22). Thus, quantitative simulation studies in education often employ *assessment tools or instruments:* collections of written items that describe individual variables and allow for the assignment of numeric values to levels of subject performance. There are many types and styles of assessment tools, and Chap. 25 is devoted entirely to this issue [18, 19].

If a given instrument is to be useful in evaluating a particular variable it must possess sufficient *validity* to be used for research purposes. In current theory, validity is not considered to be the property of a specific tool, but refers instead to the relationship between a particular decision of interest and the scores generated by that tool in a specific population and environment [18, 20, 21]. Consider a study attempting to assess the effect of a simulation-based teamwork intervention on battlefield medical crises. Many good tools addressing general teamwork skills exist, but unless they are specifically keyed to the unique situations that arise on the battlefield, the instrument may not capture important details due to its lack of sensitivity and validity in this context. This aspect of validity, which addresses the degree to which the items on the tool correspond to the conceptual framework or model that undergirds a study, is termed *content validity*. Another aspect of validity, termed *internal structure*, concerns the generalizability of the scores that the tool generates during use, and embraces common psychometrics such as inter-rater, internal consistency, and test-retest reliability. To declare a tool valid for a specific decision, various streams of evidence such as those mentioned above must be woven together to create an argument supporting its use. A number of accepted frameworks exist to support this process [20–22]. Determining a tool's validity for a given research study can be complex, and it is often better to find and utilize a previously developed tool that has been shown to be valid in the same or similar situations. If no such tool exists, however, the researcher may then be obliged to develop a new tool to measure the outcome variable of interest. Such a self-developed tool must then be validated prior to use, preferably in a pilot study. Without this step the study outcomes can be called into question. Figure 21.1 depicts the overall relationship between construct creation, assessment tool creation and validation, and a resulting interventional study. Validity will be covered in greater depth in Chap. 26.

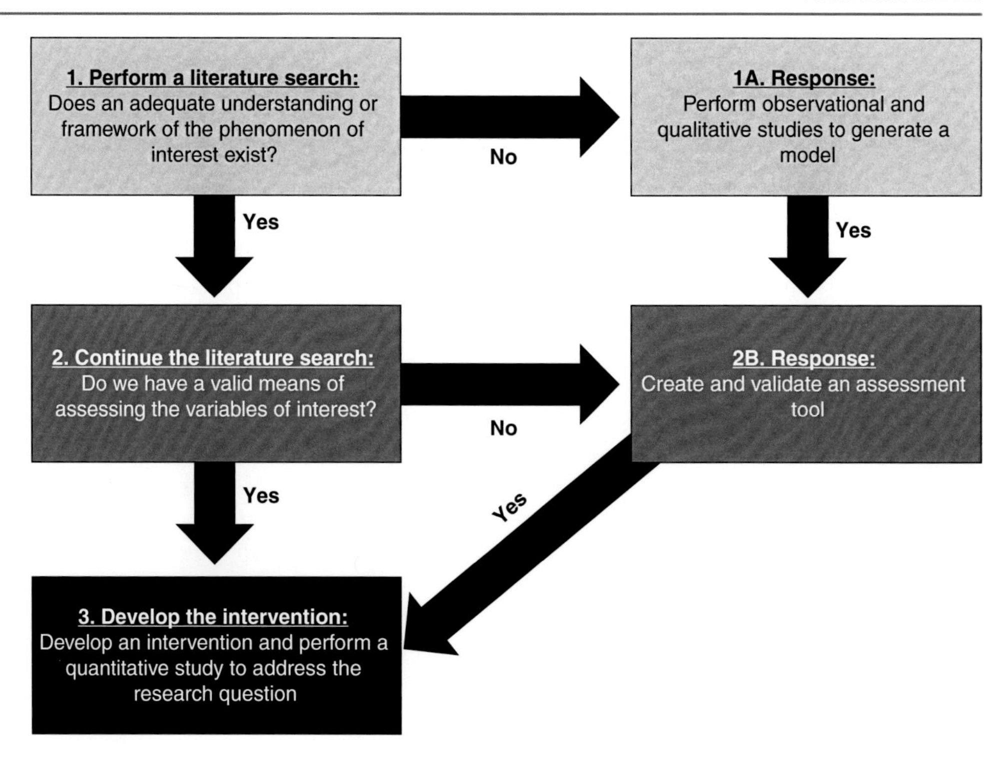

Fig. 21.1 Simplified schematic of quantitative study development. This flowchart provides a simplified pathway through the development of a quantitative interventional study beginning with the initial literature search. At each phase, questions are posed regarding necessary questions that must be answered prior to the creation of the intervention. If, at any point, important elements cannot be located, it becomes necessary to consider whether additional preliminary work is required

Basic Principles of Statistical Inference: A Review

With the exception of descriptive studies (discussed above), most quantitative studies evaluate hypotheses that postulate a specific relationship between an independent variable and an outcome/dependent variable of interest at the level of the population. Since we cannot evaluate all members of our population of interest due to its size, we instead study the effects of these variables among a smaller sample drawn from that population assuming it is representative enough of the population of interest. This raises the question of how effects seen at the level of the sample can be reliably generalized or inferred to the level of the population [23]. This is done via the use of statistical inference. While Chaps. 27, 28, 29, and 30 address these issues in depth, we offer an introductory consideration of several aspects of statistical inference that are commonly misunderstood by novice simulation researchers.

First, it is important to understand the concept of *null hypothesis testing*. Simply stated, the null hypothesis is a statement that there is no relationship between variables, and the purpose of the experiment is to gather sufficient data to reject this null hypothesis, and therefore accept the alternative hypothesis that a difference does exist. Tests of statistical inference (e.g., Student's t-test, Wilcoxon-Mann-Whitney, etc.) enable this process by analyzing experimental data and producing a probability, or p-value, which represents the likelihood that the results obtained by the experiment are due to chance and therefore that the null hypothesis is correct. In medical and educational research a conventional p-value of 0.05 or less (which translates to a 5% or less probability that the results are due to chance and the null hypothesis is true) is typically used as a threshold of statistical significance. Thus, when a given statistical test returns a p-value at or below this cutoff, the researcher can confidently reject the null hypothesis because the chance that it is true is so low (i.e. less than 5%), accept the alternative (i.e., primary study hypothesis) instead, and infer that the relationship between variables observed in the study sample is also present at the level of the population.

An important aspect of making valid inferences is an understanding of Type-1 (alpha), and Type-2 (beta) systematic error [8]. Alpha error refers to the probability that a given statistical test will indicate that the null hypothesis should be rejected when it is actually correct. Put another way, alpha error occurs when the researcher concludes that a relationship *truly exists between variables when one does not.* A researcher's tolerance for this type of error is expressed in the choice of alpha level considered significant, with the traditional criterion of 5% chance (i.e. 0.05) implying tolerance of a 5% alpha error rate. Expressed differently, if 20 such statistical tests were performed on the sample and all had a p-value of approximately 0.05, then by chance *one of these tests would constitute a false positive or Type-1 error.* Misunderstanding this possibility forms the basis of a common error that will be addressed later in this chapter.

In contrast, beta error refers to the probability a given statistical test will indicate that the null hypothesis should not be rejected when in fact it should be rejected and the alternative (study) hypothesis accepted. Practically, this error results in the researcher concluding that *no relationship exists between variables when in fact one does*. This error is tied to the magnitude of the true relationship in the population and the sample size of the study, and our tolerance for it is typically expressed as the *power* of a study (which is calculated by subtracting the acceptable beta error rate from the number one). Customary values for study power fall between 0.8 and 0.9 (i.e., a 20% to 10% chance of making a beta error) and correspond to a 1 in 5 to 1 in 10 chance that a given statistical test will yield *a false negative result or Type-2 error.* Avoiding this issue involves calculations of power and sample size. Failure to do the power and sample size calculation prior to initiating the study is a common but detrimental error. It is here that descriptive and other forms of pilot study can prove useful, as they often contain information regarding the mean or median value and standard deviation of a variable of interest within the study population, as well as its overall distribution (i.e., normal, bimodal, etc.) Such information is needed to perform power and sample size calculations.

Another important aspect of statistical inference is the appropriate selection of the test used. Each test comes with its own set of requirements and assumptions regarding the scale of measurement (i.e., nominal, ordinal, interval, and ratio) and the *data distribution* of the variables of interest. This latter characteristic refers to the overall pattern made as data produced by a given variable is ranked by magnitude and subsequently plotted on a graph [3]. Perhaps the most familiar is the Gaussian, or normal distribution, which is well described in the introductory statistical textbooks and forms a key assumption for several common statistical tests (e.g., Student's t-test, ANOVA, etc.). Unfortunately, many simulation-based studies rely on data that are not normally distributed and therefore do not meet the necessary requirements for using some statistical tests. Information regarding the data distribution of specific variables can often be derived from descriptive or pilot studies.

Finally, it is critical to recognize the difference between the statistical significance of a particular relationship between variables and the overall magnitude of that relationship in real life. It is common for researchers to misinterpret an extremely small p-value as evidence that the independent variable exerts an extremely strong effect on the dependent variable, when it only implies that the observed effect is highly unlikely to be due to chance. While p-values do depend to some extent on the strength of the relationship they also depend on other factors, such as the sample size and distribution of the data. In particular, studies with large sample sizes can often show statistical significance even when the magnitude of effect is small [24]. As an example, consider a hypothetical large multicenter study of a simulation-based intervention intended to improve communication skills that showed a significant p-value of <0.001 but only a 5% actual improvement in median scores on the assessment tool used. While the researchers could reject the null hypothesis with a great deal of confidence, and hence conclude that a real effect exists, the actual magnitude of improvement is quite meager and may not be clinically or educationally meaningful. Quantifying this magnitude thus requires some measure of *effect size*, and a number of statistical tests (e.g., Cohen's d, Pearson's r, Coefficient of Determination – r^2, etc.) exist that fill this function. Figure 21.2 depicts the relationship between statistical significance and effect size. These issues are addressed in detail in Chaps. 27 and 46 provides a case study in which the above issues are practically worked out using a conversational formal.

Common Errors in Quantitative Research

As the chapter draws to a conclusion we consider common errors seen in quantitative study designs [8, 25]. These errors find their origin in various misunderstandings of the above principles, and have been frequently encountered by the editors of this book as they review study proposals and manuscripts. By discussing these issues early in this section of the book, we hope to alert novice researchers to common pitfalls and provide the tools needed to create stronger protocols.

Insufficient Connection with the Literature or Existing Theory

As mentioned above, it is common for some novice simulation researchers to base their study interventions on personal experience or institutional need. While these are important factors, it is unfortunately easy for studies developed from these perspectives to be disconnected from the existing literature and theoretical constructs. This ultimately leads to studies that do not fit within the ongoing scholarly narrative or replicate what has already been established. This issue can be easily addressed by conducting an adequate literature search *before* the study is developed. Should this search reveal inadequate theoretical foundations for the proposed intervention, it may be necessary to perform more exploratory work prior to creating an interventional protocol.

Use of Assessments with Inadequate Validity

It is also common for quantitative study designs to incorporate novel or untested assessment tools that have not been sufficiently validated for use in research. As the strength of a researcher's conclusions is directly dependent on the reliability and validity of the outcome measures, the use of untested tools significantly limits the inferences that can be drawn from the results. To draw an analogy with biomedical research, consider a study examining the effects of a novel medication intended to reduce blood pressure. If it is later learned that the equipment used to obtain the blood pressure measurements is faulty, it becomes impossible to

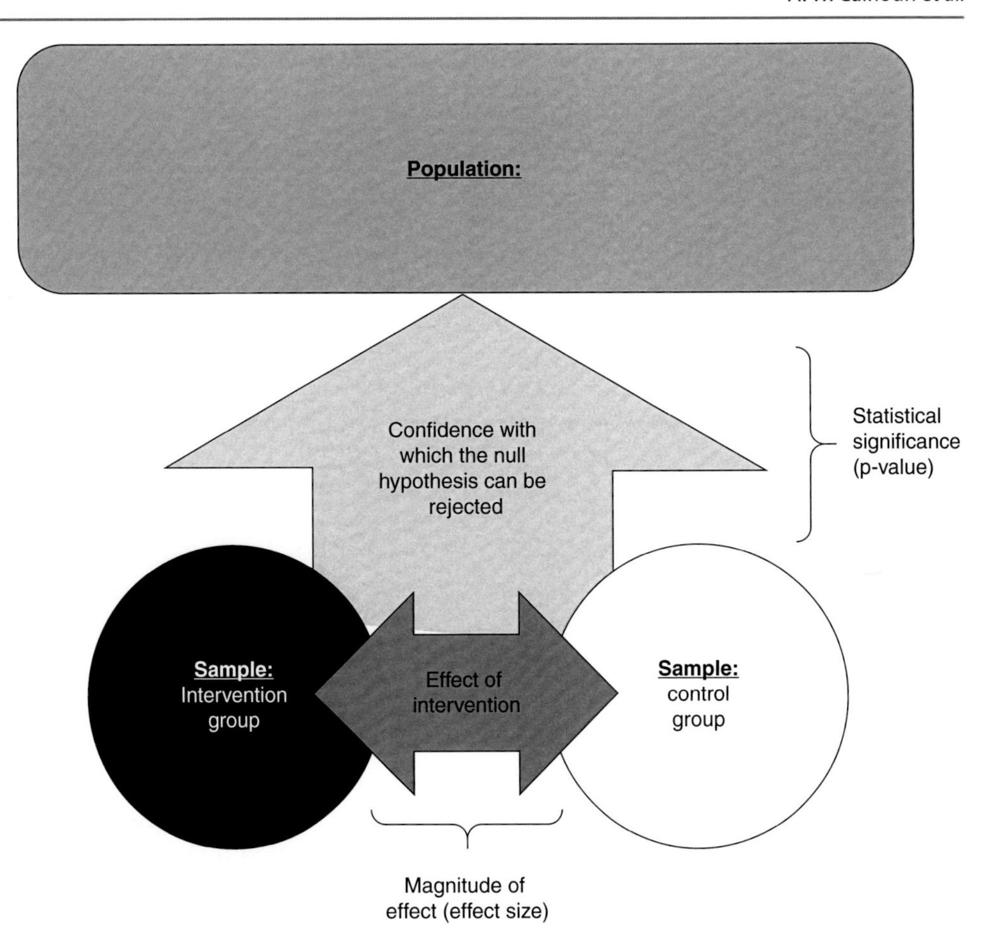

Fig. 21.2 Relationship between statistical significance and magnitude of effect. This figure depicts the relationship between the statistical significance of a relationship and the magnitude of the relationship. The two circles at the bottom of the figure depict the intervention and control groups, while the population from which the entire sample is drawn is represented by the large rectangle at the top. The horizontal two-way arrow situated between them depicts effect size, which corresponds to the overall magnitude of the difference between intervention and control groups. This overall magnitude is different, however, from the statistical likelihood that differences noted in the study are not due to chance and hence translate to an actual difference at the level of the population (i.e., the statistical significance). Statistical significance is depicted by the vertical arrow

interpret the results and the entire study is called into question. Remedies for this issue involve either the utilization of an assessment tool that has demonstrated validity for the situation in question or the simultaneous acquisition of assessment validity data within the context of the study that can then presented alongside the outcome data. Assessment tools abound in the literature (albeit with varying levels of demonstrated validity), and it is advisable to choose one with demonstrated reliability and validity as the first option whenever possible. It is critical to note, however, that even in this ideal circumstance it is incumbent upon the researcher to provide some evidence demonstrating that the tool selected possesses validity for the particular situation being assessed, especially if differences exist between the original group or situation in which the tool's validity was assessed and that of the current study. In the event that such a tool does not exist, one may need to be developed. Chapter 26 provides further information on this process.

Lack of Appropriate Power Calculations

Most educational and simulation studies utilize convenience samples, which tend to be small in size. In the event that a statistically significant effect is not detected for an outcome of interest, many authors (and journal reviewers) will then attempt to argue that this is due to a lack of sufficient sample size or the effect size of the intervention being too small. While this may be the case, the argument cannot be effectively supported without the presence of a power and sample size calculation performed before the study was executed. Possessing this information can also help guard against the temptation to argue that "trends" (i.e., changes in the dependent variable that are potentially attributable to the study intervention but that do not cross the predetermined threshold for statistical significance) are statistically meaningful results. Power calculations should be conducted prior to the initiation of the study, as calculations performed entirely post-hoc can violate the assumptions of the procedure [25]. Such after-the-fact calculations based on the specific study data can occasionally be valuable when compared with a priori calculations, however, as a check on their accuracy.

Inappropriate Use of Statistical Tests

One of the most common statistical tests used in medical research is the Student's t-test. This test, which measures the probability that a difference in means between two samples corresponds to a real difference in means for the populations represented, is quite powerful, and is indeed appropriate in many circumstances. Unfortunately, this test assumes that the data analyzed possess a normal, Gaussian distribution [23]. The assessment tools frequently used for simulation-based research rarely produce normally distributed data, however, and statistical tests that do not rely on this assumption may thus be needed. Common tests that do not require a Gaussian distribution are the Mann-Whitney-Wilcoxon test (which can be used in place of the independent samples t-test), and the Wilcoxon signed-rank test (which can be used in place of the paired t-test). Chapter 28 addresses other tests appropriate for datasets of this type in greater depth.

Inappropriate Use of Multiple Statistical Tests

The assessment tools commonly used for simulation-based research frequently contain multiple items and subscales, and it is therefore quite tempting to analyze each subscale (or each item item) separately. This leads to a cascade of p-values that often results in incorrectly interpreted data due to increased rates of alpha error. As greater numbers of statistical tests are performed, the problem expands in scope. This issue is not trivial or uncommon; the authors are familiar with studies in which p-values number in the hundreds. Ideally this should be addressed by focusing significance testing on only those variables pertinent to and/or predicted for key study outcomes [26]. Should this not be possible, however, a number of statistical techniques (such as the Bonferroni Adjustment or the use of a stricter cutoff for statistical significance) can be used to adjust the alpha level to account for the number of tests performed [27]. This concept is important particularly for researchers interested in "data mining" (the practice of statistically scanning large databases for potential associations of interest) as such studies typically involve a large number of statistical tests, all of which can produce separate p-values and collectively increase alpha error.

Use of Non-significant p-values as Proof of Lack of Relationship

In the absence of a significant p-value it can be tempting to present study data as confirming the absence of a meaningful relationship. This interpretation, however, stems from a misunderstanding of inferential statistics. Lack of a significant p-value does not imply a definite lack of relationship, but instead indicates that the null hypothesis cannot be rejected at the current power and sample size. This has a substantially different implication than the erroneous acceptance of the null hypothesis, which cannot be done in this way. Constructing a study to demonstrate true lack of difference is addressed by a very different statistical process, and requires a clear articulation of the minimal difference between groups that has clinical or educational meaning [28]. Such studies typically require large sample sizes. Chapter 30 contains a deeper discussion of these statistical approaches.

Confusion of the P-value With Magnitude of Difference

As stated above, p-values assess the likelihood that the null hypothesis explains our results, and hence allow us to infer a connection between outcomes at the level of the sample and potential outcomes among the population from which the sample was taken. P-values do not, however, provide direct information regarding the magnitude of the effect detected. It is thus quite possible (particularly when large sample sizes are present) to have a result that is significant statistically but of small or moderate magnitude of effect, and therefore of questionable value from an educational or clinical standpoint. Addressing this issue involves a presentation of the magnitude of the result as well as its statistical significance. This can be done by simply reporting the absolute difference between control and intervention groups, or by the calculation of an appropriate effect size statistic, odds ratio, or confidence intervals [24, 25].

Charting a Course

In this introductory chapter we have reviewed key aspects of quantitative simulation research, including its relationship to qualitative research, the value of theory and theoretical frameworks in developing hypotheses, and the strengths and drawbacks of common study designs. We have also introduced the concept of validity as a key aspect of assessment within research, and reviewed basic principles of inferential statics. Finally, we have presented some common errors and linked them to the information in the following chapters. It is our hope that the material that follows will assist readers' ability to produce high-quality simulation-based scholarship and prove a useful resource to those members of the simulation community embarking on quantitative research designs.

References

1. Babbie ER. The practice of social research. 12th ed. Belmont: Wadsworth Cengage; 2010.
2. Mujis D. Doing quantitative research in education with SPSS. 2nd ed. London: SAGE Publications; 2010.
3. Vogt WP. Dictionary of statistics and methodology. 2nd ed. London: SAGE Publications; 1999.
4. Kessler D, Pusic M, Chang TP, Fein DM, Grossman D, Mehta R, et al. Impact of just-in-time and just-in-place simulation on intern success with infant lumbar puncture. Pediatrics. 2015;135(5): e1237–46.

5. Calhoun AW, Sutton ERH, Barbee AP, McClure B, Bohnert C, Forest R, et al. Compassionate options for pediatric EMS (COPE): addressing communication skills. Prehosp Emerg Care. 2017;21(3):334–43.
6. Sullivan GM, Sargeant J. Qualities of qualitative research: part I. J Grad Med Educ. 2011;3(4):449–52.
7. Crandall SJ, Caelleigh AS, Steinecke A. Reference to the literature and documentation. Acad Med. 2001;76(9):925–7.
8. Picho K, Artino AR Jr. 7 deadly sins in educational research. J Grad Med Educ. 2016;8(4):483–7.
9. Cook DA, Beckman TJ, Bordage G. Quality of reporting of experimental studies in medical education: a systematic review. Med Educ. 2007;41(8):737–45.
10. Bernhard HR. Social research methods, qualitative and quantitative approaches. 2nd ed. London: SAGE Publications; 2013.
11. Evans BC, Coon DW, Ume E. Use of theoretical frameworks as a pragmatic guide for mixed methods studies: a methodological necessity? J Mix Methods Res. 2011;5(4):276–92.
12. Morgan DL. Paradigms lost and pragmatism regained: methodological implications of combining qualitative and quantitative methods. J Mixed Methods Res. 2007;1(1):48–76.
13. Tavakol M, Sandars J. Quantitative and qualitative methods in medical education research: AMEE Guide No 90: part II. Med Teach. 2014;36(10):838–48.
14. Ingham-Broomfield RA. Nurse's guide to quantitative research. Aust J Adv Nurs. 2014;32(2):32–8.
15. Neuman WL. Social research methods: qualitative and quantitative approaches. 7th ed. Edinburgh Gate: Pearson Education Limited; 2014.
16. Lopreiato JO, editor. Healthcare simulation dictionary. Rockville: Agency for Healthcare Research and Quality; 2016.
17. Issenberg SB, McGaghie WC, Petrusa ER, Lee Gordon D, Scalese RJ. Features and uses of high-fidelity medical simulations that lead to effective learning: a BEME systematic review. Med Teach. 2005;27(1):10–28.
18. Calhoun AW, Bhanji F, Sherbino J, Hatala R. Simulation for high-stakes assessment in pediatric emergency medicine. Clin Pediatr Emerg Med. 2016;17(3):212–23.
19. Calhoun AW, Donoghue A, Adler M. Assessment in pediatric simulation. In: Grant V, Cheng A, editors. Comprehensive healthcare simulation: pediatrics. Cham: Springer International; 2016. p. 77–94.
20. Cook DA, Brydges R, Ginsburg S, Hatala R. A contemporary approach to validity arguments: a practical guide to Kane's framework. Med Educ. 2015;49(6):560–75.
21. Downing SM. Validity: on meaningful interpretation of assessment data. Med Educ. 2003;37(9):830–7.
22. Messick S. Meaning and values in test validation: the science and ethics of assessment. Educ Res. 1989;18(2):5–11.
23. Vetter TR. Fundamentals of research data and variables: the devil is in the details. Anesth Analg. 2017;125:1375–80.
24. Sullivan GM, Feinn R. Using effect size-or why the P value is not enough. J Grad Med Educ. 2012;4(3):279–82.
25. Sullivan GM. Is there a role for spin doctors in Med Ed research? J Grad Med Educ. 2014;6(3):405–7.
26. Feise RJ. Do multiple outcomes measures require P-value adjustment? BMC Med Res Methodol. 2002;2(8):1–4.
27. Noble WS. How does multiple testing correction work? Nat Biotechnol. 2009;27(12):1135–7.
28. Quertemont E. How to statistically show the absence of an effect. Psychol Belg. 2011;51(2):109–27.

Research and Hypothesis Testing: Moving from Theory to Experiment

22

Mark W. Scerbo, Aaron W. Calhoun, and Joshua Hui

Overview
In this chapter, we discuss the theoretical foundation for research and why theory is important for conducting experiments. We begin with a brief discussion of theory and its role in research. Next, we address the relationship between theory and hypotheses and distinguish between research questions and hypotheses. We then discuss theoretical constructs and how operational definitions make the constructs measurable. Next, we address the experiment and its role in establishing a plan to test the hypothesis. Finally, we offer an example from the literature of an experiment grounded in theory, the hypothesis that was tested, and the conclusions the authors were able to draw based on the hypothesis. We conclude by emphasizing that theory development and refinement does not result from a single experiment, but instead requires a process of research that takes time and commitment.

Practice Points

- Theories help to organize knowledge, explain facts, and guide research.
- A good theory should suggest ways it can be tested.
- A hypothesis is a tentative statement about events that follow logically from the theory.
- An experiment represents the specific plan for testing a hypothesis.

M. W. Scerbo (✉)
Department of Psychology, Old Dominion University, Norfolk, VA, USA
e-mail: mscerbo@odu.edu

A. W. Calhoun
Department of Pediatrics, University of Louisville School of Medicine, Louisville, KY, USA
e-mail: aaron.calhoun@louisville.edu

J. Hui
Emergency Medicine, Kaiser Permanente, Los Angeles Medical Center, Los Angeles, CA, USA

Introduction

The purpose of this chapter is to discuss the role of theory in guiding research. We will offer a definition of theory, describe different types of theories and the role of theory in research, and touch on some characteristics that distinguish better theories. Next, we consider hypotheses and their relationship to theory, how they differ from more general research questions, and some statistical considerations. We also discuss theoretical constructs and how they relate to theory and hypotheses. Last, we show how the path from theory to hypotheses leads to a formal experiment. Throughout, we try to illustrate these ideas with examples drawn from science, psychology, and the healthcare simulation literature.

We cover many topics in this brief chapter and admit that we cannot delve into any in great detail (although additional and more advanced information on related topics can be found in Chaps. 23 and 24). There is an extensive literature on the philosophy of science, on theory and hypotheses, and on experimentation. We hope this primer will pique the reader's interest to dig deeper into some of the references and recommended readings cited in this chapter and in other chapters in this section of the book.

Theory

A theory is not an easy thing to describe. Most definitions of theory describe a relationship among variables. A more reasonable definition of a *theory* is: a system of ideas or set of principles that gives logical coherence to a known set of empirical relations [1, 2]. There are several important elements within this definition. First, a theory is not a single idea but instead it is based on multiple concepts that are tied together in a meaningful fashion. The relationships among the theoretical components help to organize knowledge and provide guidance for incorporating new information into our existing knowledge. Moreover, a theory provides a framework for explaining phenomena and relationships.

D. Nestel et al. (eds.), *Healthcare Simulation Research*, https://doi.org/10.1007/978-3-030-26837-4_22

A theory is not a fact. A fact is a true statement, but a theory is often based on many facts and helps to organize those facts. Also, a theory is not a law. A law is often very narrow in scope, sometimes describing a single relationship; however, a theory can encompass laws. Last, a theory is not a model, but a model may be a component of a theory. A model is a description or analogy that can help to make abstract ideas concrete; in other words, a model offers an objective representation of a theory. A theory can also be descriptive, but unlike a model it can explain why phenomena occur.

Figure 22.1 depicts graphically and describes the relationship between facts, theories, laws, and hypotheses. A theory (represented by the blue circle) is an organizing framework that contains both individual facts (represented by the yellow circles) and the potential and confirmed relationships that exist between them (represented by the black arrows). Hypotheses (represented by the red rectangle) consist of statements regarding relationships within the theory that are unconfirmed and hence still under investigation. In contrast, laws (represented by the black rectangle) are statements concerning relationships that have been confirmed as true across multiple investigations. It is important to note, however, that even confirmed laws can be superseded, resulting in those laws being viewed subsequently as special cases within a greater relationship or theory. An example of this is the manner in which Isaac Newton's law of gravitation, which still holds true in most circumstances, is now treated as a special limiting case within Albert Einstein's more comprehensive Theory of General Relativity.

Perhaps most important, theories cannot be true or false because theories represent a scheme for organizing information. Instead, as new knowledge is acquired it provides support either for or against the theory. Further, any single test of a theory cannot necessarily refute it because a negative outcome can always mean the test lacked sufficient sensitivity. Alternatively, a positive outcome may provide support for the wrong reasons [3].

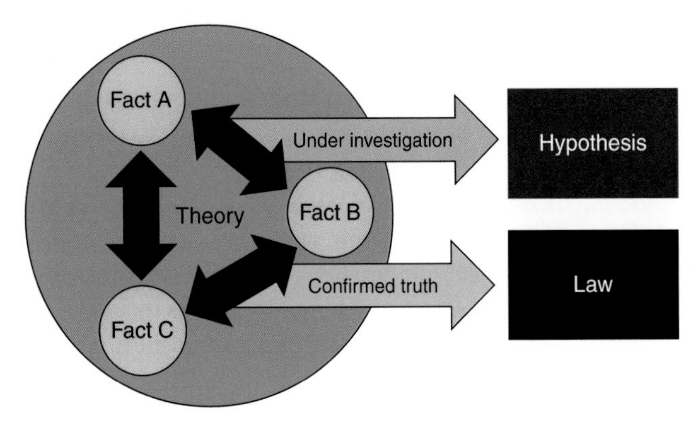

Fig. 22.1 Depiction of the relationship between facts, theories, hypotheses and laws

Types of Theories

There are many different types of theories. One important distinction is whether they are derived from empirical findings or are conceived to generate empirical research [4, 5]. *Inductive theories* represent the former. Empirical observations lead to more abstract levels of representation. For example, anesthesiologists must often monitor vital signs over extended periods of time during surgery and research has shown that their ability to maintain attention can be fragile [6]. Theories regarding the ability of individuals to focus attention over prolonged periods evolved from empirical observations of the attentional failures of military personnel to monitor radar displays [7]. On the other hand, *deductive theories* are often grand abstract ideas that drive a search for empirical support. As new data are generated they allows the theory to be verified, modified, or in some instances discarded. Freud's theory of personality was an elaborate attempt to describe human drives and motivations based on three hypothetical constructs of mind: id, ego, and superego [8]. This comprehensive theory ultimately spawned years of experimental research in an attempt to gather empirical support.

Role of Theories

Theories play an important role in research. First, they help to organize knowledge. The theory is what provides the conceptual structure to a collection of facts, laws, and/or models. Theories can be used to explain facts and laws and to predict new laws or phenomena. However, most often we rely upon theory to guide research.

A theory should suggest many possible tests. Research aimed at discovering new knowledge follows one of three approaches often attributed to the philosopher, John Stuart Mill [9]. The first is the *method of agreement*. According to this approach research is conducted to find evidence that is common among observations. This is the same approach detectives follow when they have a theory that a series of crimes may be connected and therefore were committed by the same individual. They search for clues that are common among crime scenes to determine the cause. The second approach is the *method of difference*. Here, evidence is sought to try to understand what distinguishes one set of observations from another. This is the fundamental idea behind a control group. Two groups are treated in exactly the same manner except one group receives an intervention that the control group does not. If there is a difference in outcomes between the two groups and the only factor that differed is the intervention, the intervention is then considered the likely cause. Finally, the *method of concomitant variation* is used to look for evidence that shows relationships. This may be used in very early stages of research to determine whether objects or events occur together with some regularity. For example, many in the healthcare simulation community are interested in whether learning experiences acquired in the simulation center are related to patient outcomes in their clinical settings.

Given the different roles for theory described above, how can we tell if a theory is good? Some theories are clearly better than others. Thus, theories can be evaluated according to several criteria. First, theories should show logical consistency. That is, a good theory should agree with the facts. Rarely, if ever, will a theory agree with all of the facts, but a good theory should agree with most of them (particularly when new information becomes available). Second, a good theory should generalize beyond the empirical evidence. Thus, a good theory of skill learning should generalize beyond psychomotor skills to other skills (e.g., cognitive, social, team, etc.). Third, a theory should have explanatory value. That is, it should offer reasons for *why* some phenomenon occurs. A well written theory should explain why some outcomes are expected and why others are not. For example, Ericsson's theory of deliberate practice suggests that individuals who practice skills regularly, seek and utilize feedback, and constantly pursue new challenges will show continuous improvement [10]. Therefore, if a practice regimen adopts these characteristics in an experiment, the results should show improved performance each time the trainees are assessed. On the other hand, if the experiment gives just a few opportunities for practice, with one final debrief for feedback, learners may not see their skills improve any more after the level obtained in that session. Finally, science as a discipline tends to favor theories that are parsimonious. That is, we adhere to Ockham's razor, a philosophical statement originating with William of Ockham in the thirteenth century which states that explanations containing the fewest assumptions are preferred [5].

Hypotheses and Research Questions

Many questions can stimulate curiosity and motivate someone to become engaged in research, but not all questions are based on theory. A hypothesis, however, is a specific type of research question that follows logically from a theory [5]. It is a specific prediction based on the theory and therefore is an assumption, a possible instance of the theory [11]. As such, it indicates a way for the theory to be tested empirically. A well written hypothesis identifies at least two variables and states the expected relationship between the variables. Often, a hypothesis uses the "if-then" format [11].

For example, consider a contemporary theory of stress. Lazarus and Folkman argue that stress is a mental state that arises from a situation in which an individual's perceived ability to cope is insufficient to alleviate a stimulus appraised as stressful [12]. Drawing upon this theory of stress, one possible hypothesis might be:

> If individuals must perform an unfamiliar critical task with no room for error, then they will report higher levels of stress on a rating scale than those who are familiar with the task.

In this example, the intervention, familiarity, is one variable and the other variable, stress ratings, is the outcome measure. The stated relationship, the expectation, is that the unfamiliar task should produce higher ratings of stress.

Of course, there could many other hypotheses to test this theory. Alternative hypotheses could address different degrees of familiarity, different types of tasks, different rating scales, or even different methods of assessment. An important point, however, is that each alternative hypothesis must still be tied directly to the theory. Hence, it should be clear that a single hypothesis can only add a little to our understanding of a theory. Confidence in a theory accrues or wanes from multiple hypotheses that address its breadth and depth.

There are other types of questions researchers can ask that can be quite valuable even though they are not hypotheses. Some research questions are the foundation for qualitative research approaches (see Chap. 9). These questions do not offer testable predictions, but instead allow investigators to gather information or themes that can suggest possible theories and lead to hypotheses in subsequent stages of research. For instance, several people in the healthcare simulation community have begun to ask the question; "Should we let the mannequin die?" It has generated much discussion with arguments for and against having a simulated patient die in a training scenario [13, 14]. A question such as this has implications not only for how we train providers but also for their psychological welfare. This question is not a hypothesis itself because it does not specify a prediction. The question could, however, serve as the foundation for a number of specific hypotheses (e.g., does the unexpected death of a mannequin in a simulation scenario cause elevated anxiety among medical students on a stress-anxiety instrument).

Another category of research questions follows the PICO framework (population, intervention, comparison group, and outcome). A variation of this framework exists, PICOS, which also includes setting [15]. These frameworks can be very helpful for framing a study design, guiding search strategies when conducting systematic reviews of the literature, or for gathering evidence to inform clinical decision making [16]. Again, however, this framework does not offer a testable hypothesis.

Research and the Null Hypothesis

The hypothesis derived from theory states an expectation of relationship regarding the variables under investigation. In this regard, it differs from the *null hypothesis* which is a somewhat different concept specifically tied to the way in which we frame statistical analyses [17]. Classical inferential statistics are based on the idea of rejecting the null hypothesis; that is, in a general sense, there is no difference between conditions or no relationship among variables. By contrast the *research hypothesis* states what is expected; that is, a difference between conditions or a relationship among variables. It should be noted that the research hypothesis has an important relationship to the statistical approach, as the expected effects stated within the hypothesis have direct bearing on which analyses are chosen and how they are subsequently applied. Thus, greater specificity within the hypothesis often results in improved chances of reaching statistical significance. For example, suppose one is interested in comparing an intervention with a control condition and will use a Student's t test to compare the means from two groups. If one can predict the direction of the effect (e.g., that the intervention will be greater than the control) then a one-tailed t test should be selected because it provides a greater likelihood of detecting an effect. By contrast, if one cannot predict a specific direction for the intervention (greater *or* less than the control), then one is limited to a two-tailed t test, As the t value required to achieve the identical significance level for a two-tail t test is always larger than for one-tailed t test, this may impact the outcome of the study.

Nonetheless, a researcher needs to be careful in deciding when to employ a one-tailed t test. The test is more powerful than a two-tailed test only because the two-tailed test splits the alpha level into halves, one for each tail of the distribution curve. The additional power gained actually comes at the expense of potentially missing an effect in the other direction. In certain situations, missing the effect in an opposite direction can be serious. Therefore, a researcher needs to consider the consequence of missing an effect in the opposite direction first before employing a one-tailed test. It should be noted, however, that it is inappropriate to employ a one-tailed test solely for the purpose of achieving statistical significance.

A similar situation applies to more complex experimental designs with multiple conditions. If the investigator can make specific predictions about which comparisons among three or more means are the critical ones to evaluate, they can be evaluated with preplanned comparison tests. Otherwise, the investigator is limited to one of a variety of post hoc tests (e.g., Bonferroni or Tukey tests), that may have less statistical power than preplanned comparisons because they assume all comparisons will be evaluated [3, 18]. Thus, the statement of a specific hypothesis can make a substantial difference in whether results are statistically significant.

Last, it is also important to pay attention to the balance of statistical significance and clinical significance. Not all statistically significant matters are meaningful in real life. This further illustrates the importance of having a reasonable research hypothesis.

Theoretical Constructs

Many of the issues healthcare simulation researchers wish to study are considered theoretical constructs. These are hypothetical entities that are inferred from facts, empirical observations, and even other constructs. We assume they exist and conduct research to provide supporting evidence which ultimately helps formulate or refine full theories. Their existence is often supported by multiple theories. Some examples include: intelligence, personality, situation awareness, and stress. In science, these constructs must be defined in a way that makes them both observable and measurable. We use the term *operational definition* to describe how the construct can be measured [4].

Once again, consider the construct, stress. A fundamental definition of stress is: "a relationship between the person and the environment that is appraised by the person as taxing or exceeding his or her resources and endangering his or her well-being" (p. 21) [12]. However, that definition does not specify precisely how stress is measured. We could define stress based on behavior, physiological activity, or subjective impressions. Table 22.1 shows a variety of operational definitions of stress.

Operational definitions increase the specificity of hypotheses by indicating precisely how the variables will be measured. In the example we've been using, level of stress could be measured on a subjective rating scale such as the State-Trait Anxiety Inventory (STAI) in which current levels of anxiety are scored on a scale of 20–80 with higher scores reflecting higher anxiety [19]. Therefore, stress might be defined operationally as a score of 50 or higher on the STAI.

An important benefit of precise operational definitions is that they help researchers avoid circular reasoning, which occurs when definitions are used as explanations. Consider this example:

> Members of a surgical team communicate poorly because they don't respect one another since low respect hinders communication.

Definitions do not explain behavior. Explanations require precision, and well written hypotheses incorporate the preci-

Table 22.1 Three categories of operational definitions for stress

Behavioral	Number of errors
Physiological	Heart rate in beats per minute
Subjective	Score on the State-Trait Anxiety Inventory (STAI)

sion that links variables together in a meaningful relationship. Therefore, to avoid circular reasoning, researchers use operational definitions in hypotheses. Using the previous example, communication and respect can be defined operationally in the following hypothesis, which specifies precisely how respect and communication will be measured:

> If members of a surgical team rate their colleagues lower on a rating scale of respect, then they will make fewer complimentary statements in the operating room compared to teams with higher ratings of respect.

Operational definitions are important for testing theories. The operational definition adopted by a researcher can help or hinder attempts to test a theory. For example, two researchers might conduct similar studies of stress, but one researcher chooses to measure stress by the number of errors committed and a second researcher by subjective impressions. Ideally, one would hope that both definitions would yield similar outcomes, but that does not always happen. Participants may make similar numbers of errors under two different conditions, but subsequently report that one condition was perceived to be more stressful than the other. Discrepant results such as these often require researchers to re-examine their operational definitions or the underlying theory itself.

Experiments

A hypothesis is a statement that can be true or false and suggests a test. An experiment represents the actual plan and process for testing the hypothesis. A hypothesis suggests how the theory will be tested, but not the specific details. The Method section of a research paper describes the details of the experiment and how the hypothesis was formally tested. It includes information regarding the participants, setting, equipment, assessment instruments, data collection procedures, and how the data were analyzed. A method section should include enough specificity to allow other investigators to repeat the experiment and potentially corroborate the original results.

For example, if we wanted to conduct an experiment based on the theory of stress described earlier, we must choose a task (for example, defibrillation) to test the hypothesis. Next, we have to consider the design of the experiment. We might compare groups of participants who have different levels of familiarity with defibrillation or examine the same participants when they are naïve and again when have they have acquired more experience. We also have to decide who the participants will be (e.g., medical students, residents, etc.). We need to specify the conditions where data will be gathered (in a simulation center, in situ, or with genuine patients). We also need to describe the make and model of the defibrillator to be studied and the rating scale used to measure stress. Next, we would describe the procedures followed, including how participants were recruited, what information and introductory training was provided, how the scenario unfolded, when they completed the rating scales, and what debriefing was offered. Last, we need to describe the statistical methods for analyzing the data. For example, if we chose the experimental design comparing groups with different levels of familiarity, we would analyze the results with an independent t test. Alternatively, we could look for a correlation between the level of familiarity and the stress ratings. Other possibilities may also exist. Recently, Cheng and colleagues [20] published a checklist that specifies the methodological details that should be reported for simulation-based experiments in the healthcare and we recommend researchers refer to this checklist to review their experiments.

An Example That Ties It All Together

We now offer an example that illustrates one possible pathway from theory to hypothesis to experiment, and ultimately to conclusions. Recently, Turner and his colleagues conducted an experiment to examine the abilities of standardized patients to recognize behavioral cues in learners who provided either periodic intra-session assessments or a single post-encounter assessment [21]. The investigators appealed to the theory of working memory described by Baddeley and Hitch to guide their approach [22]. They argued that when two or more tasks must: (1) be performed simultaneously and (2) utilize the same working memory subsystem it can hinder performance. Thus, if a standardized patient has to generate dialogue while at the same time monitoring the behavior of a trainee it will place greater attentional demand on the verbal part of working memory and therefore may make it more difficult to encode and maintain some of the verbal information in working memory. Furthermore, longer encounters necessitate that more information will need to be maintained in working memory. Accordingly, the investigators stated the following formal hypothesis: "…it was therefore expected that periodic evaluation would enable participants to work from a smaller subset of information in working memory at any given time throughout the scenario as a result of offloading this information more frequently. Thus, the burden on working memory would be reduced, resulting in more accurate cue recognition and improved scoring accuracy" (pg. 176) [21].

In their experiment, the investigators had standardized patients watch a 20-min video recorded encounter and complete a checklist that addressed verbal and nonverbal behaviors. One group completed the checklist at the end of the encounter. For the other group, the video was paused three times after key segments, and these SPs completed their checklists after each segment. The results showed that the SPs in the periodic encounter condition identified significantly more nonverbal cues across all segments and more critical verbal cues in the middle segment than those in the single encounter condition. The investigators concluded that their results supported the hypothesis that accuracy was better for the periodic encounters because the burden on working memory to recall behavioral information was reduced for the shorter segments compared to the full 20-min encounter. However, counter to their hypothesis, they also conceded that the effect was greater for nonverbal cues than for verbal information.

Conclusion

The goal of this chapter was to address the importance of theory to guide research. We discussed the role of theory in research, different types of theories, and how to evaluate theories. We also discussed hypotheses, how hypotheses relate to theory, the relationship between hypotheses and theoretical constructs, and how hypotheses set the stage for conducting experiments. The ideas expressed here are not exhaustive but can serve as a guide for understanding the importance of theory in research and how one moves from theory, to hypotheses, and finally to an experiment that provides the actual test of the theory.

It is important to understand that research is a process. Theory development and evaluation take time. A single test of a theory provides evidence, but rarely (if ever) conclusive evidence. A theory and an associated hypothesis can predict a specific outcome, and if the predicted outcome is obtained it provides support for the theory. There is always the possibility, however, that a *different* theory could predict the same outcome. Even a well-established theory or theories can be falsified, as is the requirement for a theory. In fact, occasionally throughout history the predominant accepted theories and laws that guide an area of science have been set aside in a paradigm shift [23]. For example, Einstein' theory of relativity challenged the well-established theories and laws of Newtonian physics. Paradigm shifts in thinking are part of the natural evolution of knowledge and led Newton-Smith to argue that any theory will be falsified within 200 years! [24].

As noted above, one of the important roles for theory is to organize knowledge and guide research. Gathering evidence that does not support a theory does not necessarily undermine its usefulness. Einstein's theory did not invalidate Newton's ideas, but instead placed boundaries on where they are and are not relevant. The value of theory rests with how it shapes our knowledge and understanding of events in our world. Within the healthcare simulation community, the value of theory rests with what we know and understand about this unique method for improving our interactions with patients and one another.

References

1. Badia P, Runyon RP. Fundamentals of behavioral research. Reading: Addison-Wesley; 1982.
2. Roeckelein JE. Elsevier's dictionary of psychological theories. Amsterdam: Elsevier Science; 2006.
3. Maxwell SE, Delaney HD. Designing experiments and analyzing data: a model comparison perspective. 2nd ed. Mahwah: Erlbaum; 2004.
4. Graziano AM, Raulin ML. Research methods: a process of inquiry. 4th ed. Boston: Allyn and Bacon; 2000.
5. Passer MW. Research methods: concepts and connections. New York: Worth; 2014.
6. Weinger MB, Herndon OW, Paulus MP, Gaba D, Zornow MH, Dallen LD. An objective methodology for task analysis and workload assessment in anesthesia providers. Anesthesiology. 1994;80:77–92.
7. Warm JS. An introduction to vigilance. In: Warm JS, editor. Sustained attention in human performance. Chichester: Wiley; 1984. p. 1–14.
8. Freud S. The standard edition of the complete psychological works of Sigmund Freud. Volume XIX (1923–1926) The ego and the id and other works. Strachey, James, Freud, Anna, 1895–1982, Rothgeb, Carrie Lee, 1925-, Richards, Angela, Scientific literature corporation. London: Hogarth Press; 1978.
9. Mill JS. A system of logic, vol. 1. Honolulu: University Press of the Pacific; 2002. p. 1843.
10. Ericsson KA, Krampe RT, Tesch-Romer C. The role of deliberate practice in the acquisition of expert performance. Psychol Rev. 1993;100:363–406.
11. Darian S. Understanding the language of science. Austin: University of Texas Press; 2003.
12. Lazarus RS, Folkman S. Stress, appraisal, and coping. New York: Springer Publishing Company; 1984.
13. Calhoun AW, Gaba DM. Live or let die: new developments in the ongoing debate over mannequin death. Simul Healthc. 2017;12:279–81.
14. Goldberg A, et al. Exposure to simulated mortality affects resident performance during assessment scenarios. Simul Healthc. 2017;12:282–8.
15. Robinson KA, Saldanha IJ, Mckoy NA. Frameworks for determining research gaps during systematic reviews. Report No.: 11-EHC043-EF. Rockville: Agency for Healthcare Research and Quality (US); 2011.
16. O'Sullivan D, Wilk S, Michalowski W, Farion K. Using PICO to align medical evidence with MDs decision making models. Stud Health Technol Inform. 2013;192:1057.
17. Christensen LB, Johnson RB, Turner LA. Research methods: design and analysis. 12th ed. Boston: Pearson; 2014.
18. Keppel G. Design and analysis: a researcher's handbook. 2nd ed. Englewood Cliffs: Prentice-Hall; 1982.
19. Spielberger CD, Sydeman SJ. State-trait anxiety inventory and state-trait anger expression inventory. In: Maruish ME, editor. The use of psychological testing for treatment planning and outcome assessment. Hillsdale: Lawrence Erlbaum Associates; 1994. p. 292–321.

20. Cheng A, et al. Reporting guidelines for health care simulation research: extensions to the CONSORT and STROBE statements. Simul Healthc. 2016;11(4):238–48.
21. Turner TR, Scerbo MW, Gliva-McConvey G, Wallace AM. Standardized patient encounters: periodic versus postencounter evaluation of nontechnical clinical performance. Simul Healthc. 2016;11:174–2.
22. Baddeley AD, Hitch GJ. Working memory. In: Bower GH, editor. The psychology of learning and motivation: advances in research and theory. 8th ed. New York: Academic; 1974. p. 47–89.
23. Kuhn TS. The structure of scientific revolutions. Chicago: University of Chicago Press; 1962.
24. Newton-Smith WH. The rationality of science. London: Routledge & Keegan Paul; 1981.

Designing Quantitative Research Studies

23

Karen Mangold and Mark Adler

Overview

There are variety of quantitative research designs that are amenable to use in educational scholarship. The design complexity will depend on available resources and the question(s) being investigated. A simulation-based medical education (SBME) quantitative study can range from an observational study to a complex, multiple group effort with or without randomization. Ensuring adequate number of participants are enrolled to have sufficient power to detect important differences is key; most educational research is underpowered. Certain designs (e.g., mastery learning) have grown in use.

Practice Points

- **Observational cohort** designs allow researchers to report natural experiments but have not investigator-initiated interventions.
- **Quasi-experimental and experimental** designs contain an intervention, but no randomization, and can infer causality on the intervention.
- **Randomized Control Trials** provide the most rigorous evidence that outcomes are secondary to the intervention, but can be expensive and cumbersome to administer.
- Educational studies are very often **underpowered**. Planning should account for acquiring sufficient participant to detect important differences.

Introduction

In the previous chapter, we discussed developing a research question or questions. In this chapter, we will focus on SBME scholarship that relies on the acquisition and interpretation of discrete data. Chapters 9, 10, 11, 12, 13, 14, 15, 16, 17, 18, 19, 20 will discuss SBME scholarship that is based on qualitative data as a primary source of information. Note that this is a false dichotomy and combined approaches (i.e., mixed methods) are often and productively used together.

SBME quantitative scholarship seeks to describe performance at the higher levels of Miller's pyramid construct ("shows how" or "does").

- How does a learner or learners perform on a given task or tasks?
- What effect does a given intervention (educational, quality improvement) have on performance comparing pretest to posttest?
- Does the instrument (checklist, rating scale) provide data that allows us to make accurate decisions?
- Do learners meet a specified passing standard?

Quantitative research begins with a question that serves as the basis for a hypothesis or hypotheses to be tested. The hypothesis serves as a foundation; a well-crafted one sets the course. The FINER mnemonic [1] provides a useful framework for developing an answerable and important question: is my idea or ideas *feasible* to implement, *interesting* to my

K. Mangold · M. Adler (✉)
Feinberg School of Medicine, Department of Pediatrics (Emergency Medicine) and Medical Education, Northwestern University, Chicago, IL, USA

Ann & Robert H. Lurie Children's Hospital of Chicago, Chicago, IL, USA
e-mail: karen-mangold@northwestern.edu; kmangold@luriechildrens.org

D. Nestel et al. (eds.), *Healthcare Simulation Research*, https://doi.org/10.1007/978-3-030-26837-4_23

audience, *novel* in that it provides new insight, *ethical* and provides *relevant* information that can be applied today. In the SBME domain, ethical issues need to be specifically considered. Will the process of collecting data expose learners to risk? Will learners fear their performance will impact academic standing or how they are viewed by peers or faculty?

Study Designs

For the following sections, we will refer to the widely cited models used by Shadish, Cook and Campbell [2]. In their terminology, "O" = observation, "X" = intervention, and "R" = randomization.

Observational Studies

Observational designs provide important insight about leaner performance. In SBME contexts, observational designs can be used when the investigator cannot or chooses not to introduce an experimental intervention. A main advantage is the lower cost and time investment compared to other designs. The lack of controls against known or unknown biases limits the strength of inferences that can be drawn compared to interventional designs. These approaches allow the investigator to describe the existing educational conditions and may be used to gather resources to support a formal intervention. Observational designs (Table 23.1) can yield data supporting association between variables but cannot establish causation.

The prospective cohort study allows for investigation of a change in practice, policy or population that is not controlled by the investigator (e.g., a "natural" experiment).

Table 23.1 Observation designs. (Reproduced with permission of Cook et al. [2])

	Past	Present	Future	Description
Cross-sectional		O		Study of a population at one specific time. The outcome(s) of interest relate to prevalence of an event or condition. *How many learners are able to perform a specific skill today? What kind of simulations do fellows in our field participate in during fellowship nationally?*
Retrospective Cohort	O			Study of a population performance over a span of time in the past. *How well did trainees perform on a specific simulated case over the past 5 years?*
Prospective Cohort			O	Study of a population performance over a span of time into future (without any investigator driven intervention). *Do trainees' skills at defibrillation change with the a hospital-wide introduction of new model of defibrillator that includes verbal auditory cues?*

Quasi-experimental Studies

A quasi-experimental study is "experimental" in that the investigator designs and controls the planned intervention. Both quasi- and experimental designs have an intervention (X) and observation(s) (O). Single-group quasi-experimental designs do not have a control group. Two (or more) group quasi-experimental designs have control arms but no randomization.

This design is often used when the investigators, for cost, efficiency, timeliness or due to ethical barriers, cannot implement a control group or random assignment. Table 23.2 summarizes uncontrolled quasi-experimental **single** group study designs typically used in educational research.

A quasi-experimental study design may be unable to determine if a learners' performance change is due to the intervention alone or due in part or in whole to other influences. If a new simulation curriculum is implemented over an extended time span, improvement in post-test scores may be due to the intervention or to a variety of other exposures implemented over the course of the year. The impact on confounding influences may be limited if one is evaluating topics not previously covered in the curriculum and have no other outside exposures. In this case, the investigator may argue that the changes seen are due to the study intervention. However, the investigator would make a stronger argument if a control group was used and only the intervention group was shown to have improved.

The study structures shown in Table 23.3 can be reframed as controlled, non-randomized, **two**-group studies.

The addition of the control improves the ability to claim causal relationship between intervention and learner performance. The pre-/post-test design is a mainstay of medical education research. The investigator seeks to demonstrate a significant and meaningful change in performance after the intervention. The pre-test established the important baseline score and the degree of variation (*dispersion* as measured by the standard deviation or inter-quartile range) for the study population. After an intervention, we expect to see higher scores and decreased dispersion. Smaller dispersion is a result of the decrease in between-subject variation as the learners improve. Most learner scores improve and outliers move closer to the group mean. This latter finding is an important and perhaps under-appreciated source of evidence for intervention effectiveness.

Table 23.2 Selected single group quasi-experimental design without control. (Reproduced with permission of Cook et al. [2])

	Past	Present	Future				Description
1 Group, Post-test only		X	O_1				Study investigating performance after an intervention only, limits inferences as baseline is not known. Is often used when the population is naïve to the topic and thus can be thought to have little or no baseline skills. Multiple post-tests (O_1, O_2, $O_{..}$) can be used. Addresses concern about impact of pre-test alone on outcome
1 Group, Pre/Post Test	O_1	X	O_2				Commonly used design. Allows for evaluation of change before/after an intervention. Without control, other factors could account for some or all of the identified performance change. Multiple pre or post-tests (O_1, O_2, $O_{..}$) can be used
1 Group, Pre/Post Test (Non-equivalent)	O_{1A}	X	O_{2A}				Design uses different pre- and post-tests. May address priming effect of the pre-test alone but introduces concern about how each test is linked to the construct studied and to each other
1 Group repeated treatment	O_1	X	O_2	O_3	X	O_4	Design used to evaluate incremental performance change with repeated interventions

Table 23.3 Selected two-group, non-randomized control quasi-experimental designs. (Reproduced with permission of Cook et al. [2])

	Past	Present	Future				Description
Post-test only		X	O_1				Design compares post-intervention performance to control group
			O_2				
Pre/Post test	O_1	X	O_2				Design compares pre/post-test change between intervention and control groups; a common design
	O_1		O_2				
Wait-list Control	O_1	X	O_2	O_3		O_4	Design adds to Pre/Post a second trial in which the wait-list control group receives the intervention. Addresses concern regarding withholding education from controls. Also provides a measure of decay in the first group
	O_1		O_2	O_3	X	O_4	

Randomized Control Trials (RCTs)

Randomization provides further support for the argument that any observed changes are secondary to the intervention and not due to other effects. The ability to make causal relationship arguments comes at the cost of increased complexity. RCT designs are often expensive and effortful. Selected RCT designs are described in Table 23.4.

The pre-test/post-test model is amongst the most commonly used both for its design strength and familiar design. The control group, however, receives no educational benefit. These control participants see less value in the study for their time and ethics review committees may take note of this deficit. An alternative approach which mitigates this concern is the **wait-list control design** (Fig. 23.1 [3]).

In the wait-list control design, investigators conduct a pre-test for all participants and then the intervention group receives the intervention followed by post-testing for all participants. Up to this point, this approach mirrors a simple pre–/post-design. To ensure all participants benefit from training, the intervention and control groups are switched and the control group then undergoes the intervention and the intervention group does not. Both groups then complete a final (second) post-test. Expected results for the experimental group will show improvement in intervention scores between the first and second test, but similar scores on the second and third tests. The control (delayed) group will not have improvement from the pre-test to the midpoint, but will then improve on the post-test, after they receive the intervention (Fig. 23.2).

An example of wait-list control simulation-based research design might be a procedural training that occurs with staff over the course of a month. All staff receive baseline training where they demonstrate the procedure on a task-trainer. Half of the staff then receive training on the procedure, while half do not, but continue on with their routine work for the month. After that month, the staff are again assessed demonstrating the procedure on a task-trainer. This design will allow for investigators to compare scores on the post-test and make inferences regarding the impact (or lack thereof) on routine exposure over that month. However, one can see how the actual practice of the procedure in the pre-test may have improved the staff members' performance of the procedure itself. Staff members may even have taken it upon themselves to have looked up the procedure and practiced during the interim.

The cross-over control study design is an extrapolation of the wait-list control design. It varies only in that there are two investigator-initiated interventions. An example of this study would be a lecture-based

Table 23.4 Selected two-group, randomized controlled experimental designs. (Reproduced with permission of Cook et al. [2])

		Past	Present	Future				Description
Post-test only	R		X	O_1				Design compares post-test data to control group; limited by lack of baseline
				O_2				
Pre/Post Test	R	O_1	X	O_2				Design compares pre/post-test change between intervention and control groups; a common design
		O_1		O_2				
Wait-list Control	R	O_1	X	O_2	O_3		O_4	Design adds to pre/post a second trial in which the control group receives the intervention after a delay. Address concern regarding withholding education from controls. Also provides a measure of decay in the first group. O_3 is not used in all implementations
		O_1		O_2	O_3	X	O_4	
Cross-over Control	R	O_1	X_1	O_2	O_3	X_2	O_4	Design features a two intervention such that each study is exposed to each intervention. O_3 is not used in all implementations
		O_1	X_2	O_2	O_3	X_1	O_4	

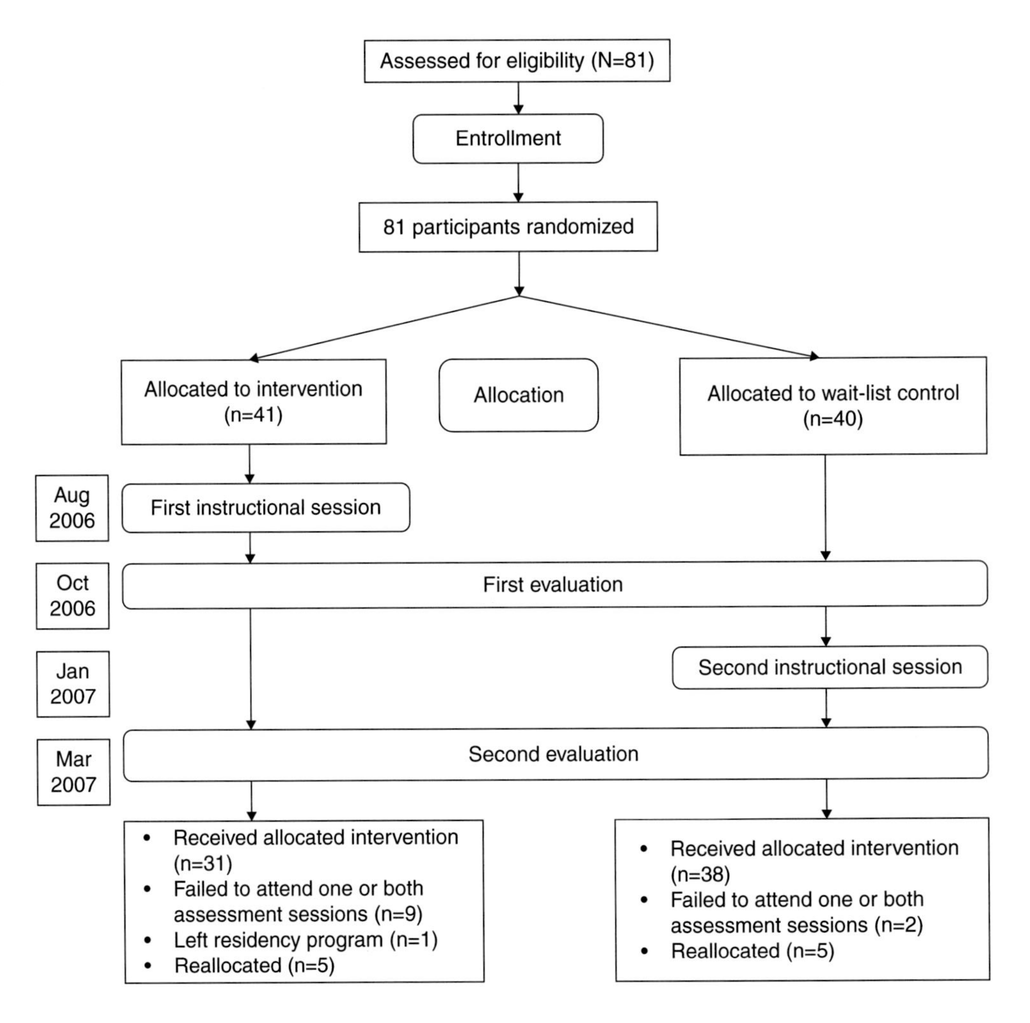

Fig. 23.1 Example of wait-list control trial design. (Reproduced with permission of Ref. [3])

curriculum versus a SBME curriculum. One group receives the standard curriculum while the other receives the new educational intervention. After the first post-test, the two groups switch to the other curriculum. Comparing the pre-test, midpoint and post-test scores allows investigators to see how the curricula compare. Care should be taken in using this model. It is reasonable to compare to two interventions to answer the question "How well do these two specific intervention compare? The comparison "Is simulation better than lecture?", however, is confounded. In this broad approach, the **medium** (simulation vs. lecture) and the **content** of each both vary in way that cannot be separated. This type of media-comparative hypothesis should be avoided [4].

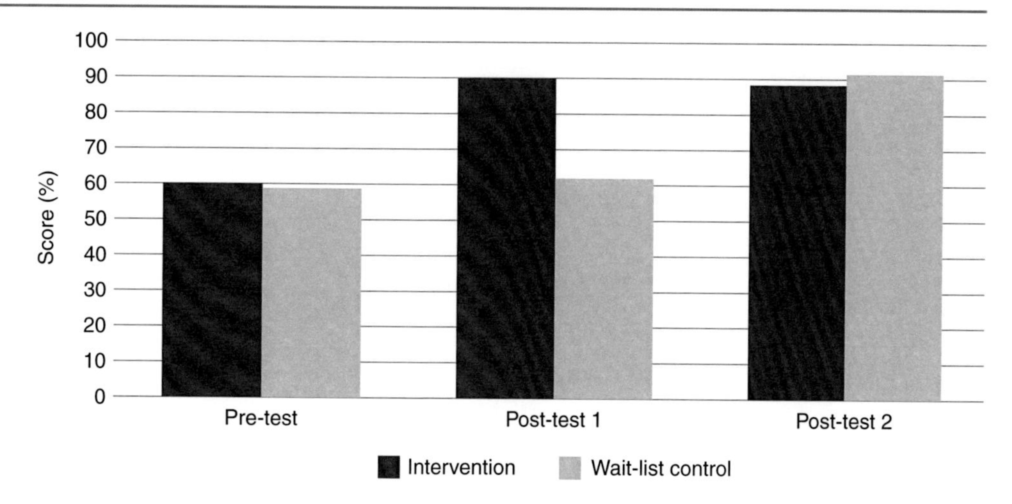

Fig. 23.2 Hypothetical example of wait-list control trial expected outcome

Considerations and Caveats

Investigators should be aware of key education research commentary as they consider their own work. Randomization, power and effect size will be considered first followed by a brief discussion or rater training and mastery learning models.

Randomization does not address or control for all sources of bias, only those arising from participants, selection bias and how the participants might change over time from a variety of causes. Table 23.5, adapted from Cook and Beckman [5], provided a broad list of threats to the validity including those not addressed by randomization.

We seek in our research to make inferences that allow us to make accurate decisions, just as we do when we assess learners. **Study power** refers to the ability to discriminate between groups given group size, the method of analysis and the predetermined study constraints (e.g., level of significance, one-or-two tailed analysis). Inadequate power risks failing to identify group differences when a difference exists (e.g., false negative). Educational investigators are often challenged to find available learners, particularly for studies beyond the undergraduate level.

Insufficient learners can seriously impact otherwise high-quality work. Cook and Hatala found that less than 30% of published health professions studies had adequate power (80%) to detect a larger group difference and essentially none (<1%) were powered to detect a smaller difference [7]. Lack of appropriate participants, however, does not provide an excuse to "throw up your hands" or to choose a study population on the basis of their availability just to supply a larger study population. A study regarding central line placement training using ultrasound which enrolls second-year medical students that the investigator has access to might show educational improvements. While a significant improvement might be shown (as there are a 100 learners or more), we learn little about those providers (residents, fellows, faculty) who actually place central lines. We need sufficient N in the population of interest.

Effect size is the quantification of the impact of a given intervention and is a unit-less metric that allows for comparison to other interventions [8]. Recall that statistical significance does not provide information about the magnitude of an outcome. Authors are expected to provide both measures of significance and of magnitude. Effect size data allows the reader to make a context-specific judgement about the reported impact and if that would be meaningful. "Our intervention group significantly outperformed the control group ($p < 0.05$) with an effect size of 0.8 which demonstrates a meaningful change".

The **mastery learning** (ML) model [9] is an educational approach that draws on traditional pre-test/post-test single group design. ML designs include a pre-test to inform both the baseline state and provide a source of psychometric data. The learners then receive an intervention, followed by a post-test. Where ML diverges from other models is how the post-test data is used. An a priori minimum passing standard is determined through a rigorous process (see Yudkowsky for more detailed information) [10]. Learners who do not meet this standard after the initial intervention timeframe are offered further time. The intervention is complete when all participants have met the standard. Unlike other designs where the outcome of the study is assessing the change in learning (e.g., learners had a 29% improvement after the intervention), ML by definition results in 100% meeting a standard. ML outcomes are directed towards translational outcomes. After a cohort achieves mastery in vitro (sim lab), does this translate to in vivo (clinically) and do patients and

Table 23.5 Threats to validity. (Reproduced with permission of Cook and Beckman [5])

Threat	Description	Randomization addresses?	Control group addresses?	Mitigation
Participants characteristics	How learners differ at outset	Y	Y	
Selection bias	Learners distributed in biased manner	Y	Y	
Maturation	Learners change over time (e.g., learn from other sources)	Y	Y	
History	"‥, all events that occur between the beginning of the treatment and the posttest that could have produced the observed outcome in the absence of that treatment" [2]	N	Y	Use a concurrent control group
Instrumentation	Change in rating tool or rater performance	N	Y	Control group
Regression to the mean	High or low performing group will tend toward average performance [6]	N	Y	Avoid choosing group assignments that are based on prior/baseline performance
Testing	Taking pre-test affects outcome – familiarity with items, prompts outside learning	N	N	Post-test only
Loss to follow-up	Participants do not complete study – losses may not be equal across groups	N	N	Collect data on those who leave; try to avoid loss
Location	Difference in group in study site or resources	N	N	Collect data on differences
Participant attitudes	Intervention group may be more motivated than control	N	N	Blinding
Implementation	Variation in instructor behavior or student adherence	N	N	Rigorous planning, collect data on how implementation was done vs. planned

the healthcare system see benefits? McGaghie has described simulation-based research as a translational science [11]. He reminds the reader that as we seek quantitative evidence of changes at the level of learner performance, we should consider and seek evidence change that impact patient and society.

Closing

Research studies are often a compromise between ideal study design and logistical reality. Control groups and randomization may not always be necessary or possible but including these features in a study improves the validity of the generated data. Designs should take into consideration how to enroll sufficient participants to be powered to detect an important difference. Appropriate study design is important both to test your hypothesis and for successful completion of your inquiry. In the next chapter, we will discuss what outcomes measures and data can be obtained after having attentively designed a simulation-based research study.

References

1. Hulley SB, Cummings SR, Browner WS, Grady DG, Newman TB. Designing clinical research. Philadelphia: Wolters Kluwer/ Lippincott Williams & Wilkins; 2013.
2. Cook TD, Campbell DT, Shadish W. Experimental and quasi-experimental designs for generalized causal inference. Boston: Houghton Mifflin; 2002.
3. Adler MD, Trainor JL, Siddall VJ, McGaghie WC. Development and evaluation of high-fidelity simulation case scenarios for pediatric resident education. Ambul Pediatr. 2007;7(2):182–6. https://doi.org/10.1016/j.ambp.2006.12.005.
4. Friedman CP. The research we should be doing. Acad Med. 1994;69(6):455–7.
5. Cook DA, Beckman TJ. Reflections on experimental research in medical education. Adv Health Sci Educ Theory Pract. 2010;15(3):455–64.
6. Bland JM, Altman DG. Some examples of regression towards the mean. BMJ. 1994;309(6957):780.
7. Cook DA, Hatala R. Got power? A systematic review of sample size adequacy in health professions education research. Adv Health Sci Educ Theory Pract. 2015;20(1):73–83. https://doi.org/10.1007/s10459-014-9509-5.
8. Salkind NJ, Rasmussen K. Encyclopedia of measurement and statistics. Thousand Oaks: SAGE Publications; 2007. 1136.
9. McGaghie WC. When I say … mastery learning. Med Educ. 2015;49(6):558–9.
10. Yudkowsky R, Park YS, Lineberry M, Knox A, Ritter EM. Setting mastery learning standards. Acad Med. 2015;90(11):1495–500.
11. McGaghie WC. Medical education research as translational science. Sci Transl Med. 2010;2(19). 19cm8.

Outcomes Measures and Data

24

Pamela Andreatta

Overview
There are two forms of disciplined inquiry that examine impacts of training or educational interventions on performance outcomes: Research and Evaluation. Although similar in many practical respects, the purposes of each are quite different. The criteria for designing and implementing controls when examining study outcomes are distinct and much more rigorous for research than they are for evaluation. *Evaluation* examines outcomes that target specific interventions in specific contexts, the results of which are not intended to be interpreted as broadly generalizable to other contexts. *Research* leads to conclusions across multiple contexts through the deliberate definition of variables and methodological controls. This chapter will focus on *Educational Research* associated with performance outcomes in healthcare contexts. Apart from the generalizability requirements, the information and processes presented in the chapter are equally applicable to program evaluation, wherein multi-site data may provide substantive information for meta-analyses. The chapter includes determining the type of research design to implement based on the outcome variables of interest (performance, clinical, financial, etc.) and their relationships to directed interventions. It also addresses identifying and operationalizing the variables of interest for study; optimizing the precision of measurement for outcomes variables; and choosing from several quantitative research designs that examine the relationships between outcomes variables of interest in healthcare education contexts: (1) Quasi-Experiment; (2) Correlational; (3) Meta-Analysis. Example studies are provided for each of the explicated research designs. Detailed methodological attributes, such as sampling, operationalization, statistical analyses, etc. are addressed elsewhere in this book and are not covered in depth.

P. Andreatta (✉)
Uniformed Services, University of the Health Sciences, Bethesda, MD, USA

Practice Points

- Define the purpose of inquiry to determine if evaluation or research protocols are appropriate.
- Identify outcomes to be examined (dependent variables) and their known and potential relationships to other factors (independent variables).
- Establish and confirm valid and reliable measurement strategies for all variables of interest.
- Identify institutional and other available sources of extant data or other metrics that are available through non-study related inquiry (patient safety, quality control, financial, utilization, turnover, etc.) to use in analyses.
- Determine optimal practical approach to facilitating scientific examination of the relationships between independent and dependent variables (or predictor/criterion variables).

Introduction

There are two types of disciplined inquiry that may examine the impacts of training or educational interventions: *Research* and *Evaluation*. Although *Research* and *Evaluation* are similar in many practical respects, their purposes are quite different and the methodological rigor required for examining *Research* study outcomes are distinct from the criteria for *Evaluation* findings. The purpose of *Evaluation* is to make data driven decisions about utilization of a particular intervention, product, or process that is expected to improve performance or practices in specific contexts [1, 2]. The data are used to determine if specific objectives are met within the context, with *no* expectation that they are transferrable to other contexts or advance understanding through the generation of new knowledge. This chapter will focus on *Research*,

D. Nestel et al. (eds.), *Healthcare Simulation Research*, https://doi.org/10.1007/978-3-030-26837-4_24

which leads to conclusions across multiple contexts through the deliberate definition of variables and methodological controls [3]. Apart from the generalizability requirements, the processes described herein are equally applicable to program evaluation and where multi-site data may provide substantive information for meta-analyses. Meta-analyses will be discussed later in the chapter.

Studies that examine the impact of interventions on performance outcomes are the epitome of educational research in healthcare, largely because all instructional and training initiatives optimally intend to target specific performance needs or gaps [4]. As such, it is essential that rigorous educational research identify and measure the salient variables that reflect the desired performance outcomes associated with the instruction or training [5, 6]. These variables may be embedded within specific learning objectives or established as having a strong association with the learning objectives. For example, a learning objective may state that the learner will be able to identify the best vein for peripheral venous access using a model or simulator, which would have a strong association with the ability to select the optimal vein for peripheral venous access in a live patient – a performance outcome. It will also likely have a strong association with the presence or absence of a hematoma in the patient after peripheral venous access is established – a clinical outcome.

The types of outcome variables (performance, clinical, financial, etc.) and their relationships to directed interventions will largely determine what type of research design to implement [7–9]. First considered in this chapter are the means by which the variables of interest for study are identified and operationalized. Second, sensitivity of measurement for outcomes variables is introduced. Third, quantitative research designs for examining the relationships between outcomes variables of interest are described: (1) Quasi-Experiment; (2) Correlational; and (3) Meta-Analysis. Details about specific methodological attributes (sampling, operationalization, statistical analyses, etc.) are covered elsewhere in this text and will not be covered in depth.

Variables of Interest

Variables are entities that represent specific factors, attributes, concepts, and contexts. For simulation supported research targeting clinical or performance outcomes, they may represent instructional, training, or practice-based interventions thought to influence a performance outcome (clinical or practice). Depending on the research questions, variables may be considered independent (predictor) variables; dependent (criterion) variables; or co-variables (intervening, moderating, control). Variables are associated with the research questions and it is essential to delineate and operationalize them accurately in order to assure the integrity of a study [10]. For example, if a researcher intends to examine the impact of simulation-based training to establish intravenous access in an adult patient (independent variable), collecting outcomes data for trained individuals performing intradermal injections (dependent variable) would not make sense. Specifying the variables of interest in a study may be quite challenging and requires researchers to fully understand the performance construct, both theoretically and empirically [11]. In the example above, the dependent variable is the ability of trained individuals to establish intravenous access in an adult patient in applied practice. However, there are important co-variables that must be measured or controlled for in the methodology and analyses. At minimum the researcher needs to consider the following variables, which have the potential to confound the study outcomes if not controlled:

1. The models, resources, and instructional methods used for both simulation and an alternate form of training.
2. Any pre-existing abilities and knowledge of the learners in the performance domain prior to instruction.
3. Post-training abilities and knowledge of learners in the performance domain.
4. Time delay between training in either context (simulation, alternate form) and applied performance in clinical context.
5. Commonality of resources during instruction and applied practice.
6. Case difficulty in applied practice context.

The first step in measuring a variable is to devise an operational definition for it. Each of the variables referenced above require operational definitions in order to collect data and control for potentially intervening factors related to the research question. Operational definitions specify the evidence the researcher is willing to accept to confirm the concepts of interest exist and are occurring – such as performance improvement as a result of instructional intervention. The more direct and measurable the operational definition is in representing the variable of interest, the more compelling the evidence will be when analyzing study outcomes [10, 11].

To operationally define a variable, first identify the concept of interest, then consider behavioral indicators that will provide evidence of that concept. Write a list of observable behaviors that researchers will accept as evidence that the concept or outcome of interest exists. Determine which indicators are most valid and which may have lesser influence. Write the operational definition as precisely as possible so that it is possible to determine measurement strategies for each variable. For observable behaviors, factors, or contexts, employing a direct approach by focusing on what is to be observed and ignoring irrelevancies will add precision to the operational definition. However, evidence of a concept of interest may not be directly observable (e.g. knowledge, compassion). An observable indicator behavior that closely aligns with the variable of interest facilitates inferential evidence for analyses. The less direct the evidence is, greater the inferential leap to answer the research question.

Indirect evidence can be bolstered by the use of multiple operational definitions and carefully collecting difference types of evidence. An effective strategy is to have multiple operational definitions and multiple methods for measuring variables, especially outcomes variables [12]. The logic of this strategy is to mitigate the weakness of an operational definition by defining other operational definitions and collecting data using different methods. Triangulation is the strategy of measuring a variable of interest from several different angles, which reduces biases and deficiencies that are inherent in any single measurement method.

The researcher is responsible for eliminating alternative explanations for the presence of evidence in research, and operationally defining all variables associated with a research question is the first step towards assuring the integrity of the research outcomes [12, 13]. Using multiple operational definitions for each variable will help determine how best to develop measurement and data collection processes.

Measurement

Once your variable operational definitions are determined, measures must be developed to capture data for analysis. The measures must reflect the variable of interest and be precise enough to capture anticipated variants within the sample. That is, the more a measure is affected by the variable of interest, the more likely the measure is going to be sensitive to differences between individuals in the sample rather than by bias, intervening variables, or random error. Sensitive measures mitigate the extent that a measure varies due to random error and not due to differences in participants. A sensitive measure involves fewer inferences, and therefore will have greater validity and reliability than an insensitive measure [14, 15].

The most important aspect of a measure is whether it allows the researcher to make the kind of comparisons necessary to answer the research question. Determining precisely what is to be measured will increase the chance that the measure is both valid and reliable. Ideally, a researcher will select a measure that is directly aligned with the characteristic to be measured [16]. Determining the form and scale of measurement should influence what measure to use and can be facilitated by considering the following questions: (1) What scale of measurement will provide information to best answer the research question? and (2) Which of the measures under consideration will provide this level of measurement? When determining a measurement scale, a higher scale of measurement will provide more information than others. The aim is to capture a wide range of scores from participants to assure as much score variability in the data as possible, and using sensitive measures avoids measuring data that produce a limited range of scores. Favoring numeric values or other concrete, observable indicators, along with considering the simplest and most direct way to measure a variable will increase both reliability and validity. However, not all numbers provide equivalent information! Numbers may represent different states (nominal scale), increments such that larger numbers represent greater amounts (ordinal scale), specific increments that represents how much greater the amount (interval scale), and equal increments with an absolute zero (ratio scale). For example, consider the following different numeric representations and associated research questions:

Nominal Scale: Do two groups to differ on ability?

Ordinal Scale: Does one group have more of an ability then another?

Interval Scale: How much more ability does one group have when compared to another group?

Ratio Scale: Does one group have more then three times as much ability that another group?

Determining the type of numbers to be used is an important part of determining a measurement. Because greater validity and reliability results when less inference is required in analyzing data, precise measurement is always favored over imprecise. All measures may be vulnerable to observer or rater biases, reliability concerns, and sensitivity of measurement. The key is to find the measure where weaknesses are less likely to confound data collection and most accurately reflect the concept or context of concern. Choose a measure that is as unbiased and reliable as possible because unreliable data resulting from random error have limited validity, which makes it difficult to obtain statistically reliable differences between groups [3, 11, 16]. To determine whether scores will actually vary, try out measures with a few participants in a pilot test before conducting a full study. Without a pilot test, measurement concerns or errors will arise as problematic after the study is completed, when it is too late to make corrections.

Example 1 (A Tale of Two Defibrillators [17]*)* Defibrillators are being replaced at a 5000-bed tertiary care hospital and level-1 trauma center. Hospital administrators have selected two models for consideration, both of which are quite different from the existing defibrillators. Administration has asked the simulation center to determine which model is easiest to use accurately across all levels of clinical staff. Optimal use of defibrillators requires accurate placement and equipment settings, and minimal time to shock delivery, so models should be intuitive to use by clinical staff members without extensive training (manual and automatic modes). The research question is: Which defibrillator is best for intuitive, accurate, and timely implementation in automatic and manual modes by all levels of clinical staff? The variables, operational definitions, and associated measurements are presented in Table 24.1.

Table 24.1 Variable operational definitions and measures for defibrillator usability study

Variables	Operational definition 1	Operational definition 2	Operational definition 3
Clinical staff level	*Professional role:* Physicians (nom 1) Residents/Interns (nom 2) Nursing (nom 3) Health professions (nom 4)	*Years of experience:* (#years)	*Self-rated defibrillation expertise:* Rating scale (low 1, 6 high)
Patient positioning	*Patient position, orientation:* Optimal (rat 2) Acceptable (rat 1) Sub-optimal (rat 0) Incorrect (rat −1)	*Proximity to defibrillator:* Optimal (int 1) Too close (int 0) Too far (int −1)	*Time to position patient, equipment:* (#sec)
Pad placement	*Pad position, adherence:* optimal (rat 2) Acceptable (rat 1) Sub-optimal (rat 0) Incorrect (rat −1)	*Defibrillator lead connection:* Accurate (rat 2) Mostly accurate (rat 1) Mostly inaccurate (rat 0) Inaccurate (rat −1)	*Time to place, connect pads:* (#sec)
Setting selections, automatic mode	*Time to access settings controls:* (#sec)	*Accuracy of setting selections:* Accurate (rat 2) Mostly accurate (rat 1) Mostly inaccurate (rat 0) Inaccurate (rat −1)	*Time to complete setting selections:* (#sec)
Activate/deliver shock, automatic mode	*Safe delivery:* Safe (ord 1) Unsafe (ord −1)	*Accurate delivery:* Accurate (ord 1) Inaccurate (ord −1)	*Time to complete shock delivery:* (#sec)
Setting selections, manual mode	*Time to access settings controls:* (#sec)	*Accuracy of setting selections:* Accurate (rat 2) Mostly accurate (rat 1) Mostly inaccurate (rat 0) Inaccurate (rat −1)	*Time to complete setting selections:* (#sec)
Activate/deliver shock, manual mode	*Safe delivery:* Safe (ord 1) Unsafe (ord −1)	*Accurate delivery:* Accurate (ord 1) Inaccurate (ord −1)	*Time to complete shock delivery:* (#sec)
Intuitiveness of equipment use	*User errors, automatic mode:* (#Errors)	*User errors, manual mode:* (#Errors)	*Staff rankings of models:* Rating scale (low 1, 6 high)
Total time to complete, automatic mode	*Sum of all automatic mode times:* (positioning, pad placement, setting selection, shock delivery)		
Total time to complete, manual mode	*Sum of all manual mode times:* (positioning, pad placement, setting selection, shock delivery)		

Notes: # (number of), *NOM* (nominal), *INT* (interval), *ORD* (ordinal), *RAT* (ratio), *SEC* (seconds)

Quasi-experimental Research

True experimental methods are largely impractical in educational research because random selection and assignment of participants to treatment conditions is rarely possible [18]. As a result, quasi-experimental methods are the most controlled form of educational research and are relatively straightforward to implement, as long as the researcher controls for confounding variables in the research processes and contexts. Researchers must have a deep understanding of the tradeoffs between external and internal validity and how to accommodate limitations for each in either the study design or analyses. Quasi-experimental procedures attempt to compensate for randomization weaknesses by exerting rigorous control of sampling, processes, procedures, and contextual elements in order to eliminate as many threats to internal validity as possible [19–21]. Quasi-experimental research designs control for selection sampling bias by confirming the extent to which comparable groups are similar at the onset of data collection, often using a pretest. If the groups are not initially equal, adjustments in the statistical analysis and interpretation must accommodate the non-equivalence.

Quasi-experimental outcomes based research is appropriately not amenable to comparisons of an untreated control group with one or more treated groups. The reason for this is largely because a true control group design – wherein one group receives no intervention and the other receives an intervention of training or instruction – logically compares the gain of something (intervention) to the gain of nothing (no intervention). If one teaches something to one group but not the other, clearly the group that receives instruction will far surpass the group that did not receive instruction when asked to apply the information or abilities that were taught. Hence, it is essential that when designing multi-group comparisons all groups receive some form of equivalent instruction. Educational studies designed to examine the relative value of simulation supported instruction must provide an equivalent learning opportunity for the comparison group, such as traditional training within the applied environment. If both interventions are documented as equivalent, apart from the salient details of interest inherent in each context, the true value of any technique, process, or knowledge gained will be irrefutable and more valid.

The interaction between study variables and other factors may also occur and these should be controlled through methodological or statistical processes to the extent possible. For example, if a treatment has a permanent impact, it will be difficult to interpret data resulting from a time series of repeated treatments if specific controls between treatments are not adhered to. Similarly, if a factor is known to impact or interact with a treatment it must be controlled for in order to mitigate its influence on study outcomes. Controlling potentially confounding factors can be achieved through a variety of techniques, including combining several quasi-experimental designs into a single study to leverage the strengths of different designs and effectively optimize validity. Similar to having multiple operational definitions of variables, varied methods of data collection improve the validity of resulting outcomes.

Example 2 (Comparison of Two Types of Training for PICC Placement [22]*)* The purpose of the study is to compare two methods of instruction for ultra-sound guided PICC placement on the acquisition of applied skills in the operative setting: in-site training in the operative context (N = 16) and simulation-based training using ultra-sound compatible vascular access phantoms (N = 16). The groups had equivalent pre-training performance abilities in the domain of interest and were randomly assigned to one of the training groups. Expert clinicians blinded to training status of the subjects scored post-training operative performance. The research question is: Are the two methods of training in ultrasound guided PICC placement equivalent for the acquisition of associated abilities in applied operative care? The variables, operational definitions, and associated measurements are presented in Table 24.2.

Table 24.2 Variable operational definitions and measures for two types of picc placement training

Variables	Operational definition
Training group	*Simulation-based training (nominal)* *Operative-based training (nominal)*
Use of ultrasound	*Operative performance score: ultrasound use;* 6-point scale, 1 = very poor to 6 = outstanding (interval)
Demonstrate vein compressibility	*Operative performance score: vein compression;* 6-point scale, 1 = very poor to 6 = outstanding (interval)
Transverse visualization of vein	*Operative performance score: vein visualization-transverse;* 6-point scale, 1 = very poor to 6 = outstanding (interval)
Longitudinal visualization of vein	*Operative performance score: vein visualization-longitudinal;* 6-point scale, 1 = very poor to 6 = outstanding (interval)
Localization of needle	*Operative performance score: needle localization;* 6-point scale, 1 = very poor to 6 = outstanding (interval)
Guide needle into vein lumen	*Operative performance score: needle guided into vein lumen;* 6-point scale, 1 = very poor to 6 = outstanding (interval)
Thread guidewire	*Operative performance score: guidewire threaded;* 6-point scale, 1 = very poor to 6 = outstanding (interval)
Exchange needle for catheter via guidewire	*Operative performance score: exchange needle/catheter via guidewire;* 6-point scale, 1 = very poor to 6 = outstanding (interval)
Dilate, advance catheter, establish access	*Operative performance score: dilate, advance catheter;* 6-point scale, 1 = very poor to 6 = outstanding (interval)
Position catheter, central superior vena cava	*Operative performance score: position catheter;* 6-point scale, 1 = very poor to 6 = outstanding (interval)
Attempts to access, position catheter in vein	*Number of attempts to access, position catheter (max 3 allowed); ratio*

Correlational Research

All research that is designed to measure the relationship between outcomes variables (performance, patient, clinical, institutional, etc.) and other factors (variables, characteristics, interventions, etc.) is subject to the same challenges of any research involving human subjects. It may be ethically impossible, morally questionable, unfeasible, or simply inconvenient to conduct any form of experimental research that requires some form of randomization of subjects to distinct treatment groups [18]. Consequently, there are many cause-and-effect research questions that are not easily examined through experimental methodologies. However, many of these questions can be addressed through qualitative or quantitative descriptive strategies, some of which may examine relationships that imply causality without directly examining a causal relationship.

Correlational research does not support causal inferences as directly as experimental methodologies; rather they demonstrate if there is a consistent relationship between a variable (e.g. treatment) and a designated outcome [23, 24]. The purpose of correlational research is to examine the extent to which changes in one variable corresponds to changes in another variable. The extent to which the variables correlate is indicated by a correlation coefficient between −1 and +1, where −1 is a perfect inverse correspondence, +1 is a perfect correspondence, and 0 is no correspondence. Correlational methodologies are non-experimental research designs that examine relationships between variables, even though they may not be causal relationships. This is especially true if the researcher demonstrates there is no relationship between treatment (e.g. simulation supported instruction) and outcome variables (e.g. ability to perform) as this provides strong evidence that the treatment did not cause the outcome. However, it is difficult to assert positive causality from positive correlation between variables.

It is relatively straightforward to perform a correlational study! It simply requires selecting a group of subjects, identifying the characteristics of interest (variables), measuring those variables for all subjects, and examining the data for relationships between the variables. Examples 3 and 4 respectively describe simple and complex correlational studies.

Example 3 (Correlation between Attitudes to Patient Safety and Motivation to Participate in Simulation-based Team Training [25]*)* The goal of the study is to examine the relationship between medical students' attitudes about patient safety and their motivation to participate in simulation-based team training in the context of surgical emergencies (Table 24.3).

Example 4 (Impact of Simulation-based Mock Code Program on Pediatric Survival Rates [26]*)* The purpose of the study is to examine the longitudinal impact of a simulation-based mock code program on pediatric and neonatal survival rates over four years. The research question is to what extent the frequency of simulation-based mock codes and training focus (cardiopulmonary arrest with and without pulse) influences the survival rates in actual pediatric and neonatal patients. The variables, operational definitions, and associated measurements are presented in Table 24.4.

Table 24.4 Variable operational definitions and measures for simulation-based mock code study

Variables	Operational definition
Study period	*Year number (interval)*
Mock code frequency	*# Mock codes/month (ratio)*
Mock code training focus	*100% Pulseless CPA (nominal 0)* *100% Pulse-present CPA (nominal 1)* *50% Pulseless CPA, 50% Pulse-present CPA (nominal 2)*
Pediatric and neonatal cardiopulmonary survival rates	*Pulseless CPA survival rates (ratio)* *Pulse-present CPA survival rates (ratio)* *Total CPA survival rates (ratio)*
Controlled variables	
Additional training	*PALS training (nominal)*
Hospital beds	*# Hospital beds (ratio)*
Patient days	*# Days services used by all patients (ratio)*
Census %	*Average number of patients per day (ratio)*
APR-DRG total pediatric inpatient case mix indices	*Calculated indices (interval)*
Discharge average LOS	*# Days patient remains in hospital (ratio)*
NACHRI patient acuity index	*Calculated indexes (interval)*
Physician staff	*# Physician staff caring for patients*
Nursing staff	*# Nursing staff caring for patients*
Facilities and equipment	*# Facility and equipment changes during study period (ratio)*

Notes: *#* (number of), *PALS* (pediatric advanced life support), *NRP* (neonatal resuscitation program), *APR-DRG* (all patient refined diagnosis-related groups), *LOS* (length of stay), *NACHRI* (national association of children's hospitals and related institutions, developed 2006), *CPA* (cardiopulmonary arrest)

There are several statistical approaches to measuring the strength of the relationship between variables, depending upon the measurement used to capture data. Pearson correlation coefficient measures the strength of linear relationships between two interval or ordinal scales. It is also possible to determine the coefficient for strength of curvilinear relationships (eta), or a linear relationship between two variables when a third variable is controlled (partial correlation). If a researcher is interested in examining a relationship between the combination of two or more variables (predictors) and a specific outcomes variable (criterion), multiple correlation analyses measure the strengths of those relationships and facilitate the development of a predictive model. Predictive modeling is beyond the scope of this chapter, however it is a powerful technique for determining both correlative and causative relationships between multiple variables of interest.

Table 24.3 Variable operational definitions and measures patient safety attitudes and training motivation

Variables	Operational definition
Attitudes about patient safety	*APSQ score: self-assessment, 26-item, 7-point Likert scale (interval)*
Motivation to participate in simulation-based team training (Surgical emergencies)	*SIMS score: self-assessment, 4-item, 7-point Likert scale (interval)*

Notes: *APSQ* (attitudes to patient safety questionnaire), *SIMS* (situational motivation scale)

Meta-Analyses

Many of the problems in interpreting and applying the results of quasi-experimental and correlational research arise because studies usually have relatively small numbers of subjects, imprecise measurements, inconsistent processes, and restricted or unique contexts in which experiments occur. Studies with relatively small numbers of subjects accumulate type II errors, which may lead to incorrectly determining negative outcomes from one treatment when a true difference exists. Power refers to the likelihood that type II error is minimized and any differences resulting between treatments are accurate. Educational research outcomes are prone to errors arising from low power, which are likely to occur with small samples (less than 70 subject). To counteract this challenge, it may seem reasonable to consider several studies on a particular topic and tally the results as a method of summarizing the most compelling evidence of outcomes. However, this would be seriously misleading! Drawing conclusions based on the outcomes from isolated studies is analogous to deriving similar conclusions from case studies of isolated learners. A better approach is to implement action research, including case studies and evaluation reports that contribute to meta-analytical integration of targeted outcomes measures [27, 28].

The major contribution of meta-analysis is that it provides a method for reviewing and synthesizing the quantitative research in the literature. Minimally, meta-analysis partially resolves concerns associated with reliability, limited operational definitions and data collection methods, threats to internal validity, instability, excessive subjectivity regarding external validity, unexamined interactions, accurate statistical conclusions, and impracticality of results. Meta-analysis effectively integrates the results of several studies using effect sizes to calculate the cost benefit ratio of implementing various educational treatments compared to their effect size benefits, thereby providing information about the practical usefulness of a treatment. An analysis of different studies using multiple operational definitions and multiple methods of measurement is strategically desirable for increasing the validity of conclusions. Errors of measurement arising from reliability are minimized because such errors tend to balance out in the long run. As the overall number of subjects in the meta-analysis increases, the likelihood of instability from type II errors reduces radically. Lastly, by considering possible contaminating factors as moderator variables, meta-analysis facilitates systematic examination of potential threats to the interpretation of outcomes. Therefore, in any meta-analysis it is important to code each study for moderator variables to determine the existence of important interactions.

Meta-analysis is a useful tool for assembling the best evidence to answer a research question when small sample sizes are in effect, however meta-analysis is not a panacea. It is important to systematically evaluate all studies for perceived weakness or differences in statistical analyses, especially for weak studies that lead to different outcomes. Reviewer bias can easily influence the nature of conclusions drawn from the review and integration. As with all quantitative research, bias can lead to false conclusions and this is no different with meta-analysis. Therefore, it is important to exercise the same methodological rigor in performing or interpreting meta-analysis as merited for other quantitative methods. Reporting of meta-analysis outcomes must include overall significance and effect sizes, interactions, qualitative data, relationship of the results to related theory, summary descriptions of operational definitions, data collection processes, threats to internal validity, and any other information involved in the coding of individual studies for analysis with regard to the independent, dependent, moderator, or control variables used in the meta-analysis.

References

1. Isaac S, Michael WB. Handbook in research and evaluation. 3rd ed. San Diego: EdITS; 1995. p. 7–11.
2. Vockell EI, Asher JW. Educational research. 2nd ed. Englewood Cliffs: Prentice Hall; 1995. p. 1–11.
3. Shavelson RJ. Statistical reasoning for the behavioral sciences. 3rd ed. Needham Heights: Allyn & Bacon; 1996. p. 1–40.
4. Shavelson RJ. Statistical reasoning for the behavioral sciences. 3rd ed. Needham Heights: Allyn & Bacon; 1996. p. 209–330.
5. Vockell EI, Asher JW. Educational research. 2nd ed. Englewood Cliffs: Prentice Hall; 1995. p. 87–120.
6. Mitchell M, Jolley J. Research design explained. 4th ed. Belmont: Wadsworth-Thompson Learning; 2001. p. 71–115.
7. Vockell EI, Asher JW. Educational research. 2nd ed. Englewood Cliffs: Prentice Hall; 1995. p. 17–30.
8. Isaac S, Michael WB. Handbook in research and evaluation. 3rd ed. San Diego: EdITS; 1995. p. 45–103.
9. Krathwohl DR. Methods of educational & social science research: an integrated approach. 2nd ed. New York: Longman; 1998. p. 21–35.
10. Vockell EI, Asher JW. Educational research. 2nd ed. Englewood Cliffs: Prentice Hall; 1995. p. 67–73.
11. Popham WJ. Modern educational measurement. Needham: Allyn & Bacon; 2000. p. 89–114.
12. Creswell JW. Research design. 3rd ed. Thousand Oaks: Sage Publication; 2009. p. 203–24.
13. Vockell EI, Asher JW. Educational research. 2nd ed. Englewood Cliffs: Prentice Hall; 1995. p. 375–96.
14. Krathwohl DR. Methods of educational & social science research: an integrated approach. 2nd ed. New York: Longman; 1998. p. 421–46.
15. Mitchell M, Jolley J. Research design explained. 4th ed. Belmont: Wadsworth-Thompson Learning; 2001. p. 120–40.
16. Popham WJ. Modern educational measurement. Needham: Allyn & Bacon; 2000. p. 278–303.
17. Andreatta P. Outcomes-based research; Performance transfer from simulation contexts to applied care contexts. [Keynote Address] Asia Pacific Meeting for Simulation in Healthcare, Singapore. 11th Nov 2016.
18. Krathwohl DR. Methods of educational & social science research: an integrated approach. 2nd ed. New York: Longman; 1998. p. 204–19.
19. Creswell JW. Research design. 3rd ed. Thousand Oaks: Sage Publication; 2009. p. 145–69.
20. Vockell EI, Asher JW. Educational research. 2nd ed. Englewood Cliffs: Prentice Hall; 1995. p. 269–90.

21. Mitchell M, Jolley J. Research design explained. 4th ed. Belmont: Wadsworth-Thompson Learning; 2001. p. 383–421.
22. Andreatta P, Chen Y, Marsh M, Choo K. Simulation-based training improves applied clinical placement of ultrasound-guided PICCs. Support Care Cancer. 2011;19:539–43. https://doi.org/10.1007/s00520-010-0849-2.
23. Vockell EI, Asher JW. Educational research. 2nd ed. Englewood Cliffs: Prentice Hall; 1995. p. 291–314.
24. Mitchell M, Jolley J. Research design explained. 4th ed. Belmont: Wadsworth-Thompson Learning; 2001. p. 426–64.
25. Escher C, Creutzfeldt J, Meurling L, Hedman L, Kjellin A, Felländer-Tsai L. Medical students' situational motivation to participate in simulation based team training is predicted by attitudes to patient safety. BMC Med Educ. 2017;17:37. https://doi.org/10.1186/s12909-017-0876-5.
26. Andreatta P, Saxton E, Thompson M, Annich G. Simulation-based mock codes significantly correlate with improved pediatric patient cardiopulmonary arrest survival rates. Pediatr Crit Care Med. 2011;12(1):33–8. https://doi.org/10.1097/PCC.0b013e3181e89270.
27. Krathwohl DR. Methods of educational & social science research: an integrated approach. 2nd ed. New York: Longman; 1998. p. 553–65.
28. Vockell EI, Asher JW. Educational research. 2nd ed. Englewood Cliffs: Prentice Hall; 1995. p. 353–74.

Designing, Choosing, and Using Assessment Tools in Healthcare Simulation Research

25

John Boulet and David J. Murray

Overview

Studies in healthcare simulation research are often based on performance scores. These scores can be used to compare provider groups, establish the efficacy of competing educational programs, and identify clinical skills deficiencies. The following chapter provides an overview of the development and use of assessment tools. Researchers need to select tools that align with the purpose of the assessment. Where human evaluators are employed, they should have sufficient expertise in the domains being assessed. Training is also necessary to ensure that evaluators are using the rubrics as intended. In the future, technology may be helpful for gathering accurate data from, and providing standardized scoring for, various simulation-based assessments. Healthcare simulation researchers who employ assessment tools need to evaluate whether the scores that are produced represent reliable and valid estimates of ability. Without some assurance of the psychometric rigor of the scores, their use in any research study could be questioned.

Practice Points

- Researchers should seek evidence to support the reliability and validity of assessment scores.
- The choice of scoring method will depend on the purpose of the assessment and the skills being assessed.
- Measurement error associated with scoring can be minimized by training evaluators and providing adequate instruction on how the assessment tool should be used.
- If aggregate assessment scores are used as an outcome measure, researchers must justify their formulation.
- When developing new scoring tools, or modifying existing ones, researchers should question whether their interpretation of the meaning of the assessment scores is reasonable for their study participants.

Introduction

Healthcare simulation research, whether based on electromechanical mannequins, standardized patients(SPs), part task trainer or hybrid simulation models, is dependent on the design of relevant content and the development of appropriate tools to assess performance [1]. The assessment tools or, more important, the scores derived from the tools, are necessary to conduct the research that is currently at the forefront of simulation-based assessment. The following chapter will outline some of the important issues to consider when designing, choosing, and using scoring tools in healthcare simulation research.

This chapter is organized into five sections. In the "Introduction", we provide a summary of assessment principles. What evidence should researchers gather, or reference, when making arguments to support the use of their assessments? Here we provide a broad overview of the criteria that one should apply when judging the quality of an assessment or, more appropriately, the utility of the scores that are produced. A more comprehensive overview of validity frameworks, including the framework proposed by Kane, is found in Chap. 26 [2]. In "Assessment

J. Boulet (✉)
Vice President, Research and Data Resources, Educational Commission for Foreign Graduates, Foundation for Advancement of International Medical Education and Research, Philadelphia, PA, USA
e-mail: jboulet@faimer.org; jboulet@ecfmg.org

D. J. Murray
Department of Anesthesiology, Washington University School of Medicine, St. Louis, MO, USA
e-mail: murrayd@wustl.edu

D. Nestel et al. (eds.), *Healthcare Simulation Research*, https://doi.org/10.1007/978-3-030-26837-4_25

Principles and Associated Validity Frameworks", we define what a score is and discuss scoring issues as they apply to simulation-based assessment and research. We concentrate on traditional metrics (e.g., checklists, global ratings), highlighting the questions that researchers should ask before choosing specific types of scoring tools. In "Why Score?" we outline the design or choice of assessment tools, focusing on the need to define what we want to measure and to align the scoring criteria with the construct, or constructs, being measured. We emphasize that the scoring tool is not simply the piece of paper or computer interface upon which the scores are recorded. Instructions on how the tool should be used, combined with rater training procedures, should also be part of the overall measurement package. In "What is a Score?" we examine some specific scoring issues, including how scores can be collected and aggregated, whether weighting should be done, and who should provide the scores. We also consider how technology could change how scores are collected in the future. Finally, in "Designing/Choosing Assessment Instruments" we argue that addressing various threats to the validity of assessment scores can lead to more meaningful research studies; ones that yield more generalizable findings.

Assessment Principles and Associated Validity Frameworks

Assessments are employed in all the health professions, often as part of certification and licensure processes [3]. There are many types of assessments that are currently in use, both "for" (formative) and "of" learning (summative) [4]. For most of these assessments, regardless of their purpose, some quantitative measure of performance is needed. Qualitative measures (e.g., comments concerning performance, stress level, confidence, etc.) can also be collected, and are helpful for providing individual feedback, understanding problems with various simulation parameters, or modifying educational interventions. Decisions concerning the quality of the assessment, however, are primarily based on quantitative measures [5]. While written remarks concerning the verisimilitude and appropriateness of the simulation scenarios can be gathered from those who are assessed, usually via questionnaires, the scores are the focus of most investigations. For much of the research conducted in the field of simulation it is these quantitative measures (scores) that are used to describe the competencies of providers, test hypotheses about knowledge, skills or abilities of individuals or groups of individuals, provide feedback or, in aggregate, to evaluate a curriculum. As such, it is important that the scores are reasonably precise and adequately represent the constructs that we are attempting to measure [6, 7].

There are several frameworks that have been used to categorize and qualify evidence to support the use of assessments or, more appropriately, the scores derived from assessments. The work by Kane, summarized in more detail in Chap. 26 [2] and by Clauser et al. [8] and Tavares et al. [9], provides a useful framework for evaluating the quality of assessments. The structure of Kane's view of validity rests on a series of assertions and assumptions that support the interpretation of the assessment scores [10, 11]. The four components of Kane's inferential chain are labeled *scoring, generalization, extrapolation, and interpretation/decision*. The *scoring* component includes evidence that the assessment was administered fairly (i.e. in a standardized way), individuals were evaluated accurately, and the scoring rules were applied consistently. The *generalization* component requires evidence that the observations (e.g., multiple choice questions, simulation scenarios) were sampled adequately from the "universe" of available observations. In addition, with respect to *generalization*, it is essential to gather evidence to indicate that the number of scenarios or encounters is large enough to provide an overall score that is a reasonably reproducible measure of ability. Here, the question of interest is whether the assessment yields reliable scores and/or decisions about the participant or team. The *extrapolation* component requires evidence that the observations represented by the assessment scores are relevant to the construct of interest. This also requires evidence that the participant or team scores (e.g., ratings, checklists) were not unduly influenced by sources of variance (rater biases, ambiguous checklist items) that were not intended when the construct and scoring was developed for the simulation scenario. The *interpretation/decision* component of the argument involves the presentation of evidence supporting the theoretical framework required for score interpretation. For example, if one can identify individuals who do not meet acceptable performance standards (as part of a simulation exercise), do they benefit more from a remediation program than those who were not identified? Likewise, where decision rules are employed (e.g., pass/fail), evidence to support the procedure and the utility of the resultant placement or categorization of those being assessed should be gathered.

Thinking about Kane's four components, there are a host of simulation-based research studies, many based on the assessment of practitioners, where attention to validity issues is certainly warranted. Studies frequently involve the comparison of different types of practitioners (e.g., physicians versus nurse practitioners) or practitioners at different stages of training (e.g., medical students, residents, practicing physicians) [12, 13]. If the more experienced practitioners score less well than those with less experience, then one should be concerned about the utility of the scores. However, even if more experienced practitioners (e.g., senior residents) outperform those with less experience (e.g., medical students), this may only constitute only "weak" score validity evidence. One must consider, based on expected ability, whether the group comparisons are really appropriate and then look at the size (meaningfulness) of any group score differences. Other studies have looked more specifically at measurement precision [14, 15]. If the scores are not precise, how can they be used as indicators of ability? Finally, there are a growing number of investigations that attempt to provide evidence to support the extrapolation argument [16]. Does performance or ability (as reflected by the score or scores) translate to performance in real-world situations? Almost all of these types of validity studies depend on ability measures (scores), and the associated tools used to collect them. As such, researchers must be diligent when they develop or choose measurement tools.

Why Score?

Scores are essential from programmatic, individual, and research perspectives. From a programmatic perspective, the use of scores assures that the curriculum has defined performance expectations and attainable course objectives. In addition, if subject matter experts agree about the expectations for performance on an exercise, or exercises, then the training is more likely to include relevant content and to meet educational objectives. This expert 'weigh in' on an expected performance (scores to be attained) results in directed experiential learning that is particularly important when procedural or cognitive skills training is used for the advanced education of health care professionals [17].

From an individual perspective, a scoring system identifies actions (diagnostic or therapeutic), expected communication and teamwork events, as well as procedural steps that are considered essential by the experts [18, 19]. The scoring system may also be based on validated clinical guidelines. If there are sequential steps or actions that are considered harmful, their recording can be useful to guide feedback. The process of defining and developing a scoring method can also help to identify those actions and tasks where expert opinion is divided about the most effective management strategy. This information can help with modifying rubrics and choosing exactly what to score. Ultimately, the availability of scores allows for the provision of feedback that can be adapted to the participant's performance. To improve performance, experienced learners require directed feedback based on their weaknesses. With a good scoring system these weaknesses can be identified and used as part of feedback that is targeted to the mitigation of performance deficits. With enough aggregate information on these deficits, the curriculum can also be adapted to the needs of the individual or team. A summary of the reasons why scores are important is provided in Table 25.1.

Table 25.1 Summary of reasons why scores are important

Reasons why scores are important
Scores require educators/content experts to define what is expected of the learner
Scores can provide information about knowledge and skill acquisition, both for individuals and groups. They can be used to quantify variation in performance among learner
Scores can be used to inform program evaluation efforts, indicating potential areas for curricular improvement
Scores, when provided as part of feedback, can help motivate learners
The analysis of scores can provide information on the quality of the assessment (reliability, validity)
Scores are needed for quantitative research studies in simulation

The development and use of scores for curriculum evaluation, to quantify individual ability, or to provide meaningful feedback naturally leads to assessment research questions. For simulation research, we may want to know whether one group performed better than another, or to identify who performed poorly so that remediation efforts can be targeted. Similarly, if we introduce a simulation training course, we might want to capture data to show that an educational intervention has had some impact. To do this, we could measure students before the educational intervention (e.g., procedural training using simulation) and then afterwards and compare their performance. Even better, to isolate the educational effects, we could randomize students to either receive or not receive intervention and measure their performance at the two time points. Regardless of the research design, at least for quantitative investigations, reliable and valid scores are needed.

What is a Score?

A score is simply a summary of the evidence contained in an examinee's (or participant's) responses to the items (or performance tasks) of an assessment related to the construct or constructs being measured. In healthcare, as applied to practitioners, the constructs of interest could be quite varied, ranging from communication skills such as history taking or breaking bad news to psychomotor skills such as central line insertion or laparoscopic surgery. Generating a score that adequately captures performance and can discriminate between novices and experts can be challenging. Where clinical guidelines exist, or an expert consensus can be achieved, the expectations about performance are likely to be more consistent. When simulation is used to recreate more complex events, the perspectives of different professions or specialists might be required to develop a scoring approach that captures the performance of interest. From a research perspective, a content review by experts to assure that the scores accurately reflect the ability of interest is essential. If the scores do not reflect the construct of interest, or do so with insufficient precision, the research study will be flawed from the onset.

There are many types of scores that are used in simulation research, the most common being "analytic" (checklists, key actions) and "holistic" (global rating scales). While some constructs can be adequately assessed using either type of tool, it is usually best to choose the measurement scale that best aligns with the construct of interest [1, 20–22]. Some constructs, or abilities, are fairly easy to reliably measure using checklists (analytic tools) [23]. History taking, physical examination, and many procedural skills can be assessed using checklists. While analytic tools (e.g., checklists) can allow for objective scoring (i.e., it is relatively easy to decide whether or not someone asked a specific question or performed a physical examination maneuver), there may be differences in opinion (subjectivity) as to what set of analytically scored items should be included to measure overall performance. As a result, one could gather objective (precise) measures of the wrong construct. For example, medical students being assessed on their history taking skills via checklists often err on the side of being more thorough, asking lots of questions to maximize the probability of obtaining a high score. Because of the scoring model, it does not matter if irrelevant questions are asked or if the sequence of questioning is illogical. For this scoring model, where only the documentation of history taking questions asked is relevant, the "shotgun" strategy could yield a higher score that does not reflect history-taking skills. Holistic measures (rating scales) overcome many of these limitations. Unfortunately, in medicine, holistic ratings are often deemed to be "subjective" measures whereas analytic scores are thought to be "objective" measures. Holistic ratings can be quite objective, however, when the construct is well defined and raters are well trained. For constructs such as communication skills, clinical decision making, and teamwork, holistic ratings often provide more reliable and valid measures of ability [24]. In simulation, one can envision many scenarios where it is not only important what the participant does, but also the order and timing of specific actions. If the sequencing of actions cannot be easily captured, or included in the analytic scoring criteria, checklists, or aggregate checklist performance, can fall short in terms of capturing overall ability. Depending on the complexity of the simulation scenario, and the ability or skill that one wants to measure, employing some form of holistic scoring tool is often preferred. A comparison of the advantages and disadvantages of analytic (e.g., checklists, key actions) and holistic (e.g., global ratings) scoring tools is provided in Table 25.2.

Table 25.2 Advantages and disadvantages of analytic and holistic scoring tools

Analytic scoring tools (Checklists, Key actions)	
Advantages	Disadvantages
Can be easier to develop scoring items, especially if clinical guidelines exist	Difficult to assess complex skills sets, including non-technical skills (e.g., communication)
Rater training is often easier to accomplish	Raters must pay attention to detail, especially if there are many scoring items
The scores provide an "objective" record of what was, and was not, done	Difficult to develop tools that account for timing and sequencing of actions
Depending on the construct being measured, non-experts can be trained to score	The scoring tools typically cannot account for egregious actions (specifying all possible harmful actions is usually not possible)
Error associated with rater leniency or stringency can be controlled	Does not indicate anything about the context in which the observations are conducted (unless written comments are added)
	Each simulation scenario may require the development of a specific scoring tool
Holistic scoring tools (Global ratings)	
Advantages	**Disadvantages**
Rely on expert judgment. Learners are often more accepting of ratings from "experts"	Experts are more difficult to recruit and train
Rater can consider overall performance, including timing and sequencing of actions	Some raters may not be objective, resulting in measurement error
Dangerous actions can be accounted for in the scoring	The construct, or constructs, being measured can be hard to define (e.g., professionalism). The numerical scales, and associated behavioral benchmarks, can be difficult to construct
The same scoring tool can often be used for different simulation stations	Providing feedback can be more difficult unless strengths and weaknesses are explicitly defined
Scores can indicate a range of achievement	Scoring criteria must be very clear and precise to avoid subjectivity of ratings

Designing/Choosing Assessment Instruments

The primary consideration when designing or choosing an assessment instrument is determining what you want to measure. A lack of specificity in defining exactly what it is you are trying to measure can have negative consequences concerning the proper interpretation of research findings. In the healthcare professions, there are many competency frameworks and associated definitions of various competencies [25]. Unfortunately, not all researchers share these definitions. More importantly, many competencies are not generic and their measurement, at least to some extent, will depend on situational factors that may or may not be modeled as part of a simulation exercise. For example, teamwork modeled in a critical care situation will be quite different from teamwork modeled in an outpatient clinic. Context-specificity, combined with inadequate definition of the construct or constructs being measured, can compromise the validity of assessment scores [26].

To design an assessment instrument, or choose a tool that has been used previously, one must be very clear on what is being, or has been, measured. Constructs such as leadership, professionalism, situational awareness, communication skills, etc., may have different meanings to different people. Likewise, optimal performance in one setting may be suboptimal in another. Differences in construct definition are often reflected in the scoring tools. In the simulation literature, there are numerous checklists and rating scales that have been developed to teach and assess communication skills [27, 28]. Checklist-based tools contain items such as "introduces self", "maintains eye contact", "does not interrupt", etc. Rating scales include domains such as "listening", "empathy", "personal manner" and "rapport". The number and content-variability of these scoring tools would suggest that common definition of communication skills does not exist. This does not necessarily invalidate the use of any of these tools. It does, however, demand that individuals who use them clearly define the construct they think they are measuring and specify the contextual factors (e.g., setting) that could impact the measurement process. The general steps necessary to develop a scoring rubric for a procedural skill are listed in Table 25.3.

For instruments that have already been developed, careful inspection of the rubrics to see if the relevant behaviors/activities have been delimited is essential. Has the instrument been employed previously? If so, with whom? It is important to know who the target examinees (persons being measured) are/were. An instrument designed to measure the history taking skills of third year medical students may not be appropriate for practicing physicians. Is there any evidence to support the reliability and validity of the scores? If yes, is it reasonable to assume that this evidence generalizes to other settings or populations? If the instrument is being adapted (adding or deleting items, changes in wording, translation into another language), do these adaptations change the nature of what is being measured? The answers to these questions can help guide decisions regarding the choice of whether to use an existing instrument or build a new one.

Table 25.3 Steps for developing a procedural skill scoring tool

1. Review the process and preparation required to perform a procedure
2. Establish the sequence of steps required
3. Develop an expert consensus concerning the content (actions) and sequence
4. Identify the major performance milestones required to perform the procedure adequately
5. Establish expectations for each step that define success
6. Define the common as well as serious failure modes
7. Develop scoring tools (checklists, rating scales) that can capture performance milestones and associated failure modes

There has been much written about developing new measurement tools. The exact specification of how to build rating scales, or design other performance data collection tools, is beyond the scope of this chapter, but is detailed elsewhere [29]. For simulation research, which often involves complex performance assessments, much effort has been dedicated to the development of scoring tools. Unfortunately, many researchers, most notably those with little experience, consider the scoring tool to be simply the piece of paper (or electronic equivalent) on which performance is recorded. While these tools may sometimes have accompanying benchmarks and guidelines, these are absent far too often, and it is not appropriate to simply give the tool (especially if a rating scale is being used) to some defined expert with the simple guidance to go and assess the medical students, residents, or other persons of interest. Training raters will orient them to construct being measured, express what poor or adequate performance looks like and, most importantly, define the criteria for scoring. Benchmarked performance videos can assist this process. Given the frequent absence of training, it is not surprising that these ratings are often deemed to be "subjective" and more dependent on the choice of the rater as opposed to the actual ability of the person being assessed [30]. If the ratings are not reasonably reflective of true ability, their use as part of a research study is suspect.

Researchers are often confronted with the question of whether "validated" tools should be used or adapted, or new tools developed instead. To inform these choices researchers need to think about what, exactly, they are attempting to measure, who is being assessed, and the potential logistical challenges of developing and validating a "new" tool while at the same time using the scores as part of a research study. The primary concern with using, or adapting, an existing tool is that any accumulated validity evidence may not applicable. For example, if the measurement tool provides valid scores for medical students it may still be inappropriate for postgraduate trainees. Likewise, if a tool is adapted (e.g., translated into a different language), the adaptation may change the meaning of the items, potentially invalidating the accrued validity evidence. While the adoption of existing tools may be appropriate, researchers need to consider whether the available validity evidence is relevant and hence whether the meaning of the scores will stay the same for a different assessment or research cohort. If a new tool is developed, appropriate steps must be taken to ensure that it yields psychometrically defensible scores [31].

Scoring Issues

There are a number of important issues that researchers must consider when employing scores. First, regardless of whether analytic or holistic rubrics are chosen, a defensible strategy for aggregating measures is needed. Often, ratings on different performance dimensions are added up and the total score is used for research. While this strategy could be appropriate, the meaning of the total score may not be clear. Second, there may be some debate on scoring rules. When checklists are employed, there could be valid arguments for weighting some items greater than others [32]. For example, one could argue that a history taking task on simulation scenario involving a depressed teenager (standardized patient) should include an item to measure the consideration of suicidal ideation. As opposed to other relevant history taking items, asking about suicidal ideation is critical and should probably be weighted. What the weight should be is a matter of opinion, but counting all items equally does not reflect their relative clinical importance. Similarly, for simulated critical care events, ignoring the timing and sequencing of actions could invalidate the scores. If one ignores timing and sequencing, and simply adds up checklist or key action performance, one could envision two people who obtain the exact same score yet, based on expert judgment, are of substantially different ability. The validity of a score can be highly dependent on how individual items, or scales, are aggregated. To the extent that validity of the scores can be compromised, any research findings based on the analysis of the scores could be questioned.

A second issue concerns who should provide the scores. For many constructs, experts are not really needed to score the performance. Standardized patients (SPs), lay people trained to model the symptom of real patients, are used to score candidates for many simulation-based assessments used for certification and licensure [33]. These individuals can be trained to rate communication skills and to document (usually via checklists) history taking and physical examination skills. Various studies have shown that well-trained SPs provide scores, at least in these domains, that are equivalent to those provided by experts [34, 35]. From both assessment and research perspectives, being able to employ non-experts decreases the cost of administering any simulation-based exercises. Nevertheless, measuring some competencies will require the judgment of experts. Assessing clinical decision making, situational awareness, and various procedural skills will require scorers/raters with expertise in the domain of interest. It should be noted, however, that employing experts does not relax the need for tool-specific rater training.

A final issue concerns how the scores are obtained. Historically, someone in the simulation environment (room, control room), and often more than one, provided the scores while the candidate/research subject did what required as part of the simulation exercise. This data collection method can be expensive and rater-candidate interactions could introduce measurement error. With the growth in fixed simulation centers, and the availability of sophisticated recording systems, real-time scoring is not essential. Recorded performances can be scored on computers, often in remote locations, using electronic databases [36]. In addition, for some simulation scenarios, the output from patient monitors can be embedded in the recording. Camera angles can be changed and raters/scorers can rewind the recordings. While this technology can make the scoring process more efficient, there may be some competencies (e.g., communication skills) where the data gathering method (e.g., rating live performances, rating recorded performances) impacts the construct being measured. For example, non-verbal communication and various procedural skills may be difficult to assess if the camera, or cameras, do not capture some important interactions.

A discussion of scoring would be incomplete without some mention of how technology could aid in the process. Regarding data collection, the use of artificial intelligence (AI) and expert systems, even as a quality assurance initiatives, could automate the scoring process [37, 38]. Computer vision and talk time analyses may eventually replace ratings of communication and interpersonal skills. Sensor-equipped mannequins and part-task trainers can provide objective data about movement and pressure. This data, if mapped to expert performance, could inform the scoring of various procedural skills. All of these technological advances offer the possibility of generating scores more efficiently and at a lower cost. While this would certainly be beneficial, the reliability and validity considerations discussed earlier still apply. From a research perspective, the application of technology to scoring simulation-based assessments will open the door for a host of feasibility and comparative studies.

Gathering Evidence to Support the Use of Assessment Scores

There are many ways to gather evidence to evidence to support the validity of assessment scores. As noted previously, Kane's framework serves a useful guide for exploring and categorizing different validation strategies. While a complete overview of validation strategies is beyond the scope of this chapter, guiding principles are described Chap. 26 [2]. Those embarking on simulation-based research, at least where quantitative analyses are employed, need to pay attention to scoring. If the scores are not sufficiently precise, then how will we really know that any comparisons of scores between groups, or longitudinally within a group, are meaningful? If we do not have evidence to support the validity of the scores then any inferences we make about the abilities of individuals or groups based on the scores may be flawed.

When simulation-based research involves scores (and most studies will), gathering evidence to support their psychometric adequacy is essential.

Conclusion

Scores are a fundamental element of most simulation research. They are used to provide feedback, evaluate training programs, compare groups of individuals, and determine competence. They can be obtained in numerous ways, employing both analytic and holistic frameworks. They need to be precise and adequately measure the construct of interest. These properties are a function of the tool development process, the training of the people who provide the scores, and the assessment administration conditions. Researchers should be familiar with the basic psychometric principles needed to evaluate the adequacy of the scoring tools and, more specifically, the scores they generate.

References

1. Boulet JR, Murray DJ. Simulation-based assessment in anesthesiology: requirements for practical implementation. Anesthesiology. 2010;112(4):1041–52.
2. Hatala RA, Cook D. Reliability and validity. In: Nestel D, Hui J, Kunkler K, Calhoun A, Scerbo M, editors. Healthcare simulation research: a practical guide. Cham: Springer.
3. Holmboe E, Rizzolo MA, Sachdeva AK, Rosenberg M, Ziv A. Simulation-based assessment and the regulation of healthcare professionals. Simul Healthc. 2011;6(7 SUPPL):58–62.
4. Epstein RM. Assessment in medical education. N Engl J Med 2007;356(4):387–396. Cox M, Irby DM, editors.
5. Boulet JR, Mckinley DW, Whelan GP, Hambleton RK. Quality assurance methods for performance-based assessments. Adv Health Sci Educ. 2003;8(1):27–47.
6. Cizek GJ. Defining and distinguishing validity: interpretations of score meaning and justifications of test use. Psychol Methods. 2012;17(1):31–43.
7. Cook DA, Hatala R. Validation of educational assessments: a primer for simulation and beyond. Adv Simul. 2016;1(1):31.
8. Clauser BE, Margolis MJ, Swanson DB. Practical guide to the evaluation of clinical competence. In: Holmboe ES, Durning SJ, Hawkins RE, editors. Practical guide to the evaluation of clinical competence. 2nd ed. Philadelphia: Elsevier; 2017. p. 22–36.
9. Tavares W, Brydges R, Myre P, Prpic J, Turner L, Yelle R, et al. Applying Kane's validity framework to a simulation based assessment of clinical competence. Adv Health Sci Educ. 2017;23(2):1–16.
10. Kane MT. Current concerns in validity theory. J Educ Meas. 2001;38(4):319–42.
11. Kane MT. Validating the interpretations and uses of test scores. J Educ Meas. 2013;50(1):1–73.
12. Blum RH, Muret-Wagstaff SL, Boulet JR, Cooper JB, Petrusa ER, Baker KH, et al. Simulation-based assessment to reliably identify key resident performance attributes. Anesthesiology. 2018;128(4):821–31.
13. Henrichs BM, Avidan MS, Murray DJ, Boulet JR, Kras J, Krause B, et al. Performance of certified registered nurse anesthetists and anesthesiologists in a simulation-based skills assessment. Anesth Analg. 2009;108(1):255–62.
14. Watkins SC, Roberts DA, Boulet JR, Mcevoy MD, Weinger MB. Evaluation of a simpler tool to assess nontechnical skills during simulated critical events. Simul Healthc. 2017;12(2):69–75.
15. Kreiter CD, Gordon JA, Elliott ST, Ferguson KJ. A prelude to modeling the expert: a generalizability study of expert ratings of performance on computerized clinical simulations. Adv Health Sci Educ. 1999;4(3):261–70.
16. Mcgaghie WC, Issenberg SB, Barsuk JH, Wayne DB. A critical review of simulation-based mastery learning with translational outcomes. Med Educ. 2014;48(4):375–85.
17. Griswold-Theodorson S, Ponnuru S, Dong C, Szyld D, Reed T, McGaghie WC. Beyond the simulation laboratory: a realist synthesis review of clinical outcomes of simulation-based mastery learning. Acad Med. 2015;90:1553–60.
18. Weinger MB, Banerjee A, Burden AR, Mcivor WR, Boulet J, Cooper JB, et al. Simulation-based assessment of the management of critical events by board-certified anesthesiologists. Anesthesiology. 2017;127(3):475–89.
19. Wiggins LL, Morrison S, Lutz C, O'Donnell J. Using evidence-based best practices of simulation, checklists, deliberate practice, and debriefing to develop and improve a regional anesthesia training course. AANA J. 2018;86(2):119–26.
20. Boulet JR, Swanson DB. Psychometric challenges of using simulations for high-stakes assessment. In: Dunn WF, editor. Simulators in critical care and beyond. Des Plains: Society of Critical Care Medicine; 2004. p. 119–30.
21. Jonsson A, Svingby G. The use of scoring rubrics: reliability, validity and educational consequences. Educ Res Rev. 2007;2(2): 130–44.
22. Vu NV, Barrows HS, Marcy ML, Verhulst SJ, Colliver IA, Travis T. Six years of comprehensive, clinical, performance-based assessment using standardized patients at the southern illinois university school of medicine. Acad Med. 1992;67(1):42–50.
23. Ilgen JS, Ma IWY, Hatala R, Cook DA. A systematic review of validity evidence for checklists versus global rating scales in simulation-based assessment. Med Educ. 2015;49(2):161–73.
24. Boulet JR, McKinley DW. Criteria for a good assessment. In: McGaghie WC, editor. International best practices for evaluation in the health professions. London: Radcliffe Publishing, Inc; 2013. p. 19–43.
25. CanMeds. The Royal College of physicians and surgeons of Canada: CanMEDS framework [Internet]. 2017 [cited 2018 Jul 16]. Available from: http://www.royalcollege.ca/rcsite/canmeds/canmeds-framework-e.
26. Durning SJ, Artino AR, Boulet JR, Dorrance K, van der Vleuten C, Schuwirth L. The impact of selected contextual factors on experts' clinical reasoning performance (does context impact clinical reasoning performance in experts?). Adv Health Sci Educ. 2012;17(1):65–79.
27. Scalese RJ, Obeso VT, Issenberg SB. Simulation technology for skills training and competency assessment in medical education. J Gen Intern Med. 2008;23(1 SUPPL):46–9.
28. Ryan CA, Walshe N, Gaffney R, Shanks A, Burgoyne L, Wiskin CM. Using standardized patients to assess communication skills in medical and nursing students. BMC Med Educ. 2010;10(1):1–8.
29. McDowell I. Measuring health: a guide to rating scales and questionnaires. 3rd ed. New York: Oxford University Press, Inc.; 2006.. 768 p
30. Athey TR, McIntyre RM. Effect of rater training on rater accuracy: levels-of-processing theory and social facilitation theory perspectives. J Appl Psychol. 1987;72(4):567–72.
31. Cheng A, Auerbach M, Hunt EA, Chang TP, Pusic M, Nadkarni V, et al. Designing and conducting simulation-based research. Pediatrics. 2014;133(6):1091–101.

32. Boulet JR, Van Zanten M, De Champlain A, Hawkins RE, Peitzman SJ. Checklist content on a standardized patient assessment: an ex post facto review. Adv Health Sci Educ. 2008;13(1):59–69.
33. Boulet JR, Smee SM, Dillon GF, Gimpel JR. The use of standardized patient assessments for certification and licensure decisions. Simul Healthc. 2009;4(1):35–42.
34. Ben-David MF, Boulet JR, Burdick WP, Ziv A, Hambleton RK, Gary NE. Issues of validity and reliability concerning who scores the post-encounter patient-progress note. Acad Med. 1997;72(10 Suppl 1):S79–81.
35. Boulet JR, McKinley DW, Norcini JJ, Whelan GP. Assessing the comparability of standardized patient and physician evaluations of clinical skills. Adv Health Sci Educ Theory Pract. 2002;7(2):85–97.
36. Boulet JR, Errichetti AM. Training and assessment with standardized patients. In: Riley RH, editor. Manual of simulation in healthcare. 2nd ed. Oxford: Oxford University Press; 2016. p. 185–207.
37. Wu JT, Dernoncourt F, Gehrmann S, Tyler PD, Moseley ET, Carlson ET, et al. Behind the scenes: a medical natural language processing project. Int J Med Inform. 2018;112:68–73.
38. Hodges BD. Learning from Dorothy Vaughan: artificial intelligence and the health professions. Med Educ. 2018;52(1):11–3.

Reliability and Validity

26

Rose Hatala and David A. Cook

Overview

The choice of outcome measure for simulation studies is a crucial element of the research design, for without careful forethought and planning how can we be confident that we are measuring what we intend to measure? In this chapter, we follow on from the concepts introduced in Chap. 25, outlining the key elements in developing and examining the validity argument for the outcome measure used in a simulation research study, with an emphasis on Kane's framework.

Practice Points

- Explicitly state the intended decisions or conclusions you will be trying to make from the study data. This should guide your choice of outcome measure for the study.
- Use a framework (we suggest Kane's framework) to guide the planning and examination of the validity evidence for your chosen outcome measure.
- Appraise existing validity evidence and collect new evidence as needed.
- Ultimately, a judgment must be made as to whether the validity evidence supports the intended use of the outcome measure.

R. Hatala (✉)
Department of Medicine, St. Paul's Hospital, The University of British Columbia, Vancouver, BC, Canada

D. A. Cook
Mayo Clinic Multidisciplinary Simulation Center, Office of Applied Scholarship and Education Science, and Division of General Internal Medicine, Mayo Clinic College of Medicine and Science, Rochester, MN, USA
e-mail: cook.david33@mayo.edu

Background and Key Concepts

As outlined in Chap. 25, the choice of outcome measure for simulation studies is a crucial element of the research design, for without careful forethought and planning how can we be confident that we are measuring what we intend to measure? Put another way, will our outcome measure allow us to draw meaningful conclusions from the study? In this chapter, we follow on from the concepts introduced in Chap. 25, outlining the key elements in developing and examining the validity argument for the outcome measure used in a simulation research study.

First, a note regarding some common and confusing terminology. The word "validity" is sometimes used to refer to the methodological rigor of a study. This convention is firmly entrenched in clinical research and we do not oppose it, but simply note the potentially conflicting usage of this word. The simulation community also speaks frequently of validating simulation devices or training activities, but we believe there are better and less confusing terms to refer to such studies. We reserve the term "validation" to refer to the collection of evidence to evaluate outcome measures and their associated interpretations and decisions. In this chapter, we are focussed on validity as it pertains to the evidence supporting the intended use or interpretation of an outcome measure.

Finally, we have yet to mention reliability (and we commonly hear "reliability and validity" said together, much like "peanut butter and jelly"!). Using Kane's framework for gathering and evaluating validity evidence, the classical concepts of reliability are part of the Generalization inference and will be discussed under that inference in this chapter.

Why Do Researchers Need to Pay Attention to Validity?

After all the careful planning we put into designing a study, we will be unable to draw meaningful conclusions if the selected outcome measure does not accurately measure what

D. Nestel et al. (eds.), *Healthcare Simulation Research*, https://doi.org/10.1007/978-3-030-26837-4_26

it is intended to measure. The most important step in guiding the validation process is to begin by explicitly stating the decision we are trying to make. Validation (and validity) ultimately refers to evidence that supports a specific interpretation or use of study data to support a specific decision. As outlined in the previous chapter, we have the choice in our research studies of selecting an existing outcome measure or creating a new measure. In either case, validity evidence must be gathered to support the interpretations that will be drawn from the data.

Validation refers to the process of collecting validity evidence to evaluate the appropriateness of the interpretations of the study results (i.e. do the results from our outcome measure support the decision we wish to make?) [1, 2]. This definition highlights that validation is a process, not an endpoint, and the interpretations are specific to the decision at hand and the context in which they were collected. Thus, if we implement a previous outcome measure in a new context (like a new research population), then validity evidence will need to be gathered to support our interpretations in this new context. Labeling an outcome measure as "validated" means only that evidence has been collected in a specific context (learner group, learning objectives, educational setting) to support a specific decision. The process of validation is vitally important – but that process will vary for every outcome measure, and elements of the process will need to be repeated with each new implementation of the outcome measure. If an outcome measure is developed *de novo,* then more validity evidence will be necessary to support data interpretation, as no prior validity evidence exists for this tool. The process of collecting validity evidence can be done in a separate validation study (ahead of the main research study), or collected concurrently in the main study. There is a risk with collecting validity evidence concurrently, for if the evidence suggests that the outcome measure is not measuring what was intended then the main study results will be inconclusive. From a pragmatic perspective, however, with finite time and money, this is often the tactic that researchers employ.

Validation of an Outcome Measure

As was outlined in the previous chapter, it is helpful to use a framework to guide the planning and examination of validity evidence for a chosen outcome measure. Historically, classical validation frameworks focused on three different types of validity: content, construct and criterion. However, more modern frameworks such as those defined by Messick [3] and Kane [4] take a more unified view of validity, namely that it is a hypothesis that requires testing. While validity can never be proven, evidence is gathered to support or refute the hypothesis (validity argument).

Messick's framework rests on collecting and/or examining five sources of validity evidence for a given outcome measure: content evidence (relationship between the content of the measure and the construct being measured), internal structure (relationship between the items within the measure), relationship with other variables (relationship between current measure and other related or unrelated measures), response process (relationship between observed performance and the record of that performance e.g., checklist or global rating score), and consequences (impact of the assessment and decisions made as a result of the assessment) [3]. While it is helpful to have these categories for collecting and evaluating validity evidence, Messick's framework does not provide guidance on how to prioritize or weight the evidence that might be most impactful for a particular decision.

In contrast, Kane's approach asks us to begin by outlining the intended-use argument (IUA) for the outcome measure. This argument, akin to a research hypothesis laid out before the study begins, addresses the decisions or conclusions that we would make from the data. Knowing the decision or conclusions that we wish to make, we can then outline the types of evidence that would be required to support these conclusions. After collecting evidence, we return to the IUA, compare the empiric findings against those we hypothesized, and then make a judgment as to whether the evidence supports the intended use. This final judgment, together with the evidence collected, is called the validity argument. Kane identifies four sources of validity evidence that can inform the IUA and validity argument, organized into four categories or inferences that link the original performance to a decision based on the outcome measure. We begin with an observation of a performance (e.g., hand motion in laparoscopic surgery, team performance in a cardiac arrest scenario, central line-associated bloodstream infections) that is documented as a score (e.g., checklist, global rating, computer-generated metric). Going from an observation to a score assumes that the score is an accurate reflection of the original performance; we call this the Scoring inference. Combining scores from several observations yields an overall test score that we assume fully represents performance of this task across all possible conditions in the study setting; we call this the Generalization inference. Performance in the study setting is assumed to reflect performance in the real world (Extrapolation inference), and this in turn is further assumed to serve as the basis for making a meaningful decision based on the study's intent (Implications/Decision inference). To summarize: performance generates a score; several scores yield a test score; the test score is presumed to reflect real-life performance; and this presumption is used to make a decision. The process of gathering the evidence under each of these inferences, relevant to the study's context and purpose, is called validation [2, 4].

Throughout the rest of this chapter, we have chosen an example of a quantitative outcome measure to illustrate the key concepts in validation. However, the process we outline would be the similar had we chosen a qualitative measure. When using a qualitative approach our data become words rather than numbers, but the process of validation remains the same. Interested readers can learn more about this in the following publication [5].

A Practical Approach to Validation

We will now describe an eight-step approach to validation, and illustrate this approach using a hypothetical quantitative research study in which we will assess medical students' suturing skills. Let's imagine that we are designing a research study to address the question of whether blocked versus spaced practice of suturing, using a low fidelity simulator, leads to better suturing skills for medical students during their surgical clerkship rotation. The intended decision is whether students will be permitted to suture patients in the operating room (OR) under direct supervision by the senior resident. After the educational intervention, we plan to assess our learner participants across 4 suturing stations, with each station presenting a different suturing challenge (varying the suture, instruments, and visibility of the surgical field). A senior colleague suggests that we consider using as our outcome measure the Objective Structured Assessment of Technical Skills (OSATS) [6] wherein trained assessors complete a checklist and assign a global rating to the learner's suturing performance at each station [7]. Our colleague points out that "This instrument has been well validated."

While we respect our colleague, we know that validation is a process and that "well validated" does not indicate whether the validity evidence is appropriate or sufficient to support using the OSATS for our study purposes. Moreover, we remain uncertain regarding what additional validity evidence might be ideally gathered during the course of our study. In this section, we will walk through the steps outlined in Table 26.1, which presents a practical approach to the validation process for our selected outcome measure [1].

Table 26.1 A practical approach to validation of a study outcome measure. (From [1])

1. Define the construct and proposed interpretation
2. Make explicit the intended decision(s) or conclusion(s) that your data will need to address
3. Define the interpretation-use argument, and prioritize needed validity evidence
4. Identify candidate outcome measures and/or create/adapt a new instrument
5. Appraise existing evidence and collect new evidence as needed
6. Keep track of practical issues including cost
7. Formulate/synthesize the validity argument in relation to the interpretation-use argument
8. Make a judgment: does evidence support the intended use?

Step 1. Define the construct and proposed interpretation.

The first step is to explicitly articulate the outcome—knowledge, skill, behavior or patient effect—we intend our study intervention to address. It is imperative that our outcome measure assesses what the study intervention targets.

This step also highlights the issue of surrogate outcomes. While the ultimate goal of our simulation-based educational interventions is better patient care outcomes, measuring patient outcomes in most studies is not feasible. Thus, we often must choose surrogate outcome measures that may either directly or indirectly reflect the intended patient outcomes. If validity evidence exists that supports the link between our surrogate outcome and real patient outcomes (typically validity evidence under the Extrapolation and/or Implications inferences), this will further strengthen the validity argument for our outcome measure [8].

In our study, we are interested in suturing skills. Since the OSATS is intended to assess a variety of surgical skills, it may be a good tool to choose. While it would be ideal to measure patient-related behaviors, in this study we will be satisfied with measuring skills.

Step 2. Explicitly state the intended decisions/conclusions.

One of the most crucial steps in the validation process is to clearly outline the decisions/conclusions that we ultimately want to draw from our study. A clear statement of the conclusions we anticipate supporting with our outcome measure will frame all the subsequent steps in the validation process. Without this, we will be unable to craft a coherent validity argument.

In our study, we want to draw conclusions about the suturing skill of our participants and whether they can safely suture under supervision on a real patient during their clerkship rotation.

Step 3. Define the interpretation-use argument, and prioritize needed validity evidence.

Once we have outlined the decisions or conclusions we wish to draw from our study, we need to outline the assumptions underlying those decisions in order to determine what validity evidence we need to collect. We cannot emphasize strongly enough the importance of this up-front work to create the interpretation-use argument [4]. Crafting the IUA is akin to developing a research hypothesis for a study and outlining what type of evidence is needed to examine that hypothesis. Once we have run the study and have the validity evidence in hand, then examining how and why the evidence supports or refutes our IUA forms the validity argument for our outcome measure.

Even before we collect or review the first piece of validity evidence, we can outline the evidence that we would hope to find under each of the inferences in Kane's framework:

(a) Scoring (for more details see Chap. 25): the observation of performance is correctly transformed into a consistent numeric score. Evidence will ideally show that the checklist items are relevant to suturing performance and that raters understood how to use the instrument.
(b) Generalization: scores in a study setting fully and accurately represent the task (i.e., across the full breadth of desired variations in the patient, context, or other conditions). The generalization inference emphasizes two main issues: [1] sampling (Have we adequately sampled suturing performance across a sufficient number of tasks and a variety of conditions?) and [2] reliability (Are scores reproducible?). Many traditional psychometric analyses, such as inter-rater reliability and Generalizability studies (i.e., G-studies), support the Generalization inference. Evidence will ideally demonstrate adequate sampling in addition to reproducibility of scores from the OSATS.
(c) Extrapolation: scores in the study setting relate to real-world performance. Measures of real world performance might include global ratings assessed during procedures with real patients, measures of experience such as procedural logs or year of training, patient adverse events, or end-of-rotation ratings. Associations with other measurements obtained in a test setting can also support this inference (the argument being that if two independent measures correlate as expected it would support, but not confirm, that they are measuring the intended real-life construct). Evidence under this inference will ideally examine the association between multiple other measures and the OSATS.
(d) Implications: the decisions and actions based on the outcome assessment have intended favorable effects and negative effects are minimal. These conclusions or decisions may be about the effectiveness of a particular simulation approach or simulation-based intervention on individual health care providers, their institutions or more directly on patients themselves. Foregrounding the conclusion or decision can guide the validation process [9].

Evidence under this inference will ideally demonstrate that students feel more prepared following the assessment, that suturing complications in the OR related to student errors decline and that students who were held back from suturing in the OR until they underwent additional training feel this time was well spent.

Step 4: Identify candidate outcome measures and/or create/adapt a new instrument.

It is unlikely that any single study could gather all of this validity evidence, and this is particularly true when score validation is only one component of a much larger study. Rather than start entirely from scratch, we advise to look carefully for previously described assessments that measure the same or a similar construct (knowledge, skill). If existing assessments aren't very good, the researchers can improve upon it. If nothing can be found to measure the precise task in question, the researchers can learn much from measures designed either to measure a related task or to measure a distinct task using a similar approach.

In our case, we look to the literature to examine what outcome measures are available and what evidence already exists to support their use. Guided by our senior colleague, we find two systematic reviews relevant to the use of OSATS (checklists and global ratings) for technical skill assessment [7, 10]. Furthermore, several of the procedures in the original OSATS stations (e.g., abdominal wall closure, control of haemorrhage) include specific checklist items related to suturing that would be directly applicable to our current study [6].

Step 5. Appraise existing evidence and collect new evidence as needed.

In this step, we review and appraise the existing validity evidence for our outcome measure, compare this against our IUA to decide what new evidence we will need to collect, and then collect that evidence.

(a) Intended Use

Reviewing the literature, we learn that the OSATS was initially developed to provide feedback on specific technical skills but subsequently has been used for promotion decisions [7]. Our intent of using the OSATS to determine if students are competent to attempt suturing in the OR under direct supervision is in line with this intended use. We further realize that the OSATS should facilitate providing feedback to students on which suturing elements require improvement, and decide to include sharing the checklist scoring with the participants as an intentional part of our study design.

(b) Scoring

Under the Scoring inference, we learn that checklists and global ratings both have validity evidence to support their use (i.e., they both appropriately translate the observation of performance into a numeric score) but that by necessity checklists need to be created and rigorously evaluated anew for each specific task and context. By contrast, the global rating scale is transferrable across specific skills [7, 10]. We also identify that most studies have not paid attention to rater training in the use of the checklists and global ratings, and hence there is insufficient validity evidence in this area to support the Scoring inference [7]. We thus decide to have all raters participate in a multi-faceted rater training intervention [11].

(c) Generalization

Under the generalization inference we find that some studies, including the original OSATS reports, have intentionally sampled across tasks and conditions. This seems to substantially strengthen the validity argument supporting the OSATS. This is in contrast to many simulation studies where the 'scenario' is only applied once or twice during the study and thus sampling is quite limited.

Another aspect of generalization is the reproducibility or reliability of scores. Reliability can be a confusing topic, perhaps due to the number of different types of reliability—which include internal consistency, inter-rater, inter-station, test-retest, and parallel-form reliability. All of this can be simplified by focusing on the concept of reproducibility: if the observation were repeated again would we draw the same conclusion? Importantly, whenever we repeat the observation, at least one or more test conditions have changed, and the different "types" of reliability simply attach names to what might have changed. If the observations are different items on a multiple-choice test, we speak of internal consistency (or synonymously, inter-item) reliability. If the observations are scored by two different raters we speak of inter-rater reliability. If the whole test is identical except for a lapse in time we speak of test-retest reliability.

For the OSATS, we find that a number of studies have gathered evidence for reliability in terms of inter-rater reliability (showing it is generally acceptable and higher for global rating than checklist), and a few have also examined internal consistency, inter-station and inter-item reliabilities [7]. We plan to measure inter-rater reliability during rater training and, if acceptable, use single raters during the study. Although the 4 OSATS stations we intend to use in our study may seem like too few stations for adequate sampling, we are comfortable that we have sampled across the settings that a medical student would be commonly be exposed to.

(d) Extrapolation

Under the extrapolation inference, the most common type of validity evidence gathered in simulation-based research studies is expert-novice differences [12]. However, there may be multiple reasons why expert-novice differences exist for the outcome measure that are unrelated to the construct of interest [13]. While *not* finding expert-novice differences would suggest a serious limitation in the validity argument for that outcome measure, the *presence* of expert-novice differences contributes little to the overall argument.

Fortunately, for the OSATS there is a reasonable body of evidence under the extrapolation inference beyond expert-novice differences [7]. Previous studies have demonstrated improvement in OSATS scores following training, which suggests that OSATS scores in our study will be able to capture differences in performance between the two training approaches. Additional studies have demonstrated correlation of OSATS with other technical skills measures. For our suturing study, we will further augment this evidence base by collecting the student operative performance logs completed as part of their clerkship rotation and correlating performance on the OSATS to their clinical performance scores.

(e) Implications

Evidence to support (or refute) the Implications inference is arguably the most important validity evidence, and yet is typically the most difficult to collect. In our experience, having reviewed multiple outcome measures used in simulation and other health-professions education fields, this evidence is almost never collected [2, 7, 14, 15]. Yet if we cannot provide evidence that the decisions made based on our outcome measure are having an overall favourable effect, our validity argument will be lacking the evidence most relevant for clinical practice.

Turning to our suturing study, we are surprised to learn that there are no studies that examine the consequences of the OSATS on the learner, training program or patients [7]. Although this will add substantial work to our study, we decide to collect data relevant to the Implications inference. To do this we will collaborate with our qualitative research colleagues to conduct exit interviews with a sample of participants, purposively sampling from those who completed the study and went on to suture in the OR and those who required additional training before real clinical practice. Our questions will focus on the perceived impact of the assessment on their learning, whether the decision stemming from the assessment felt correct and whether they were or were not prepared to suture in the OR.

Step 6: Keep track of practical issues including cost.

Education research has traditionally ignored cost and other practical issues related to intervention or assessment implementations, but such information is invaluable to others trying to decide whether to change their educational practices. For example, we may find that while the OSATS is an outcome measure that supports our validity argument, the cost and feasibility issues will make it difficult for future researchers to implement on a large scale. Keeping track of these and other practical issues related to development, implementation and interpretation of the outcome measure will be helpful to other researchers.

Step 7. Formulate/synthesize the validity argument in relation to the interpretation-use argument.

Having laid out the IUA and decided upon which new evidence we intend to collect, we are ready to start our study. When the study is complete, and our data are in hand, we will need to examine the validity evidence that we collected under each of Kane's inferences. We will determine what evidence we were able to collect, judge whether the evidence supports or refutes the validity of proposed inferences, and contrast these findings against the IUA we proposed in Step 3. This process of laying out and examining the collected data constitutes the validity argument for our outcome measure.

Step 8. Make a judgment: does evidence support the intended use?

Ultimately, we need to make a judgment whether the evidence supports or refutes our validity argument. If we decide the OSATS scores are not well supported it will call into question the results of our entire study; hence the importance of identifying up front as much evidence as we can! When writing up the study for publication, we will specifically include a section in the Methods where we outline our a priori interpretation-use argument, a section in the Results where we present the collected validity evidence, and a section in the Discussion where we synthesize for the reader whether the validity argument was supported or refuted and what evidence will be required in future studies. Alternatively, it would also be reasonable to present all of this information (IUA, evidence, and synthesis) in the Methods, particularly if the incremental contribution of the validity evidence is small relative to previously existing evidence and/or the overall aims of the research study.

Practical Problems with Validation

In our experience as associate editors across some prominent medical education and simulation journals, we have found that researchers commonly use outcome measures that lack adequate supporting validity evidence. Researchers often seem to gather the lowest-hanging fruit (validity evidence that is easy to capture, like expert-novice differences, but does not provide the strongest support for the data interpretation that the study requires), or ignore the validation process altogether. Not only does this leave future researchers with incomplete validity evidence for these outcome measures, but it brings into question the interpretation of the study results (for how else can we be confident that the outcome measured what it was intended to measure?).

There are many simulation-based outcome measures available that would be relevant and adaptable to most research study interests, and it is the rare study that requires a new outcome measure. If we adapted existing tools and gathered the relevant validity evidence for our study purposes, we would be working together as a field to strengthen the quality of our outcome measures. A 2011 study by the authors found 417 studies that examined validity evidence for simulation-based outcome measures, and yet only 11 outcome measures had been examined in 5 or more studies [14]. That's a lot of potential outcome measures with limited validity evidence to support their interpretations!

Another common limitation in simulation-based studies occurs when researchers gather validity evidence but do not organize it or interpret it using a validity framework. This forces the reader to do the work of crafting a validity argument and examining the study's evidence against this argument themselves. Most readers lack the time or skill required to make such judgments. As researchers, it behoves us to prospectively outline for ourselves (and ultimately our readers) the conclusions or decisions that we want to make using the study data and the validity evidence required to support those decisions. It then falls to us to collect, organize, and interpret that evidence.

Conclusions

In this chapter, we have focused on a practical approach to the process of validation for the outcome measures used in our studies. We wish to emphasize three key points. First, validation is a process, not an outcome, and must be considered afresh with every use of an outcome measure. Second, outlining the decision or conclusion that our outcome measure is intended to support is a crucial step in the process; all subsequent steps rest on this this decision. Third, although difficult, it is imperative that simulation researchers collect and examine Implications evidence, so that the consequences of the decisions or conclusions on learners, systems and patients are empirically demonstrated.

References

1. Cook DA, Hatala R. Validation of educational assessments: a primer for simulation and beyond. Adv Simul. 2016;1:31. https://doi.org/10.1186/s41077-016-0033-y.
2. Cook DA, Brydges R, Ginsburg S, Hatala R. A contemporary approach to validity arguments: a practical guide to Kane's framework. Med Educ. 2015;49(6):560–75. https://doi.org/10.1111/medu.12678.
3. Messick S. Validity. In: Linn RL, editor. Educational measurement. 3rd ed. New York: American Council on Education and Macmillan; 1989. p. 13–104.
4. Kane MT. Validating the interpretations and uses of test scores. J Educ Meas. 2013;50(1):1–73.
5. Cook DA, Kuper A, Hatala R, Ginsburg S. When assessment data are words: validity evidence for qualitative educational assessments. Acad Med. 2016;91:1359–69. https://doi.org/10.1097/ACM.0000000000001175.
6. Martin JA, Regehr G, Reznick R, MacRae H, Murnaghan J, Hutchison C, Brown M. Objective structured assessment of technical skill (OSATS) for surgical residents. Br J Surg. 1997;84(2):273–8.
7. Hatala R, Cook DA, Brydges R, Hawkins R. Constructing a validity argument for the Objective Structured Assessment of Technical Skills (OSATS): a systematic review of validity evidence. Adv Health Sci Educ. 2015;20:1149–75. https://doi.org/10.1007/s10459-015-9593-1.
8. Brydges R, Hatala R, Zendejas B, Erwin PJ, Cook DA. Linking simulation-based educational assessments and patient-related outcomes. Acad Med. 2015;90(2):246–56. https://doi.org/10.1097/ACM.0000000000000549.
9. St-Onge C, Young M, Eva KW, Hodges B. Validity: one word with a plurality of meanings. Adv Health Sci Educ. 2016;22(4):853–67. https://doi.org/10.1007/s10459-016-9716-3.
10. Ilgen JS, Ma IWY, Hatala R, Cook DA. A systematic review of validity evidence for checklists versus global rating scales in simulation-based assessment. Med Educ. 2015;49(2):161–73. https://doi.org/10.1111/medu.12621.
11. Holmboe ES, Hawkins RE, Huot SJ. Effects of training in direct observation of medical residents' clinical competence: a randomized trial. Ann Intern Med. 2004;140(11):874–81.
12. Cook DA, Zendejas B, Hamstra SJ, Hatala R, Brydges R. What counts as validity evidence? Examples and prevalence in a systematic review of simulation-based assessment. Adv Health Sci Educ Theory Pract. 2014;19(2):233–50.
13. Cook DA. Much ado about differences: why expert-novice comparisons add little to the validity argument. Adv Health Sci Educ. 2015;20(3):829–34. https://doi.org/10.1007/s10459-014-9551-3.
14. Cook DA, Brydges R, Zendejas B, Hamstra SJ, Hatala R. Technology-enhanced simulation to assess health professionals: a systematic review of validity evidence, research methods, and reporting quality. Acad Med. 2013;88(6):872–83. https://doi.org/10.1097/ACM.0b013e31828ffdcf.
15. Hatala R, Sawatsky AP, Dudek N, Ginsburg S, Cook DA. Using In-Training Evaluation Report (ITER) qualitative comments to assess medical students and residents: a systematic review. Acad Med. 2017;92(6):868–79. https://doi.org/10.1097/ACM.0000000000001506.

Statistical Analysis: Getting to Insight Through Collaboration and Critical Thinking

27

Matthew Lineberry and David A. Cook

Abbreviation

NHST Null hypothesis statistical testing

Overview
Statistical analyses are key for deriving important insights from quantitative data in educational research. While the technical aspects of statistics can seem daunting, and expert consultation is well-advised, we do not advise thinking of analysis as a task to be "handed off" to a statistician after data is collected. Instead, analyses are best completed in close collaboration with a broad research team beginning early in the conceptualization of the research. This chapter outlines foundational concepts in statistics with the goal of familiarizing non-statisticians with key terms and concepts. We also share tips and tricks for running analyses and writing about your findings. Our hope is that readers who consider themselves statistically "challenged" will appreciate that their critical thinking about analyses can be essential to conducting sound and insightful research, provided a basic understanding of content and a team-centered approach.

M. Lineberry (✉)
Zamierowski Institute for Experiential Learning, University of Kansas Medical Center and Health System, Kansas City, KS, USA
e-mail: mlineberry@kumc.edu

D. A. Cook
Mayo Clinic Multidisciplinary Simulation Center, Office of Applied Scholarship and Education Science, and Division of General Internal Medicine, Mayo Clinic College of Medicine and Science, Rochester, MN, USA
e-mail: cook.david33@mayo.edu

Practice Points

- Statistical analyses are a key part of how our theories and practices change following a research effort.
- Statistical analyses are best performed as part of an early and ongoing team effort among all investigators, rather than as a task "handed off" late to a statistician disconnected from the broader research project.
- Understanding foundational concepts and terms can help non-statisticians contribute to the research team's critical thinking about how to best analyze the data.

Introduction

The word "statistics" evokes anxiety and confusion for many educators and scholars. If you are among this group, take a deep breath; this chapter is for you. Our goal is to reframe the topic and perhaps ease that anxiety and confusion! In the chapter we first consider *why* we conduct analyses, as well as with *whom*, and *when* they should be conducted. We also share guidance on how to understand *what* your data mean in terms of foundational statistical concepts. Finally, we discuss tips and "pearls" for *how* you can conduct and report analyses in sound and effective ways.

Why Analyze

"New insight" is the primary pursuit of research efforts. New insights begin by asking a good research question, followed by robust study design, rigorous data collection, and sound data interpretation. Sometimes new insights will reinforce and refine current ideas or approaches to education. At other times new insights will challenge preconceptions and point to innovative ways we can improve our theories and our

D. Nestel et al. (eds.), *Healthcare Simulation Research*, https://doi.org/10.1007/978-3-030-26837-4_27

work. Data analysis (including but not limited to statistical analyses) is only a means to the end of insightful and defensible interpretation, helping us to decide how data should change our thoughts and practices.

In the analysis phase, though, research teams often pursue things other than "new insight", slipping instead into the *avoidance* of feared negative consequences. First, there can be fear of breaking statistical "rules", which may seem infinite and incomprehensible. Second, teams often ask research questions in which they have an emotional stake in the direction of the results, resulting in studies designed primarily to *justify* an educational approach that the authors developed [1]. Third, even when a research team is open to results in any direction there may be fear that if no interesting statistical patterns are discernible the study may be difficult to publish [2]. Some concern for avoiding mistakes is healthy, but not when it becomes debilitating, nor when it threatens to bias the research team – for instance, making them unwilling to publish data that do not affirm prior beliefs.

With Whom and When to Analyze

One way to overcome these fears is to abandon the notion that any *individual* should be doing the analysis alone. First, experienced scientists will approach the same dataset in different ways and may draw different conclusions; a team approach facilitates consideration of multiple alternatives [3]. Further, analysis choices sometimes reflect researchers' biases, [4] and transparency and collaboration with divergent thinkers can be very useful in moderating individuals' biases. Additionally, biostatisticians may seem to possess comprehensive statistical expertise to education scholars but there are actually "specialties" within statistics just as there are in medicine. For example, most biostatistics training programs include little training in *psychometrics* (the statistical science of measuring and analyzing psychological constructs like procedural skill or inter-professional attitudes – i.e., the constructs of interest in simulation-based education research). A team approach brings together diverse expertise.

So, a great way to start analysis is the assembly of an analysis *team* that contains divergent thinkers with different types of expertise (statistical and otherwise). While the most statistically-proficient member is likely to run the analyses, they should also explain what they are doing so that everyone understands the concepts involved. Everyone on the team should feel empowered to ask questions and explore alternatives. If you are unable to recruit a trained statistical expert to join the team, consider recruiting a statistical trainee eager to learn and develop new skills. Those with less statistical expertise (which often includes the lead education scholar) can prepare to be effective and essential partners in that teamwork by learning and applying the concepts and principles outlined in this chapter.

Early team thinking about analyses can guide design modifications before you start collecting data, so assemble the team as soon as you have an initial sense of your research question and intended data collection. For example, if you can only sample ten residents for a simulation-based experience you would like to assess and that sample would give only a 4% chance of getting an interpretable answer to your research question, the research team must recognize early that this presents a problem and consider alternative research questions.

What to Analyze: Understanding Statistical Basics

Here we share a quick introduction to foundational concepts that can help you get started as a collaborator in statistical analysis. You can go deeper by studying any of several short primers on statistics, such as "PDQ Statistics" [5] or "Essential Biostatistics." [6]

Variables: The "Raw Materials" of Analysis

Variables are the attributes of people, moments in time, or other objects of measurement that you collect data about, like "age in years" or "exam scores". Variables play different roles in a study, depending on your research questions and assumptions.

- A **dependent variable**, or "outcome", is an attribute of interest that you expect might change or vary and wish to understand how and why it changes. "30-day mortality" is one example.
- An **independent variable** is one that you suspect causes or is associated with a change in a dependent variable. Independent variables are also sometimes called "factors" or "predictors". Sometimes researchers manipulate independent variables, as when assigning participants to one of several experimental conditions. However, an independent variable doesn't always have to be manipulated. For instance, you can't assign people to have certain hand sizes, but you could still see how hand size (an independent variable) predicts surgical performance.
- Sometimes a variable of interest is **descriptive**, and you are not focused on whether it is affected by or affects another variable. Instead, you are simply interested in the values that it takes. Demographic variables often fall under this category if you simply use them to describe the sample and do not relate them to your main independent or dependent variables.

Each variable consists of two or more levels. The variable "learner handedness" might have levels "right-handed", "left-handed" and "ambidextrous"; the variable "training intervention" might have levels "high feedback" and "low feedback"; and "knowledge test score" might have levels 0 to 100.

Scales of Measurement

For each variable, the **scale of measurement** is the set of permissible values that variable can have and what each value means. Different statistical analyses are designed to work with data having particular scales.

- **Nominal** variables' values have no quantitative meaning; for instance, "color" is a variable, but "red" is not twice as much color as "green".
- **Ordinal** variables have values with an order of increasing or decreasing magnitude, but limited information about the intervals between values. For example, the ordinal results of a race would indicate who finished first, second, and third, but would not indicate if all three finished within one second of each other, or if the second-place runner lagged three seconds behind the first.
- **Interval** variables have order, *and* the intervals between values are all equal. For example, on a well-designed multiple-choice test, the difference between 65% and 66% is similar to the difference between 66% and 67%. It would be reasonable to treat these scores as interval data.
- **Ratio** variables are interval data for which the zero point really means "absence of the attribute" and is thus more than a simple point on the scale. For temperature, zero degrees in Celsius doesn't mean "absence of temperature" so it is an interval scale; but zero degrees Kelvin means "no thermal motion", so it is a ratio scale.

By the way, nominal and ordinal variables are both called **discrete**, which means the levels cannot be subdivided. For example, a runner can take first place or second place, but not 1.25th place. Interval and ratio data are called **continuous** and can have decimal points as far out as is justified by the precision of your measurement. Numbers are often used to represent nominal or ordinal variables, but these are merely placeholders and shouldn't be read to imply quantity.

Describing Variables' Shapes

If you measure a set of objects (or people) using a variable such as "height", you will have a data **frequency distribution**. This represents the way those objects' values plot when graphed along the variable's scale. For instance, you may already know about the "normal" distribution: a bell-shaped distribution with one big "hump" of frequently-observed values in the middle and "tails" of progressively less frequent values on either side (Fig. 27.1).

A few key values can help describe a distribution's shape concisely.

- **Central tendency** statistics (a.k.a. "location" statistics) point out the most frequent values in a distribution.
  - The *mean* is a simple average of all observed values. One issue with the mean is that it can be strongly affected by big *outliers* – values that are extremely high or low relative to the rest of the observations.
  - The *median* is the number that would cleanly divide the data in half, with equal numbers of values above and below it. It is less influenced by outliers.
  - The *mode* is the value that appears most often. (There might be more than one mode; for example, in the sequence of numbers "12222334444566", the modes are 2 and 4 because each appears four times).
- **Dispersion** or **variability** statistics indicate how spread out data are.
  - The *range* lists the lowest and highest values (e.g., "1 to 6" in the number sequence above). The *interquartile range* is a narrower range defined by dividing the ordered values into four sections (quarters) of equal size, and reporting the range of values that define the second and third quarters ("quartiles").
  - *Variance* is defined as the average squared difference between the mean and each observed value. For practical purposes it can be viewed as an overall measure of how spread out the data points are, accounting for every point. Like the mean, it is disproportionately influenced by extreme outliers. The *standard deviation* is the square root of the variance, which is handy since that puts the statistic back in line with the original scale of measurement. (It would be strange to report that the variance in some class' test scores is "227 squared points"!)
  - Finally, some statistics point out additional information about a distribution's shape, most importantly *skew*, which indicates if the data is "pushed" over to one side or another.

Types of Analyses: Descriptive Versus Inferential Statistics

If you intend only to describe a distribution of your data in your specific context, and don't intend to apply these results to a different setting or future replication, then you are just using **descriptive** statistics. An example might be reporting the evaluation of a local workshop to your Dean.

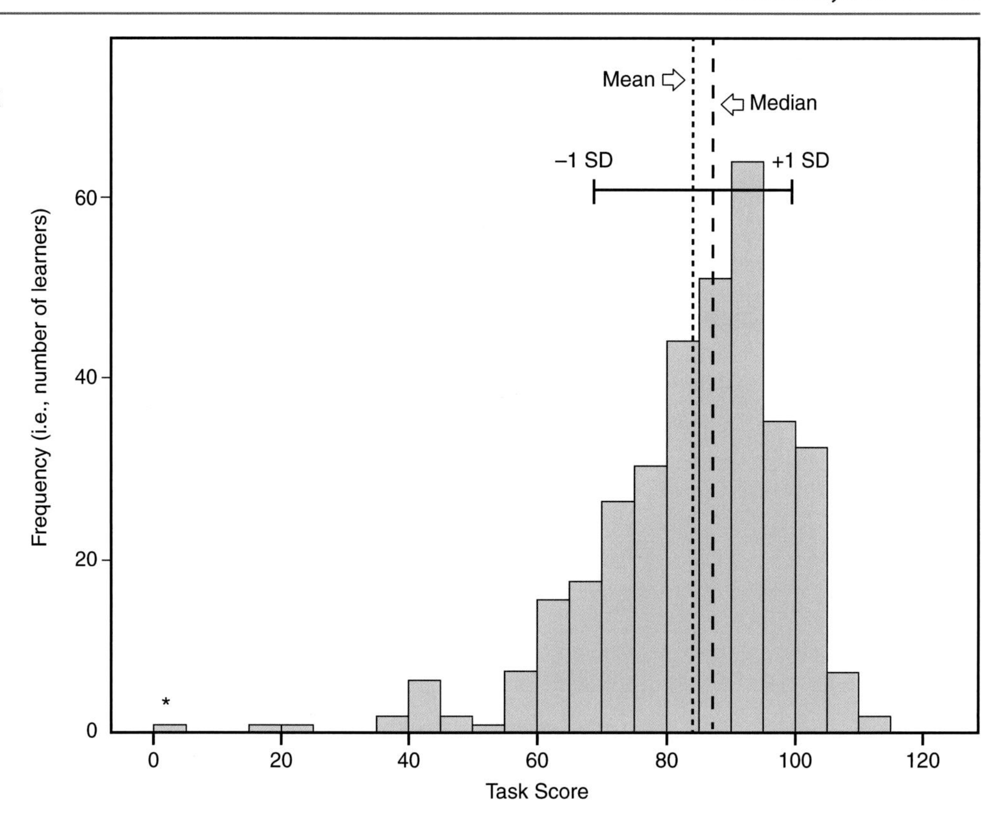

Fig. 27.1 Example frequency distribution of learner's scores on one variable. *Note.* SD = standard deviation. These scores show a mostly *normal* distribution – that is, there is a bell shape that is somewhat symmetrical. There is a small degree of negative *skew*, with a few learners scoring very low who might be considered *outliers* relative to the rest of the observed values. Because of this, the *median* and especially the *mean* are "pulled" to the left, relative to the most frequently observed scores. Vertical bars indicate how wide a single *standard deviation* is in each direction above and below the mean. While this specific display is referred to as a *frequency distribution*, the general type of graph is termed a *histogram*

However, if you wish to generalize your findings beyond the study (which is true for most research) – for example, to anticipate what might happen if you repeated the study using another group and/or in a new context – that calls for **inferential** statistics. By extension, inferential statistics also give you an idea of how well these results reflect the "truth" about a broader population. If you did repeat the study, it is unlikely that you would observe the exact same results. Instead you would see at least small differences due to participation by different people, or differences in performance or response. In this situation inferential statistics can help you understand how much variation you might expect to see. When people talk about "statistical analysis" they usually mean inferential statistics.

Inferential statistics look either for **differences** between groups on an outcome variable or **associations** between variables. Comparing test scores for a group of learners from before vs. after a course would be difference-focused, whereas seeing how closely their pre- and post-course scores correlated with each other would be association-focused.

An **effect size** simply describes the size or strength of a difference or association. An *unstandardized* effect size statistic is given in the units of the original scale of measurement, e.g., "an improvement of 2.3 points on a 7-point scale". *Standardized* effect sizes are unit-less, and thus more readily comparable across studies. The standardized mean difference is a commonly-used standardized effect size for reporting differences, calculated as the difference between means divided by the standard deviation. Cohen's d is one way of calculating the standardized mean difference. Stated differently, Cohen's d converts the mean difference into standard deviation units; if $d = 0.53$, then the difference was exactly 0.53 standard deviations. Effect sizes in studies of associations include r (the correlation coefficient) and R^2.

The differences or associations you observe in a study might reflect real underlying phenomena, but they also reflect the particular learners and conditions you sampled. Inferential statistics can help you draw more robust statistical conclusions by accounting for **random sampling error** in your studies.

A closely related concept is the **precision** of estimation for statistics like the mean or variance. A precisely-estimated statistic wouldn't change much if you repeated a given study many times. Imprecision due to random sampling error is often reported as the **standard error of the estimate** for a given statistic – a special "standard deviation" for the estimate. In research reports it is not uncommon to see a **95% confidence interval** provided for a statistic. This is based on the standard error of the estimate. When interpreting confidence intervals, it is common but technically wrong to say, "we are 95% confident that the 'real' underlying value is somewhere in this range". A more correct interpretation is "If we ran the experiment again 100 times, our results would probably fall in this range 95% of the time." [7]

Inferential analyses can be either **parametric** – which simply means that they use some simplifying statistic about variables' distributions, like a "mean" or "variance" – or **non-parametric,** meaning that the analysis does not use such simplifying statistics. Parametric approaches are generally preferred if the data distributions meet certain requirements, such as a normal distribution as defined above. Chapters 21 and 28 go into more detail about when to use each.

The "Machinery" of Analysis

Nearly all basic statistical approaches in use today involve **null hypothesis statistical testing** (NHST). The detailed mechanics of it are beyond what we can and should cover in this chapter, but you'll know the approach when you see it: it spits out "*p*-values" at the end. You also might hear researchers saying they used an "alpha (or "α") of .05". Given that random sampling error is an issue, NHST is designed to help you balance your risks of making opposing mistakes in your statistical inferences.

One mistake would be to say, "There's a real difference between these groups for this outcome!" when really there is not, and the observed difference was just due to random error. This is called a "**Type I**" error. In theory, **alpha** (α) tells how probable you are to make that mistake. Specifically, an alpha of .05 theoretically means there is a 5% chance that, if there were no real difference, you would erroneously declare that there was one.

The other mistake would be to say, "There is no difference between these groups", when in fact there *is* an underlying difference, and random sampling error obscured that difference. That's called a "**Type II**" error, and it has a probability as well, called **beta** (β) – although it is mentioned in reports a lot less than alpha. As alpha goes down, beta goes up, meaning you cannot avoid making mistakes; you only get to decide which ones to make more often. Ideally, researchers would think critically about how relatively "undesirable" Type I and Type II errors are for their research context and then pick an alpha/beta balance that is suitable. Unfortunately there is a strong norm in research to unthinkingly set alpha at .05 and ignore beta [8, 9]. This is not always the best approach. For instance, early in a program of research you might want to err on the side of not missing a potentially important association by using a higher alpha (such as .1), which would allow a lower beta error. You would aim to identify *possible* effects, and plan to follow with further research to refine which ones are likely to be real. Conversely, experts in biomedical research are suggesting that alpha levels in most research studies are too high, and that levels of .01 or .005 might be more appropriate for studies with widespread impact [10].

If beta is your chance of missing real effects, then 1-beta is your chance of finding real effects when they are present. This is also known as the statistical **power** of the study. Statistical power is discussed more in Chaps. 21 and 29, but suffice it to say you always want the power of a study to be high enough to give you a good chance of finding **practically-significant effect sizes.** That is, if it would be practically significant to reduce a rate of infections from "five per month" to "three per month", a study should be designed with enough power to detect a difference of two per month. It is also important to distinguish statistical significance and practical (i.e., clinical or educational) significance. Statistical significance is established by the results of inferential statistics, typically using the *p*-value in relation to the alpha level. Practical significance is a judgment call by the person interpreting the study results, and whether a given result is accepted as practically significant will vary across individuals, study settings, and research questions. Although there are no rigid cutoffs, some general guidelines are commonly accepted for different standardized effect sizes [11].

When you run an NHST-based analysis you'll get a *p*-value, and if that value is less than your chosen alpha, you will say that the tested effect is "statistically significant". Actually, it would probably be more accurate to say that the effect was "**statistically discernible**" as being more than just random chance, since people usually take "significant" to mean "important", and that's not necessarily the case. Even accomplished scientists often think *p*-values mean much more than they really do [12, 13]. For instance, a significant *p*-value doesn't mean that it is very likely that the study finding would be replicable if you ran a study again. It is also dangerous to think that very small *p*-values mean your findings are "especially significant" or "highly significant", or that findings a bit higher than your alpha (e.g., $p = .07$) are "marginally significant" [12]. Best just to say, "we did (or did not) see a discernible effect, given our chosen alpha", and move on. Conversely, when a *p*-value is not statistically significant that does NOT mean that there is "no effect"; you just were not able to discern one. Basically, there are only two types of results from NHST analyses: "We think there is an effect" or "We are not sure if there is an effect."

How to Analyze: Tips and Pearls

Know Your Variables

If your study incorporates advice from earlier chapters, such as those on crafting hypotheses, choosing assessment tools, and evaluating reliability and validity, then you hopefully have a good sense of what the variables in your dataset *mean*. Keep a single, rich summary of that information in a *codebook*. An especially powerful approach is to rough-draft the codebook very early in the project and update it as the study design progresses. For each *variable* in your data the codebook should include certain fundamental bits of information, like variable names and basic measurement

Table 27.1 Example codebook

Short variable name (for statistical software)	Variable label	Scale of measurement	Value labels	Data source	Measurement considerations & concerns	Role in the present study
ACLS_Cert	ACLS certification status	Ordinal	0 = No, 1 = Yes	Residency program database	This is measured for everyone within a team individually	Descriptive only (demographics)
DebriefCondition	Randomly-assigned debrief condition	Nominal	0 = learner-directed, 1 = instructor-directed	Random assignment key document	This is assigned per team (not per individual). We are assuming that the random assignment key document was followed correctly for each session	Independent variable
CompressStartMinutes	Minutes from simulated patient cardiac arrest to start of compressions	Ratio	Minutes, in decimal format (e.g., 1′15″ is "1.25 min")	Observers reviewing session video	One value per team. As with other observer-scored variables, we had two observers score 10% of videos to check their coding consistency	Outcome variable
PctCorrectDepth	Percentage of compressions between 2 and 2.4 inches	Ratio	Percentage values from 0% to 100%	Zoll defibrillator accelerometer	One value per team. Accelerometer does not compensate for patient bed compressibility, so compression depth measurements prior to backboard placement may be biased	Outcome variable
(cont'd)	…	…	…	…	…	…

ACLS advanced cardiac life support

details. We suggest several other details that are useful to specify as well (and which are not often found in codebooks). An example is given in Table 27.1.

Look Before You Leap

One common tendency among researchers is to jump straight to running inferential analyses. This typically springs from the excitement of finally getting an answer to a fascinating research question. It is more sensible (and often just as fascinating), however, to simply look at the data first. Literally just look at the data; scan through the raw numbers and words, and be curious. Are there any strange numbers you didn't expect, like that second-year surgery resident who reported she had previously performed 11,111 unsupervised laparoscopic cholecystectomies? When data is missing, why might that be? If you asked for open-ended comments on a survey, is there anything in those comments that should make you reconsider what the other responses mean (for instance, if a response hints that a participant may have interpreted your questions in an idiosyncratic way)?

Along with looking at the raw data, ask your team's analyst to prepare numeric summaries and simple graphs showing distributions and relationships in the data. Are there strange patterns, such as a test on which a sizeable proportion of learners got zero points? When two variables' values are plotted together (i.e., a "scatterplot"), do the dots line up in a row, do they look like a disorganized swarm of birds, or do they form some sort of curve? Add to your codebook any insights gained through this visual inspection of the data.

Choose Suitable Statistical Tests

With a clear picture of what your variables mean, the effects that you are interested in, and how your data are distributed, it becomes much easier to select the most suitable statistical analyses. For instance, if you have two different groups of learners and you wonder if their average scores on a 15-item checklist-based assessment are different, you might choose to run a parametric test such as an "independent samples *t*-test" if the scores meet the assumptions of that test. If they do not, you might decide instead to use a non-parametric test that makes fewer assumptions. Flow-chart-style guidance for this decision making is available in most textbooks and other online resources. Understanding the concepts discussed earlier will make these much easier to use. You can also let your best statistical "expert" guide this process; just make sure they are able to clearly explain their proposed analysis choices in a way that aligns with your understanding of what the variables mean and what the relationships of interest are. Remember that the research question should drive decision-making about analyses, not the other way around.

Share Your Findings

Chapter 42 addresses the topic of "Writing for Publication" in detail; here, we will share a few important pearls for reporting the results of statistical analyses:

1. Report key numeric results, not just the *p*-value. These will typically include the group means or proportions, and additionally (depending on the analysis) the difference between means, the correlation coefficient, the regression coefficient, or the odds ratio.
2. Explain what numbers mean. Rather than saying "The score in group A was 45 (5)", say instead "The mean (standard deviation) score in group A was 45 (5) seconds." When reporting a regression coefficient, explain what the coefficient (i.e., the slope or beta) means (e.g., "the regression coefficient was 1.7, indicating a 1.7-point increase in test scores for a 1-point increase in baseline motivation score").
3. Report the number of data points, (usually participants) included in each analysis. This number often varies slightly from one analysis to another due to missing data, and thus reporting the total number enrolled is insufficient. This information is particularly important if a reader wants to include your study in a systematic review or meta-analysis. Also, report the numerator and denominator when giving a percentage (e.g., "43/50 (86%)", not "86%" or "N = 43 (86%)").
4. Report the exact *p*-value, unless it is very small ($p < .001$) (e.g., report "$p = .03$", not "$p < .05$").
5. Report confidence intervals, which give readers a concrete sense of the precision of your findings. When comparing two groups (or the change in one group), report the difference and the confidence interval around the difference, such as "We found a mean difference of 4% (95% confidence interval: [1%, 7%]; $p = .02$.")". This provides useful information about the range of plausible results, and is particularly important for results that do not reach statistical significance (so-called "negative" studies). For example, suppose you find a between-group difference of 2.8% ($p = .23$), and that you have determined that a difference of 7% is "educationally significant." A 95% confidence interval of [−2.7%, 8.3%] would allow for the possibility of educationally significant effects (since the upper limit 8.3% is greater than your 7% threshold); these results are inconclusive. On the other hand, a 95% confidence interval of [−0.5%, 6.1%] would be more definitive since the upper confidence limit is less than your threshold of 7.
6. Report only the number of significant digits justified by your sample size. For samples <100, two significant digits is usually enough (e.g., "2.3" not "2.31"; "93" not "93.2"). *P*-values should be reported to 2 decimal places or, if <.01, with 1 digit (e.g., "$p = .23$" not "$p = .231$"; "$p = .002$" not "$p = .0023$").
7. In tables and figures, explain all abbreviations; special use of italics, parentheses, and dashes; special symbols; and empty cells. Keep abbreviations consistent with the main text, and define all abbreviations using footnotes (so that the table or figure can stand alone).
8. Do not use the word "trend" to describe statistical results that approach but do not reach statistical significance (e.g., do not say "There was a trend toward significance ($p = .06$)."). The word "trend" has special meaning in statistics, namely a tendency of the data values to move up or down across a series of repetitions (i.e., over time).

Parting Thoughts

As educators and scholars, you may often think of statistical analyses as a highly technical and somewhat-bewildering part of the research process. We encourage thinking of statistical analysis instead as another aspect of research that calls us to be interdisciplinary and collaborative, to apply critical thinking, and most importantly to be curious and eager to find new insights – all characteristics that are part of successful and joyful scholarship. As methodologist Paul Meehl said, there is no "automatic 'inference-machine'" [14]. Drawing transformational insight from data demands our best critical and collaborative efforts, and we wish you the best in those efforts!

References

1. Cook DA, Bordage G, Schmidt HG. Description, justification and clarification: a framework for classifying the purposes of research in medical education. Med Educ. 2008;42(2):128–33.
2. Rosenthal R. The file drawer problem and tolerance for null results. Psychol Bull. 1979;86(3):638–41.
3. Silberzahn R, Uhlmann EL, Martin D, Anselmi P, Aust F, Awtrey EC, et al. Many analysts, one dataset: making transparent how variations in analytical choices affect results. PsyArXiv [Internet]. 2017 24 [cited 2018 Jan 19]. Available from: https://psyarxiv.com/qkwst/.
4. Gelman A, Loken E. The garden of forking paths: why multiple comparisons can be a problem, even when there is no "fishing expedition" or "p-hacking" and the research hypothesis was posited ahead of time. Unpublished Manuscript [Internet]. 2013. Available from: http://www.stat.columbia.edu/~gelman/research/unpublished/p_hacking.pdf.
5. Norman GR, Streiner DL. PDQ statistics. 3rd ed. Hamilton: People's Medical Publishing House; 2003. p. 218.
6. Motulsky H. Essential biostatistics [Internet]. Oxford: Oxford University Press; 2016 [cited 2018 Aug 30]. Available from: https://global.oup.com/ushe/product/essential-biostatistics-9780199365067?cc=us&lang=en&.
7. Morey RD, Hoekstra R, Rouder JN, Lee MD, Wagenmakers E-J. The fallacy of placing confidence in confidence intervals. Psychon Bull Rev. 2016;23(1):103–23.

8. Lineberry M, Walwanis M, Reni J. Comparative research on training simulators in emergency medicine: a methodological review. Simul Healthc J Soc Simul Healthc. 2013;8(4):253–61.
9. Murphy K. Using power analysis to evaluate and improve research. In: Rogelberg SG, editor. Handbook of research methods in industrial organizational psychology. Malden: Blackwell Publishers; 2002. p. 119–37.
10. Ioannidis JPA. The proposal to lower p value thresholds to.005. JAMA. 2018;319(14):1429–30.
11. Cohen J. Statistical power analysis for the behavioral sciences. 2nd ed. New York: Taylor & Francis; 1988.
12. Cohen J. The earth is round (p<. 05). Am Psychol. 1994;49(12):997–1003.
13. Miller J. What is the probability of replicating a statistically significant effect? Psychon Bull Rev. 2009;16(4):617–40.
14. Meehl PE. High school yearbooks: a reply to Schwarz. J Abnorm Psychol. 1971;77(2):143–8.

Nonparametric Tests Used in Simulation Research

28

Gregory E. Gilbert

Overview
This chapter discusses tests for data often seen in simulation studies – data that are not normally distributed (A statistical distribution is the way the data are shaped. If we plotted the data values on the *x-axis* and frequency with which they occurred on the *y-axis* we would have the distribution of data.). This chapter briefly discusses types of data found in simulation studies and types of normality tests, prior to examining nonparametric tests useful for testing data that are not normally distributed.

Practice Points

- Unless a researcher is very experienced he or she should involve a statistician in their research study.
- Normality testing should always be done using multiple methods.
- There exist nonparametric methodologies corresponding to most parametric methodologies.
- Sample size estimates for nonparametric tests can be accomplished by doing sample size calculations for a parametric procedure and multiplying by 1.15 and rounding up.
- If a nonparametric procedure cannot be found, researchers can apply parametric procedures to the ranked data.

What Are Nonparametric Tests and Why Do We Use Them?

An important assumption in statistics is that data are normally distributed – the shape data subscribe to or resembles is a bell-shaped curve (Fig. 28.1). It was once common practice to ignore this assumption if the sample was "large" (meaning greater than 30 observations). This was so because the Central Limit Theorem [1] specifies, that a (sufficiently "large") random sample from any distribution will be normally distributed, no matter what the original distribution looks like.

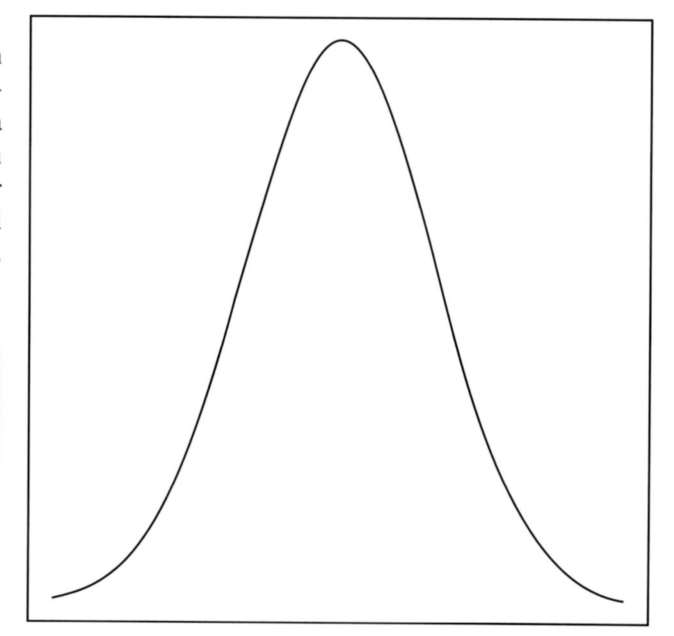

Fig. 28.1 Standard normal distribution

The importance of the normal distribution is undeniable since it is an underlying assumption of many statistical procedures. It is also the most frequently used distribution in statistical theory and applications. Therefore, when carrying out statistical analysis using parametric methods, validating the assumption of normality is of fundamental concern for the analyst. An analyst often concludes the distribution of the data "are normal" or "not normal" based on graphical exploration (Q-Q plot, histogram or box plot) and formal testing of normality (e.g., Anderson-Darling, Shapiro-Francia, Shapiro-Wilk). Even though graphical methods are useful in checking the normality of sample data, they are unable to provide

G. E. Gilbert (✉)
SigmaStats© Consulting, LLC, Charleston, SC, USA

D. Nestel et al. (eds.), *Healthcare Simulation Research*, https://doi.org/10.1007/978-3-030-26837-4_28

formal, conclusive evidence that data are normally distributed. Graphical methods are subjective as what appears to be a normal distribution to one may not necessarily appear to be a normal distribution to others. In addition, statistical experience and knowledge are required to interpret graphs properly. In most cases, formal statistical tests are required to confirm the conclusion drawn from graphical methods.

Why Do We Need Nonparametric Tests?

Checklist data. Often, those teaching or assessing simulation participants use checklists.[1] Checklists are very useful in learning procedural steps and are useful in reducing errors in healthcare and sentinel events. However, as is illustrated in Fig. 28.2, checklist data have a tendency not to be normally distributed. In addition, checklist data are discrete in nature. The data are not continuous; hence checklist data are not appropriately analyzed using parametric methodology.

Scores. Similar to checklist data, simulation results are often expressed as scores and are discrete in nature. Scores may range from zero to the maximum possible score. The data will be discrete in nature and will most likely be negatively (or left) skewed (Fig. 28.3). Therefore, score data are difficult to assess using parametric tests or methods.

Percentages. Those testing or assessing simulation participants often convert raw scores to percentages.

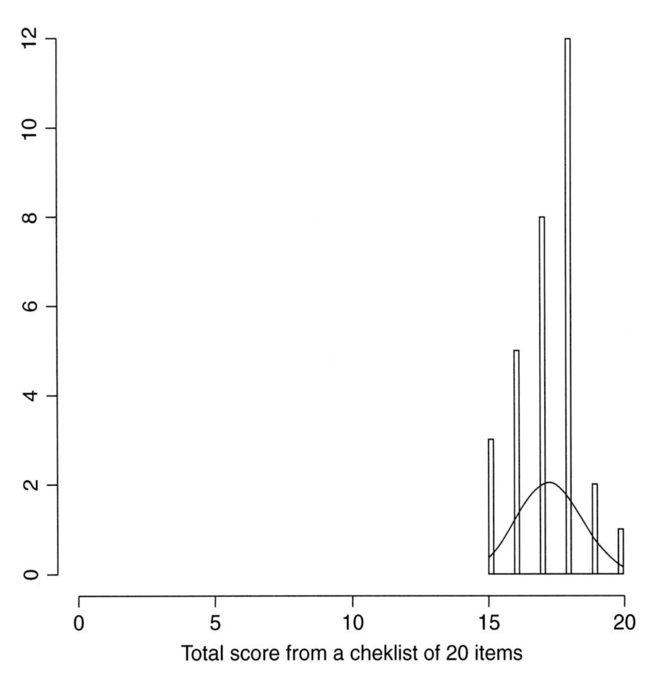

Fig. 28.2 Hypothetical simulation checklist data with normal distribution superimposed

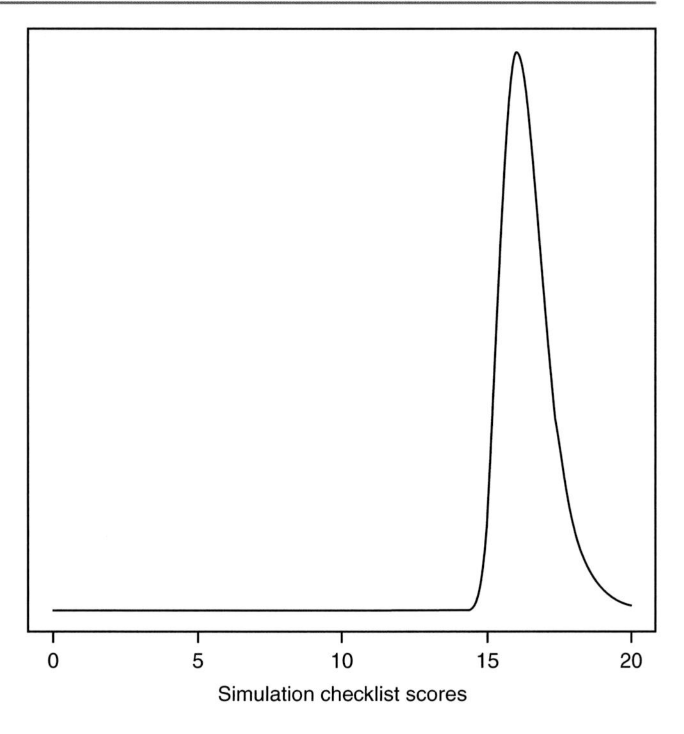

Fig. 28.3 Distribution of hypothetical negatively (left) skewed simulation checklist data

Percentages are not normally distributed; they subscribe to a binomial distribution.[2] One solution is to transform the data (mathematically change the outcome values); however, this causes problems with interpretation because researchers must either reverse transform the results (apply the opposite transformation) or interpret the results with respect to the transformed data. Interpreting the transformed results is not intuitive and often makes no sense. The most common transformation for percentages is the arcsine transformation.[3] [2] The arcsine transformation is complicated, and back transformation is not easy because the transformation is the arcsine of the square root of a number y (arcsine$\sqrt{y}$). Further complicating the matter, is the transformed result[4] is in radians, so researchers would have to draw conclusions in terms of radians if they did not back transform the data.

The point of this discussion is that simulation data, when used as percentages, are not normally distributed and requires a mathematical transformation *or* the use of nonparametric

[1]Hopefully these checklists have demonstrated psychometric reliability and psychometric validity.

[2]A binomial distribution is a statistical distribution of any number ($n > 1$) of binary trials (an experiment with only two outcomes – often success or failure like a coin flip) where there is the same probability of success.

[3]The arcsine transformation is appropriate because the percentages are derived from count data – the number of correct answers.

[4]For example, for 17 of 20 checklist items correct (or 85%) the arcsine transformation would be 1.016 radians. This is the value you would use for your analyses – not very intuitive!

statistical analysis for proper interpretation. Analysis using nonparametric statistics offers the simpler, more straightforward solution.

How Do We Know Data Need a Nonparametric Test?

There are different methods of examining normality in continuous distributions. All have advantages and disadvantages. Researchers can examine distributions graphically and compare how observed data are distributed to a normal distribution, use a goodness-of-fit test, a normality test based on regression and correlation, tests based on the distribution of the data (empirical distribution tests), moment[5] tests, or a number of other methodologies. Of these, only some of the more common normality tests will be discussed. In doing any research, it is best to use a variety of methods in order to test normality to triangulate results.

Quantile-Quantile plot. One way to examine normality is to use quantile-quantile or Q-Q plots [3]. A Q-Q plot graphically compares the degree to which observed data resemble a normal distribution. In this way it can be thought of as a graphical empirical distribution function test (see "Empirical distribution function tests" below; Fig. 28.4). However, this method should be the jurisdiction of a statistician or experienced researcher, as the degree to which the data subscribes to a normal distribution is quite subjective.

The Chi-square (CSQ) test for goodness-of-fit for normality. The oldest and most well-known normality test is the chi-square goodness of fit test. A goodness-of-fit test compares how well the observed data match a theoretical distribution. However, the CSQ test is not highly recommended for continuous[6] distributions (interval or ratio data) since the CSQ uses only the counts of observations in each cell in computing the test statistic rather than the observations themselves. Another caveat of the CSQ goodness-of-fit test is that it requires grouping of data. These groupings are arbitrary, thus influencing the results of the test. Conover also points out the CSQ test is not overly powerful [4]. (See the section on "Power of Nonparametric Tests" for a definition of power).

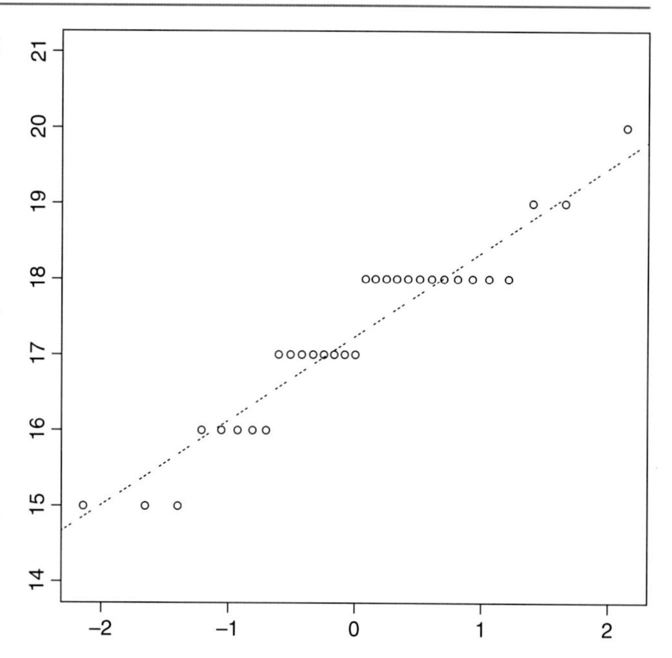

Fig. 28.4 Normal probability (Q-Q) plot of hypothetical simulation checklist data

Tests based on regression and correlation (Shapiro-Francia, Shapiro-Wilk, and Ryan-Joiner tests). The Shapiro-Francia [6] test is the squared correlation between the ordered sample values after they have been ranked and the (approximated) expected ordered quantiles[7] from the standard normal distribution. Because it uses correlation it belongs to this class of tests. The Shapiro-Wilk statistic is calculated by summing the differences of the ordered data values from one another multiplied by coefficients derived by Shapiro and Wilk and squaring that value divided by the sums of squares of the ordered observations. The Shapiro-Wilk [7] test is the most powerful test for all types of statistical distributions and sample sizes, whereas the Kolmogorov-Smirnov test (to be discussed in the next section) is the least powerful test [8]. The Ryan-Joiner test is the correlation between the sample data and the corresponding percentage point of the normal distribution. The Ryan-Joiner [9] test is very similar to the Shapiro-Wilk test; however, it has the advantage of being easier to implement in statistical software. It also has the advantage of being easier to explain to researchers how the calculations are done. Because it is the most powerful test for all distributions and sample sizes, the Shapiro-Wilk test is the recommended test for normality.

[5]The *moment* of a statistical distribution describes one aspect of the shape of the distribution. In statistics the most common moments are the first four moments – the mean (describing where observations clump), the variance (second moment; describing the spread of the distribution), skewness (the third moment; describing the symmetry of the distribution), and kurtosis (the fourth moment; describing the peakedness or flatness of the distribution). Moment-based tests use these quantities in their calculations.

[6]Continuous data are data measured on a scale that is potentially infinite. Data can take on almost any value within its range and is only limited in value by the precision of the measuring instrument or the convenience of rounding. Example of continuous data would be length or weight is the chi-square test for goodness-of-fit, first presented by Karl Pearson [4, 5].

[7]Quantiles consist of cut points dividing a range of data into equal contiguous groups. For example, if we divide data into 100 equal parts we call those quantiles percentages, if we divide data into 10 equal parts those are deciles, if we divide data into four equal parts those are called quartiles. If we divide data into two equal parts those quantiles are termed "halves".

Empirical distribution function (EDF) tests (Kolmogorov-Smirnov, Cramer-von Mises, Anderson-Darling, and Lilliefors tests). An empirical distribution[8] is how observed data are distributed. Empirical distribution tests compare observed data to the theoretical normal distribution to determine if there is agreement between the theoretical distribution and the distribution of observed data. There are a number of empirical distribution function (EDF) tests. The most popular of these tests is the Kolmogorov-Smirnov test [10, 11]. Other popular tests are the Cramer-von Mises, Anderson-Darling, and Lilliefors tests [12–16]. The Kolmogorov-Smirnov test is preferred when sample sizes are small, but otherwise has no advantage over the other tests [17, 18]. The Kolmogorov-Smirnov and Cramer-von Mises tests are similar enough in approach that there is no preference in the literature concerning their use [4]. However, the Shapiro-Wilk test is preferred over the Kolmogorov-Smirnov test [19]. The performance of the Anderson-Darling test is similar to the Shapiro-Wilk test and is preferred over the Kolmogorov-Smirnov test [8]. The Lilliefors test also presents a better alternative to testing normality than the Kolmogorov-Smirnov test [20]. In summary, if using an EDF test, the best tests to use are the Anderson-Darling or Lilliefors test unless there is a very small sample size then the Kolmogorov-Smirnov test should also be used.

Tests based on moments (skewness, kurtosis, D'Agostino-Pearson K2, and Jarque-Bera tests). Normality tests based on moments[9] include the skewness ($\sqrt{b_1}$), the kurtosis (b_2), the D'Agostino-Pearson K^2 and the Jarque-Bera tests. The procedures for the skewness and kurtosis tests can be found in D'Agostino and Stephens [21] and D'Agostino et al. [22]. Discussion of these two tests is not included here as they are not available in major statistical software and are not commonly used. The D'Agostino-Pearson K^2 test [22] is an omnibus test (similar to the F test[10]) analyzing data to determine skewness and kurtosis. It calculates how much the skewness and kurtosis values of the observed data differ from the values expected from a normal distribution. A *P* value is computed from the sum of the squares of these discrepancies. The D'Agostino-Pearson K^2 test has the advantage of not being affected by tied data, whereas the Shapiro-Wilk test is affected by ties. The Jarque-Bera test [23, 24] is a goodness-of-fit test testing whether the skewness and kurtosis of the sample data are significantly different from the skewness and kurtosis values of the normal distribution. The Jarque-Bera test statistic is based on the number of observations, the sample skewness, the sample kurtosis, and the number of independent variables (or regressors).

Summary: Normality testing. When dealing with simulation data it is highly recommended normality testing always be done. When testing it is suggested some graphical assessment be done under the direction of a statistician or experienced researcher. In addition to graphical analysis, it is recommended a small array of normality tests be used. Of the tests presented, due to their performance and availability in software, it is recommended the Shapiro-Wilk (or Ryan-Joiner), Anderson-Darling, and D'Agostino-Pearson K^2 tests be used to get a complete picture of whether data subscribe to a normal distribution. If dealing with a very small sample size the Kolmogorov-Smirnov test should be considered.

Descriptive Statistics

In dealing with data that are not normally distributed the mean (or arithmetic average) and standard deviation are not good measures of central tendency and variance. The median[11] and interquartile range should be reported, instead, when using nonparametric tests. However, in the literature, the mean (or arithmetic average) and standard deviation are more commonly reported. This is appropriate when dealing with normally distributed data. If data are not normally distributed – for example, in the case of checklist data – the arithmetic average is less descriptive in terms of central tendency. When dealing with data that are not normally distributed, the median and the interquartile range[12] should be

[8]An empirical distribution is the distribution represented by *observed* data not theoretical data. Therefore, an empirical distribution function is a mathematical function that most closely models or describes the observed data.

[9]Recall, the *moment* of a statistical distribution describes one aspect of the shape of the distribution. In statistics the most common moments are the first four moments – the mean (describing where observations clump), the variance (second moment; describing the spread of the distribution), skewness (the third moment; describing the symmetry of the distribution), and kurtosis (the fourth moment; describing the peakedness or flatness of the distribution). Moment-based tests use these quantities in their calculations.

[10]An omnibus test is a statistical test used to test an overall hypothesis. It tends to find general significance when examining parameters of the same type. For example, an F test examines whether there are significant differences between more than two groups. If significant, one can conclude differences exist, but does not know where the exact differences exist (i.e. Do the differences exist between groups 1 and 2, between groups 1 and 3, or between groups 2 and 3?).

[11]The median (Q2 or $\hat{\theta}$) is the data value dividing the sample equally in half. If the sample size has an odd number of observations the median is the middle value; for an even number of observations the median is the average of the middle *two* observations statistics associated with nonparametric tests.

[12]Data can be divided into fourths, called quartiles. The value dividing the lowest quarter and the second lowest quarter of the group is called Q1, the next value, dividing the bottom 50% and top 50% is called the median (Q2), the value dividing the bottom 75% and the top 25% is called Q3. The interquartile range (IQR) is the difference between the third quartile and the first quartile: Q3–Q1.

reported because they better describe the distribution of the data.

Pearson's Chi-Square (χ^2) Test

Pearson's chi-square test is appropriate for nominal data that are not normally distributed.[13] However, the data must be mutually exclusive.[14] One of the most common uses for the chi-square (χ^2) test is to determine if data are independent (or mutually exclusive). In other words, are the data related? For example, in a high-fidelity simulation is passage of bag-valve mask (BVM) ventilation skills (yes or no) dependent upon hand dominance (left-handed or right-handed) or gender (male, female, nonbinary)? If the *P* value is significant for the chi-square (χ^2) test, we would conclude the data are *not* independent of one another and thus are related, i.e. left-handed people *or* right-handed people (one group or the other) has an unfair advantage when it comes to passing the BVM test.

"Nonparametric Student's *t* Test": The Wilcoxon-Mann-Whitney Test

The Wilcoxon-Mann-Whitney (WMW) test was independently developed by Frank Wilcoxon and Henry Mann and Donald Whitney [25–27]. It is the nonparametric equivalent of Student's *t* test, where the two groups being sampled are unrelated. The hypothesis being tested is that it is equally likely a randomly selected value from one sample will be less that or greater than a randomly selected value from another sample. If the WMW test is significant it would indicate the samples are significantly different. For example, the WMW test would be appropriate to test whether medical students score significantly differently than nursing students on a nasogastric tube insertion checklist. The WMW test performs almost as well on normal distributions as it does on samples that are not normally distributed [28]. In statistical terms, the WMW test is very efficient.[15]

"Nonparametric Paired *t* Test": Wilcoxon Signed-Rank Test

In the same paper where Frank Wilcoxon proposed the Wilcoxon rank-sum test, he proposed the Wilcoxon signed-rank test. When dealing with normally distributed data that are dependent[16] a paired *t* test must be used to account for the data not being independent. The reason it is necessary to use a paired test is because the observations are related (i.e., highly correlated). The paired test (paired *t* test or Wilcoxon signed-rank test) is more powerful[17] than their independent counterparts (Student's *t* or WMW test) because the paired tests leverage this information. Wilcoxon's signed-rank test determines whether the population mean ranks differ [25]. It assesses whether two dependent samples, selected from the same population have the same distributions. The classic simulation example involves giving participants a pretest, introducing an intervention, and giving the participants a posttest to determine if anything was learned. If the Wilcoxon signed-rank test is significant, it could be concluded participants' pretest scores were significantly different from their posttest scores.

Kruskal-Wallis One-Way Analysis of Variance (ANOVA)

This nonparametric ANOVA is named after William Kruskal and Allen Wallis [29]. Like most nonparametric methods, it is based on ranks. Instead of testing whether the means of groups are equal, it tests whether the groups are from the same distribution [4, 28, 29]. In the same way that ANOVA is an extension of Student's *t* test for more than two groups, Kruskal-Wallis one-way ANOVA can be thought of as an extension of the WMW test for more than two groups. Like the omnibus F test[18] which indicates at least two means are significantly different, a significant Kruskal-Wallis test indicates at least one sample comes from a different distribution than the other samples. For

[13]Nominal data are categorical in nature such as types of simulation, colleges, universities, or genders.

[14]Data that are *mutually exclusive* are data belonging to only one category. For example, one cannot be functioning as a medical student and a nursing student at the same time. The categories are said to be *mutually exclusive*.

[15]A test that is more statistically efficient means that is takes fewer observations to reach a given power. For example, for nonnormally distributed data it takes fewer observations for the WMW test to achieve 80% power than it would for Student's *t* test.

[16]Two samples that are related in some way, such as measurements on the same or matched participants, are termed *dependent* and must be assessed using different statistical methods than independent or unrelated samples.

[17]Power is the probability of correctly rejecting the null hypothesis (H_0) when the alternative hypothesis (H_A or H_1) is, in reality, true.

[18]An omnibus test is a statistical test used to test an overall hypothesis. It tends to find general significance when examining parameters of the same type. For example, an F test examines whether there are significant differences between more than two groups. If significant, one can conclude differences exist, but does not know where the exact differences exist (i.e. Do the differences exist between groups 1 and 2, between groups 1 and 3, or between groups 2 and 3?).

example, if a significant *P* value is found when testing medical school faculty, medical students, nursing faculty, nursing students, and residents, it could be concluded that at least one of those groups was significantly different from the others.[19] However, where those differences lie would be unknown. Similar to the F test, Kruskall-Wallis ANOVA does not identify which group or how many pairs of groups are significantly different. Instead of using Bonferroni's test for multiple comparisons [30, 31] or Tukey-Kramer's method (a better alternative to Bonferroni [32–34]) to determine which groups are different (a technique often used in parametric ANOVA), Dunn's test [35] or the more powerful, but less well known, Conover-Iman test [36] would help identify which groups are significantly different. In the example above, it would help determine whether medical students are significantly different from residents, or nursing students are significantly different from nursing faculty.

Friedman's Two-Way ANOVA

Like most nonparametric tests, the Friedman test is named after the person first proposing it, Milton Friedman, and is rank-based [37–39]. Friedman's test is used for nonparametric two-way ANOVA and involves ranking each row (or block) together and then examining the values of ranks by columns. Friedman's test might be applicable for a simulation involving medical students, nursing students, nurses, and physicians (the blocks) and three different simulation treatments such as learning using a task-trainer, low-fidelity simulation, and high-fidelity simulation. Using Friedman's test we could not only assess whether there was a significant difference between the types of learners, but also whether there was a significant difference between the types of simulation. Like the other omnibus tests discussed, a significant value for Friedman's test indicates that at least one difference exists. To determine what the specific differences are post hoc procedures would have to be used. Not all statistical software supports post hoc analysis for Friedman's test; however, R (Vienna, AT) and SPSS (Armonk, NY) are two that have methods for post hoc testing when Friedman's test is used.

Power of Nonparametric Tests

All things being equal, nonparametric tests are less powerful,[20] than parametric tests. In other words, to reach the same level of power they need a greater sample size. This being true, however, nonparametric tests never require more than 115% of the sample size required by their parametric alternative, assuming the sample size is greater than 25 and the distribution of data is not "unusual".[21] When planning to use a nonparametric test, compute the sample size required for a parametric test, round up to the nearest integer, multiply the parametric sample size estimate by 1.15, and round up to the nearest integer again.[22] [40]

Advanced Methods

Outlined above are the more common nonparametric statistical methods used. With a moderate amount of time and experience, understanding and confidence could be gained to perform these methods, although it is always better to consult a statistician. However, situations exist where research questions cannot be answered using these methods. For example, perhaps there is a need for analysis of covariance (ANCOVA) when using nonnormally distributed data. The Kruskal-Wallis test or Friedman's two-way ANOVA cannot accommodate a covariate; therefore, nonparametric regression must be used. The same is true if it were necessary to examine outcomes over time for decidedly nonparametric data. To do this, nonparametric longitudinal data analysis would be needed. In both cases, researchers should consult a statistician.

One way to approach nonparametric testing is to use an approach by Conover (1999) [4]. For a nonparametric analysis, Conover suggests that any parametric method can be used if it is applied to ranked data. In other words, suppose you administer a clinical practice exam (CPX) and want to predict scores on a licensure exam adjusting for demographic variables such as age, gender, race, and cumulative grade point ratio. Given the data are not distributed normally, how would you do this? Conover suggests ranking all the CPX scores (1 = best, …, *n* = worst) and use linear regression to predict the nonnormally distributed licensure exam ranks. Of course, you would have to interpret the

[19] In ANOVA, if the omnibus F test is significant, a researcher knows the means of at least two of the groups are significantly different. However, it is not known which groups are significantly different. To find these differences each group must be tested against the other using post hoc (or "afterwards") tests. In other words, *multiple comparisons* must be done. For example, we must compare medical students vs. nursing students, medical students vs. residents, and nursing students vs. residents to find which one (or all) of the groups are significantly different.

[20] Power is the probability of correctly rejecting the null hypothesis (H_0) when the alternative hypothesis (H_A or H_1) is, in reality, true.

[21] "Unusual" means the distribution does not have infinite tails.

[22] For example, if you want to use the WMW test, calculate the sample size for Student's *t* test (80% power, $\alpha = .05$, $\mu_1 = 15$, $\mu_2 = 18$, $\sigma = 3$) the sample size for each group should be about 16. Multiply this by 1.15 to get a sample size estimate for the WMW test $16 \times 1.15 = 18.4 \approx 19$ participants per group.

results in terms of ranks, but that is better than not carrying out the study!

The methods discussed thus far fall into an area of statistical practice or methodology classified as frequentist statistics or frequentist inference. These practices are earmarked by conclusions being drawn rely on sample data emphasizing the outcome frequency. Any of the tests could also be performed using Bayesian statistics or Bayesian inference. This field of statistics is based on a probability theory related to Bayes' Theorem [41]. Bayesian theory allows statisticians to describe the probability of an outcome based on prior knowledge of variables related to that outcome. For example, if the passage rates of a CPX are known, Bayes' theorem can use the probability of a certain score on the CPX to more accurately predict the score on the licensure exam. Whether addressing a two-sample problem or a more complex research question, a statistician should be consulted if a researcher wishes to use a Bayesian approach.

Summary

In this chapter we have discussed approaches to testing simulation data that are not normally distributed. Some methods for testing whether data are normally distributed have been examined and some of the more popular nonparametric (also known as assumption-freer) statistical methods available in statistical packages have been described. Calculating a priori[23] power for nonparametric procedures was also mentioned – calculate power for the parametric method, multiply by 1.15, and round up. Finally, some more advanced approaches to statistical methods were noted, including a valid approach to nonparametric analysis when no appropriate methods are available.

References

1. de Moivre A. The doctorine of chances: a method of calculating the events of play. 3rd ed. London; 1756. https://www.ime.usp.br/~walterfm/cursos/mac5796/DoctrineOfChances.pdf.
2. Sokal RR, Rohlf FJ. Biometry. New York: Freeman; 1995.
3. Wilk MB, Gnanadesikan R. Probability plotting methods for the analysis of data. Biometrika. 1968;55(1):1–17.
4. Conover WJ. Practical nonparametric statistics. 3rd ed. New York: Wiley; 1999.
5. Pearson KX. On the criterion that a given system of deviations from the probable in the case of a correlated system of variables is such that it can be reasonably supposed to have arisen from random sampling. Philos Mag Ser 5. 1900;50(302):157–75.
6. Shapiro SS, Francia RS. An approximate analysis of variance test for normality. J Am Stat Assoc. 1972;67(337):215–6.
7. Shapiro SS, Wilk MB. An analysis of variance test for normality (complete samples). Biometrika. 1965;52(3–4):591–611.
8. Razali NM, Shamsudin NR, Maarof NNNA, Hadi AA, Ismail A. A comparison of normality tests using SPSS, SAS, and MINITAB: an application to health related quality of life data. In: IEEE Staff, editor. International conference on statistics in science, business, and engineering (ICSSBE) Langkawi. Kedah: Institute for Electronics and Electrical Engineers (IEEE); 2012.
9. Ryan TA, Joiner BL. Normal probability plots and tests for normality. Happy Valley; 1976.
10. Kolmogorov A. Sulla determinazione empirica di una legge di distribuzione. G dell' Inst Ital degli Attuari. 1933;4:83–91.
11. Smirnov N. Table for estimating the goodness of fit of empirical distributions. Ann Math Stat. 1948;19(2):279–81.
12. Cramer H. On the composition of elementary errors. Skand Aktuarietidskr 1928;11:13–74, 141–180.
13. von Mises RE. Wahrscheinlichkeit, Statistik und Wahrheit. Vienna: Julius Springer; 1928.
14. Anderson TW, Darling DA. A test of goodness-of-fit. J Am Stat Assoc. 1954;49(268):765–9.
15. Lilliefors HW. On the Kolmogorov-Smirnov test for the exponential distribution with mean unkown. J Am Stat Assoc. 1969;64(325):387–9.
16. Lilliefors HW. Corrigenda: on the Kolmogorov-Smirnov test for normality with mean and variance unknown. J Am Stat Assoc. 1969;64(328):1702.
17. Slakter MJ. A comparison of the Pearson chi-square and Kolmogorov goodness-of-fit tests with respect to validity. J Am Stat Assoc. 1965;60(311):854–8.
18. Slakter MJ. Corrigenda: a comparison of the Pearson chi-square and Kolmogorov goodness-of-fit tests with respect to validity. J Am Stat Assoc. 1966;61(316):1249–52.
19. Ghasemi A, Zahediasl S. Normality tests for statistical analysis: a guide for non-statisticians. Int J Endocrinol Metab. 2012;10(2):486–9.
20. Razali NM, Wah YB. Power comparisons of Shapiro-Wilk , Kolmogorov-Smirnov, Lilliefors and Anderson-Darling tests. J Stat Model Anal. 2011;2(1):21–33.
21. D'Agostino RB, Stephens MA. Goodness-of-fit techniques. New York: Marcel Dekker; 1986.
22. D'Agostino RB, Belanger A, D'Agostino RB Jr. A suggestion for using powerful and informative tests of normality. Am Stat. 1990;44(4):316–21.
23. Jarque CM, Bera AK. Efficient tests for normality, homoscedasticity and serial independence of regression residuals. Econ Lett. 1980;6(3):255–9.
24. Bera AK, Jarque CM. Efficient tests for normality, homoscedasticity and serial independence of regression residuals. Econ Lett. 1981;7(4):313–8.
25. Wilcoxon F. Individual comparisons of grouped data by ranking methods. Biom Bull. 1945;1(6):80–3.
26. Mann HB. Nonparametric tests against trend. Econometrica. 1945;13(3):163–71.
27. Mann HB, Whitney DR. On a test of whether one of two random variables is stochastically larger than the other. Ann Math Stat. 1947;18(1):50–60.
28. Hollander M, Wolfe DA. Nonparametric statistical methods. 2nd ed. New York: Wiley; 1999.
29. Kruskal WH, Wallis WA. Use of ranks in one-criterion variance analysis. J Am Stat Assoc. 1952;47(260):583–621.
30. Dunn OJ. Estimation of the medians for dependent variables. Ann Math Stat. 1959;30(1):192–7.

[23]"A priori" refers to calculating power prior to beginning a study or experiment. One should never calculate power after conducting an experiment. If nonsignificant results are found it is because the study was underpowered. See Gilbert and Prion (2016) more a more complete discussion [42].

31. Dunn OJ. Multiple comparisons among means. J Am Stat Assoc. 1961;56(293):52–64.
32. Kirk RE. Experimental design: procedures for the behavioral sciences. 4th ed. Thousand Oaks: Sage Publications, Inc; 2013.
33. Tukey JW. The problem of multiple comparisons. Unpublished manuscript. In: The collected works of John W Tukey multiple comparisons: 1948–1983. 8th ed. New York: Chapman & Hall; 1953.
34. Kramer CY. Extension of multiple range tests to group means with unequal numbers of replications. Biometrics. 1956;12(3):307–10.
35. Dunn OJ. Multiple comparisons using rank sums. Technometrics. 1964;6(3):241–52.
36. Conover WJ, Inman RL. On multiple-comparisons procedures. Los Alamos; 1979.
37. Friedman M. The use of ranks to avoid the assumption of normality implicit in the analysis of variance. J Am Stat Assoc. 1937;32(200):675–701.
38. Friedman M. A correction: the use of ranks to avoid the assumption of normality implicit in the analysis of variance. J Am Stat Assoc. 1939;34(205):109.
39. Friedman M. A comparison of alternative tests of significance for the problem of m rankings. Ann Math Stat. 1940;11(1):86–92.
40. Lehmann EL, D'Abrera HJM. Nonparametrics: statistical methods based on ranks, Revised. Upper Saddle River: Prentice Hall; 1998.
41. Bayes T. An essay towards solving a problem in the Doctrine of Chances. Philos Trans. 1763;53:370–418.
42. Gilbert GE, Prion SK. Making sense of methods and measurement: the danger of the retrospective power analysis. Clin Simul Nurs. 2016;12(8):303–4.

Contemporary Analysis of Simulation-Based Research Data: P Values, Statistical Power, and Effect Size

29

Emil R. Petrusa

Overview
Many quantitative researchers proceed from the assumption that statistical significance as represented by the p-value is enough to explain their results. In reality, though, p-values form only a part (albeit an important one) of the logic of hypothesis testing. The purpose of this chapter is to explain this logical flow by addressing core concepts such as the null hypothesis, alpha and beta error, and statistical power. Particular attention should be paid to the concept of effect size, which is a quantitative means of expressing the magnitude of an observed effect. Only when a p-value is correctly interpreted in the context of the power and effect size of a study can the results be given the most appropriate interpretation and meaningful conclusions be derived.

Practice Points

- P-values, which represent the probability that we are incorrectly rejecting the null hypothesis, are inadequate for the full interpretation of study results.
- Statistical power, defined as the likelihood of rejecting the null hypothesis when it is false, is dependent on the sample size of a study as well as the levels of alpha and beta error the investigator is willing to tolerate.
- The effect size of a study represents the magnitude of the difference noted between study groups relative the total variability, and provides critical context for interpreting the results.
- Meaningful conclusions can be derived only through careful consideration of these concepts and how they relate to the specific study results.

E. R. Petrusa (✉)
Department of Surgery and Learning Lab (Simulation Center), Harvard School of Medicine, Massachusetts General Hospital, Boston, MA, USA
e-mail: epetrusa@mgh.harvard.edu

Introduction

A common misconception in quantitative research is that the p-value is the primary determinant of whether a study's results are meaningful. P values (e.g., $p = 0.05$), however, are not enough to achieve this goal. While p values are important, **effect size** is fast becoming a second standard by which to judge a study's result [1–3]. An effect size is an index of how substantial one's results are. By "substantial" I mean the statistical result divided by the variance in the study. A common effect size coefficient is Cohen's *d*. For an independent groups t-test, Cohen's *d* is the difference between the 2 means divided by the pooled standard deviation from the 2 groups. Other statistical results have different formulas for calculating their respective effect size. This chapter will provide both an explanation of effect size, but also how it relates to statistical power. Statistical power is the ability to detect a real result when it exists. Often, we think about statistical power as having enough subjects in our study. This is indeed an important part, but number of subjects is contingent upon the effect size we hope to obtain. I will make brief mention of strategies to increase the likelihood of a larger effect size, but the focus is on understanding p values, power <u>and</u> effect size. What are these? Why you should care about them? How should these guide the planning or interpretation of study results? This chapter is a conceptual introduction to how these questions can be rigorously addressed. Four elements interrelate: effect size, statistical power, alpha value, and the sample size (i.e., the number of people needed to find the results we believe or hope we will find). We will show how these inter-relate before and after a study. Before answering these questions, it will be useful to review the logic of hypothesis-testing studies.

D. Nestel et al. (eds.), *Healthcare Simulation Research*, https://doi.org/10.1007/978-3-030-26837-4_29

The Logic of Hypothesis-Testing Studies

Whether it is straight research or an evaluation of some intervention, quantitative studies often have one or more research questions and/or hypotheses to be supported. Here are some examples.

- Training with rapid cycle, segmented practice will be more efficient and effective than whole-event training for conducting a code.
- Those with less clinical training will speak up more often after participating in sensitivity training than those without sensitivity training.
- Actual infection rates in the intensive care unit will be lower after at least 60% of ICU clinicians are trained to a specified level of competency for simulated central line placement.
- There will be a strong relationship between amount of video game playing and the time it takes to learn laparoscopic cholecystectomy on a simulator.

A positive answer to any of the above will add to the collective understanding about simulation (i.e., it could be published if the results of the statements/hypothesis are supported).

Classical inferential statistics begins with the assumption that there is no effect, no relationship, no finding. This is called the ***null hypothesis*** – the operationalized assertion that there is no effect, no relationship, no findings. In scientific writing this null hypothesis is understood and is not specifically written. Instead, we write our research hypothesis, which is an alternative to the null. We then conduct a study to find evidence to support this alternative statement/hypothesis. This scientific assumption of "no findings" is operationalized in our studies using the conditions, subjects and variables we are studying. An example of an operationalized null hypothesis is, "there will be no difference between mean performance scores of a group of 4th year medical students who are given detailed verbal instructions about how to do a cricothyrotomy and another group of 4th year medical students who watch a silent video of a properly done cricothyrotomy, when both groups are assessed on performance of a cricothyrotomy on a mannequin." This operationalized assumption is often unstated, but it is always present. For the above example, our research hypothesis might be, "those medical students receiving only a correct verbal description about performance of a cricothyrotomy will have significantly higher performance scores than those watching a silent video of a properly performed cricothyrotomy." The evidence we collect in our study must be sufficiently different from "no result" to allow us to *reject the null* and accept our alternative hypothesis.

Another scientific principle is that nothing is ever proven in an absolute sense. Instead, results are accepted within a certain probability; that is, the likelihood that we are wrong in rejecting the null hypothesis. By convention, scientists use 5% as the largest likelihood that we will be wrong in rejecting the null hypothesis. This "alpha value" is set *prior* to analyzing our results. One output from the statistical analysis is the actual probability that we are incorrectly rejecting the null hypothesis, the "p value." Said another way, the likelihood of finding a difference or relationship as large as we did by chance is 5%. We are thus 95% confident that we are correct in rejecting the null hypothesis (i.e., the likelihood of wrongly rejecting the null hypothesis is 5%). Researchers also call this mistake a Type I error. The smaller the p value (e.g., 1%) the less likely we are to mistakenly reject the null. By convention, however, the p value is written as a decimal number (e.g., 0.05 or 0.01). We use the $<$, $=$, or $>$ symbols to indicate that the actual p value is different from the traditional alpha level of 0.05 or 0.1. Thus, a $p < 0.5$ is read as "the probability is less than 5%." Statistical programs often produce an exact p value such as $p = 0.0259$.

There is also an opposite mistake – accepting the null hypothesis (saying there is "no effect") when it should have been rejected. Mistakenly accepting the null hypothesis is called a Type II error. In this situation, the size of our result is insufficient to reject the null. Had we organized our study differently (more subjects, more reliable outcome measure, *stronger* intervention, etc.) we would have found something reportable; something strong enough to reject the null hypothesis and support our alternative one. While not usually reported in a manuscript, a traditional likelihood for a Type II error is 20%. This likelihood of falsely accepting the null is called *beta* (β).

Statistical Power

Statistical power is defined as "the likelihood of rejecting the null hypothesis when it is false." Stated differently, power is the likelihood of accepting our alternative hypothesis when it is true [4–7]. When designing our study, we should arrange it to maximize statistical power so that we have increased as much as possible the chances we will find the results we hypothesize are true. With all the work involved (getting the institutional review board (IRB) to approve our study, getting permission to recruit and then recruiting subjects, running the study, perhaps compensating subjects for their time, collecting, analyzing and reporting data, etc.) we will want to do everything we can to maximize the likelihood of correctly rejecting the null and accepting our result. Since beta is the probability of incorrectly accepting the null hypothesis, then $1-\beta$ is the probability of correctly rejecting the null. If we set beta to be 20% (0.20), then 1–0.20 = 0.80 or 80%. This probability applies to any statistical calculation, whether a t-test, chi square or correlation. When planning a study we can set our power to any level. As mentioned, with all the effort we have already made to conduct the study, we wouldn't want a 50/50 chance of finding a true result (i.e., a power of 50%). Ideally, higher is better.

Choosing a value for statistical power should be one of the first steps in planning a study. However, there is another approach to calculating power – after the data are analyzed. We call this a post hoc (after the fact) approach for calculating statistical power, especially when we do not find the results statistically significant as we had hoped. For this analysis we know exactly how many subjects we have, the exact statistical result and the exact effect size. We can use these to calculate our exact power coefficient. It is quite possible that this after-the-fact power coefficient is different from the one we calculated before our study. Usually calculations before the study are based on estimates. After the study, we know the components exactly to calculate our exact power.

There are many aspects of a study that increase the likelihood of finding a statistically significant result. We could include as many subjects as possible. We could add items or cases or repetitions to the assessment of our dependent/outcome variable in an attempt to increase reliability of the outcome measure. We could choose subjects that have greater differences from each other when creating comparison groups. We could choose a different outcome measure with higher reliability for our dependent variable. In all this, the heart of the matter is our expected effect size (which we will discuss in detail shortly). Estimating the effect size forces us to quantify the size of the difference or relationship we expect to find. For example, consider a study assessing two approaches to studying for a test, in which two groups of learners each use a different method. The scores on a final exam are then used as a dependent variable. To most accurately calculate power, we need to estimate the size of the difference between the means that we believe will be meaningful. To continue with the example, say we chose a difference of 1 point on a 100-point test as the difference we want to find statistically significant. A critic might claim that a difference of 1 test item correct is trivially small and is thus unimportant to report. Suppose, however, we decided that a difference of 15% would be important (notice that this is a *judgment made by you,* the investigator) and that such a large difference would persuade us to adopt one of the study methods in the future. A 15% difference could arise from means of 70% and 80.5% ([.85–0.7]/0.7 = 0.15 or a 15% increase over 70%) or from means of 51.75% and 45% ([.5175–.45]/.45 = 0.15 or 15% increase over 45%). Notice this is not the same as a 15-percentage point difference, although we could choose that. For a 15-percentage point difference, the means might be 60% and 75% or 25% and 40%.

There are other ways to increase statistical power, such as using a dependent measure that has high reliability. Adding more well-written items should increase reliability. Both of our study methods should reduce the variability of test scores for the two groups. Smaller standard deviations result in a smaller pooled standard deviation which in turn supports higher statistical power. Having a potent intervention for a treatment group will support statistical power. A weak intervention might be an on-line module that subjects just glance at or do not engage at all. Another change that supports statistical power is the choice of a different probability for rejecting the null hypothesis. Changing the threshold of significance from p = 0.01 to p = 0.05 increases the probability that we are likely to mistakenly reject the null, but also increases the power to detect the effect of the intervention.

Effect Size

The final concept for this chapter is "effect size." What is an effect size? An effect size is a number that represents the substantiality or magnitude of our result [1–3, 8, 9]. Put another way, effect size gives a sense of how large the statistical result is relative to the total variability in the study. Cohen is credited for the original effect size concept and calculation [9, 10]. It is important to understand that effect size is only relevant to the numeric results (i.e., the size of the difference or relationship we want to find, the variability within groups, and the number of subjects in each group, etc.). An effect size is NOT, however, about the importance of our result. The importance is determined by the context in which we obtain our result. Is a difference of four points an important increase in the performance of central line placement after simulation training? Experts should determine whether this difference is an important one based on external clinical and/or educational considerations. Effect size is typically stated as small (up to 0.2), moderate (up to 0.5) or large (0.80 and larger). Each statistical method such as a t-test, analysis of variance, chi square, regression equation or similar non-parametric method has its respective effect size formula. Wikipedia (https://en.wikipedia.org/wiki/Effect_size) has a nice description of the effect size calculations for these approaches, and organizes the presentation into 3 categories of statistical analysis: correlations, differences and categories. A partial summary of the Wikipedia information is presented in Fig. 29.1. There are many online calculators for effect sizes with different statistical methods. A web site for Psychometrica [10] has several calculators listed in Fig. 29.2. These calculators are intuitive. Another resource is

Class or family of analysis	Specific statistic	Effect size indicator	Name	Formula	Comments
Correlations	Pearson r	r^2		R^2 or r^2	The correlation coefficient squared is the proportion of variance in one variable accounted for by one or more others
	Regression R	R^2	Coefficient of determination		
		η^2	Eta squared	$\eta^2 = \frac{SS_{Treatment}}{SS_{Total}}$	Similar to r and R, but more specifically is the variance in a dependent variable by a predictor variable when other predictor variables are controlled
Differences	Independent groups t-test	*d*	Cohen's *d*	$d = \frac{(\bar{x}_1 - \bar{x}_2)}{S}$	The difference between the means of 2 independent groups dividied by the pooled standard deviation (both groups)
		Δ	Glass's Δ	$\Delta = \frac{(\bar{x}_1 - \bar{x}_2)}{S_2}$	Difference between means divided only by the standard deviation of the control or comparison group
	ANOVA	ω^2	Omega squared	$\omega^2 = \frac{SS_{treatment} - df_{treatment} \cdot MS_{error}}{SS_{Total} + MS_{error}}$	For between-groups ANOVA only and where all groups have equal N
	Dfference between 2 correlation coefficients	*q*	Cohen's *q*	$q = \frac{1}{2}\log\frac{1+r_1}{1+r_1} - \frac{1}{2}\log\frac{1+r_2}{1-r_2}$	For testing the difference between 2 Fisher-transformed correlation coefficients
	ANOVA		Root mean square standardized effect	$\psi = \sqrt{\frac{1}{k-1} \cdot \frac{\Sigma(\bar{x}_j - \bar{X})^2}{MS_{error}}}$	This looks at the overall difference of the entire model adjusted by the root mean square. This formula is for a one-way ANOVA. Other formulas exist for more complex ANOVAs
Categorical data	Chi square	phi	Phi coefficient	$\phi = \sqrt{\frac{X^2}{N}}$	Phi is the square root of the chi square value dividied by the total N
		V	Cramer's V	$\phi_c = \sqrt{\frac{X^2}{N(k-1)}}$	V is the square root of the chi sqare value divided by the total N times the smaller of rows or columns
		h	Cohen's *h*	$h = 2$ (arcsin $\sqrt{p_1}$ –arcsin $\sqrt{p_2}$)	Formula for comparing two independent proportions or probabilities. "Arcsin" is the arcsine transformation.
	Odds ratio			Larger odds/Smaller odds	This is the larger odds (a ratio) from one group divided by the smaller odds (also a ratio) of the other group.
	Risk difference			(Risk treatment group - Risk controlgroup)	Simply the numerical difference between the risk (probability) of an event in a treatment group and the risk (probability) in the control group. This is especially useful for comparing the effectiveness of an intervention.
	Relative risk or risk ratio			Probability of success in treatment group / Probability of success in control group	Similar to Odds Ratio except that probabilities are used instead of odds
Ordinal data	Mann-whitney U	*d*	Cliff's delta	$d = \frac{2U}{mn} - 1$	U is the value of the Mann-Whitney computation and *m* and *n* are the Ns of the 2 groups

Fig. 29.1 Selected statistical analyses and respective effect size formulas. (Source: Wikipedia https://en.wikipedia.org/wiki/Effect_size)

1. Comparison of groups with equal size (*Cohen's d, Glass Δ*)
2. Comparison of groups with different sample size (*Cohen's d, Hedges' g*)
3. Effect size for pre-post-control studies with the correction of pretest differences
4. Effect size estimates in repeated measures designs
5. Calculation of d from the test statistics of dependent and independent t-tests
6. Computation of d from the F-value of Analyses of Variance (ANOVA)
7. Calculation of effect sizes from ANOVAs with multiple groups, based on group means
8. Increase of success through intervention: The Binomial Effect Size Display (BESD) and Number Needed to Treat (NNT)
9. Risk Ratio, Odds Ratio and Risk Difference
10. Effect size for the difference between two correlations
11. Effect size calculator for non-parametric Tests: Mann-Whitney-U, Wilcoxon-W and Kruskal-Wallis-H
12. Computation of the pooled standard deviation
13. Transformation of the effect sizes *r, d, f, odds* Ratio and *eta square*
14. Computation of the effect sizes *d, r* and η^2 from X^2 - and *z* test statistics
15. Table for interpreting the magnitude of d, r and eta square according to Hattie (2009) and cohen (1988)

Fig. 29.2 List of calculators on the Psychometrica website (https://www.psychometrica.de/effect_size.html) [10]

called G*Power, which is a freeware application for calculating effect size, power and sample size [11, 12]. This application can be downloaded and run on any computer platform. Figure 29.3 shows a screen shot of a calculation needed for sample size with a given effect size (0.50), alpha (0.05) and power (0.95). The result is that 88 subjects are needed in each group. The web site has a manual and a publication describing the application [13].

To illustrate the use of effect size in interpreting results, we will use an independent groups t-test. Figure 29.4 shows the formula for Cohen's *d* for an independent group t-test. It is the difference between the means divided by the pooled standard deviation from both groups. A "pooled standard deviation" is the average of the standard deviation from each of the two groups. Figure 29.4 also has the formula for calculating the pooled standard deviation for two independent groups. There are different effect size formulas for different statistical methods. As mentioned, Wikipedia has a nice summary of these with links for more information.

To further illustrate use of effect size, we will use a fictitious study with 2 groups of learners doing a simulated procedure. One group receives a new training approach, while the other group is our comparison group and has no special training. The outcome measure is the number of actions done correctly. An independent groups t-test is the statistic for testing the difference between two such groups, assuming the outcome variable has a normal, Gaussian distribution. After discussion, the investigators feel that a 20% increase in scores would be an educationally useful difference. We have 15 subjects per group. After performing the study, the results show a Group 1 mean of 10.67 (sd = 3.11) and a Group 2 mean of 8.89 (sd = 4.74). The increase of Group 1's mean over Group 2's is (10.67–8.89)/8.89 = 20%. The t-test result was t = 1.216, df = 28, p = 0.23. So, we obtained our 20% difference, but it did not reach statistical significance. What is the effect size? Using the G*Power application, d = 0.46 which is a moderate effect size. However, the power to detect a 0.05 level of significance associated with those t-test results and effect size was only 0.34. Recall that power is the ability to detect a real difference if it exists. Our study was under powered to detect a moderate effect size (0.46) at p = 0.05. Using the G*Power application again, but this time for prediction of the sample size needed to detect a difference at p = 0.05 with effect size = 0.46 and power = 0.80, we find that we need 59 subjects in each group. Notice that we use the same means and standard deviations, so we retain the 20% higher mean score for Group 1 that we thought would be educationally useful. If we had used 59 subjects in each group our t-test result would now be statistically significant at p = 0.05. After all the work to set up and run the study it would be disappointing to get a result that is not statistically significant. Although the effect size is moderate, it is unlikely that a journal will publish a study with a t-test result of p = 0.23.

An effect size is also helpful in interpreting a statistically significant result. Let's consider the same study as above but with different results where t = 1.98, p = 0.05 and effect size = 0.28. This looks like a much better result; a more modest effect size but with a significant p value. Let's see the details. Group 1 mean = 10.67 and sd = 3.11 and Group 2 mean = 9.59 and sd = 4.47 with 100 subjects in each group. Despite the statistical results, the increase in mean score for Group 1 over Group 2 is 11% – just over 1 point (1.08). Is this still an educationally useful difference?

In the context of this example, we found a statistically significant difference with our t-test indicating that there is a less

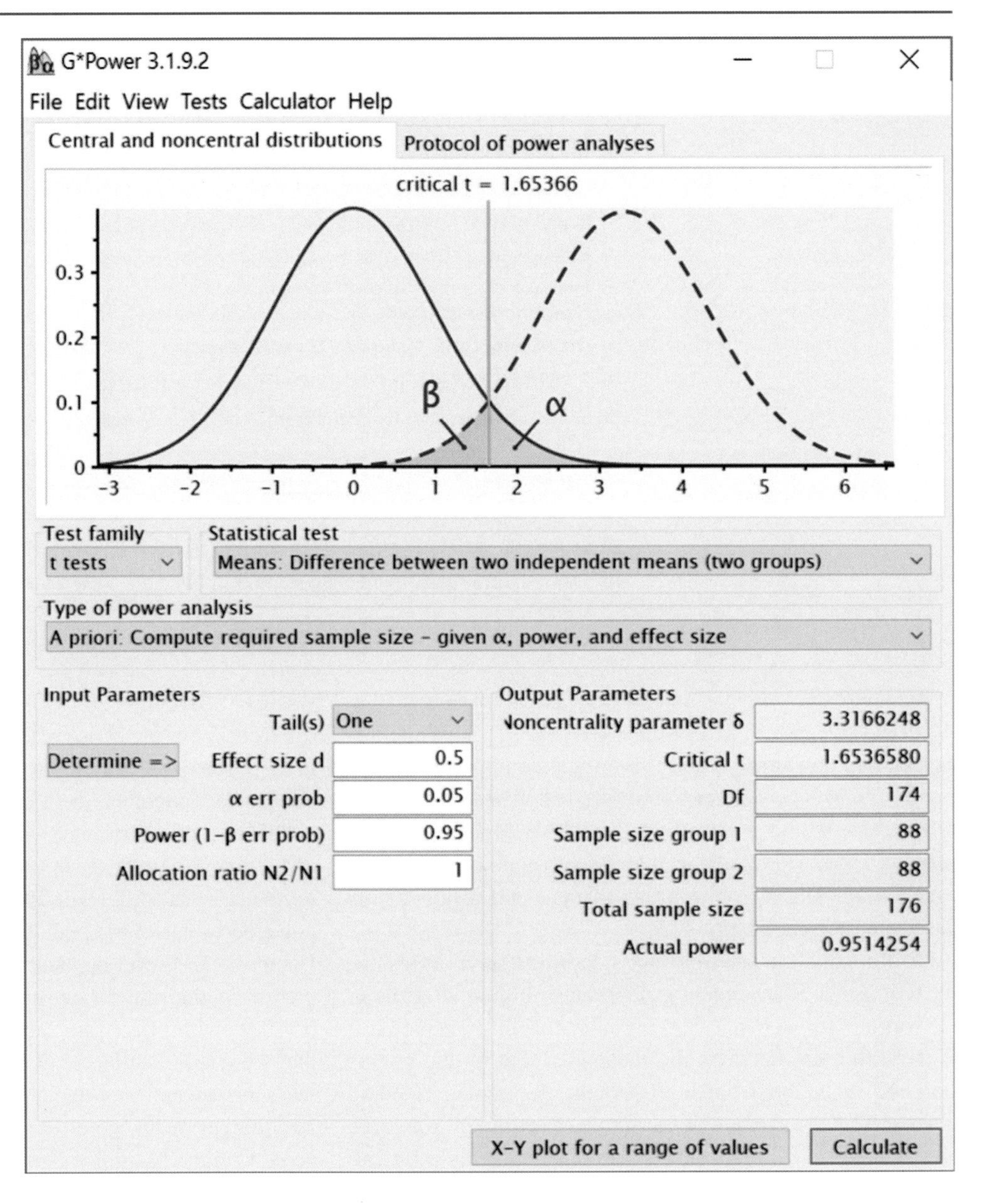

Fig. 29.3 Screen shot of G∗Power calculation of needed sample size

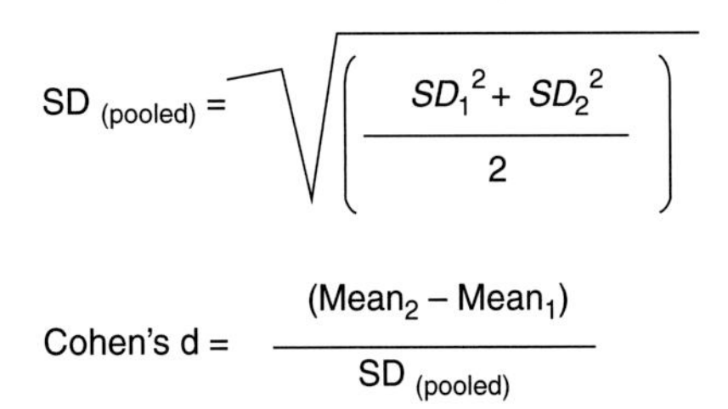

Fig. 29.4 Formulas for Cohen's *d* and the pooled standard deviation for a t-test

than a 5% chance that we are rejecting the null hypothesis incorrectly. Cohen's *d* indicates that we have a small effect. Even if we had a p value of 0.01, the effect size could still be small. We say there is a small effect size – a small magnitude difference – despite a statistically significant p value of 0.05. The effect size helps to qualify our result beyond what the p value means. It is not uncommon for some to misinterpret a 0.01 p value as indicating a bigger or better result than a value of 0.05 and, as discussed above, the p value does indicate that the statistical result has a higher probability of being a true result. This does not, however, translate to a greater magnitude of effect. In fact, we could have a 0.01 probability and our result would still be "small" if the effect size is unchanged. In the final manuscript, such a result would be reported as "a small effect significant at p = 0.01."

Another statistical concept called "degrees of freedom" also plays a role in determining the best interpretation of the p value. Degrees of freedom is a concept in social science statistics that relates to how many numbers in a set can change and still retain the same mean value. Consider the following set of numbers: 2, 4, 6, 8 and 10. The sum is 30 and the mean is 6.0. Four of the five numbers can vary, but after those four are set, there is only one number that will allow the mean to be 6.0. We say that this data set has 4 degrees of freedom. In a study, the degrees of freedom relate to the number of subjects in the study, assuming each subject has an outcome. Consider again our example. For each of our 2 independent groups, the degrees of freedom are equal to the number of subjects in each group minus 1. If there are 9 subjects in each group, the total degrees of freedom = ((9–1) + (9–1)) = 16. For independent t-tests, the number of subjects in each group may be different. As a further example, consider a similar study containing 6 subjects in one group and 18 in another. For this variant the degrees of freedom are ((6–1) + (18–1)) = 22. The same t-value has a smaller risk of falsely accepting the null hypothesis with more degrees of freedom. Figure 29.5 contains a table of critical values for the t-test. A critical value is the result from the t-test calculations that must be met or exceeded to claim a particular level of statistical significance. If we have 10 degrees of freedom and a t-test value = 1.725, our result is not statistically significant. However, if we had 100 degrees of freedom, our result would have been significant at $p < 0.05$.

Even with careful estimates for a pre-study effect size the actual results of the study may be different. If you do not achieve statistical significance with your study data, consider doing a post-study calculation of the obtained effect size. The calculations are the same as those used for pre-study estimates except that you will use data from your study. You can use this post-study analysis to estimate the additional number of subjects needed to achieve statistical significance. Of course other changes mentioned earlier may also increase your statistical power, including a higher number of subjects in our study, a stronger intervention, a more reliable outcome measure, groups created with greater initial differ-

df	.25	.20	.15	.10	.05	.025	.02	.01	.005	.0025	.001	.0005
1	1.000	1.376	1.963	3.078	6.314	12.71	15.89	31.82	63.66	127.3	318.3	636.6
2	0.816	1.061	1.386	1.886	2.920	4.303	4.849	6.965	9.925	14.09	22.33	31.60
3	0.765	0.978	1.250	1.638	2.353	3.182	3.482	4.541	5.841	7.453	10.21	12.92
4	0.741	0.941	1.190	1.533	2.132	2.776	2.999	3.747	4.604	5.598	7.173	8.610
5	0.727	0.920	1.156	1.476	2.015	2.571	2.757	3.365	4.032	4.773	5.893	6.869
6	0.718	0.906	1.134	1.440	1.943	2.447	2.612	3.143	3.707	4.317	5.208	5.959
7	0.711	0.896	1.119	1.415	1.895	2.365	2.517	2.998	3.499	4.029	4.785	5.408
8	0.706	0.889	1.108	1.397	1.860	2.306	2.449	2.896	3.355	3.833	4.501	5.041
9	0.703	0.883	1.100	1.383	1.833	2.262	2.398	2.821	3.250	3.690	4.297	4.781
10	0.700	0.879	1.093	1.372	1.812	2.228	2.359	2.764	3.169	3.581	4.144	4.587
11	0.697	0.876	1.088	1.363	1.796	2.201	2.328	2.718	3.106	3.497	4.025	4.437
12	0.695	0.873	1.083	1.356	1.782	2.179	2.303	2.681	3.055	3.428	3.390	4.318
13	0.694	0.870	1.079	1.350	1.771	2.160	2.282	2.650	3.012	3.372	3.852	4.221
14	0.692	0.868	1.076	1.345	1.761	2.145	2.264	2.624	2.977	3.326	3.787	4.140
15	0.691	0.866	1.074	1.341	1.753	2.131	2.249	2.602	2.947	3.286	3.733	4.073
16	0.690	0.865	1.071	1.337	1.746	2.120	2.235	2.583	2.921	3.252	3.686	4.015
17	0.689	0.863	1.069	1.333	1.740	2.110	2.224	2.567	2.898	3.222	3.646	3.965
18	0.688	0.862	1.067	1.330	1.734	2.101	2.214	2.552	2.878	3.197	3.611	3.922
19	0.688	0.861	1.066	1.328	1.729	2.093	2.205	2.539	2.861	3.174	3.579	3.883
20	0.687	0.860	1.064	1.325	1.725	2.086	2.197	2.528	2.845	3.153	3.552	3.850
21	0.686	0.859	1.063	1.323	1.721	2.080	2.189	2.518	2.831	3.135	3.527	3.819
22	0.686	0.858	1.061	1.321	1.717	2.074	2.183	2.508	2.819	3.119	3.505	3.792
23	0.685	0.858	1.060	1.319	1.714	2.069	2.177	2.500	2.807	3.104	3.485	3.768
24	0.685	0.857	1.059	1.318	1.711	2.064	2.172	2.492	2.797	3.091	3.467	3.745
25	0.684	0.856	1.058	1.316	1.708	2.060	2.167	2.485	2.787	3.078	3.450	3.725
26	0.684	0.856	1.058	1.315	1.706	2.056	2.162	2.479	2.779	3.067	3.435	3.707
27	0.684	0.855	1.057	1.314	1.703	2.052	2.158	2.473	2.771	3.057	3.421	3.690
28	0.683	0.855	1.056	1.313	1.701	2.048	2.154	2.467	2.763	3.047	3.408	3.674
29	0.683	0.854	1.055	1.311	1.699	2.045	2.150	2.462	2.756	3.038	3.396	3.569
30	0.683	0.854	1.055	1.310	1.697	2.042	2.147	2.457	2.750	3.030	3.385	3.646
40	0.681	0.851	1.050	1.303	1.684	2.021	2.123	2.423	2.704	2.971	3.307	3.551
50	0.679	0.849	1.047	1.299	1.676	2.009	2.109	2.403	2.678	2.937	3.261	3.496
60	0.679	0.848	1.045	1.296	1.671	2.000	2.099	2.390	2.660	2.915	3.232	3.460
80	0.678	0.846	1.043	1.292	1.664	1.990	2.088	2.374	2.639	2.887	3.195	3.416
100	0.677	0.845	1.042	1.290	1.660	1.984	2.081	2.364	2.626	2.871	3.174	3.390
1000	0.675	0.842	1.037	1.282	1.646	1.962	2.056	2.330	2.581	2.813	3.098	3.300
z^*	0.674	0.841	1.036	1.282	1.645	1.960	2.054	2.326	2.576	2.807	3.091	3.291

Fig. 29.5 Critical values for the t-test. Circled values show the difference in critical values for df = 10 and df = 100. To be statistically significant, the calculated t value must be the same or larger than the critical value. With 100 degrees of freedom, the critical value is only 1.66. However, for 10 degrees of freedom, it is 1.812. The same numerical difference between 2 means could thus be statistically significant at $p = 0.05$ with 100 degrees of freedom, but not significant with 10

ences related to our intervention (such as medical students and final year residents) and a decision to accept a p value of 0.05 instead of 0.01 for our minimal level of significance.

Conclusion

In summary, it is important to understand the interrelationships between the logic of hypothesis testing, the risk of falsely rejecting the null hypothesis (alpha), the risk of falsely accepting the null hypothesis (beta), the likelihood of detecting a real difference when one exists (statistical power), sample size (and degrees of freedom) and the substantiality of statistical results (effect size). The internet has many resources for understanding, calculating and interpreting these components of statistical analysis.

References

1. Sullivan GM, Feinn R. Using effect size – or why the p value is not enough. J Grad Med Educ. 2012;4(3):279–82.
2. Ferguson CJ. An effect size primer: a guide for clinicians and researchers. Prof Psychol Res Pract. 2009;40(5):532–8.
3. Cumming G. Understanding the new statistics: effect sizes, confidence intervals, and meta-analysis. New York: Taylor & Francis Group, L.L.C; 2012.
4. Cohen J. Statistical power analysis for the behavioral sciences. 2nd ed. Hillsdale: Lawrence Erlbaum Associates; 1988.
5. Cohen J. A power primer. Psychol Bull. 1992;112(1):155–9.
6. Cohen J. The statistical power of abnormal-social psychological research: a review. J Abnorm Soc Psychol. 1962;65:145–53.
7. Murphy KR, Myors B, Wolach A. Statistical power analysis: a simple and general model for traditional and modern hypothesis tests. 4th ed. New York: Taylor & Francis Group; 2014.
8. Ellis PD. The essential guide to effect sizes: statistical power, meta-analysis and the interpretation of research results. New York City: Cambridge University Press; 2010.
9. Grissom RJ, Kim JJ. Effect sizes for research: univariate and multivariate applications. 2nd ed. New York: Taylor & Francis Group, L.L.C.; 2012.
10. Lenhard W, Lenhard A. Calculation of effect sizes. 2016. Available: https://www.psychometrica.de/effect_size.html. Dettelbach (Germany): Psychometrica. https://doi.org/10.13140/RG.2.1.3478.4245.
11. Faul F, Erdfelder E, Lang AG, Buchner A. G*Power 3: a flexible statistical power analysis program for the social, behavioral, and biomedical sciences. Behav Res Methods. 2007;39:175–91.
12. Faul F, Erdfelder E, Buchner A, Lang AG. Statistical power analyses using G*Power 3.1: tests for correlation and regression analyses. Behav Res Methods. 2009;41:1149–60.
13. http://www.gpower.hhu.de/ [Web site for the G*Power application].

Advanced Statistical Analyses

30

Miguel A. Padilla

Overview
Statistical models offer much flexibility and many fall under several general umbrellas. The linear mixed model is one such model, and it can be specified to answer vastly different research questions. Three such linear mixed model specifications (or methods) are the hierarchical linear models used when there are clustered data structures; generalizability theory used for evaluating the reliability or consistency of a measurement process; and equivalence testing used for investigating the similarity between conditions. Here, each method is presented in the context of healthcare simulation with a worked-out example to highlight its central concepts while technical details are kept to a minimum.

Practice Points

- A linear mixed model (LMM) is an umbrella model that can be formulated to answer a variety of research questions.
- A LMM formulated to account for clustered (nested) data structures is called a hierarchical linear model.
- Generalizability theory is a form of a LMM formulated to evaluate the consistency of measurement (assessment).
- Equivalence testing is another form of a LMM formulated to measure the equivalence of groups.
- Any statistical package with a LMM routine can obtain the basic results for the three methods discussed.

M. A. Padilla (✉)
Department of Psychology, Old Dominion University, Norfolk, VA, USA
e-mail: mapadill@odu.edu

Introduction

Healthcare simulation is a rapidly growing field due to advancements in technology and research methodology. Statistical methods are a major part of research methodology, and three advanced statistical methods are presented here. Statistical models have been developing for over a century, and many of them can fall under several general umbrellas. The linear mixed model (LMM; or just mixed model) is one such umbrella model [1]. It is called a LMM because it can model any combination of random and fixed effects. A random effect is when the levels of a variable (or factor) can be thought of as being sampled from a corresponding population. For example, if data are collected from different medical centers and "center" is in the model, then "center" can be thought of as a random effect. By contrast, an effect is called fixed if the levels in the study represent all the possible levels of a variable (or factor). Some examples of fixed effects include gender (male, female) and treatment method (treatment, placebo). The modeling flexibility of LMMs allows them to be specified to answer a variety of research questions. LMMs have been widely used in medical research to study longitudinal change, the consistency of measurement (assessment), and to establish bioequivalence. The methods presented here are examples of each one of these instances. Therefore, these methods should be adaptable to research in healthcare simulation. Specifically, three methods are discussed: hierarchical linear models, generalizability theory, and equivalence testing.

Before moving forward, a disclaimer is needed. The statistical methods here are advanced and are being presented within the context of a general model (i.e., LMM). To keep the discussion concise, some statistical notation and equations are used. However, the models are presented in their simplest forms through examples. Therefore, the general concepts are accessible to all academics and researchers.

D. Nestel et al. (eds.), *Healthcare Simulation Research*, https://doi.org/10.1007/978-3-030-26837-4_30

Hierarchical Linear Models

An intuitive form of the LMM is the hierarchical linear model (HLM) [2, 3]. A key distinction of HLMs is that they are specifically formulated to account for a clustered (nested) data structure. The simplest clustered data structure is when the units of analysis are nested within a cluster. Such clustered structures can occur in organizations and individual change. An organizational example is when students (units) are nested within medical schools (clusters). An example of individual change is when repeated measures are made of each individual in a study. In this example, the repeated measures (units) are nested within individuals (clusters). These two separate clustered structures can also be combined. Suppose an obesity study is being conducted at multiple clinics in the country in which participants are weighed multiple times over the duration of the study. In this situation, the repeated weight measures are nested within each participant, and the participants are nested within the clinics.

Consider data in which airway management skills (AMS) are measured over time (a_t; 4 occasions, 2 months apart) for paramedic trainees that received one of two training methods: modified simulation (ms; $n_{ms} = 16$) or standard simulation (ss; $n_{ss} = 11$). In such a situation, a split-plot ANOVA is the standard way to analyze AMS with time as the within-subjects factor, and training method as the between-subjects factor. Table 30.1 presents the results indicating significant method and time main effects. These effects would typically be investigated with post hoc tests. However, an alternative is to approach the whole analysis through HLM.

HLM specifies models by breaking them up into levels that account for the clustered structure of the data. A level is added for every clustering in the data. For this reason, HLM is also commonly referred to as multilevel modeling. The current example constitutes a two-level model in which time (units) is nested within paramedic trainees (clusters) and can be captured through a random-coefficient regression model (or random coefficient model).

The level-1 model can model time linearly through regression for each trainee and takes the following form:

$$y_{ti} = \pi_{0i} + \pi_{1i}a_{ti} + e_{ti}. \quad (30.1)$$

The model has an intercept (π_{0i}), slope (π_{1i}), and residual (e_{ti}) for each trainee. The intercept is the AMS at start for each trainee. The slope is the 2-month AMS change for each trainee; i.e., how much does AMS change every 2 months. The residual is assumed be independently normally distributed with constant variance σ^2. The level-1 model essentially models where the trainees started and how much they changed over the course of the study.

There are two things to point out about the level-1 model. First, this is the simplest form the level-1 model can take for time. It can be expanded to include higher order terms as needed. Second, the spacing between measurements can be different for the individuals; i.e., trainees do not have to be measured exactly every two years.

The level-2 model takes the following form:

$$\pi_{0i} = \beta_{00} + \beta_{01}ms_i + r_{0i} \quad (30.2)$$

$$\pi_{1i} = \beta_{10} + \beta_{11}ms_i + r_{1i}. \quad (30.3)$$

Notice that now the intercept (π_{0i}) and slope (π_{1i}) from level-1 are each modeled through regression. The model can now be described in terms of fixed (βs) and random effects (r_{0i} and r_{1i}). The first set of fixed effects are the average AMS for the standard simulation (β_{00}) and the average distance difference for the modified simulation (β_{01}) at start. The second set of fixed effects are the average 2-month distance slope for the standard simulation (β_{10}) and the slope difference for the modified simulation (β_{11}).

The random effects are captured with r_{0i} and $r_{1i,}$ which are assumed to be normally distributed with variances τ_{00} and τ_{11}, respectively. Here, τ_{00} captures the variability in π_{0i} (i.e., how much the trainees vary in AMS at start), and τ_{11} captures the variability in π_{1i} (i.e., how much the trainees vary in their change). An additional component not explicitly shown in the models above is the covariance τ_{01} between the intercept (r_{0i}) and slope (r_{1i}) random effects. Now the relationship between where trainees start and how much they change can be estimated.

The fixed effects are presented in Table 30.2. First, there is no significant AMS difference between the modified simulation and standard simulation at start (p-value = .088). Second, AMS for the standard simulation significantly increases over the time of the study (p-value < .001).

Table 30.1 ANOVA table for Airway Management Skills (AMS)

Source	SS	df	MS	F	p-value
Between					
Method (M)	140.465	1	140.465	9.292	.005
Error	377.915	25	15.117		
Within					
Time (T)	209.437	3	69.812	35.347	<.001
M × T	13.993	3	4.664	2.362	.078
Error	148.128	75	1.975		

Table 30.2 Fixed effects of random-coefficient regression model

Fixed effects	Estimate	SE	t-test	p-value
AMS at start				
Avg. ss AMS (β_{00})	21.21	0.61	34.77	<.001
Avg. ms AMS difference (β_{01})	1.41	0.79	1.78	.088
Slope for 2-Month AMS change				
Avg. ss AMS slope (β_{10})	0.48	0.10	4.80	<.001
Avg. ms AMS slope difference (β_{11})	0.30	0.13	2.31	.026

Table 30.3 Random effects of random-coefficient regression model

Random effects	Estimate	SE	z-test	*p*-value
Residual variance (σ^2)	1.72	0.33	5.21	<.001
Intercept variance (τ_{00})	2.91	1.14	2.55	.006
Slope variance (τ_{11})	0.02	0.03	0.67	.243
Intercept-slope covariance (τ_{10})	−.01	0.15	−.07	.956

However, AMS for the modified simulation method significantly increases at a faster rate than the standard simulation (*p*-value = .026).

The random effects are presented in Table 30.3. First, the level-1 residual variance is significant, indicating that there is unexplained variance in the model. Perhaps adding a quadratic term that can model time curvilinearly at level-1 can help explain more variance and improve the fit of the model. Second, the intercept variance is significant, indicating that trainee AMS varies at start. Third, slope variance is not significant suggesting that trainees do not vary in their rate of AMS change. Lastly, the intercept-slope covariance is not significant, so there is no relationship between AMS at start and its rate of change.

In summary, the modified simulation method is more effective than the standard simulation at improving AMS. Specifically, trainee AMS improves under both the modified simulation and standard simulation over time, but improves at a faster rate under the modified simulation. In addition, trainee AMS is similar at the start of the study for both methods, and trainee AMS improved over time regardless of their AMS at the start of the study. For HLM examples see Gadde et al. [4] and Elobeid et al. [5].

Generalizability Theory

Another form of the LMM is generalizability (G) theory [6]. However, the question(s) addressed here pertain to the consistency of measurement, and hypothesis testing is of little to no interest. Measurement is an important process in any of the sciences as it is the foundation by which data are generated. This is as true for establishing the efficacy of a medical intervention as it is for simulation-based training. Measurement is a discipline itself, but all the ideas fall into one of two equally important concepts: validity and reliability (see Chap. 26). Here, the focus is on reliability as it relates to G theory. However, G theory formulates and extends the classical true score model using a LMM. Even so, classical test theory (CTT) reliability is discussed first.

The classical true score model from CTT formulates the observed score for a measurement as

$$x = \tau + u \tag{30.4}$$

where x is the measured data point (observed score), τ is the true score, and u is random measurement error. The idea is that every time a data point is measured, it has an element of truth (τ) plus an element of error (u). The model can be used to form the following reliability index

$$\rho = \frac{\sigma_\tau^2}{\sigma_\tau^2 + \sigma_u^2} \tag{30.5}$$

which is the proportion of true score variance to true score variance plus measurement error variance. The ideal situation is when there is no error (i.e., $u = 0$) as then the data point is equal to truth, and reliability would be perfect (i.e., $\rho = 1$). However, this is extremely rare in behavioral/social science research. Depending on the assumptions, the reliability index can take on different forms. If the assumption of tau-equivalence (or essentially tau-equivalence) is at least satisfied, one form that the reliability index can take is coefficient (or Cronbach's) alpha [7].

Coefficient alpha is the most common reliability index reported for a measurement instrument in many fields, including medicine and nursing [8, 9]. Coefficient alpha owes its popularity to three key features [7]. First, it is computationally simple, requiring only the number of items in the measurement instrument and the corresponding covariance matrix. Second, it can be computed for continuous, ordinal, or dichotomous items. Third, it only requires a single administration of the corresponding measurement instrument. Coefficient alpha is defined as

$$\rho = \alpha_C = \frac{k}{k-1}\left(1 - \frac{\sum_i \sigma_{ii}}{\sum_i \sum_j \sigma_{ij}}\right) \tag{30.6}$$

where k is the number of items, $\sum_i \sigma_{ii}$ is the sum of all the k item variances, and $\sum_i \sum_j \sigma_{ij}$ is the sum of all the item variances and covariances.

For example, suppose researchers are interested in how well a set of 3 emergency medicine simulation scenarios scored by 2 raters measures knowledge of emergency medicine in junior residents. In the study, a sample of 13 junior residents participate in every scenario scored by every rater. A standard way to investigate the reliability of this design is to compute coefficient alpha for the scenarios and raters. The following covariance matrices are obtained for scenarios and raters, respectively:

$$\hat{\Sigma}_S = \begin{bmatrix} 2.59 & 2.50 & 1.29 \\ 2.50 & 4.17 & 1.75 \\ 1.29 & 1.75 & 1.56 \end{bmatrix} \quad \text{and} \quad \hat{\Sigma}_R = \begin{bmatrix} 6.86 & 3.71 \\ 3.71 & 5.14 \end{bmatrix}.$$

The corresponding coefficient alphas are $\hat{\alpha}_C = 0.86$ for scenarios, and $\hat{\alpha}_C = 0.76$ for raters. The issue here is that scenarios and raters interacted with one another as part of one design (or measurement process) and the coefficient alpha for each ignores this aspect of the design; e.g., coefficient alpha for raters ignores the impact of scenario and vice versa. This highlights a limitation of the CTT reliability methods: they can only assess one form of measurement at a time. Therefore, a method that can simultaneously assess multiple forms of measurement is required. This is precisely what G theory does.

Before moving forward, some G theory terminology must be briefly presented. In G theory anything that is used to measure is considered a source of measurement error and called a facet. The variance associated with the facets and anything they interact with is considered error variance. On the other hand, what is being measured is called the object of measurement, and the associated variance is the universe score variance (i.e., G theory's version of true score variance). In the current example, scenarios (s; $n_s = 3$) and raters (r; $n_r = 2$) are facets, and junior residents (p; $n_p = 13$) are the objects of measurement. Lastly, G theory breaks the entire analysis into two pieces: a generalizability (G) and decision (D) study. In the G study, researcher(s) obtain estimates of all the relevant variance for the measurement process. The D study is where researcher(s) obtain reliability estimates for the measurement process.

In a G study, the variability in the measurement process is captured by reformulating the classical true score model as a LMM with each facet and objects of measurement as terms in the model. As such, G theory has the same modeling flexibility as a LMM in that it can have any combination of fixed and random effects. Continuing with the current example, every junior resident participated in every scenario scored by every rater. In G theory, this constitutes a completely crossed design ($p \times s \times r$) that can be captured with the following model

$$x = \mu + p + s + r + ps + pr + sr + u \qquad (30.7)$$

where x is the observed score, μ the grand mean, p are the junior residents, s are the scenarios, r are the raters, and u is the error (residual). Although this is a LMM with all random effects, the main interest in G theory is the variability via the variance component (VC) associated with each of the terms (or sources) and their corresponding interactions (e.g., $p \times s$). The variance sources can be presented through an ANOVA table.

Table 30.4 is the ANOVA table for the knowledge of emergency medicine example. There are a few important differences to note from traditional ANOVA. First, there are no F-tests and accompanying p-values because these are not of interest in G theory. Second, what is considered the sample size in traditional ANOVA methodology is now an important source in the model (p) as it captures true differences in knowledge, skill, etc. (i.e., true score variance). Third, there is no variance for the highest order interaction ($p \times s \times r$) because there are no *df* to estimate it. This is because the sample size (i.e., p) is now a term in the model (see previous second point). Thus, the variance for the highest order interaction and error are confounded and cannot be disentangled, and together they are referred to as the residual.

Table 30.4 ANOVA table for two-facet $p \times s \times r$ design

Source	SS	df	MS	VC $(\hat{\sigma}^2)$	% of total variance
Junior residents (p)	38.82	12	3.24	0.38	0.36
Scenario (s)	1.56	2	0.78	0.01	0.01
Rater (r)	5.65	1	5.65	0.13	0.12
$p \times s$	11.10	24	0.46	0.10	0.10
$p \times r$	9.18	12	0.77	0.17	0.16
$s \times r$	0.54	2	0.27	0.00	0.00
Residual (u)	6.13	24	0.26	0.26	0.25

Note. All facets are random, $u = (p \times s \times r) + \text{error}$

The ANOVA table is the G study and first piece of a G theory analysis. The last column in Table 30.4 contains the relative percentage of each VC. As such, the largest VC (0.36) is for junior residents (p) indicating that the most variance is universe score variance (i.e., true score variance). In terms of error, the second largest VCs are for raters (r) and its interaction with junior residents ($p \times r$). This indicates that raters are not as consistent as scenarios and are vary in their scoring of junior residents; i.e., the raters are scoring the junior residents differently. However, the largest error VC is for the residual (0.25), suggesting that something not accounted for by the G study design is impacting the measurement process. Once the VCs are estimated, they can then be used to estimate G theory reliability analogs [6].

In a D study, G theory offers two reliability analogs. The first index is the generalizability coefficient defined as

$$E\rho^2 = \frac{\sigma_\tau^2}{\sigma_\tau^2 + \sigma_\delta^2} \qquad (30.8)$$

where σ_τ^2 is universe score variance and σ_δ^2 is relative error variance defined as

$$\sigma_\delta^2 = \frac{\sigma_{ps}^2}{n_s} + \frac{\sigma_{pr}^2}{n_r} + \frac{\sigma_{psr,e}^2}{n_s n_r}. \qquad (30.9)$$

The second is the index of dependability defined as

$$\Phi = \frac{\sigma_\tau^2}{\sigma_\tau^2 + \sigma_\Delta^2} \qquad (30.10)$$

where σ_Δ^2 is the absolute error variance defined as

$$\sigma_\Delta^2 = \frac{\sigma_s^2}{n_s} + \frac{\sigma_r^2}{n_r} + \frac{\sigma_{ps}^2}{n_s} + \frac{\sigma_{pr}^2}{n_r} + \frac{\sigma_{sr}^2}{n_s n_r} + \frac{\sigma_{psr,e}^2}{n_s n_r} \qquad (30.11)$$

Continuing with the current example using the estimated VCs, then $\hat{\sigma}_\delta^2 = 0.16$ and $\hat{\sigma}_\Delta^2 = 0.23$. Using these quantities with $\hat{\sigma}_\tau^2 = \hat{\sigma}_p^2 = 0.38$, then $E\hat{\rho}^2 = 0.70$ and $\hat{\Phi} = 0.62$.

In the behavioral/social sciences, a typical criterion for adequate reliability is .70 or higher [10]. Because the current measurement process with 3 items and 2 raters is at or below .70, its reliability is questionable. While there does not appear to be such a criterion for G theory reliability indices, the .70 criterion does give a good benchmark from which to start. Of course, the criterion is contingent on the discipline and purpose of the measurement process. Even though both G theory reliability indices were demonstrated, each has a specific use when making decisions about the objects of measurement [6]. The generalizability coefficient is only appropriate when making relative decisions, and the index of dependability when making absolute decisions. Relative decisions are based on comparing the objects of measurement to one another; i.e., how a person compares with other people. Absolute decisions are based on comparing objects of measurement to the pre-established criterion for what is being measured; i.e., does a person meet a certain skill level. For G theory examples see McBride et al. [11] and Nadkarni et al. [12].

Equivalence Testing

The last method discussed is equivalence testing which originated in pharmacokinetics for establishing practical similarity (bioequivalence) between groups [13]. A typical situation in pharmacokinetics is a pharmaceutical company wanting to determine if a generic drug is as effective as the current drug. As such, establishing statistical equivalence is growing in popularity across the sciences outside of pharmacokinetics in a variety of settings.

Equivalence testing is the simplest form the LMM can take but is probably the most difficult to grasp. This is because the same model and corresponding results are used to test seemingly opposing hypotheses than traditionally done in null significance hypothesis testing (NSHT). NSHT and equivalence testing are flip sides of the same coin. Each method sets up two opposing hypotheses: the null (H_0) and alternative (H_A) hypothesis. Both methods assume H_0 to be true unless data provide sufficient evidence to reject it. The difference between the methods lies in how these hypotheses are stated. Consider the situation involving two means. In a typical NSHT scenario, H_0 states that the mean difference is equal to zero and H_A that the mean difference is not equal to zero. On the other hand, equivalence testing states H_0 as the mean difference surpassing or being equal to Δ and H_A as the mean difference being within Δ, where Δ is a content specific value chosen by the researcher(s) using literature, prior knowledge, or expertise.

To illustrate, consider a study in which medical students are trained in patient-centered communication through an online interactive tool. The students (T) and professionals in the field (C) are then asked to view a 6-min video of a clinical scenario and assess the care providers' communication behavior. The following assessment estimates are obtained: $n_T = 30$, $\hat{\mu}_T = 25.7$, $\hat{\sigma}_T = 1.8$; $n_C = 32$, $\hat{\mu}_C = 24.6$, $\hat{\sigma}_C = 2.1$. The standard way to compare the means from the two conditions is an independent-samples t-test which is the simplest form of the LMM with only one fixed effect. In traditional NSHT, the idea is to test for a difference between the two conditions. Here, the corresponding hypotheses can take on the following forms

$$\begin{aligned} H_0 &: \mu_T - \mu_C = 0 \\ H_A &: \mu_T - \mu_C \neq 0 \end{aligned}. \quad (30.12)$$

The hypotheses above set up a two-sided test with $df = 60$ that gives a critical value of $t_{crit} = \pm 2.00$ using $\alpha = .05$. The corresponding independent-samples t-test is $t = 2.21$.

A two-sided test provides two options for proceeding with hypothesis testing. The first option uses the following criteria: if a test statistic surpasses the critical value, reject H_0. For the current example, H_0 can be rejected because $t = 2.21 > t_{crit} = 2$. The second option considers the following criteria: if zero is not within the confidence interval (CI), reject H_0. For the current example, the 95% CI is [0.103, 2.097], and H_0 can be rejected because zero is not within the CI. In either case, it can be concluded that students gave more favorable assessments than the professionals.

By contrast, suppose the researcher wants to test for equivalence between students and professionals. In addition, based on prior assessment studies, the researcher specifies an assessment difference of $\Delta = 2.5$ as not meaningful. This is a situation for equivalence testing via two-one-sided t-tests (TOST) [14, 15]. Here, the corresponding hypotheses take on the following forms

$$\begin{aligned} H_0 &: |\mu_T - \mu_C| \geq \Delta \\ H_A &: |\mu_T - \mu_C| < \Delta \end{aligned}. \quad (30.13)$$

TOST sets up two composite hypotheses based on H_0. The lower H_0 takes the following form

$$H_{0L} : \hat{\mu}_T - \hat{\mu}_C \leq -\Delta \quad (30.14)$$

with corresponding independent-samples t-test $t_L = 7.22$. The upper H_0 takes the following form

$$H_{0U} : \hat{\mu}_T - \hat{\mu}_C \geq \Delta \quad (30.15)$$

with corresponding independent-samples t-test $t_U = -2.81$. These are two one-sided t-tests that are corrected for Type I error by dividing α by the number of tests (i.e., Bonferroni procedure). With $df = 60$, the critical values are $t_{crit} = \pm 2.00$ using $\alpha/2 = .05/2 = .025$.

The TOST procedure also provides two options for proceeding with hypothesis testing. The first option considers

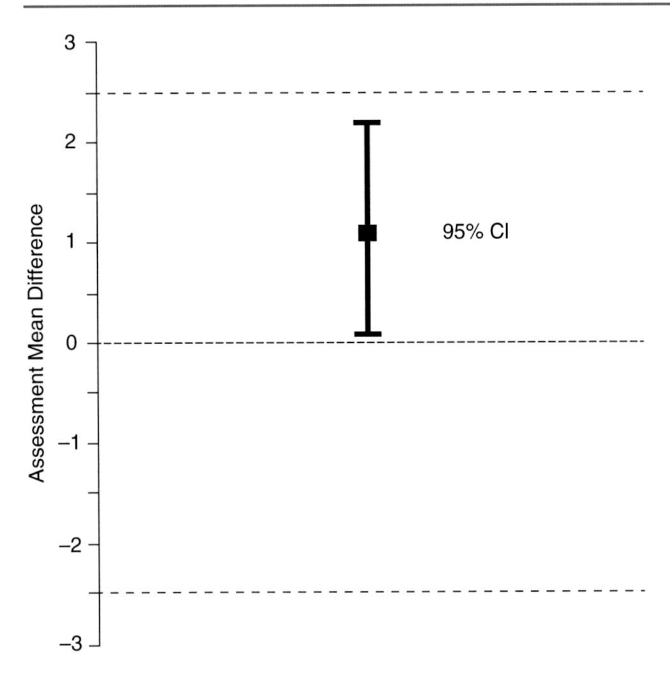

Fig. 30.1 Assessment mean difference 95% confidence interval (CI) with equivalence bounds (±2.5)

the following criteria: if the lower test statistic (t_L) is greater than a positive critical value and the upper test statistic (t_U) is less than a negative critical value, reject H_0. For the current example, H_0 can be rejected because $t_L = 7.22 > t_{crit} = 2$ and $t_U = -2.81 < t_{crit} = -2$. The second option considers the following criteria: if the CI is within ±Δ, reject H_0. For the current example, the 95% CI is [0.103, 2.097], and H_0 can be rejected because the CI is within ±2.5. In either case, the assessment of the students is practically equivalent to the professionals. The idea behind equivalence testing can be succinctly presented in a graph. Figure 30.1 presents the 95% CI along with the equivalence bounds (±Δ) where it is clear that the CI is within the equivalence bounds. For an equivalence testing example see Anderson-Montoya et al. [16].

Conclusion

HLM, G theory, and equivalence testing were briefly presented. However, this brief presentation does not do any of these methods justice and the reader is referred to the corresponding references for further details. Additionally, the methods were presented in the context of a LMM to show that, even though these methods answer different questions, they share a general statistical framework. As such, any of the standard statistical packages (e.g., SAS, SPSS, R, etc.) through their LMM routines can run any of the models. However, the packages only compute the G theory VCs but not the corresponding reliability indices. The VCs can be used to hand-compute or use software (e.g., MS Excel) to get the required reliability estimates ($E\rho^2$, Φ). To get all the G theory estimates, then GENOVA [6] or EduG [17] can be used. For equivalence testing, a simple t-test routine from the packages can be used.

In conclusion, three statistical methods commonly used in medical research were presented. Each method answers different research questions and hence have different applications. Therefore, each method was presented through an application. In each application, the advantage of the method is demonstrated by contrasting it with the traditional method of analysis. Through this process it was made clear that HLM and G theory offer more flexibility and provide a richer analysis than traditional ANOVA and coefficient alpha, respectively. By contrast, equivalence testing is not necessarily richer than a NSHT, but it does demonstrate how the same results can be used to answer seemingly opposing hypotheses. Although the hypotheses are opposing, both have their place in research. It is hoped the presentation here has sparked the curiosity of healthcare simulation researchers and given ideas as to how to adapt the methods in their research.

References

1. Muller KE, Stewart PW. Linear model theory: univariate, multivariate, and mixed models. Hoboken: Wiley-Interscience; 2006. xiv, p. 410.
2. Raudenbush SW, Bryk AS. Hierarchical linear models: applications and data analysis methods. 2nd ed. Thousand Oaks: Sage Publications; 2002. xxiv, p. 485.
3. Snijders TAB, Bosker RJ. Multilevel analysis: an introduction to basic and advanced multilevel modeling. 2nd ed. Los Angeles: Sage; 2012. xi, p. 354
4. Gadde KM, Franciscy DM, Wagner HR, Krishnan KRR. Zonisamide for weight loss in obese adults – a randomized controlled trial. Jama-J Am Med Assoc. 2003;289(14):1820–5.
5. Elobeid MA, Padilla MA, McVie T, Thomas O, Brock DW, Musser B, et al. Missing data in randomized clinical trials for weight loss: scope of the problem, state of the field, and performance of statistical methods. PLoS One. 2009;4(8):e6624.
6. Brennan RL. Generalizability theory. New York: Springer; 2001. xx, p. 538
7. Padilla MA, Divers J, Newton M. Coefficient Alpha bootstrap confidence interval under nonnormality. Appl Psychol Meas. 2012;36(5):331–48.
8. Cortina JM. What is coefficient alpha? An examination of theory and applications. J Appl Psychol. 1993;78(1):98–104.
9. Hogan TP, Benjamin A, Brezinski KL. Reliability methods: a note on the frequency of use of various types. Educ Psychol Meas. 2000;60(4):523–31.

10. Peterson RA. A meta-analysis of cronbach's coefficient alpha. J Consum Res. 1994;21(2):381–91.
11. McBride ME, Waldrop WB, Fehr JJ, Boulet JR, Murray DJ. Simulation in pediatrics: the reliability and validity of a multi-scenario assessment. Pediatrics. 2011;128(2):335–43.
12. Nadkarni LD, Roskind CG, Auerbach MA, Calhoun AW, Adler MD, Kessler DO. The development and validation of a concise instrument for formative assessment of team leader performance during simulated pediatric resuscitations. Simul Healthc. 2018;13(2):77–82.
13. Hauck WW, Anderson S. A new statistical procedure for testing equivalence in two group comparative bioavailability trials. J Pharmacokinet Biopharm. 1984;12(1):83–91.
14. Schuirmann DJ. A comparison of the two one-sided tests procedure and the power approach for assessing the equivalence of average bioavailability. J Pharmacokinet Biopharm. 1987;15(6):657–80.
15. Wellek S. Testing statistical hypotheses of equivalence and noninferiority. 2nd ed. Boca Raton: CRC Press; 2010. xvi, p. 415.
16. Anderson-Montoya BL, Scerbo MW, Ramirez DE, Hubbard TW. Running memory for clinical handoffs: a look at active and passive processing. Hum Factors. 2017;59(3):393–406.
17. Cardinet J, Johnson S, Pini G. Applying generalizability theory using EduG. New York: Routledge; 2010. xviii, p. 215.

Part V

Mixed Methods and Data Integration

Applying Mixed Methods Research to Healthcare Simulation

31

Timothy C. Guetterman and Michael D. Fetters

Overview
Mixed methods has the potential to add values to simulation research by informing healthcare simulations, developing assessments and measures, and evaluating the effectiveness of simulations. However, it seems under-utilized relative to other single methodology designs. This chapter provides an introduction to mixed methods in simulation research and evaluation. We introduce mixed methods and cover major designs. Each form of research brings unique strengths. We highlight the integration of qualitative and quantitative research as a central feature of mixed methods and discuss integration strategies. To illustrate the application of mixed methods to simulation, we discuss studies that have used mixed methods to: (1) evaluate simulations by integrating qualitative and quantitative data, (2) use qualitative methods to develop simulations and its features, and (3) develop assessments and surveys for use in simulation research, such as developing models of learner experiences. Finally, we provide recommended criteria that apply to writing and reviewing for publication or funding proposals.

Practice Points

- Consider mixed method to address complex research questions and aims that require integrating quantifiable outcomes or measures along with the qualitative nuanced and contextual information.
- Identify your core mixed methods design—convergent, explanatory sequential, and exploratory sequential. Core designs offer a way to think about the entire process and how to integrate the two forms of research and build to more complex applications such as intersecting other designs such as randomized controlled trials or case studies.
- The integration of qualitative and quantitative research is essential when conducting mixed methods.
- Include critical aspects of mixed methods whenever writing proposals or manuscripts. Use recommendations (Box 31.2) as a self-check of writing.

Overview of Mixed Methods Research and Evaluation

Mixed methods research and evaluation has emerged as a rigorous and value added methodology. In brief, mixed methods is an approach to research and evaluation that involves collection, analysis, and integration of qualitative and quantitative data within a single study or a closely connected series of studies. It is a methodological approach used across disciplines, including health sciences [1], education [2], and interdisciplinary research topics. It has also seen increasing use in healthcare simulation research. The major advantage of mixed methods lies in its ability to address complex research questions and aims — understanding both quantifiable outcomes and measures along with the nuanced and contextual information that qualitative research yields. Given that healthcare simulation research involves nuanced topics, such as (but not limited to) patient outcomes, education, and human behavior of learners in addition to the simulation methodologies themselves, mixed methods may be ideally suited for simulation research. However, mixed methods appears to be under-utilized in simulation research based on several recent systematic reviews [3, 4].

T. C. Guetterman (✉)
Creighton University, Omaha, NE, USA
e-mail: timguetterman@creighton.edu

M. D. Fetters
Department of Family Medicine, University of Michigan, Ann Arbor, MI, USA
e-mail: mfetters@umich.edu

D. Nestel et al. (eds.), *Healthcare Simulation Research*, https://doi.org/10.1007/978-3-030-26837-4_31

Despite the relatively limited adoption of mixed methods, investigators are actually conducting innovative simulation research using this methodology. Applications of this approach include evaluating simulations as educational interventions [5, 6], comparing the effectiveness of simulations while measuring emotional engagement and stress [7], and developing learning models and theories [8]. The premise behind mixed methods is that it has the potential to provide a more complete understanding of research questions by leveraging the strengths of both qualitative and quantitative methods [9]. Therefore, the purpose of this chapter is to provide an introduction to mixed methods research and evaluation as applied to healthcare simulation. We define the key characteristics of mixed methods with particular focus on major mixed methods designs and integration. Based on our experiences of conducting mixed methods simulation research and our review of studies published in the past ten years, we discuss example studies and potential applications, and provide guidance for conducting mixed methods simulation research and evaluation.

Definition of Mixed Methods Research and Evaluation in Healthcare Simulation

We distinguish mixed methods from multimethod research. Multimethod research can refer to other combinations of research approaches. For example, multimethod research includes the use of quantitative and qualitative research that is not integrated within a study. It may also refer to the use of more than one quantitative or qualitative approach (e.g., both grounded theory and ethnography) within a study [10]. Our focus for this chapter is mixed methods research with the key distinction being the integration of qualitative and quantitative research in a study.

Mixed methods can be defined by its major features: (1) the collection and analysis of both quantitative and qualitative data; (2) the use of rigorous and systematic qualitative and quantitative research procedures (e.g., for sampling, data collection, and data analysis approaches); (3) the employment of a mixed methods design to guide the entire process of research; and (4) the integration of the qualitative and quantitative research. Integration is the point at which qualitative and quantitative research interacts, and can occur during analysis, after analysis (by integrating results), or by using the results of one form or research to inform a follow-up phase [10]. Finally, investigators might rely on a philosophical, theoretical, or conceptual framework to guide the research (e.g., crafting research questions or next hypothesis, informing analysis and interpretation).

In healthcare simulation, a mixed methods approach has numerous advantages centered around the idea that it may generate a more complete understanding. For instance, through a mixed methods study, investigators can develop cases and content for simulations, and evaulate assessments or other outcome measures. Furthermore, integrating qualitative data brings learner perspectives to the forefront, which is critical to ensure authentic learning and understand usability issues in learners' own words in order to both evaluate and refine simulations. Nevertheless, managing the complexity of mixed methods brings challenges in terms of resources and skills. It tends to be more resource intensive, meaning it can take more time given the collection and analysis of both qualitative and quantitative data. It also requires skillsets within the research team of quantitative, qualitative, and often psychometric methods in addition to mixed methods specific skills [11]. Psychometric methods include developing instruments and writing items, examining the reliability of assessment instruments, and conducting validation studies. Therefore, in considering whether it is the best approach to address research questions or aims, we urge researchers to fully consider these practical issues, such as resources to conduct mixed methods, the time needed, and the skills required.

Mixed methods designs. Investigators use mixed methods designs to *guide* the process of research. Although scholars have developed numerous different terms and typologies for mixed methods designs, they can be characterized by the timing of qualitative and quantitative strands, their relative emphasis, and the level of integration (at a certain point or at multiple points through the study), [12]. Creswell and Plano Clark (10) offer a parsimonious set of three core mixed methods designs—convergent, explanatory sequential, and exploratory sequential—as a foundation upon which more complex designs can develop. In the *convergent design*, qualitative and quantitative data are collected and analyzed in about the same time frame, and then the results are integrated (i.e., brought together) for analysis and comparison. Qualitative and quantitative research are often equally emphasized in this design. The *explanatory sequential* design begins with a quantitative phase, which is often emphasized more. Then, based on those results, investigators conduct a follow-up qualitative phase for the purpose of explaining the initial quantitative results. Integration occurs in this design when moving from the quantitative to qualitative phase and then again when interpreting how the qualitative findings helped to explain the quantitative. In the *exploratory sequential* design, investigators begin with a qualitative phase [10]. Based on those findings, integration occurs as researchers systematically build to a quantitative phase (e.g., develop a survey and administer, develop an intervention and implement quantitatively).

More complex designs arise by building on these core designs, employing multiple core designs and intersecting a mixed methods design with another methodology, such as a case study or an intervention. Using an intervention mixed methods design, researchers might develop, test, and refine a simulation. For example, a research team used a series of qualitative and quantitative phases to develop an virtual human simulation to train communication skills [13] and then evaluated the intervention through a mixed methods randomized controlled trial, collecting and integrating multiple data sources [6]. In complex mixed methods studies, integration occurs at multiple points. Researchers conducting studies that involve multiple phases should carefully ensure that each phase informs the next. Beginning with a qualitative exploration, investigators can use those findings to inform important outcomes to measure in a subsequent trial as well as which assessment instruments or surveys to use. Similarly, results from a quantitative component could systematically inform a qualitative follow-up phase designed to unpack mechanisms, elaborate, and explain the quantitative results.

Integration is well-established as the cornerstone of mixed methods research and is critical to the analysis of data in a mixed methods study [14]. Integration is the point at which quantitative and qualitative research procedures interact [10], forming the basis for final mixed methods metainferences or conclusions drawn from the study [14]. *Metainferences* are new inferences generated from the integration of qualitative and quantitative results beyond what either strand alone would generate [15]. Metainferences arise during interpretation of results. For example, Kron et al. [6] integrated results of a quantitative survey of learner attitudes during a communication simulation with qualitative themes and examined the extent to which the quantitative and qualitative results were congruent. Another example of generating metainferences involves qualitatively comparing the differential experiences of high, medium, and lower performers in a simulation. In this example, the new insight might be how the learner attitudes or motivation towards the simulation affected performance, which could in turn inform ways to modify the simulation. Though integration can occur in many ways, as a starting point, we often think first about integration with respect to the different mixed methods designs [10, 16].

In a **convergent design**, integration occurs through merging the qualitative and quantitative data for comparison or by transforming data from one type to another for further analysis (e.g., counting and quantifying qualitative codes to model statistically) [2, 10]. The intent of integration within convergent designs is to expand understanding or examine concordance between the two strands. Investigators then draw metainferences about the extent to which the results were concordant, discordant, or expanded understanding. Next, integration in **explanatory sequential designs** occurs by systematically connecting to a qualitative phase with the intent of explaining the initial quantitative results [2, 10]. Specific procedures may include using the initial quantitative results to determine what results need further explanation, developing specific questions for an interview or focus group protocol, or selecting a purposeful qualitative follow-up sample. Metainferences arise by considering how the qualitative follow-up phase helped to explain the initial quantitative results. Finally, integration in the **exploratory sequential design** occurs as researchers build from the initial qualitative phase to a follow-up quantitative feature, such as systematically using the qualitative codes and themes to develop an assessment instrument or to develop simulation features [2, 10, 16]. Investigators conclude with metainferences about how well the quantitative feature was contextualized, or perhaps how well the qualitative findings generalized.

Complex mixed methods designs employ multiple strategies for integration. For example, in an intervention mixed methods design, integration might occur by using an initial qualitative phase to inform the intervention or assessments. Integration may happen during the intervention in a randomized controlled trial by collecting qualitative data (e.g., observations of the experience) about the process, and integration might occur post-intervention through exit interviews with learners. As with the other designs, researchers consider the additional metainferences generated through the integration of both forms of data.

Integration is reported through narrative writing and through joint displays that are a means to visually facilitate and represent integration [17]. Across the designs and integration strategies, a joint display can be used to explicitly link the qualitative and quantitative components. Ideally, a joint display will include qualitative and quantitative results and a mixed methods metainference consistent with the integration strategy used. We strongly recommend careful attention to integration throughout the entire study when designing mixed methods to achieve meaningful integration when conducting and disseminating the study [18].

Potential Applications of Mixed Methods to Simulation in Healthcare

As noted in Box 31.1, qualitative and quantitative research each bring unique strengths to healthcare simulation research. Further potential lies in the integration of the two types of research within a single project. In the following section, we focus on three potential applications of mixed methods. These applications are most common in the mixed methods healthcare simulation research we reviewed, but other applications are possible.

Box 31.1 Reasons for integrating qualitative and quantitative data in simulation research

Potential goals of a qualitative component	Potential goals of a quantitative component
• Conduct a needs assessment to inform a simulation • In communication simulation, find actual language to build into the simulation • Assess learner experiences in detail and in their own words • Begin qualitatively to use findings to systematically develop instruments • Develop a theory, model, or framework explaining the simulation • Understand trainee reflections of the experience (pedagogical and research advantage) • Explain results of simulation trial and potential mechanisms of action • Examine the authenticity of the simulated learning experience	• Assess primary outcomes quantitatively • Test effectiveness using a randomized controlled trial • Gather validity evidence and reliability of measures • Assess quantitatively learner attitudes or self-assessment of skills • Assess the effects of an intervention on clinical outcomes

Mixed Methods to Evaluate Healthcare Simulation Interventions

As an educational intervention, simulation needs evidence of effectiveness. An evaluation assesses the merit, worth, or value of a simulation [19], often by assessing outcomes and process. A randomized controlled trial (RCT) may be ideally suited for establishing causal inference as to the effectiveness of educational intervention [2, 20]. Moreover, a mixed methods randomized trial permits the evaluation of outcomes and processes, allowing contextual factors and learner experiences to be explored regarding the effectiveness in other educational settings outside of the trial. However, the integration of qualitative methods can help to identify relevant outcomes measures pre-trial, inform recruitment procedures, or inform the intervention components itself. For example, narratives could be collected from patients to include in a history taking simulation to enhance its authenticity. On the other hand, integrating qualitative data post-trial, such as conducting follow-up interviews with a sample of participants, can elucidate mechanisms of action and help to explain RCT results.

Example of Using Mixed Methods in an RCT

Silberman and colleagues conducted a mixed methods randomized controlled trial of a high fidelity human simulation for physical therapy students [5]. The primary outcome of the study was self-efficacy in acute care. Their data collection included demographic information, academic achievement, the Acute Care Confidence Survey to assess self-efficacy, and a semi-structured focus group with the experimental group to understand their perceptions of the value of the simulation. Finally, they integrated the qualitative themes with the self-efficacy results by merging the two. The quantitative results were consistent with the qualitative themes about increased confidence, and the qualitative themes expanded their understanding of the nonthreatening simulation environment and of how the simulation fostered skills development.

Mixed Methods to Develop Simulation Features

An innovative application of mixed methods is to use qualitative data to inform the development of simulations. The general idea is to begin with an exploration, such as interviewing learners, clinicians, or experts, and based on those findings to systematically develop a simulation or its features. For example, Kron and colleagues developed a virtual human simulation to provide training in empathic communication skills [13]. As prototypes developed of the simulation, they were tested with learners and with experts to refine the prototype. Similarly, qualitative quotes might be the basis for actual language used in communication simulations. Investigators might also use qualitative findings to develop an educational model and use that to build key components of the simulation that accounts for clinical skills and learner behavior.

Example of Using Mixed Methods to Develop Simulation Features

Waznonis used mixed methods to examine nursing simulation debriefing practices, as an important but often overlooked simulation feature, in order to inform future simulation debriefings [21]. The researcher employed a survey with 62 closed-ended and open-ended items. We should note that open-ended questions does not typically yield data that is as rich as other sources, such as interviews or focus groups [22]. In this study, authors surveyed faculty who teach in accredited nursing programs about their (1) background and specialty area, (2) training debriefing, (3) debriefing practices (e.g., when debriefing occurred, who was present, what type of debriefing was used), (4) challenges in debriefing, and (5) evaluation of debriefing. Another major qualitative source was one to two paragraph descriptions by faculty of their debriefing approaches. The integration of numeric and text-based questions yielded more understanding about existing debriefing practices and evaluation in addition to themes about debriefing as a semi-structured process and the eclectic approaches used [21]. Although aspects, such as challenges to debriefing, might have been gathered through a checklist-type item, the benefit of an open-ended item is that it did not restrict participants in their responses. The integrated findings produced recommendations to have skilled debriefing facilitators, create a safe environment, delineate facilitator roles to observe the simulation and engage the learner in debriefing, use a structured format, and focus on learning objectives. The investigator plans a future qualitative strand to further elaborate on issues such as training needs, student engagement, and evaluation [21].

Example of Using Mixed Methods to Develop Surveys or Assessments

Mixed methods can be used in simulation research to develop assessments of learning outcomes and to develop surveys to investigate learning principles. As noted, an exploratory sequential design begins with a qualitative phase and builds to a quantitative feature (e.g., an instrument). For example, qualitative research might be developed to better understand or develop a model of relevant health skills. Then, that information can be used to identify major constructs to measure. The subsequent assessment might be a self-assessment or a robust objective structured clinical examination (OSCE) with a standardized patient. One approach is to use qualitative themes to identify major scales or sections in the assessment, use codes to identify variables within each item, and use participant quotes to inform item language [10]. The goal is to ensure that the instrument is contextually relevant for the target participants. Beginning qualitatively with the target participants ensures the instrument is grounded in the perspectives of the target participants.

Tripathy et al. used mixed methods to generate, test, and refine a model of student responses to mannequin death in simulation [8]. Their aim was to develop the model of the effects on psychological and educational factors to inform future simulations. They began with a qualitative grounded theory approach using data collected from focus groups with learners. Based on the themes and theoretical framework created, they developed a survey, which they sent to learners to test and refine the theory. By using mixed methods, the investigators were able to both develop a theory of learner experiences—qualitatively and then disseminate a survey to refine the theory—quantitatively in a single research study.

Recommended Practices for Applying Mixed Methods to Healthcare Simulation

Box 31.2 presents our recommendations of critical aspects to include when writing or reviewing mixed methods simulation literature, grants for funding, and other proposals. The recommendations are derived from our review of literature.

Box 31.2 Recommendations for writing publications of mixed methods simulation and evaluation research

- ☐ Identify a rationale for using mixed methods to study the research problem
- ☐ Ensure that research questions or aims reflect what you hope to understand using mixed methods
- ☐ Identify a mixed methods design
- ☐ Include a procedural diagram of the mixed methods design
- ☐ Detail rigorous procedures for the management of both quantitative and qualitative strands
- ☐ Specify the approach to integration and describe procedures for integration
- ☐ Report the integrated results in narrative
- ☐ Report integration using joint displays
- ☐ Specify the value-added from mixed methods relative to a single methods approach

Conclusion

Mixed methods research and evaluation involves the collection, analysis, and integration of qualitative and quantitative research. The integration of the two distinguishes mixed methods and assists researchers in developing metainferences beyond what the two approaches alone could generate. Investigators might begin by considering which core design—convergent, explanatory sequential, or exploratory sequential—applies to their research questions. Thinking through the design in addition to potential data sources should inform thinking about integration. In brief, the quantitative and qualitative results could be merged for comparison or related in convergent designs. In an explanatory sequential design, quantitative results could connect to a follow-up qualitative phase by using the quantitative results to identify who to sample and what questions to ask. In the exploratory sequential design, qualitative findings could build to specific items on a survey on instrument. More complex designs also exist, such as interventions that employ multiple core designs and integration approaches.

Mixed methods research has the potential to add value to healthcare simulation research by yielding a more comprehensive understanding of research questions or aims than either qualitative or quantitative approaches alone. However, to achieve the value-added of mixed methods, rigorous qualitative and quantitative research needs to be conducted and integrated. Failing to integrate the two forms, when collected and when the research questions call for both, is likely a missed opportunity. As simulation researchers continue to use mixed methods, we encourage both empirical and methodological articles to disseminate important example and lessons learned about how to leverage the methodology.

References

1. Curry L, Nunez-Smith M. Mixed methods in health sciences research: a practical primer. Thousand Oaks: SAGE; 2015.
2. Creswell JW, Guetterman TC. Educational research: planning, conducting, and evaluating quantitative and qualitative research. 6th ed. Boston: Pearson; 2019.
3. Aura SM, Sormunen MS, Jordan SE, Tossavainen KA, Turunen HE. Learning outcomes associated with patient simulation method in pharmacotherapy education: an integrative review. Simul Healthc. 2015;10(3):170–7.
4. Alanazi AA, Nicholson N, Thomas S. The use of simulation training to improve knowledge, skills, and confidence among healthcare students: a systematic review. Internet J Allied Health Sci Pract. 2017;15(3):2.
5. Silberman NJ, Litwin B, Panzarella KJ, Fernandez-Fernandez A. High fidelity human simulation improves physical therapist student self-efficacy for acute care clinical practice. J Phys Ther Educ. 2016;30:14–24.
6. Kron FW, Fetters MD, Scerbo MW, White CB, Lypson ML, Padilla MA, et al. Using a computer simulation for teaching communication skills: a blinded multisite mixed methods randomized controlled trial. Patient Educ Couns. 2017;100:748–59.
7. Ignacio J, Dolmans D, Scherpbier A, Rethans JJ, Chan S, Liaw SY. Comparison of standardized patients with high-fidelity simulators for managing stress and improving performance in clinical deterioration: a mixed methods study. Nurse Educ Today. 2015;35:1161–8.
8. Tripathy S, Miller KH, Berkenbosch JW, McKinley TF, Boland KA, Brown SA, et al. When the mannequin dies, creation and exploration of a theoretical framework using a mixed methods approach. Simul Healthc. 2016;11(3):149–56.
9. Creswell JW. A concise introduction to mixed methods research. Thousand Oaks: Sage; 2015.
10. Creswell JW, Plano Clark VL. Designing and conducting mixed methods research. 3rd ed. Thousand Oaks: SAGE; 2018.
11. Guetterman TC. What distinguishes a novice from an expert mixed methods researcher? Qual Quant. 2017;51:377–98.
12. Leech NL, Onwuegbuzie AJ. A typology of mixed methods research designs. Qual Quant. 2009;43(2):265–75.
13. Fetters MD, Guetterman TC, Scerbo MW, Kron FW. A two-phase mixed methods project illustrating development of a virtual human intervention to teach advanced communication skills and a subsequent blinded mixed methods trial to test the intervention for effectiveness. Int J Mult Res Approaches. 2018;10(1);296–316.
14. Bazeley P. Integrating analyses in mixed methods research. London: SAGE; 2018.
15. Teddlie C, Tashakkori A. Major issues and controveries in the use of mixed methods in the social and behvioral sciences. In: Tashakkori A, Teddlie C, editors. Handbook of mixed methods in social & behavioral research. Thousand Oaks: SAGE; 2003. p. 3–50.
16. Fetters MD, Curry LA, Creswell JW. Achieving integration in mixed methods designs—principles and practices. Health Serv Res. 2013;48:2134–56.
17. Guetterman TC, Fetters MD, Creswell JW. Integrating quantitative and qualitative results in health science mixed methods research through joint displays. Ann Fam Med. 2015;13(6):554–61.
18. Fetters MD, Molina-Azorin JF. The journal of mixed methods research starts a new decade: the mixed methods research integration trilogy and its dimensions. J Mixed Methods Res. 2017;11(3):291–307.
19. Stufflebeam D. Evaluation models. N Dir Eval. 2001;2001(89):7–98.
20. Shadish WR, Cook TD, Campbell DT. Experimental and quasi-experimental designs for generalized causal inference. Boston: Houghton Mifflin; 2002.
21. Waznonis AR. Simulation debriefing practices in traditional baccalaureate nursing programs: national survey results. Clin Simul Nurs. 2015;11:110–9.
22. Stake RE. Qualitative research: studying how things work. New York: Guilford Press; 2010.

Making Use of Diverse Data Sources in Healthcare Simulation Research

32

Jill S. Sanko and Alexis Battista

Overview

This chapter provides readers with an introduction to some of the diverse data sources available to simulation researchers. This chapter compliments previous chapters (e.g., Chap. 4, *Starting your Research Project*) by drawing on research examples to highlight data generated from healthcare simulation-based encounters that may help the reader determine which data forms will best answer their research questions. In keeping with the goals of this text and the diversity of simulation in healthcare, the chapter also draws on examples from a variety of health professions domains, including nursing, medicine, and allied health, while highlighting a variety of simulation modalities, such as live simulations and augmented reality. The chapter ends with advice and suggestions as well as pointers to other complementary chapters in this book.

Practice Points

- Simulation-based learning contexts and activities provide researchers with diverse data forms, including participant performance or self-reported outcomes, faculty and simulated persons (includes standardized patients, confederates; see definition in the Appendix A at the end of the chapter) perspectives, biologic data and visual forms of data, such as video, to name a few.
- Keep an open mind and carefully consider what data will best support your research purpose and research questions.
- Reach out for help and guidance when working with data forms you are less familiar with. This may include adding study team members who specialize in specific research methods or statistical analyses approaches.
- Be strategic: starting with study design, consider what types of technical support may you need to support your research project. This may include the need for support from your information technology staff, institutional review board or, in some cases, advice from your legal department.

The opinions and assertions expressed herein are those of the author(s) and do not necessarily reflect the official policy or position of the Uniformed Services University or the Department of Defense.

J. S. Sanko (✉)
School of Nursing and Health Studies, University of Miami, Coral Gables, FL, USA
e-mail: j.sanko@miami.edu

A. Battista
Graduate Programs in Health Professions Education, The Henry M. Jackson Foundation for the Advancement of Military Medicine, Bethesda, MD, USA
e-mail: alexis.battista.ctr@usuhs.edu

Introduction

A widely read 2004 article, *The Future Vision of Simulation in Healthcare* by David Gaba, highlighted eleven diverse dimensions on which simulation may be applied and categorized [1]. Gaba aptly noted that using simulations to improve healthcare practice is a complex effort, stating,

> A fundamental part of the vision for the future is that clinical personnel, teams, and systems should undergo continual systematic training, rehearsal, performance assessment, and refinement in their practice. This vision is inspired in part by the systems in place in various high reliability organizations, particularly commercial aviation, but it is not slavishly copied from their experiences. Needless to say, using simulation as part of the process of revolutionizing healthcare is more complex than merely attempting to stick simulation training on top of the current system.

The first Research Consensus Summit of the Society for Simulation in Healthcare (SSH) was held in January 2011, with the goal of providing guidance for future research efforts in healthcare simulation [2]. In keeping with Gaba's

D. Nestel et al. (eds.), *Healthcare Simulation Research*, https://doi.org/10.1007/978-3-030-26837-4_32

notions of complexity, the summit also highlighted the diverse applications of simulation in healthcare, including the use of simulation to support systems analysis, team training and the vital role of feedback and reflection.

Both Gaba and findings from the 2011 research summit highlight the notion that simulation can support a variety of goals within healthcare, ranging from individual level education and training to analyzing systems (this idea highlights the need for diverse research efforts to grow). They also acknowledge that addressing the complex needs of healthcare requires thoughtful consideration rather than simple mimicry of what has been done before or what has been done in other fields.

In keeping with these perspectives, this chapter highlights several examples of how researchers have sought to thoughtfully study healthcare simulation. We emphasize some of the diverse data types that researchers use with the goal of helping the reader consider how various data types may be used to support research endeavors. This chapter also complements previous chapters (e.g., Chap. 4, *Starting your Research Project*) by drawing on research examples to highlight the use of data generated from healthcare simulation-based encounters that may help the reader determine which data forms might best answer their posed research questions. We also address some of the practical issues and challenges of working with complex, and potentially uncommon, data sources and offer suggestions and advice about how to overcome issues should they arise.

Taking Advantage of the Rich Data Forms in Simulations

One of the unique aspects of research with and about simulation is the multitude of data types that may be generated from a single simulated encounter. Studies discussed in this chapter, demonstrate the use of learner-generated self-reported data (e.g., reflections, surveys, questionnaires, psychometric measures, and self-assessments), biological or physiological data, visual data (e.g., video, audio, simulated electronic charting records) and simulator-generated data (e.g., simulator activity logs and simulation timelines). These data sources can be helpful when exploring the impact of a simulation-based encounter on an individual's or team's performance when seeking to understand how individuals or teams interact within a healthcare system or with a new medical device or when seeking to evaluate a simulation program or curriculum. Table 32.1 presents a summary of some of the common data types generated and utilized by simulated researchers while highlighting some of their advantages, disadvantages and common analyses approaches.

Table 32.1 Summary of diverse data forms available for simulation-based research

Data type	Advantages	Disadvantages/Limitations	Common analyses approaches
Self-report measures	Can be easy to collect and replicable across studies or study locations Data collection can be embedded within the simulation session (e.g., included as a part of the pre- or post-simulation phase) Can include qualitative and quantitative data forms	Identifying or creating *high quality* questionnaires or surveys can be time consuming Subject to various forms of bias (e.g., acquiescence) Subject to demonstration of reliability and validity and should have psychometric analysis completed prior to use	Varied forms of quantitative analyses (e.g., descriptive, regression) Open or categorical qualitative coding
Simulated person – based data	Can be easy to collect if simulated persons are already part of the simulation Collection can be done with limited interruptions or impact to the simulation encounter Data collection can be embedded within the simulation session (e.g., included as a part of the pre- or post-simulation phase)	Requires trained observers May require time to train the individuals collecting the data (e.g. simulated participants) Like self-reported data may be subject to forms of bias Real-time observations may be at risk for missing data Difficulties related to asking a human to do two things at once (e.g., play role, make observations and record/collect data)	Varied forms of quantitative analyses (e.g., descriptive, [a]regression) Open or categorical qualitative coding

Table 32.1 (continued)

Data type	Advantages	Disadvantages/Limitations	Common analyses approaches
Biologic and physiologic data	Gives an individualized view of the impact of the simulation on aspects of learner biology such as stress, etc Depending on the method used (e.g., salivary cortisol, heart rate, blood pressure), biologic and physiologic data can help researchers detect real-time fluctuations of biologic and physiological processes, such as stress	May be perceived as invasive by study participants which could impact willingness to participate May increase the time commitments of study participants which may also impact willingness to participate Can be expensive depending on method used Requires specialized equipment (e.g., access to a laboratory, holter monitors) which requires specific upkeep, and maintenance to ensure accuracy of data collected Requires the inclusion of study team members with the expertise and knowledge to interpret the findings	Varied forms of quantitative analyses (e.g., descriptive, [a]regression)
Digital video and audio	Readily available because video recording is commonly done as a part of the day-to-day operations of simulation-based learning Can be viewed or analyzed multiple times from multiple sites and by multiple people	Video files can be large and require advanced database planning, ample storage space, thoughtful security and consent practices May require team-members with expertise in analysis or data management Can be difficult to anonymize, and therefore poses additional challenges when seeking institutional review board approval May require additional software to facilitate data collection (e.g. software designed for collecting observational data) Data collection equipment can fail or researchers may forget to press record, thus the use of more than one camera or audio recorder is needed	[a]Content analysis, [a]Semiotics, [a]Discourse analysis
Systems based interface data	Regularly collected or available as a part of Simulation-based education because many simulators and medical devices have this capability built-in Can be analyzed multiple times from multiple sites and by multiple people May help improve the accuracy of measuring the length of time for a procedure or skill or for when a designated activity is performed	Files may be large and require ample database space and planning Requires researchers to ensure that saving the data files is completed and stored appropriately May require team-members with expertise in analysis or data management	[a]Content analysis, [a]Semiotics, [a]Human-computer interaction analysis
Simulator- and sensor generated data	Easily collected as many simulators have this capability built-in Can allow for the automated observation or collection of data without having to have a human present. This can be especially helpful when desiring to collect data during unusual time-periods Can easily collect large amounts of nuanced data	If the simulator does not have the innate capability, it requires specialized equipment that may require programing by individuals with knowledge in computer programing May require designing specific devices to collect data depending on your research purpose or questions Equipment can be expensive; however, advances in technology continue to drive the costs of these technologies down, making it less costly Equipment may require specific upkeep, such as calibration to ensure accurate data collection Usually requires team-members with expertise in analysis or data management	Varied forms of quantitative analyses (e.g., descriptive, [a]regression)

Note: Simulated person refers to a person who portrays a patient (simulated patient), family member or healthcare provider in order to meet the objectives of the simulation [3]

[a]See definitions provided in the Appendix A at the end of the chapter

Data Types, Their Uses and Examples

Self-Report Measures

Survey, questionnaire, or testing data are commonly used to assess changes in knowledge or attitudes or participants perspectives and experiences following a simulation encounter. Simulation participant perspectives or knowledge may be gathered as a sole measure or can be combined with other survey and testing data. For examples of high quality studies that employ some of these measures, (e.g., knowledge, procedural skills, self-confidence, self-efficacy), we refer you to several meta analyses [4, 5].

For simulation-based research, surveys, questionnaires, tests, or assessments are frequently used within a pre/post research design. This means that the participant encounters the questions both before an intervention (simulation) and then again following the intervention (simulation). While the use of self-report measures within a pre – post intervention research design is common, it also has some limitations (See Table 32.1), including the lack of subject randomization, inability to control for potentially important confounding variables, and presence of only a single statistical approach to demonstrate causality. Limitations noted, the pre – post intervention research design was a common research approach during the early days of simulation-based education when there was a need to establish simulation-based learning (SBL) as a valid and effective teaching modality. Today, this approach is still viable for new or novel uses of simulation or as part of a set of data of various forms being collected together, where correlation or triangulation will be used to assess the impacts of the intervention.

When considering this design it is important to keep in mind that, as the field of healthcare simulation matures, simulation-based interventions must be novel enough for a self-report pre- post design to be considered for publication. If you are considering this approach, we recommend pairing the self-report data with other types of data, data analysis techniques, or perspectives that enhance your ability to draw meaningful conclusions. For example, a researcher might be interested in looking at changes in attitudes following participating in a simulation encounter. In this case of pre and post design, the researcher may consider using a tool that measures attitudes changes, coupled with opened-ended reflection questions focused on perceptions of the experience and the perceived value of learning opportunity.

Simulated Person-Based Data

This category includes data collected by individuals (sometimes called confederates or standardized patients) that portray/simulate patients, family members or healthcare providers within the simulated context [3]. These individuals are often trained to assess learner performance, support participant learning during practice and provide feedback. Although the main job of these individuals is to play a human role in a simulation, they may also be tasked with making observations or assessments for the purposes of research [6, 7].

This approach works especially well because simulated persons are often trained to make assessment-related decisions and provide feedback. Additionally, their presence within the simulated encounter allows them to be able to collect the data unobtrusively (a hidden in plain sight approach). An example of this tactic is to have the simulated person record how many learners wash their hands prior to conducting an examination and before leaving the room, and this person was actually employed by one of the authors during simulation encounters enbedded in a patient safety course. Another example is the 2014 study by Falcone and colleagues in which simulated persons assessed student ability to communicate or perform certain aspects of an assessment while taking part in objective structured clinical examinations [8].

Biologic and Physiologic Data

The use of biologic and physiologic data includes data from biological sources (e.g., saliva, blood, and stress hormones cortisol or dehydroepiandrosterone [DHEA]) and physiological data (e.g., heart rate, blood pressure, respiratory rate). Among studies in healthcare simulation, the use of biologic markers may also be used along with physiological markers, self-report measures or on their own.

Studies involving the use of physiological and biological data in healthcare simulation research are not yet widespread. They are becoming more common, however, in protocols that examine the impacts of simulation on individuals while they are performing a procedural skill or engaging in a scenario. Measuring biologic markers, like cortisol levels, can help researchers better understand the impacts of highly realistic settings and encounters on learners biologic stress response.

An example of collecting biological and physiological data is the study by DeMaria *et al.* They studied the impact of simulated patient death on medical students' stress response and learning during an American Heart Association (AHA), Advanced Cardiac Life Support (ACLS) course. Heart rate, salivary cortisol, and DHEA levels were used to examine the stress responses of medical students encountering a patient who either died or survived a simulated resuscitation [9]. While the researchers did not note differences in medical students' long-term knowledge or skills, students encountering the death of the simulated patient had higher heart rates compared to baseline than those participants where the simulated patient survived [9].

In another study, Bong. Lightdale, Fredette and Weinstock examined differences in physiological stress among physicians participating in simulation-based training (SBT) compared to those training with traditional tutorial-based interactive education (IET) training [10]. Not surprisingly, those in the SBT condition demonstrated increases in heart rate while those the IET group had a decrease in heart rate from baseline [10]. The physicians in the SBT group also showed increases in serum cortisol levels, whereas the physicians in the IET group had decreases in serum cortisol [10].

Additionally, the approach of using physiological data coupled with the use of other data types (observation or self-appraisals for example) may help researchers better understand the impacts of stress on human behaviors and decision making in a given scenario. Harvey, Nathens, Bandiera, and LeBlanc's 2010 study of the impact of stress on cognitive appraisal during simulated trauma resuscitations is a good example of this dual data approach [11]. The findings from this study demonstrated that stress levels (assessed using both cortisol levels and self-reports of situational demands) were higher for those individuals interacting in high stress trauma resuscitations compared to those individuals participating in low stress trauma resuscitations [11]. Further, the findings of the Harvey and colleagues' study suggested that high-acuity events should include interventions targeting stress management skills [11].

Findings from studies using biologic and physiologic data can promote the development of strategies designed to keep learners safe during simulated learning encounters. They may also help researchers to better understand stress responses during actual clinical crisis responses, for example, when simulation is used as a surrogate for actual encounters. Additionally, when coupled with the use of one of the other data types (e.g., observation, self-report measures), this approach can help simulation stakeholders better understand the impact of stress on human behaviors and decision making in a given scenario or situation.

Video and Audio Data

An alternative to live observational analysis of data is the use of video or audio recordings. Both data forms are often widely available because both are frequently collected as a part of normal everyday operations in simulations labs and centers. Video and audio data in healthcare simulation can include video recordings captured during the simulated session, audio recordings captured during post-simulation debriefing or reflection, audio analysis of video recordings, and recordings made by various simulators or medical devices such as defibrillators or ultrasound devices.

Video and audio analysis supports researchers' efforts to conduct in-depth analysis of social contexts and can help them examine how individuals interact in a designated setting [12]. Video and audio data are well suited to help researchers examine process oriented research questions such as examining how individuals communicate or how individuals think about specific task while performing it.

Digital Video

Although video is often used to support the reflection phase of simulations, it can also be used in healthcare simulation research. These uses include in-depth analyses of simulation participants' behaviors [13–15] and video-assisted assessment [16]. For example, Battista categorized and quantified common learning behaviors using recorded videos of nursing students participation in four scenario-based simulations [13]. In another example, Sadideen and colleagues conducted a comprehensive video analysis of participant performance in interprofessional simulations to examine and categorized which leadership behaviors were more successful than others [14]. Furthermore, Sanko and Mckay used video analysis to compare behaviors associated with medication administration between groups who received a prior pharmacology enhanced curriculum and groups participating in a traditional lecture-based pharmacology course [15].

Video-assisted assessment can also be used to allow assessors or researchers to review performance at a time that is more convenient to them, to enable more complex evaluations of a participant's performance and to allow for blinded assessment. For example, Kowalewski and colleagues video-recorded student participants pre and post-test performance of laparoscopic skills performance and shared that video with designated independent raters to support blinded assessment of participant performance [16].

Audio Data

These data can be used to study a participant experience or perspective related to participation in simulation, a use similar to other research domains (e.g., health professions education, human factors engineering). In these situations, audio data are collected during interviews with individuals or focus groups. We refer you to Chaps. 12, 13 and 17 of this text for more information about these kinds of data collection procedures.

Researchers have also utilized audio data to conduct in-depth analysis of participant's reflections or experiences [17] or to study clinical or diagnostic reasoning processes [18, 19]. For example, Partin and colleagues asked nursing students to audio-record their reflections following a simulated encounter to help researchers understand how student's experienced learning in simulation [17]. Forsberg and colleagues asked experienced nurses to *think aloud* while partaking in a virtual patient encounter to investigate how they made clinical decisions during complex virtual patient scenarios [18]. Thinking aloud while performing a task or viewing a video can provide rich information about how an individual thinks while problem solving [20]. Finally, Tschan and colleagues asked teams of two to three physicians to engage in a simulated encounter where the diagnosis of the patient was intentionally complex and ambiguous to explore how teams of physicians made diagnostic decisions [19]. To study these processes, the researchers coded the transcribed utterances of participants' simulated performance and conducted behavioral analysis of their activities and actions using video [19].

The use of video and/or audio in these examples highlight how these data forms enable researchers to engage with simulated activity beyond the scheduled simulation event while also supporting in-depth analysis of participant activity or thinking. Studies that employ video or audio recordings should be carefully planned before they start. The section below addresses some special considerations related to the use of video and audio data.

Special Considerations When Using Video and Audio-Based Data

When considering the use and collection of digital video and audio data, there are several operational and regulatory issues that must be kept in mind. For example, it is important to carefully consider the full life-cycle of video data. These steps can include: (1) the selection of video or audio equipment that will be used to collect your data, (2) the determination of the number of cameras needed and the angles or perspectives that will be most useful to your research, (3) the consent of participants (i.e. beyond simply seeking permission to video-record), (4) how participant privacy will be maintained and how video data will be de-identified (if possible), (5) data security, including how videos will be transported, stored and shared and (6) longer term considerations of how video data will be managed until the end of its useful life. For in-depth discussions on what constitutes quality in video research we refer you to *Guidelines for video Research in Education: Recommendations from an Expert Panel* [21]. For greater detail related to visual analysis in SBL and SBR, please also see Chap. 15 of this text.

Systems-Based Interface Data

Systems-based interface data includes data captured by the systems that learners interact with during the simulated encounter. These data forms may include simulated Electronic Health Records (EHRs), adverse event reporting systems, and participant notes and documentation. Other forms of systems-based interface data include data derived from medical equipment's stored history or data logs. Examples of such equipment include intravenous pumps, medical imaging devices, ventilators, medication dispensing stations, and defibrillators. These data types enable simulation researchers and evaluators to study how and when learners use such systems, and can assist researchers in answering a variety of questions about how humans interact with medical equipment. Clinical record keeping databases can also be used to provide more accurate details about when a participant engaged in a designated task such defibrillation or ventilator manipulation. Systems-based data is also particularly useful when researchers want to study how humans engage with and navigate clinical settings and utilize clinical equipment. Of growing interest within the simulation community, this approach often draws from human factors psychology and frequently uses human-computer interaction analysis to analyze the data. These approaches have the explicit goal of making systems safer.

The 2013 study by Al-Rasheed provides an excellent example of using simulator generated data in simulation research [22]. This study examined the impacts of real-time manikin compression feedback on chest compressions during cardiopulmonary resuscitation during a 6-min simulation encounter [22]. The real-time feedback improved chest compression performances [22].

Simulator- and Sensor-Generated Data

This category includes information captured by the internal sensors and/or software of the simulation equipment itself, and can be extended to include data captured through sensors worn by participants or placed on specified medical devices (e.g., blood bank door) located in the simulation environment. Similar to systems-based data, simulator-and sensor-based data help researchers examine participant, team or system processes, or support automated data collection during complex scenarios.

Simulator-Based Data

Not only have simulators (full bodied, task trainer, and computerized) become more realistic, but many contain robust tools for collecting and generating data as an information source. For example, the vital sign (e.g., blood pressure, heart and respiratory rates, oxygen saturations) changes used to provide participants with context clues can also be recorded and collected as data. These data can then be used to track measures such as how long it took for participants to recognize a change in the patient's condition, how long it took to complete a task or support analysis of the process or steps participants engaged in.

In addition to vital sign data, many simulators have features that allow for the recording of behavior during various procedures (e.g., compressions depth or rate during cardiopulmonary resuscitation (CPR) and ventilation patterns). Simulators with these features record the rate, depth, number, and duration of compressions or ventilations that can be used to look at a variety of issues. Ashton and colleagues and Bjorshol and colleagues both utilized a simulator with this capability to explore the effects of rescuer fatigue on performance of continuous external compressions [23, 24]. In addition to these metrics, some simulators have still more advanced features that allow for measuring tidal volumes, drug levels, and gas exchange/levels (CO_2 and O_2).

Sensor-Based Data

As technology has advanced, so has the ability to generate new and interesting data using wearable devices and other electronic sensing gadgets. Data captured through sensors are often attached to learners or to key places in the simulated or clinical environment (in the case of in situ simulations), allowing researchers to explore ways in which humans interact with patients, patient environments, clinical equipment, and with other members of the healthcare team. Included in this category are Radio Frequency Identification (RFID), proximity beacons/sensors, depth sensing cameras, Real-Time Location Systems (RTLS).

Some of the questions researchers have examined using these devices include exploration of the interactions between individuals and teams within a designated environment (e.g., clinical or simulated). For example, a team at the University of Miami used a depth sensing camera (similar to those cameras used by the X-Box, which allows players to interact with the game system kinetically) to create and test a new system that reminded healthcare providers to wash their hands [25]. In another example Barrat and colleagues used wearable proximity sensors to explore the contact patterns between individuals in order to better understand how communicable diseases are passed from person to person (see Fig. 32.1) [26].

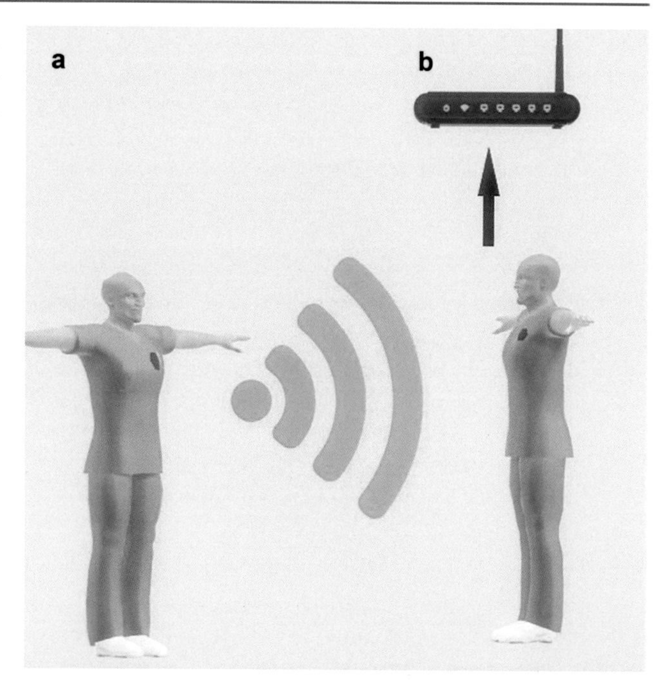

Fig. 32.1 Schematic of the sensing infrastructure of a Radio Frequency Identification Device (RFID). RFID badges are worn by individuals: (a) Illustration of RFID badge function. When two badge wearing individuals enter into the pre-defined proximity of each other a signal is sent to the RFID readers located in the environment. (b) Example of a RFID device [26]

Conclusions

In this chapter we have presented most common and unique forms of data frequently available for use by researchers in simulation, provided examples of how some researchers have utilized them and discussed some of the practical issues and challenges that individuals or research teams may encounter. As the field of healthcare simulation continues to grow, it is important for those involved in simulation-based education and simulation-based research to thoughtfully consider which data forms may best help answer their research questions. We encourage researchers to keep pondering new and creative ways to leverage the unique and interesting data available to advance the field. Box 32.1 provides some final practical practice points for individuals and research teams to consider.

Box 32.1: Practical Considerations for Researchers

- Be willing to think "Outside the Box" when planning your research study: Don't fall into an "autopilot" mode in which you default to using data that you are most familiar with or that is commonly presented in the current literature.
- Begin with the end in mind: Identify and consider which data types will help you best address your research project's purpose and research goals.
- Plan ahead: While designing the study, consider what types of research or technical support you may need if you decide to work with data forms or methods or analysis that you are less familiar with. You may need to recruit or collaborate with individuals who possess the expertise to help guide you.
- Develop a Data Management Plan: This plan should address all aspects of the data lifecycle, including considerations on how to collect, store access, analyze, share and protect your data. Work with your institution's information technology, legal, and research divisions to make sure the system is consistent with your local organizational policies or needs. When working with novel data such as digital video there may also be new or additional regulatory steps that need to be addressed (e.g., consent for all participants, participant privacy and confidentiality, data sharing and access).
- Look Beyond Simulation-based Learning Research: Consider examining studies beyond simulation and even health professions education to gain new ideas for study design and methodological approaches.

Appendix

Definitions of Terms Found in Chapter

Content analysis: is a widely used qualitative research technique to analyze data to uncover themes occurring in a set of data. It is not considered a single technique, but rather three distinct approaches that assist in interpretation of meaning from text data. [Hsiu-Fang, Shannon, 2005].

Discourse analysis: is analysis of language. It is used to study chunks of language as it flows together, this type of analysis allows for the interpretation language considering its context, and taking into consideration the environment, and activity.

Simulated person: any human who plays a live role in a simulation encounter; may include the role of a family member, healthcare provider (confederate, embedded simulated person, simulation actor), or patient (standardized patient [when standardized and provides feedback to learners] or simulated patient [when not standardized and/or does not provide feedback to learners]).

Regression: is a set of statistical methods used to estimate relationships between variables. Regression does not demonstrate causality, but rather allows one to see how closely related two or more variables are.

Semiotics: is the study of meaning-making from signs and symbols.

Human-computer interaction analysis: is a type of analysis aimed at understanding how humans interact with computers in terms of functionality, and usability.

References

1. Gaba DM. The future vision of simulation in health care. BMJ Qual Saf. 2004;13(suppl 1):i2–10.
2. Dieckmann P, Phero JC, Issenberg SB, Kardong-Edgren S, Østergaard D, Ringsted C. The first research consensus summit of the society for simulation in healthcare: conduction and a synthesis of the results. Simul Healthc. 2011;6(7):S1–9.
3. Lopreiato JO, Downing D, Gammon W, Lioce L, Sittner B, Slot V, Spain AE, The Terminology & Concepts Working Group. Healthcare Simulation Dictionary. 2016. Retrieved from http://www.ssih.org/dictionary.
4. Cook DA, Hatala R, Brydges R, Zendejas B, Szostek JH, Wang AT, Hamstra SJ. Technology-enhanced simulation for health professions education. JAMA. 2011;306(9):978–88.
5. McGaghie WC, Issenberg SB, Cohen MER, Barsuk JH, Wayne DB. Does simulation-based medical education with deliberate practice yield better results than traditional clinical education? A meta-analytic comparative review of the evidence. Acad Med J Assoc Am Med Coll. 2011;86(6):706.
6. Swanson DB, van der Vleuten CP. Assessment of clinical skills with standardized patients: state of the art revisited. Teach Learn Med. 2013;25(sup1):S17–25.
7. Petrusa ER. Status of standardized patient assessment: taking standardized patient-based examinations to the next level. Teach Learn Med. 2004;16(1):98–110.
8. Falcone JL, Claxton RN, Marshall GT. Communication skills training in surgical residency: a needs assessment and metacognition analysis of a difficult conversation objective structured clinical examination. J Surg Educ. 2014;71(3):309–15.
9. DeMaria S, Silverman ER, Lapidus KA, Williams CH, Spivack J, Levine A, Goldberg A. The impact of simulated patient death on medical students' stress response and learning of ACLS. Med Teach. 2016;38(7):730–7.
10. Bong CL, Lightdale JR, Fredette M, Weinstock P. Effects of simulation versus tutorial-based training on physiologic stress levels among clinicians: a pilot study. Simul Healthc. 2010;5(2):272–8.
11. Harvey A, Nathens A, Bandiera G, LeBlanc V. Threat and challenge: cognitive appraisal and stress responses in simulated trauma resuscitations. Med Educ. 2010;44(6):587–94.
12. Derry SJ, Pea RD, Barron B, Engle RA, Erickson F, Goldman R, Hall R, Koschmann T, Lemke JL, Sherin MG, Sherin BL. Conducting video research in the learning sciences: guidance on selection, analysis, technology, and ethics. J Learn Sci 2010;19(1):3–53. Derry SJ. Guidelines for video-research in education: recommendations from an expert panel. Retrieved 28 Dec 2017 from https://pdfs.semanticscholar.org/1fd7/7f96cd217b18d71105686d4997d46731f816.pdf.

13. Battista A. An activity theory perspective of how scenario-based simulations support learning: a descriptive analysis. Adv Simul. 2017;2(1):23.
14. Sadideen H, Weldon SM, Saadeddin M, Loon M, Kneebone R. A video analysis of intra-and interprofessional leadership behaviors within "The Burns Suite": identifying key leadership models. J Surg Educ. 2016;73(1):31–9.
15. Sanko JS, Mckay M. Impact of simulation-enhanced pharmacology education in prelicensure nursing education. Nurse Educ. 2017;42(5S Suppl 1):S32–7.
16. Kowalewski TM, White LW, Lendvay TS, Jiang IS, Sweet R, Wright A, Hannaford B, Sinanan MN. Beyond task time: automated measurement augments fundamentals of laparoscopic skills methodology. J Surg Res. 2014;192(2):329–38.
17. Partin JL, Payne TA, Slemmons MF. Students' perceptions of their learning experiences using high-fidelity simulation to teach concepts relative to obstetrics. Nurs Educ Perspect. 2011;32(3):186–8.
18. Forsberg E, Ziegert K, Hult H, Fors U. Clinical reasoning in nursing, a think-aloud study using virtual patients–a base for an innovative assessment. Nurse Educ Today. 2014;34(4):538–42.
19. Tschan F, Semmer NK, Gurtner A, Bizzari L, Spychiger M, Breuer M, Marsch SU. Explicit reasoning, confirmation bias, and illusory transactive memory: a simulation study of group medical decision making. Small Group Res. 2009;40(3):271–300.
20. Fonteyn ME, Kuipers B, Grobe SJ. A description of think aloud method and protocol analysis. Qual Health Res. 1993;3(4):430–41.
21. Derry SJ. Guidelines for video-research in education: recommendations from an expert panel. Retrieved 28 Dec 2017 from https://pdfs.semanticscholar.org/1fd7/7f96cd217b18d71105686d4997d46731f816.pdf.
22. Al-Rasheed RS, Devine J, Dunbar-Viveiros JA, Jones MS, Dannecker M, Machan JT, Jay GD, Kobayashi L. Simulation intervention with manikin-based objective metrics improves CPR instructor chest compression performance skills without improvement in chest compression assessment skills. Simul Healthc. 2013;8(4):242–52.
23. Ashton A, McCluskey CL, Gwinnutt AM, Keenan AM. Effect of rescuer fatigue on performance of continuous external chest compressions over 3 min. Resuscitation. 2002;55(2):151–5.
24. Bjorshol CA, Sunde K, Myklebust H, Assmus J, Soreide E. Decay in chest compression quality dues to fatigue is rate during prolonged advanced life support in a manikin model. Scand J Trauma Resusc Emerg Med. 2011;19(46):1–7.
25. Birnbach D, Fitzpatrick M, Thomas R, Ramirez J, Sanko J, Rosen L, Shekhter I. Testing a hand hygiene compliance monitoring system utilizing a depth-sensing camera in a simulated clinical environment. Presented at the 13th annual international meeting on simulation in healthcare. Jan 2013, Orlando.
26. Barrat A, Cattuto C, Tozzi AE, Vanhems P, Voirin N. Measuring contact patterns with wearable sensors: methods, data characteristics and applications to data-driven simulations of infectious diseases. Clin Microbiol Infect. 2014;20(1):10–6.

Part VI

Professional Practices in Healthcare Simulation Research

33 Writing a Research Proposal for Sponsorship or Funding

Debra Nestel, Kevin Kunkler, Joshua Hui, Aaron W. Calhoun, and Mark W. Scerbo

Overview

A main reason for writing a research proposal is to obtain funding or other non-monetary support. For this chapter, the purpose and predominant theme of writing a research proposal is for obtaining funds to support the research project. Many of our readers already know there is no one research proposal "formula" that can be applied to all funding applications. Many of the opportunities and application processes have similar steps and may request much of the same information, however one should not "cut and paste" a previous proposal into another organization's or agency's application process as there are significant differences. Proposals will need some degree of repackaging to properly align to what the organization or funding agency is seeking. This chapter will offer just a glimpse of some of those common steps, and address items that make healthcare simulation unique from a research application process. It will explore some common pitfalls that can result in a research proposal's rejection.

D. Nestel
Monash Institute for Health and Clinical Education, Monash University, Clayton, VIC, Australia

Austin Hospital, Department of Surgery, Melbourne Medical School, Faculty of Medicine, Dentistry & Health Sciences, University of Melbourne, Heidelberg, VIC, Australia

K. Kunkler (✉)
School of Medicine – Medical Education, Texas Christian University and University of North Texas Health Science Center, Fort Worth, TX, USA
e-mail: k.kunkler@tcu.edu

J. Hui
Emergency Medicine, Los Angeles Medical Center, Kaiser Permanente, Los Angeles, CA, USA

A. W. Calhoun
Department of Pediatrics, University of Louisville School of Medicine, Louisville, KY, USA

M. W. Scerbo
Department of Psychology, Old Dominion University, Norfolk, VA, USA

Practice Points

- A research proposal is a formal request for financial support or sponsorship.
- It is important to know your audience when structuring and applying research proposals.
- It is important to read the opportunity or application thoroughly, and from more than one perspective to appreciate what the organization or funding agency is seeking.
- Thoroughly complete the structured steps requested by the sponsor/funder, and design the proposal to complement these steps with an innovative and creative research plan.
- An understanding of relevant historical data and examples can assist you in detecting and addressing potential failure points in your proposal prior to submission.
- It is also important to have colleagues with different perspectives and backgrounds review your proposal prior to submission.

Introduction

A research proposal is a formal request for an individual, organization, or agency to consider sponsoring for funding or other non-monetary support. The research proposal is the tool and mechanism by which investigators convince the reviewers from the organization or funding agency that the proposal is worthy of funding and that it should be approved [1, 2]. It may also serve as a reference guide, particularly for project management during the actual research. Some proposals are limited in length and content, while others may require an extensive amount of information. This depends on the funding

D. Nestel et al. (eds.), *Healthcare Simulation Research*, https://doi.org/10.1007/978-3-030-26837-4_33

agency. There are, however, several common components that almost all research proposals will need to contain. These are:

1. The research question(s) (i.e., what do you want to research)
2. The outlined objective(s), goal(s), and/or aim(s)
3. The plan and/or methodology that will be used to pursue the research question(s)
4. The metrics/tools that will be used to evaluate the outcome of the research
5. The estimated cost [budget] of the project
6. The projected length and timeline of the project
7. The potential impact and anticipated results of the research once completed
8. Key features that differentiate this project from others that the funding agency may receive
9. The proposed deliverable(s)
10. The project's key personnel

Other components may need to be included depending on the type of research and the organization's or funding agency's instructions. Examples of the latter include human and animal subject protection, testing and evaluation conditions and protocols, policies and regulations [compliance], and standard operating procedures/processes. Furthermore, some agencies will have additional processes, such as the submission of pre-proposals or letters of intent, which must occur prior to submitting the full proposal. Always, always read the offering or request for proposal thoroughly, as failure to follow the directions of the agency or organization providing the support or funds can lead to summary rejection. It may also be helpful to correspond with the organization or funding agency identified point of contact to ask them direct questions about the opportunity or application process if allowed.

Some of the Variabilities Involved in Research Proposal

Research proposal writing is not unique to healthcare simulation and health professional education. In fact, nearly every market, industry, academic institution, and non-for-profit organization writes research proposals at some point in order to receive sponsorship and support. There are many publications that describe the steps necessary to write a research proposal [3–8].

When first considering the preparation of such a proposal, it is important to keep in mind just how varied the offerings may be. For example, some research proposals are for small amounts of funding given over a limited time while others are for enormous amounts that are disbursed over many years. The source of funding may also vary between international or federally funded grants, businesses or venture capitalist, and private foundations, just to name a few. This also applies to the type of research supported. Some organizations and funding agencies span the entire breadth of research, while most do not. Table 33.1 is just one resource that explains how one funding agency defines different stages of research. Research funding is often broken into very basic and theoretical research versus advanced and applied research. Some research dollars also focus on gathering the proverbial last piece of data that finalizes our understanding of a particular question (such as occurs in clinical trials). Some of the later stages of research, such as during the manufacturing stages, are typically not funded through international or federally funded sources; yet they remain a part of the overarching research process. While the array of potential funding opportunities is vast, the "catch", so-to-say, is that there are usually some form of rules, regulations, and expectations that are intertwined with receipt of funding or support. It is the agreement or contractual piece, often referred as "red-tape" and/or "bureaucracy," that may repel potential investigators from applying. These "contractual" obligations may range from submission of reports to issues concerning intellectual property, penalties, or even ownership of the outcomes and product. It is imperative that investigators read the entire opportunity and application process. This chapter will not venture into the rules, regulations, and expectations. It is critical, however, to know that they exist.

Table 33.1 Range of research funding per the U.S. Department of Defense [9]

Basic research (6.1)	Directed toward the greater knowledge or understanding of the fundamental aspects of phenomena and of observable facts
Applied research (6.2)	Study to understand the means to meet a recognized and specific need
Advanced technology development (6.3)	Development of subsystems and components and efforts to integrate subsystems and components into system prototypes
Demonstration and validation (6.4)	Evaluate integrated technologies, representative modes, or prototype systems in a high fidelity and realistic operating environment
Engineering & manufacturing development (6.5)	[Conduct] engineering and manufacturing development tasks aimed at meeting validated requirements prior to full-rate production
RDT&E management support (6.6)	Management support for research, development, test, and evaluation efforts and funds to sustain and/or modernize the installations or operations
Operational system development (6.7)	Development efforts to upgrade systems that have been fielded or have received approval

Research Proposal Steps

Investigators are motivated to create and write research proposals. This motivation will somewhat drive the direction, focus, objectives, goals, and anticipated outcomes of the research. For the purposes of this chapter it does not matter if these motivational factors are financial incentives, academic progression (e.g. tenure), academic prominence/prestige, job security, self-esteem (e.g. relevance), expressing one's creativity and ingenuity, or scientific prowess [10]. Regardless of the motivation, what remains are the rules, regulations, and expectation of the organization or funding agency. A significant amount of strategy is necessary to improve the chances of success; it is important to remember that success is never assured even when this advice is followed. The concept of research proposal strategy is discussed in another chapter, but can briefly be summarized as well as some of the research proposal steps:

- Follow the instructions and understand ALL of the requirements!
- Know who the audience is and what they want in return
- Outline the sections and deconstruct components to provide detailed information
- Create a clear title. Sometimes a clever title goes a long way!
- Emphasize the impact and significance of the research. Think about its impact at multiple levels such as the institution, the agency/organization offering the funds or support, the immediate community who will benefit, and the community at large
- Emphasize realism and feasibility while simultaneously addressing the distinguishing factors of the research concept and what your team offers
- Make sure the abstract is clear, impactful, and addresses the central question(s)
- Provide and support proposed methodologies and/or techniques
- Provide preliminary data/information through citation of articles, prior research performed, or other evidenced-based sources
- Provide and support proposed objectives, goals, metrics and measuring tools that will assist in determining "successes" versus "failure"
- Propose timelines and project management
- Propose a budget
- Identify key personnel and bibliographies/curricula vitae

From my perspective the creative 'why', the strategy, the outline, the methodologies, the internal and external metrics and measurement tools, the balance between pushing the boundary and realism, the impact, and the value proposition are some of the most important areas on which to concentrate. There is no question that hard work and extensive research are key when addressing these issues. Partnerships play a role, especially if there are specialized areas that the prime investigator or institution lack expertise or do not support very well. Others have their own thoughts regarding what constitutes successful research proposals [2, 11, 12], and so one should use multiple resources to formulate what will work best when submitting towards an opportunity or funding agency.

Always have colleagues review and edit the research proposal. By bringing multiple perspectives to bear such a review can result in significant improvements from the overall research concept, to project management, to the grammatical level, to budgetary, and to the impact and deliverables, just to name a few. Provide plenty of time, when appropriate, for the development of the research proposal and minimize the "all-nighter" approach when possible. There are, of course, occasional opportunities that one hears about late, and in these circumstances the investigator and team must weigh the pros and cons of application carefully before pursing.

In concert with the steps discussed above, it is important that the research proposal maintain an appropriate "feel" or "ethos". This can be difficult to maintain at times, especially since writing a research proposal is no easy task. Unfortunately and additionally, it is inevitable that something will not go according to the original plan during the writing of the research proposal. During the writing and preparation phase, do your best to maintain a positive and enthusiastic attitude among research team members. Careful attention to the following considerations can assist in this process:

- Be passionate about the proposed research
- Be as realistic as best as possible, even when striving for the stars
- Have an overarching strategy. Minimize components within the research project that have limited connectivity to the overarching strategy
- Be ethical, morally responsible, and truthful
- Understand what stage (or cycle) the various components of the research resides. The proposal may be superb in its content, methodology, etc.; but if the research is at an applied phase and the funding opportunity is for basic research, it most likely will not be accepted
- Follow the organization and funding agency's guidelines and application process
- Ask questions!

Uniqueness and Options

Healthcare simulation research has many potential roles [13] and a wide array of options for funding exist, including sources that are open to the ideas and focus on: (1) education and training; (2) processes and methodologies; (3) product development; and (4) clinical applications or clinical outcomes. For example, organizations such as the Food and Drug Administration [14] are increasing their level of interest in simulation and modeling. Recent developments such as the 510 K approval of the Microsoft Hololens™ [15, 16] for use in pre-operative surgical planning is an example of how simulation technology can be utilized for tomorrow's healthcare providers and patients.

Because of this range and diversity of healthcare simulation roles there is not a one-size fits all approach when it comes to writing research proposals. Each of the focus areas mentioned above have distinct differences in terms of the research project's delivery and outcomes so it is somewhat logical that the contents and structure of the relevant applications will vary not only between organizations and between agencies, but even within the same organization or agency. When compounded with the range of research types possible (from basic research through operationalization studies) and the multitudes of potential organizations and funding agencies, it should be no surprise that there is not a one-size fits all formula. As stated above, it is vital that the investigator and team thoroughly read the opportunity or application process so that all of the rules, requirements, and expectations from the organization or funding agency are clearly understood.

Not Accepted: Some Common Reasons

So why might a research proposal be rejected? Why do so many seem to "fail"? First, there may be ethical issues, such as plagiarism, within the proposal. These are addressed more thoroughly in Chap. 34. It is also important to remember that there is a limited pool of funds and so worthy projects (including those with very good to excellent scores) are simply and often not funded. Receiving a rejection when the reason is that the organization lacks funds does not alleviate the "sting" yet it is a reality within research and remains, unfortunately, a basic fact.

There is variation in the percentage of proposals that do not receive support between funding agencies. Not all organizations report and provide their acceptance or rejection rates. Several of the US federal funding agencies do document their rejection rates and some report an over 80% rejection rate [17]. Data produced by the National Institutes of Health reporting branch in 2017 revealed rejection rates ranging from 0–100% for specific funding opportunities, with an overall average rejection rate of nearly 78.8% [18]. In my past role on the funding agency side, it was not uncommon to see 33% to over 95% rejection rates depending on the specific program. Reasons as to why individual proposals were rejected or declined ranged from incomplete research proposals (e.g. sections were missing such as budgets), lack of details of how the research will be accomplished, outright contradiction on the proposed direction or inconsistent information, and proposals that had almost no alignment with the funding opportunity topic.

Given this, it is imperative that you do not allow simple non-compliance issues within the research proposal to be the reason why the proposal was not accepted [6]. If the organization or funding agency requests a table of contents, then create, format, and insert a table of contents into the proposal. If the funding agency requires an abstract, then create, properly format, and insert an abstract into the proposal. Addressing the basic steps of the opportunity or application process is essential when writing and submitting a research proposal. Attention to these details will not assure success, but they can assure that your proposal will be read, reviewed, and considered and not summarily dismissed.

Conclusion

In summary, follow the instructions that the organization or funding agency has published. Address the needs and compliance items of the organization or funding agency. Provide the data and information the organization or funding agency seeks in addition to the proposed content, objectives, methodologies, techniques, etc. that you want to stress in the research proposal. Provide yourself enough time to properly pursue the support or funding opportunity whenever possible. Perform due diligence in the outline and background research. Be passionate and minimize discouragement. Have others assist you with editing the proposal and try to recruit people with different talents in order to receive different perspectives during the editing process. Writing research proposals takes motivation, creativity, innovation, time, patience, perseverance, organization, hard work, research, and, most of all, a team. Best of luck!

References

1. Al-Riyami A. How to prepare a research proposal. Oman Med J. 2008;23(2):66–9.
2. Bradshaw CJA. Twenty tips for writing a research proposal. https://conservationbytes.com/2015/05/04/twenty-tips-for-writing-a-research-proposal/.
3. How to write a research proposal. https://www.wikihow.com/Write-a-Research-Proposal.
4. How to write your research proposal. https://www.westminster.ac.uk/study/postgraduate/research-degrees/how-to-apply/entry-requirements/how-to-write-your-research-proposal.
5. Patel A. Research in simulation: research and grant writing 101. https://www.laerdal.com/usa/sun/presentations/phoenix/day1/8-patel.pdf.
6. Study Guide and Strategies. How to write a research proposal. http://www.studygs.net/proposal.htm.
7. Sudheesh K, et al. How to write a research proposal? Indian J Anaesth. 2016;60(9):631–4.
8. Center for innovation in research and teaching. Developing a research proposal. https://cirt.gcu.edu/research/developmentresources/tutorials/researchproposal.
9. Defense Acquisition University. Research, development, test, and evaluation (RDT&E) funds. Created 19-Apr-2015; last reviewed 20-June-2018. https://www.dau.mil/acquipedia/Pages/ArticleDetails.aspx?aid=e933639e-b773-4039-9a17-2eb20f44cf79.

10. von Hippel T, von Hippel C. To apply or not to apply: a survey analysis of grant writing costs and benefits. PLoS One. 2015;10(3): e0118494. https://doi.org/10.1371/journal.pone.0118494.
11. Bourne PE, Chalupa LM. Ten simple rules for getting grants. PLoS Comput Biol. 2006;2(2):e12.
12. Proctor EK, et al. Writing implementation research grant proposals: ten key ingredients. Implement Sci IS. 7 96. 12 Oct 2012 doi:https://doi.org/10.1186/1748-5908-7-96.
13. Kunkler K. The role of medical simulation: an overview. Int J Med Robot. 2006;2(3):203–10.
14. Morrison T. How simulation can transform regulatory pathways. 9-Aug-2018. https://www.fda.gov/ScienceResearch/AboutScienceResearchatFDA/ucm616822.htm.
15. Microsoft. Surgeons use Microsoft HoloLens to 'see inside' patients before they operate on them, 8-Feb-2018. https://news.microsoft.com/en-gb/2018/02/08/surgeons-use-microsoft-hololens-to-see-inside-patients-before-they-operate-on-them/.
16. Novarad. OpenSight augmented reality system is the first solution for Microsoft HoloLens 510(k) cleared by the FDA for medical use, 24-Oct-2018. https://www.businesswire.com/news/home/20181024005714/en/Novarad%E2%80%99s-OpenSight-Augmented-Reality-System-Solution-Microsoft.
17. Harris M. Why did my research proposal fail. https://granttraining-center.com/blog/research-proposal-fail/.
18. National Institutes of health, success rates. https://report.nih.gov/success_rates/.

Writing an Ethics Application

34

Gabriel B. Reedy and Jill S. Sanko

Overview

Ethical research practice should be the aim of all simulation researchers. However, this intention does not always translate into the successful completion of an ethics application. In this chapter, we discuss considerations for conducting ethical research in simulation, and the relationship between these principles and the requirements for ethical approval. We will also give practical advice about the common requirements of ethics panels and institutional review boards (IRBs), to enable you to proceed as smoothly as possible through the process of gaining ethical approval for your simulation study.

Practice Points

- Ethical research practice requires you to demonstrate clear plans for your research activity, to show you are a careful custodian of research data, and to consider and justify your research decisions.
- An ethics application requires detailed information on all elements of the research process, from the specifics of how data will be stored, to CVs from the investigators.
- Often, ethics applications can be supported by using pre-existing templates for documentation and material you have written about your research project for other purposes.
- Always involve either colleagues with experience of ethics applications or establish who can provide advice from your research ethics office. Some departments employ people to work with investigators on prospective projects.
- Expect an ethical review process to take time, to involve potential changes to your research plan, and to require you to make careful choices about how you conduct your research.

Introduction

For the simulation researcher, the process of an ethical review of your proposed research might seem daunting and confusing. This is especially true because often those of us working in simulation-based education have training in clinical or technical fields, and may not have experience in conducting scholarly research. Ethical research practice is also of particular importance in simulation because of the potential for learners to feel as if aspects of the simulation can lead them to perform poorly or make choices different to those they would make in clinical practice. The perceived hurdle of gaining ethical approval for research can dissuade colleagues from conducting research in the field. However, the potential benefits of simulation research are great, and not just for the field, but also for learners, patients, and society at large. In this chapter, we explore the principles behind ethical simulation research, applying for and obtaining an ethics board approval, and some considerations you will want to have in mind when you design and plan your research. We discuss some of the common aspects of the ethics board approval processes and how you can avoid common issues that may hamper getting the ethics board approval needed to move forward.

G. B. Reedy (✉)
Faculty of Life Sciences and Medicine, King's College London, London, UK
e-mail: Gabriel.Reedy@kcl.ac.uk

J. S. Sanko
School of Nursing and Health Studies, University of Miami, Coral Gables, FL, USA
e-mail: j.sanko@miami.edu

D. Nestel et al. (eds.), *Healthcare Simulation Research*, https://doi.org/10.1007/978-3-030-26837-4_34

Ethics in Practice: What is an Ethical Researcher?

Ethical research is an important element in conducting quality research. Historically, in the absence of rules and regulations for the ethical conduct of research, research participants (or *subjects*, as they were often referred to) suffered grave atrocities in the name of science (see Box 34.1 for some noteworthy examples). These mistakes, while tragic, helped to bring to light cases and instances that should never occur again, and also paved the way for the development of now widely practiced processes, policies, laws, and codes of research conduct. Together, these serve to minimize future harm to participants while maximizing the potential for research benefit.

Ethics in research is not only important at the individual level, but also for society and the larger body of individuals who may potentially be impacted by research and research outcomes. The maintenance of the established norms for conducting research assist in upholding ethical standards, the promotion of accountability, bolstering public support, and aligning with commonly held moral and social values [1].

Norms for conduct promote the aims of research, including knowledge, truth, and error mitigation [2]. Cooperation and coordination of a diverse group of researchers is a fundamental element when carrying out even simple research endeavors. Garnering the collective cooperation and coordination of all those individuals involved on a research team using ethical standards as foundational principles helps to promote the values that are essential to collaborative work in research [3].

Accountability is another primary aspect and vital part of ethical research. Researchers must operate under the basic assumption that they are accountable to their participants, their research colleagues, their institutions, themselves, and the public at large. Various mechanisms exist to structure this accountability. These include organizational ethics boards, as well as larger national and international regulatory agencies.

Research findings have intended, and sometimes unintended, consequences and impacts for a discipline and also for society. Therefore, public trust that research findings are accurate and sound are important for support of research. Gaining trust means people are more likely to participate in, and pay attention to, research, if they have faith in the process and its outcomes. Finally, research norms help to uphold basic moral and social values such as social responsibility, human rights, and health and safety [1]. Ethical and accountable research ensures that values and morals are upheld, even when it means that holding to this high moral standard may be to the detriment of the research itself.

Prior to engaging in research, it is important to seek out a colleague with research experience and ask for advice. Become familiar with general ethical research principles and the process of conducting research at your institution and across the broader academic, social, and cultural contexts in which you work. These efforts will help to prepare you to practice in an ethical manner. Ethical researchers prepare themselves with the information needed to ensure that they are conducting ethical research—minimizing harm and risk to participants while generating necessary knowledge—and are upholding the most basic rule of healthcare to "do no harm" [4] . It is also helpful to remember that most simulation research is not seamlessly analogous to interventional medical research, and therefore some very different issues can be at play [5, 6].

Why You Need an Ethics Application

The ethics application process is the oft-dreaded, first step for beginning research. The ethics application can be a point of frustration for many new researchers. Producing the written application materials may also seem like an impossible hurdle, however, we argue that these steps are not, and should not, become a restrictive barrier for new researchers. It is unreasonable to expect the process to be easy or seamless. With some guidance and understanding, the process can become easier. Hopefully, this chapter will assist by sharing some of the insights that we have learned from successful ethical approvals for simulation research.

The first step in becoming proficient at ethics applications is developing a basic understanding of why ethical approval for your research is almost always required. At the most fundamental level, obtaining ethical approval says to the public that a study has undergone review by an unbiased source that has deemed the intent, procedures, and methods to be appropriate, sound, and developed in a way that mitigates potential and undue risk to research participants. Additionally, the process of submitting an ethics application allows research involving humans (human subjects research) to be continuously monitored by an impartial body, responsible for keeping research participants safe and overseeing ethical research practice. Approval also assists in providing a direct path for accountability.

The contemporary process of ethical review and oversight grew out of the regulations that were put into place to keep research participants safe after a few noteworthy cases where research resulted in harm to research subjects. The Belmont Report summarizes the ethical principles and guidelines for research involving human subjects [7].

Box 34.1: Noteworthy examples of research that helped to identify a need for stricter regulations and oversight

- Tuskegee Syphilis Experiment
- Nazi Concentration Camp Medical Research, including:
  - Joseph Mengele's Twin Experiments
  - Nerve, Muscle, and Bone Transplantation Experiments
  - Hypothermia Experiments
  - Fertility and Sterilization Experiments
- Chester Southam's Live Cancer Studies
- Henrietta Lacks and the HeLa Cell Line
- Holmesburg Prison Experiments
- Stateville Penitentiary Malaria Study
- Willowbrook State School Hepatitis Experiments

Is an Ethics Application Required?

For most simulation researchers, the first consideration is whether or not ethical approval is even required. Some institutional or legal contexts classify some types of work to be research (and thus subject to ethical review), while similar projects might be classed as evaluation or quality improvement and not subject to ethical review. Even if a project is considered to be research, if the work does not involve close interaction with research participants, it may not be subject to ethical review or approval. However, most simulation research involves interaction with research participants, and therefore does require the approval of an institutional ethics board. Indeed, much simulation research is qualitative in nature, and thus focuses on human beings and their interactions and behavior in simulated environments. As such, this work is necessarily subject to ethical review.

Even if your proposed study is not human subjects research, you may still need institutional approval. The process of obtaining approval for research not involving human subjects or participants is usually a comparatively simple and expedited one. Institutional guidelines, checklists, and decision trees will help you make the determination of what is required for your project. Pugsley and Dornan provide a helpful list of questions to ask when planning medical education research; these can also be useful when exploring and conceptualizing ethical simulation research [8].

Consent and Consenting

Many research studies require that participants provide informed consent. That is, the process whereby research participants are fully informed about the research aims, procedures, and processes, including the nature of the data that is being collected and how it will be used, and in light of that information, provide their consent to participate in the research.

Informed consent is most often conducted formally and recorded with a signed document. Interventional studies are usually required to gather formal signed consent, but in some cases (such as evaluation of simulation-based aspects of university courses) it may not be necessary.

Other types of consent include implied consent and verbal consent. Implied consent is often associated with studies where surveys are used to collect data. A prospective participant is given information about the study and what is being asked of them. Their subsequent completion of the survey implies that they consent to participating. Verbal consent is another mechanism that may be used for obtaining consent from a research participant. This is sometimes used when having a participant sign a consent puts them at risk for discovery or harm (e.g., if a researcher were studying women who work as sex workers and signing consent may make them vulnerable to abuse, or legal action). Ethics boards usually offer templates for informed consent. Care must be taken to adapt the form to your project.

Special Considerations for Studies Using Simulation or About Simulation

Vulnerable Populations and Relative Power

Vulnerable populations in research are those who may be at heightened levels of potential risk associated with their participation in research. Some examples are patients, children, prisoners, pregnant women, those with cognitive impairments, or members of socially or culturally disenfranchised groups. Additionally, and most commonly in simulation research, students, trainees, and employees may also be considered vulnerable in certain circumstances.

This uneven power gradient (especially if you are an academic at the institution) may make students may feel undue pressure to participate if they believe participation or non-participation could impact their grades, progression, or ability to graduate [9]. Therefore their freedom and capacity to protect themselves from risk may not be the same as someone who is not a student. This does not mean you cannot use students as potential research subjects, but rather that extra care and precaution in 'distancing' yourself as an investigator (where possible) should occur.

Such precautions may include having someone who is not directly involved in teaching or administrating (i.e. an administrator) sending out communications, and or consenting potential participants. Precautions will absolutely involve having language indicating to participants that their decision to participate in the study will not impact their grades, ability to graduate, future employment, or future admission to the school included in the consent language. The obligatory statement about being able to withdraw from the study at any time without penalty should also be included. It is important to note that sometimes a participant can withdraw from a project but their data may not be able to be withdrawn, since it has been identified and pooled.

There are also other ways for students to participate in simulation research that may not require the need for consent. Examples include using data collected as a routine part of a course, such as test scores, post-course evaluations, or video footage (if this is routine at your simulation center). This type of data can potentially be used both prospectively or retrospectively, following ethics review board approval. When exploring using these types of data sources, you should determine whether or not the data is protected. In the United States, student data may be protected under the Family Educational Rights and Privacy Act (FERPA). In the European Union, the General Data Privacy Regulation (GDPR) applies. In other countries, specific laws will govern the use of student information.

In addition to dealing with students as potential 'vulnerable' populations of interest, you may also be conducting simulation research in a clinical setting where patients and employees may be involved. Patients may not automatically be listed as a vulnerable population, but in some circumstances, they could be considered vulnerable (e.g., children, pregnant women, and prisoners). However, like students, their position in relation to you as an investigator may also make them feel undue pressures to participate in research; this is especially true if they feel that it could impact the care that they receive. Therefore, again, care and thoughtful consideration for mitigating this potential influence needs to be given when conducting research that may utilize patients or patients' data. Again, language used as part of the consenting process will need to be cautiously phrased. Additionally, you will need to ensure that the information is not protected under federal laws. In the United States, the Health Insurance Portability and Accountability Act (HIPPA) protects much of the patients' data. In European Union countries, the General Data Protection Requirement (GDPR) applies not just to patients but to any data generated or collected which is personally identifiable. Prior to generating or collecting data you will need to determine the need for written permissions, what is and is not protected, and what rules and regulations your institution has regarding use of patients' data. You can find out more information below about these in the resources section.

Finally, if you are planning on asking employees to participate, or using employees' data in your simulation research, the same careful considerations mentioned above are needed. Distancing yourself as a researcher should be done when possible, and proper language will need to be included in the consent process making it known that participation is voluntary and will not impact pay, pay raises, promotions, lay-offs, firing, etc. As always, please check with your organization and refer to organizational and governmental policies regarding specific rules and regulations in place.

The Process of Submitting an Ethics Board Approval Application

Because of the potential for harm to research participants, there is by extension legal and reputational risk to the institutions where research is conducted. There are therefore often a number of hurdles that simulation researchers must overcome when seeking approval for their work. For most simulation researchers, the first step for ethics review will be a departmental level contact or committee, which may have either a formal or informal gatekeeping function. If the gatekeeping function is a formal one, then a complete application using the appropriate local format will likely be required; this will typically be reviewed and feedback provided before the application is sent on for a higher level review. If the gatekeeping function is formative or formal signoff is not required, you may only need to have a preliminary plan for your work that can form the basis of discussion or approval locally. In some cases, this local process forms the basis for a required peer review of your study before it can go forward to an ethics review panel.

Ethical review panels at universities are frequently comprised of a combination of specialist administrators with a background in research administration, and academics who represent the disciplines and fields of the applications which they review. It is always advisable to make contact with colleagues working in the research ethics departments to inquire about the process. In almost all institutional contexts, transparency about the process is encouraged, and most research ethics colleagues view it as part of their job to assist researchers to successfully and safely conduct their desired research. In addition to the basic process and timeline questions, you may want to ask:

- How is the ethical review panel formed?
- Who is on the panel and what is their background?
- What are the guidelines that panel members use to make their decisions?
- What are the common reasons why research proposals are rejected? What special considerations should be taken into account?

- What templates or guides are available that would be helpful when developing an application?
- What is the process if the panel requires further information or amendments, or initially rejects an application? Will feedback be provided to help with a resubmission?

Because simulation researchers' work often spans boundaries between hospitals or hospital systems and universities, additional considerations might apply between the two contexts. For instance, in the UK, research where hospital, social care, or prison service employees or patients are involved will almost certainly require approval via the Integrated Research Application System (IRAS). In other places, there may be national, state or provincial-level processes that apply. The application is then routed through to the appropriate panel within hospitals or local trusts for review. If the researcher is a university student or academic, the research must also be approved subsequently or concurrently by the university ethics review panel. In the US, the process is similar. There may also be circumstances where the academic institution can serve as the institutional review board of record if it is accredited as a Human Research Protection Program (AAHRPP).

Completing an Application

First looking at an online application system, or downloading an empty form, can be a daunting task even for an experienced simulation researcher. Approaching the ethical review application as another part of the research process helps to put this feeling into context: it's just one part of any research project. Further, in most cases, the written information required for an ethical approval process is not something you need to generate from scratch. You will almost certainly have explored these issues in the process of either applying for grant funding or getting your work approved by a peer-review process or as part of an academic project. Additionally, work put into the process of obtaining ethics review board approval can often be used for sections of any written dissemination of your research.

Remember that the ethical review process requires balancing potential benefits and risks, and so the application is an opportunity to explain to a review panel how important your work is and how you have considered, and will take steps to minimize, potential risk to those who participate in your research. As you put yourself in the place of panel members who will review your application, and you think about how to write your application, keep this balance in mind. Research always has the potential to both help and hurt those who are a part of it. As you begin this process, consider how your written ethical application can present your work—and even help you adjust your research design—to maximize the potential value while minimizing the potential impact.

As always when writing, *remember your readers*: make it as clear and easy as possible for them to read and review your application and to achieve their goals. Panel members almost certainly review applications in addition to other aspects of their work, so construct your application using clear and simple language, while demonstrating how your project will be conducted ethically.

Research Aim and Rationale

Most applications will ask for the purpose of the study, and potentially the rationale of the research. The research aim or research question may suffice here, but with some further explanation of why the research is being conducted. Whether it is stated as a requirement or not, consider writing these aspects of your form in plain language rather than technical academic jargon. Even if those on the panel are social scientists, ethicists, or quality improvement scientists, they may not be familiar with your particular approach or conceptual framework. Making your work as easy to understand as possible, while retaining the essential explanation and rigor, is central to this effort. When detailing the rationale of a simulation research project, remember to highlight how your project will have an impact on or make a difference to simulation learners, have an impact on patient safety, or improve the quality of patient care, even if this is one or two steps removed (as is often the case with simulation research). The generation of any type of scientific knowledge is typically not directly and immediately applicable, but if you can illustrate the chain of potential impact, then you have helped to establish the case for how your planned research addresses the balance between benefit and risk.

References and Citations

Most ethical application processes will require academic references to show that your work is grounded in existing literature (even if it addresses a gap in the current knowledge) and that you are basing your research on well-known and accepted research practice. Keeping in mind your readers on a review panel, be economical and concise with your references and use any citations to show that you are aware of the potential issues at play in your work.

Methodological Approach and Research Methods

In most cases, this is the part of the application that will receive the most scrutiny. Reviewers will be looking for a clear and unambiguous articulation of how you will interact with research participants, what data you will seek to generate based on those interactions, and how you will protect their rights, privacy, and integrity. Your research methodology should always follow logically from your aims and research questions, and you should clearly explain the methodological tradition within which your research is situated. Further, your research methods should be clearly aligned with your chosen methodological approach.

If you are using instruments such as surveys or checklists as part of your research, you should plan on including them in your application—a review panel will almost certainly want to have a clear sense of what you are proposing to use with your participants. Similarly, if you are planning to conduct interviews, you will need to include interview schedules, even if they are only topical guidelines for semi-structured interviews. Observation schedules or frameworks will be an important addition if you are conducting live observations, and coding frameworks if you are analyzing video recordings. If you are generating and collecting measurable and quantifiable data, your methods will need to be fully explicated and your tools outlined. If your work is exploratory in nature, you will want to explain methodologically how this is warranted and the nature of the data you expect to generate; panels will want to be confident that you are not generating or collecting data that is either unwarranted or inappropriate for the aims of your study.

Whether it is required by the process or not, a visual flow diagram of the research process can be helpful both for you as a researcher and for the review panel. By representing your work diagrammatically, you may be able to see parts of your planned research process where problems might occur and you can make adjustments to your process prior to beginning your work. In any case, a clear outline of the process will help reviewers make sense of your research plan and satisfy them that you have taken a thoughtful and considered approach to your work.

Potential for Risk or Harm

Most application processes ask researchers to articulate the potential risks for participants. In this area, like most aspects of the process, taking advice from your research ethics office and from more experienced colleagues is invaluable. Many ethics panels take the perspective that *all* research has inherent risk for participants, and that denying any potential for risk to participants in an ethics application shows that the researcher is not taking seriously their duty to thoughtfully consider the impact of the research process. As always, the analysis for research ethics review panels is balancing the potential for valuable outcomes of the work with potential for harm to participants. Highlighting that there is a potential for risk to participants, even if it is small, and that the risk has been mitigated by the research design, shows that you are engaging meaningfully and thoughtfully with the process and with your obligations as an ethical researcher.

For example, in research involving interviews with simulation participants, there is a small risk that an interview could evoke an emotionally sensitive topic. Being aware of, and having access to, trained and qualified support services for your participants is often an expectation of ethics review panels (and quite sensible research practice). In research involving biofeedback markers, there may be a very small risk of an allergic reaction to a sensor; highlighting this potential and the ways you might respond if a participant has a reaction is important.

Special Considerations for Privacy

Depending on where your research is conducted, there may be special laws or regulations to protect the privacy of research participants, and this may be either addressed separately in the process or through sections on the potential for risk and harm. Although you should take advice about these, it is important to consider the social and cultural as well as the legal contexts within which your research is conducted. For instance, what is the potential harm that could come to a participant if your research data is compromised? What steps would you take to mitigate that risk, and to recover the situation for your participants if it occurs? Respective legal contexts (e.g. FERPA and HIPAA in the USA; GDPR in European Union countries) will also apply Box 34.2.

Box 34.2: Tips for completing your ethical review process

- Always remember that the goal of the research ethics review process is to balance the potential risks of the research with the potential for benefit. Your job is to show how the benefits outweigh the risks, and that you have thought about the risks and attempted to mitigate them when possible.
- Be careful about promising more than you can practically deliver. Design your data collection or generation methods so that you can actually complete your ethical obligations as a researcher.

- Ask your colleagues and your contacts at your institutional research ethics office for any boilerplate language or templates that can help you in the process. As an example, informed consent paperwork typically has a format that should be closely followed. Successful applications using particular methods can provide the basis for subsequent research designs and applications.
- Thoughtfully designed research proposals, whether produced for academic purposes or for grant funding, often contain much of the justification for your work that is also required for the ethical review process. Use what you've already written as the basis for your application when you can.
- Write your application with your reader in mind: use plain English; help the reviewer understand your thinking; make it easy for the reviewer to find what they need to approve your application.
- Ask experienced colleagues to review your application and give you feedback before you submit it for formal review, even if peer review is not a required part of the process.

Conclusion

Although writing an ethics application can sometimes seem like additional work without a significant reward, the importance of ethical research conduct cannot be overstated. Most publications of peer-reviewed research, including the major simulation journals, require evidence of approval by the relevant authority (e.g. your ethical review process or institutional review board) before publishing the research. Just like our patients, our simulation learners rely on us as professionals and trust us to be guided by ethical principles in our work. An ethical simulation researcher is one who commits to the principles of ethical research behaviour while also ensuring that any required processes are completed and followed. An ethical simulation researcher holds the needs of patients, learners, service users, families and carers in high regard and does not let the generation of new knowledge take priority. An ethical simulation researcher recognises that the benefits from sound and ethically-conducted research are greater—both to the field and to society—than those obtained by unethical expediency. Research can help us understand the power of simulation-based learning, but to do so it must be sound and conducted ethically.

References

1. World Medical Association. World medical association declaration of helsinki: ethical principles for medical research involving human subjects. JAMA. 2013;310(20):2191–4. https://doi.org/10.1001/jama.2013.281053.
2. Resnik DB. What is ethics in research and why is it important? [Internet]. 2015. Available from: https://www.niehs.nih.gov/research/resources/bioethics/whatis/index.cfm
3. Maggio LA, Artino Jr AR, Picho K, Driessen EW. Are you sure you want to do that? Fostering the responsible conduct of medical edu- cation research. Acad Med. 2017. https://doi.org/10.1097/ACM.0000000000001805.
4. Miles SH. The Hippocratic Oath and the ethics of medicine. Oxford: Oxford University Press; 2004.
5. Eva KW. Research ethics requirements for medical education. Med Educ. 2009;43:194–5. https://doi.org/10.1111/j.1365-2923.2008.03285.x.
6. ten Cate O. Why the ethics of medical education research differs from that of medical research. Med Educ. 2009;43:608–10. https://doi.org/10.1111/j.1365-2923.2009.03385.x.
7. The Belmont Report. Belmont report: ethical principles and guidelines for the protection of human subjects of research. 1979; Report of the national commission for the protection of human subjects of biomedical and behavioral research.
8. Pugsley L, Dornan T. Using a sledgehammer to crack a nut: clinical ethics review and medical education research projects. Med Educ. 2007;41:726–8. https://doi.org/10.1111/j.1365-2923.2007.02805.x.
9. Boileau O, Patenaude J, St-Onge C. Twelve tips to avoid ethical pitfalls when recruiting students as subjects in medical education research. Med Teach. 2018;40(1):20–5. https://doi.org/10.1080/0142159X.2017.1357805.

Additional Resources

Australian Code For the Responsible Conduct of Research. https://nhmrc.gov.au/about-us/publications/australian-code-responsible-conduct-research-2018.

Collaborative Institutional Training Initiative (CITI) certification. https://about.citiprogram.org/en/homepage/.

Family Educational Rights and Privacy Act (FERPA). https://www2.ed.gov/policy/gen/guid/fpco/ferpa/index.html.

General Data Protection Regulation (GDPR). https://eugdpr.org.

Health Insurance Portability and Accountability Act (HIPAA) Privacy Rule and Research. https://privacyruleandresearch.nih.gov/clin_research.asp.

Research Ethics Timeline (1939 – present). https://www.niehs.nih.gov/research/resources/bioethics/timeline/index.cfm.

The Belmont Report. https://www.hhs.gov/ohrp/regulations-and-policy/belmont-report/index.html. *The Belmont Report is one of the leading publications on the ethics of healthcare research. This report helped to establish the fundamental principles we follow as healthcare researchers including: (a) respect for persons, (b) protection of autonomy, and (c) informed consent.

United Kingdom National Health Service (NHS) Health Research Authority (HRA) Guidance on the General Data Protection Regulation. https://www.hra.nhs.uk/hra-guidance-general-data-protection-regulation/.

Strategies in Developing a Simulation Research Proposal

35

Sharon Muret-Wagstaff and Joseph O. Lopreiato

Overview

This chapter focuses on steps that investigators can take before starting to write a grant proposal in order to increase the likelihood of securing funding.

Practice Points

- First, ask a good research question.
- Consider the fit of your research question with your expertise and goals, your unique research environment, and a potential funder's mission, goals, and priorities.
- Choose an effective group of collaborators.
- Develop a realistic timeline for proposal development.

Introduction

Today's emphasis on healthcare quality and patient safety coupled with rapid advances in simulation technology create a substantial need and opportunities for simulation-based research. However, grant-writing to secure necessary research resources can be daunting. Researchers can increase their likelihood of success by conducting strategic background work before starting to write a grant proposal. We recommend six considerations.

S. Muret-Wagstaff
Department of Surgery, Emory University School of Medicine, Atlanta, GA, USA
e-mail: smuretw@emory.edu

J. O. Lopreiato (✉)
Uniformed Services University of the Health Sciences, Bethesda, MD, USA
e-mail: joseph.lopreiato@usuhs.edu

First, Ask a Good Question

Is your research question significant, feasible, clear, ethical, novel, timely, high-impact, and translatable into practice or policy? Is it likely to move the field forward? Can you make a convincing argument for the need for this project? In other words, should this question be answered? [1–4] Ultimately, the single most common reason that reviewers recommend acceptance of medical education manuscripts is that the study addresses an "important, timely, relevant, critical, prevalent problem." [5] A question that addresses the impact on clinical outcome measures is particularly powerful [6].

Simulation-based research questions typically fall into two broad categories: [1] research on the efficacy of simulation as a training methodology; [2] research using simulation as an investigative methodology [7]. Both capitalize on the use of simulation to replicate clinical tasks and scenarios, make rigorous assessments, and repeat these cycles rapidly and continuously without risk to patients.

A researcher can identify and refine a compelling question in several ways. A in-depth systematic literature review can reveal gaps in the current knowledge base, promising theories and conceptual frameworks, and recommendations for future research topics. Published research agendas or policy statements from groups such as professional societies, advocacy organizations, and governmental agencies also can be revealing for the same purpose.

Of course, others may be considering questions similar to yours, and projects that are not yet published may be underway. The National Institutes of Health provides the Research Portfolio Online Reporting Tools (RePORT) which can help you learn about these potential efforts. RePORT includes a searchable public database with details of past and current federally funded projects (https://projectreporter.nih.gov/reporter.cfm). Perusing society conference abstracts also can be fruitful. Finally, participation in scientific meetings such as the International Meeting on Simulation in Healthcare offers the opportunity to learn firsthand about studies that

D. Nestel et al. (eds.), *Healthcare Simulation Research*, https://doi.org/10.1007/978-3-030-26837-4_35

are not yet published and to discuss your ideas and potential research questions with others in the field of healthcare simulation.

How Does Your Research Question Fit with Your Expertise and Goals?

Once you have identified a potential and exciting research question that fits most of the aforementioned criteria of what consitutes a good research question, consider its fit with your level of expertise and your career goals and trajectory.

Grant reviewers will be eager to learn about the suitability of your background for your proposed project—insufficient time and insufficient training are common perceived barriers to generating scholarship among clinician educator faculty [8]. Your relevant training, experiences in the field, academic level and advancement, prior funding and publication records, and recognitions are important considerations. As noted below, various types of grants are best suited for researchers who are at different points in their career trajectories. For example, a career development grant is an excellent fit for those in early career, but inappropriate for a senior researcher.

If the question that you want to answer appears to be beyond your current reach, consider a different but related question that could be a key stepping-stone toward your ultimate goal or partner with a senior researcher who have obvious track records in the field.

How Does Your Research Question Fit with Your Unique Research Environment?

For early career investigators, availability of a qualified research mentor is critical in planning a research endeavor. In choosing and engaging a mentor, consider the most useful mentoring model for you and clarify specific elements and responsibilities of a mentoring arrangement that are workable for both you and your mentor [9].

Does your research environment include a sufficient number (based on sample size calculation) of willing research participants with the needed characteristics? How long do you estimate it will take to recruit participants? How will you protect the rights and welfare of your subjects, whether they be patients, trainees, practicing clinicians, or others? What are the requirements of your local institutional review board?

If your study involves the use of simulation center, you will want to take advantage of particular capabilities and differentiators offered by your simulation facility. Simulation center accreditation by organizations such as the Society for Simulation in Healthcare or the American College of Surgeons adds to confidence that the proposed research can be carried out as planned. You will want to gain a full understanding of details such as your simulation facility's equipment and functionality, staffing, assessment tools and capabilities, video system, data capture and management, and ability to execute the study with fidelity [10]. It will be important to choose the simulation modality that is best suited to answer your research question. If barriers should arise, how will you mitigate these?

Lastly, you must garner local leadership support and institutional commitment in advance. Funders want assurance that your leaders are willing to provide the space, time, and resources needed as a foundation for your success.

How Does Your Research Question Fit with Your Potential Funder's Mission, Goals, and Priorities?

As you seek a funder with interests that are aligned with yours, consider intramural awards (particularly for pilot testing and gathering preliminary data), patient advocacy organizations, philanthropists, foundations, professional societies, industry, and federal agencies [11]. During your literature review, notice how published projects similar to yours have been funded.

Read about a potential funder's mission, goals, priorities, requirements, and current strategic plan. What types of projects have they recently funded and who are the recipients? Would your project be feasible within the funder's budget allowances and project period? Seek out grant-writing guidance from the organization as well as the manuscript elements required for the journals in which you aspire to publish your findings. For example, the Agency for Healthcare Research and Quality (AHRQ) offers Tips for Grant Applicants [12]. Talking with the program officer to better understand the potential fit between your research question and the agency can be helpful.

A single funder may offer various types of grants. For example, a foundation might offer both modest seed grants and larger grants for more mature projects. The National Institutes of Health (NIH) offer various types of grants including career development awards, exploratory or developmental research grants, investigator-initiated research project grants, center grants, and small business innovative research grants for technology development.

Once you have narrowed your selection of potential funders, and before you start writing, consider how you might respond to each of the organization's peer review criteria, and how you might bolster your proposal in this regard [13, 14]. (Note: Research methodology and study design considerations are addressed in other chapters of this book.) Many non-governmental organizations follow or adapt the NIH research review criteria that query the significance, investigator background, innovation, approach, environment, and overall impact of a proposal, along with additional review criteria and considerations that may apply to a specific funding announcement [15].

Advice in the NIH "Before You Start Writing" document [16] offers suggestions and a plain language checklist that are applicable to any grant-writing endeavor. For example, you could talk with seasoned reviewers at your institution, make a list of questions that are likely to be on the minds of peer reviewers, and plan that each section of your proposal will answer one or more of these questions with clarity. Remember that reviewers may not be subject matter experts in your precise area of research.

How Will You Form an Effective Group of Collaborators?

Your choice of collaborators contributes substantially to the probability of success of your project. Make certain that your co-investigators have experience and training that are appropriate to the research question, and that complementary and integrated expertise among team members is evident. Statistical expertise is particularly important; a recent analysis showed that the top reason for rejection of an education research manuscript was "statistics inappropriate, incomplete, or insufficiently described." [5]

Additionally, consider how you will lead, organize, manage, and communicate with your collaborative team in advance. Clarify various roles, expectations, time commitments, and ways decisions will be made regarding authorship and scholarly responsibilities to ensure transparency and avoid later misunderstandings. These commitments, arrangements, and understandings must be spelled out clearly in letters of support.

What Is a Realistic Timeline to Develop, Review, Revise, and Submit Your Proposal?

Think backward from your proposal deadline and create a Gantt chart listing realistic timeframes to carry out each process. Examples of elements to include in your timeline in addition to writing and revision are:

- Conduct literature review and the preliminary considerations and arrangements discussed above
- Request review of your specific aims by your mentor and senior colleagues
- Start cross-institutional arrangements as required for multi-site projects
- Develop budget and budget justification and request review by your departmental and institutional administrator
- Request internal review and feedback on an early draft of the proposal by your mentor, collaborators, and colleagues using the peer review criteria, followed by your revision
- Assemble materials, bio-sketches, and letters of support
- Conduct final proof-reading
- Submit a proposal to your institutional review board
- Meet all institutional administrative requirements
- Submit your grant proposal early according to your institution's sponsored programs office process
- Allow time for unexpected delays

Developing an effective and persuasive simulation-based research proposal takes time and persistence. Investigators can increase likelihood of securing simulation-based research funding by addressing these six essential considerations before starting to write the research grant proposal as well as creating the timeline to execute the proposed study successfully. No funding agency will fund a study that could not be finished on time with the budget it asks for. Finally, the writing process of the grant application is not entirely linear and the successful investigator may revisit these considerations throughout the pre-writing phase.

References

1. SSH research tools. Society for simulation in healthcare. https://www.ssih.org/ssh-resources/research-tools. Accessed 26 Nov 2018.
2. Morrison J. Developing research questions in medical education: the science and the art. Med Educ. 2002;36:596–7.
3. Blanco MA, Lee MY. Twelve tips for writing educational research grant proposals. Med Teach. 2012;34:450–3.
4. Prideaux D, Bligh J. Research in medical education: asking the right questions. Med Educ. 2002;36:1114–5.
5. Bordage G. Reasons reviewers reject and accept manuscripts: the strengths and weaknesses in medical education reports. Acad Med. 2001;76:889–96.
6. Chen FM, Burstin H, Huntington J. The importance of clinical outcomes in medical education research. Med Educ. 2005;39:350–5.
7. Cheng A, Auerbach M, Hunt EA, Chang TP, Pusic M, Nadkarni V, Kessler D. Designing and conducting simulation-based research. Pediatrics. 2014;133(6):1091–101.
8. Arnsten JH, Grossman P, Townsend JM. A scholarship-generating project for clinician educators. Med Educ. 2005;39(5):532–3.
9. Kashiwagi DT, Varkey P, Cook DA. Mentoring programs for physicians in academic medicine: a systematic review. Acad Med. 2013;88(7):1029–37.

10. McIvor WR, Banerjee A, Boulet JR, Berkhuis T, Tseytlin E, Torsher L, DeMaria S Jr, Rask JP, Shotwell MS, Burden A, Cooper JB, Gaba DM, Levine A, Park C, Sinz E, Steadman RH, Weinger MB. A taxonomy of delivery and documentation deviations during delivery of high-fidelity simulations. Simul Healthc. 2017;12:1):1–8.
11. Wisdom JP, Riley H, Myers N. Recommendations for writing successful grant proposals: an information synthesis. Acad Med. 2015;90:1720–5.
12. AHRQ. Tips for grant applicants. Agency for healthcare research and quality, Rockville. https://www.ahrq.gov/funding/process/grant-app-basics/apptips.html. Accessed 26 Nov 2018.
13. Bordage G, Caelleigh AS, Steinecke A, Bland CJ, Crandall SJ, McGaghie WC, Pangaro LN, Penn G, Regehr G, Shea JS. Joint task force of academic medicine and the GEA-RIME committee. Review criteria for research manuscripts. Acad Med. 2001;76(9):897–978.
14. Blanco MA, Gruppen LD, Artino AR Jr, Uijtdehaage S, Szauter K, Durning SJ. How to write an educational research grant: AMEE guide No. 101. Med Teach. 2016;38:113–22.
15. National Institutes of Health. Review criteria at a Glance. 2018. https://grants.nih.gov/grants/peer/guidelines_general/Review_Criteria_at_a_glance.pdf. Accessed 26 Nov 2018.
16. NIH. Before you start writing. National Institutes of Health. https://www.nih.gov/institutes-nih/nih-office-director/office-communications-public-liaison/clear-communication/plain-language/before-you-start-writing. Accessed 26 Nov 2018.

Identifying and Applying for Funding

36

Kevin Kunkler

Overview
In the current environment of dwindling national level funding, the need to properly identify and successful apply for funding is ever more critical. Armed with your research strategy, assuring that the funding source aligns with your strategy and not vice-versa is critical. Diligence in reading the announcement or solicitation carefully is important to assure qualification status, time-lines, budget allowance, and many other critical factors. Involvement of a program manager and editorial team is important with increasing chances of success.

Practice Points

- Understand the various funding sources.
- Aligning your strategy with the goals and objectives of the funding source.
- Read the announcement or solication and then re-read the announcement or solication.
- Know your team's roles, responsibilities, and understand team's dynamics and dependencies.

Introduction

In the current environment of dwindling national level funding, the need to properly identify and successful apply for funding is more critical than ever. Having a great research strategy assists with prioritizing where to look for funding, assist in saving precious time, and will assist with the type of terms you or your organization are willing to negotiate with the funding source.

K. Kunkler (✉)
School of Medicine – Medical Education, Texas Christian University and University of North Texas Health Science Center, Fort Worth, TX, USA
e-mail: k.kunkler@tcu.edu

Furnished with your research strategy, assuring that the funding source aligns with your strategy and not vice-versa is critical. Diligence in reading the announcement or solicitation carefully is important to assure qualification status, time-lines, budget allowance, and many other critical factors. Involvement of a program manager and editorial team is important with increasing chances of success.

There are several perspectives to consider when thinking about specific research strategies and how it aligns with funding. Where is your project within the research lifecycle; is it at Basic Research, Advanced Research, Applied Research, etc.? What are the fundamental building blocks that allows you to link your funding whether in terms of breadth, depth, or paradigm shifting? It is best, when possible to have as much flexibility in your research project to allow multiple funding resources and opportunities yet has enough definition and focus that you are not spending more of your time researching the respective funding opportunities rather than focusing on your research. In an attempt to make your area of research as broad as possible but allows you to entertain many funding opportunities, you need to see your research from different perspectives and different levels. An example of this concept may use chemistry, biochemistry, and biology. Think of what of your research may constitute as electrons, protons, and neutrons; so very basic particles. Next think of how those particles form elements, such as Hydrogen, Carbon, Oxygen, Nitrogen, and Sulphur. Keeping it somewhat simple, yet flexible, these elements form basic building blocks called amino acids. From amino acids, proteins can be formed and from proteins enzymes, hormones, cellular structure, antibodies, etc. There are other examples that you can use, regardless of the example you select there are keys to your research and potential funding sources and how to best align which type of research is at which level and which direction is the best route to provide the most and/or best benefits.

As just one example of a national or federal sponsor organization and the depth and breadth of topics they support, The National Institutes of Health, as of the writing of this chapter, has twenty-seven **(27) Institutes and Centers** (Fig. 36.1).

D. Nestel et al. (eds.), *Healthcare Simulation Research*, https://doi.org/10.1007/978-3-030-26837-4_36

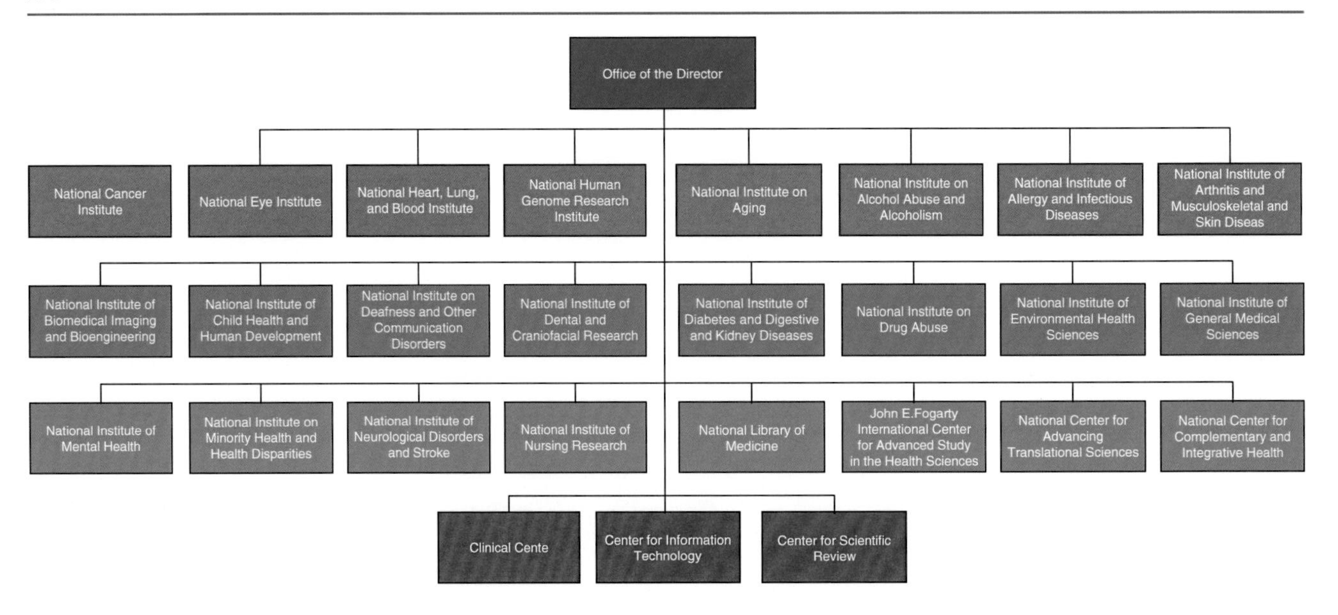

Fig. 36.1 National Institute of Health (NIH) org chart

Each Centers has a specific research agenda, often focusing on particular diseases or body systems and sometimes with some overlap with other Institutes and Centers.

Levels of Research

Many of the Federal Funding Institutes (NIH, DoD, NSF as examples) have the concept of tiers of funding dependent on the type of research. Most organizations who have multi-tier alignments have the concept of 'Basic Research' and other tiered research. Whether an organization uses a definition such as, "fundamental knowledge about the nature and behavior of living systems" (NIH); "systematic study directed toward greater knowledge or understanding of the fundamental aspects of phenomena and/or observable facts without specific applications" (DoD); to "build a foundation for generating sustainable, science-based solutions" (Gates Foundation), understand what the definition of words as 'Basic', 'Advanced', 'Applied', 'Development', 'Trials', or 'Test and Evaluation' and then align it with the respective components of your research. [https://phpartners.org/grants.html; https://www.rand.org/].

Just as a researcher needs to align their respective work with the grant or request for proposal opportunity, one further needs to align their proposed work with the level or funding or the program tier that is offered (Table 36.1).

Grant or Contract

When a researcher sees that funding is available, considering whether the award mechanism is in the form of a grant, a cooperative agreement, a contract, or potentially some other form of agreement may be one of the last things a researcher takes the time to understand. "Funding is funding? Isn't it?" Grants and Cooperative Agreements are the most similar with one very large difference between the two: cooperative agreements allows the agency providing the funds to have substantial involvement with the organization carrying out the research activity. Both Grants and Cooperative Agreements are non-repayable funds disbursed by an organization to a recipient to provide services, such as perform research or to build a proof of concept. Both Grants and Cooperative Agreements fundamentally are to have outcomes for public purpose and results released to the public.

Contracts, on the other hand, are a very different beast. Contracts typically benefit the organization providing the funds. Often, but not always, the organization providing the funds has the rights for the outcome be delivered to the respective organization, or the organization may state that they have first rights of refusal. The scope of work is typically defined by the funding organization. This is very important as the organization may actually edit, merge, delete, etc. items within the terms that were not proposed by the organization original proposal. Reporting and documentation typically is stricter as well. [Grants.gov; FedBizOps].

Due to the differences between Grants, Cooperative Agreements, and Contracts, researchers might automatically dismiss applying for Requests for Proposals to avoid contracts. Obviously, this is the discretion of the Principal Investigator and the organization to which the contract is awarded, however based upon the progression of where project or the type of research in the research cycle, the respective PI and organization should not be so quick to automatically dismiss contracts. These funding mechanisms allow for further advancement of a particular project or product than most

Table 36.1 Examples of range & types of funding categories (Open source – publicly available web site) https://grants.nih.gov/grants/funding/ac_search_results.htm

Code/ Category	Title	Description
R00/ Research projects	Research transition award	To support the second phase of a Career/Research Transition award program that provides 1–3 years of independent research support (R00) contingent on securing an independent research position. Award recipients will be expected to compete successfully for independent R01 support from the NIH during the R00 research transition award period
R01/ Research projects	Research project	To support a discrete, specified, circumscribed project to be performed by the named investigator(s) in an area representing his or her specific interest and competencies
R03/ Research projects	Small research grants	To provide research support specifically limited in time and amount for studies in categorical program areas. Small grants provide flexibility for initiating studies which are generally for preliminary short-term projects and are non-renewable
R15/ Research projects	Academic research enhancement awards (AREA)	Supports small-scale research projects at educational institutions that provide baccalaureate or advanced degrees for a significant number of the Nation's research scientists but that have not been major recipients of NIH support. The goals of the program are to (1) support meritorious research, (2) expose students to research, and (3) strengthen the research environment of the institution. Awards provide limited Direct Costs, plus applicable F&A costs, for periods not to exceed 36 months
R18/ Research projects	Research demonstration & dissemination projects	To provide support designed to develop, test, and evaluate health service activities, and to foster the application of existing knowledge for the control of categorical diseases
R21/ Research projects	Exploratory/ Developmental grants	To encourage the development of new research activities in categorical program areas. (Support generally is restricted in level of support and in time.)
R24/ Research projects	Resource-related research projects	To support research projects that will enhance the capability of resources to serve biomedical research
R25/ Research projects	Education projects	For support to develop and/or implement a program as it relates to a category in one or more of the areas of education, information, training, technical assistance, coordination, or evaluation
R28/ Research projects	Resource-related research projects	To support research projects contributing to improvement of the capability of resources to serve clinical research
R34/ Research projects	Planning grant	To provide support for the initial development of a clinical trial or research project, including the establishment of the research team; the development of tools for data management and oversight of the research; the development of a trial design or experimental research designs and other essential elements of the study or project
R35/ Research projects	Outstanding investigator award	To provide long-term support to an experienced investigator with an outstanding record of research productivity. This support is intended to encourage investigators to embark on long-term projects of unusual potential
R41/ Research projects	Small business technology transfer (STTR) grants – phase I	To support cooperative R&D projects between small business concerns and research institutions, limited in time and amount, to establish the technical merit and feasibility of ideas that have potential for commercialization. Awards are made to small business concerns only
R42/ Research projects	Small business technology transfer (STTR) grants – phase II	To support in – depth development of cooperative R&D projects between small business concerns and research institutions, limited in time and amount, whose feasibility has been established in Phase I and that have potential for commercialization. Awards are made to small business concerns only
R43/ Research projects	Small business innovation research grants (SBIR) – phase I	To support projects, limited in time and amount, to establish the technical merit and feasibility of R&D ideas which may ultimately lead to a commercial product(s) or service(s)
R44/ Research projects	Small business innovation research grants (SBIR) – phase II	To support in – depth development of R&D ideas whose feasibility has been established in Phase I and which are likely to result in commercial products or services. SBIR Phase II are considered "Fast-Track" and do not require National Council Review
R61/ Research projects	Phase 1 exploratory/ Developmental grant	As part of a bi-phasic approach to funding exploratory and/or developmental research, the R61 provides support for the first phase of the award. This activity code is used in lieu of the R21 activity code when larger budgets and/or project periods are required to establish feasibility for the project
R90/ Research projects	Interdisciplinary regular research training award	To support comprehensive interdisciplinary research training programs at the undergraduate, predoctoral and/or postdoctoral levels, by capitalizing on the infrastructure of existing multidisciplinary and interdisciplinary research programs. For trainees who do not meet the qualifications for NRSA authority

any grants will allow. Think of some of the alternatives to Federal contracts, such as equity investors or venture capitalists; the terms and conditions offered through some of these respective options may be considerable more damaging to the future direction and control of the project or product versus a contract with a Federal government agency. [https://techcrunch.com/2016; https://techcrunch.com/2017].

The Application

For novices, this process may be intimidating. For experts, this process still might be daunting. For many the application process is 'red tape', 'bureaucracy', 'paper-pushing', shear frustration, or expressions of expletive language that is not fit for print. However, the application process is inevitable when applying for funding.

There are numerous decisions a researcher must consider when presented with a funding opportunity. As mentioned earlier, research strategy is very important. By knowing your strategy, often the best decision to make when reviewing funding opportunities is to say 'No'. Just because there are funding opportunities, does not mean one should always apply. There are numerous reasons why one should not apply and will only mention a few: (1) The type of funding and where your research in the lifecycle may not correlate: the research may have a more basic research component and the opportunity is specifically requesting an advanced stage. It may be even worse when your project is at an advanced stage and the public funding in your area of research is only for basic. (2) The terms and conditions of the agreement may be asking too much, such as first rights of refusal or the ability of the funding agency to request the focus of the project to be in an area that you believe is not in the best interest of the project or the organization. (3) Do not go chasing funding opportunity after funding opportunity. One ends up with many tangential projects that may be nearly impossible to piece together. Perception by funding agencies may be that you and your organization have no strategic plans; then when a great opportunity appears, you may not receive the funding. So do not be afraid to say 'No' to certain funding opportunities.

Other common items that occur to researchers. Some researchers appear to be afraid to contact the public funding organization. Researchers may receive vast differences in response when approaching an agency, but the researcher should attempt to contact. There are times where the program officer, or equivalent, will listen to you and assist which components of your research might have stronger interests to the specific topic. They should also be able to indicate if the type of research you are pursuing is duplicative with other sponsored research. There are some public funding organizations, however, who do not provide much guidance as the interpretation of strict and fair competition may limit what and how much information they can provide you.

Always, always, always read, and then re-read, and then even re-read again the funding opportunity. Just skimming through the opportunity and believing what you think the funding organization wants to hear in an application is a poor approach. There are many components to an application and if you or someone from your team does not read the opportunity well, it may lead to many late nights near deadlines just to satisfy the application. Or worse the application could be disqualified because you and your team didn't follow the instructions.

Resources

There are too many resources to name within this chapter, but there are many for you and your organization to use as potential resources. The field of medical or healthcare simulation and modeling spans several different areas such as Education and Training; Computer Science; Computational Modeling and Mathematics; Bioengineering; Material Properties; Healthcare Processes and Polices; Systems and Organ Tissues (such as Heart, Lung, and Blood or Diabetes and Digestive Diseases); Standards; Science Foundations; Medical and Healthcare Societies or respective Colleges. The list is extensive! Collaboration opportunities with different inter-professional experts (such as nursing, pharmacy, social work, physical therapy, pastoral, respiratory therapists, etc.); engineers (computer, bio-medical, chemical, electrical, mechanical, systems, aeronautical, etc.); education; theatre and communications; and other industries (entertainment, technology, informatics; devices, pharmaceuticals, genomics, etc.) are almost limitless as well. Due to the vast collaborations, variables, directions, etc., no wonder identifying the best possible funding opportunities at the most optimal time with the most unique and opportunistic groups makes applying for funding such as challenge. Yet, due to so many opportunities, it should also encourage that many and diverse funding opportunities are available, if you just know where and how to look for them. Systematic and disciplined search approaches typically assist researchers more than the random and sporadic approach. For example, setting-up quarterly meetings with a grants team, collaborators, etc. should assist in allocating specific time to comb through funding opportunities. Do not forget to look within your own institution as sometimes there are funding opportunities especially to begin projects to acquire initial data or to begin the proof-of-concept process. Table 36.2 provides some of the many funding agencies and organizations that offer opportunities. Do note that some of the agencies have limitations of who are considered. Example – some are limited to United States organizations, academic institutions, businesses, etc.

Table 36.2 Examples of different agencies and some of their funding opportunities (All information obtained is open source – publicly available from the respective organization/agency web site)

Organization	Web address	Overview
National Institutes of Health – main page	https://grants.nih.gov/funding/index.htm	NIH offers funding for many types of grants, contracts, and even programs
National Cancer Institute (NCI)	https://www.cancer.gov/grants-training/grants-funding/funding-opportunities	NCI conducts and supports research that will lead to a future in which can prevent cancer before it starts, identify cancers that do develop at the earliest stage, eliminate cancers through innovative treatment interventions, and biologically control those cancers that cannot currently eliminate
National Eye Institute (NEI)	https://nei.nih.gov/funding	"Conduct and support research, training, health information dissemination, and other programs with respect to blinding eye diseases, visual disorders, mechanisms of visual function, preservation of sight, and the special health problems and requirements of the blind."
National Heart, Lung, and Blood Institute (NHLBI)	https://www.nhlbi.nih.gov/grants-and-training/funding-opportunities-and-contacts	Research, training, and education program to promote the prevention and treatment of heart, lung, and blood diseases and enhance the health of all individuals
National Human Genome Research Institute (NHGRI)	https://www.genome.gov/10000884/funding-opportunities/	The study of the ethical, legal and social implications of genome research
National Institutes on Aging (NIA)	https://www.nia.nih.gov/research/grants-funding	Research on the biomedical, social, and behavioral aspects of the aging process; the prevention of age-related diseases and disabilities; and the promotion of a better quality of life
National Institutes on Alcohol Abuse and Alcoholism (NIAAA)_	https://www.niaaa.nih.gov/funding-opportunities	Research focused on improving the treatment and prevention of alcoholism and alcohol-related problems to reduce the enormous health, social, and economic consequences of this disease
National Institute of Allergy and Infectious Diseases (NIAID)	https://www.niaid.nih.gov/grants-contracts/opportunities	Strives to understand, treat, and ultimately prevent the myriad infectious, immunologic, and allergic diseases that threaten millions of human lives
National Institute of Arthritis and Musculoskeletal and Skin Diseases (NIAMS)	https://www.niams.nih.gov/grants-funding	Supports research into the causes, treatment, and prevention of arthritis and musculoskeletal and skin diseases
National Institute of Biomedical Imaging and Bioengineering (NIBIB)	https://www.nibib.nih.gov/research-funding#quicktabs-funding_tabs=1	Leading the development and accelerating the application of biomedical technologies. The Institute is committed to integrating the physical and engineering sciences with the life sciences
Eunice Kennedy Shriver National Institute of Child Health & Human Development (NICHD)	https://www.nichd.nih.gov/grants-contracts/FOAs-notices	Research on fertility, pregnancy, growth, development, and medical rehabilitation strives to ensure that every child is born healthy
National Institute on Deafness and Other Communication Disorders (NIDCD)	https://www.nidcd.nih.gov/funding/all-nidcd-funding-opportunities	Biomedical research and research training on normal mechanisms as well as diseases and disorders of hearing, balance, smell, taste, voice, speech, and language
National Institute of Dental and Craniofacial Research (NIDCR)	https://www.nidcr.nih.gov/grants-funding/funding-opportunities	Research program designed to understand, treat, and ultimately prevent the infectious and inherited craniofacial-oral-dental diseases and disorders
National Institute of Diabetes and Digestive and Kidney Diseases (NIDDK)	https://www.niddk.nih.gov/research-funding/current-opportunities	To disseminate science-based information on diabetes and other endocrine and metabolic diseases; digestive diseases, nutritional disorders, and obesity; and kidney, urologic, and hematologic diseases
National Institute on Drug Abuse (NIDA)	https://www.drugabuse.gov/funding/funding-opportunities	Advance science on the causes and consequences of drug use and addiction and to apply that knowledge to improve individual and public health

(continued)

Table 36.2 (continued)

Organization	Web address	Overview
National Institute of Environmental Health Sciences (NIEHS)	https://www.niehs.nih.gov/funding/index.cfm	Discover how the environment affects people in order to promote healthier lives
National Institute of General Medical Sciences (NIGMS)	https://www.nigms.nih.gov/grants/Pages/Funding.aspx?tab=All	Investigate how living systems work at a range of levels, from molecules and cells to tissues, whole organisms and populations: affect multiple organ systems
National Institute of Mental Health (NIMH)	https://www.nimh.nih.gov/funding/index.shtml	Understanding, treating, and preventing mental illnesses through basic research on the brain and behavior
National Institute on Minority Health and Health Disparities (NIMHD)	https://www.nimhd.nih.gov/funding/nimhd-funding/index.html	Research to improve minority health and eliminate health disparities
National Institute of Neurological Disorders and Stroke (NINDS)	https://www.ninds.nih.gov/Funding/Find-Funding-Opportunities	Supports and conducts basic, translational, and clinical research on the normal and diseased nervous system
National Institute of Nursing Research (NINR)	https://www.ninr.nih.gov/researchandfunding/desp/oep/fundingopportunities	Basic research and research training on health and illness across the lifespan to build the scientific foundation for clinical practice, prevent disease and disability, manage and eliminate symptoms caused by illness, and improve palliative and end-of-life care
National Library of Medicine (NLM)	https://www.nlm.nih.gov/ep/Grants.html	Supports research in biomedical communications; creates information resources for molecular biology, biotechnology, toxicology, and environmental health
National Science Foundation (NSF)	https://www.nsf.gov/funding/azindex.jsp	Promote the progress of science; to advance the national health, prosperity, and welfare; to secure the national defense
National Institute of Standards & Technology	https://www.nist.gov/about-nist/funding-opportunities	
NIST – BioScience	https://www.nist.gov/oam/funding-opportunities	The development of quantitative analytical measurement tools for nucleic acids, proteins, metabolites, and cell systems to aid industry in the deployment of innovative biotechnologies and advanced biomaterials
NIST – Health	https://www.nist.gov/oam/funding-opportunities	Measurement assurance for biomedical stakeholders through the development of measurement tools for medical devices, clinical diagnostics, imaging tools and the characterization of complex biotherapeutics
NIST – Information Technology	https://www.nist.gov/oam/funding-opportunities	Advances the state-of-the-art in IT in such applications as cybersecurity and biometrics
NIST – Materials	https://www.nist.gov/oam/funding-opportunities	Develops testbeds, defines benchmarks and develops formability measurements and models for a variety of emerging materials
NIST – Nanotechnology	https://www.nist.gov/oam/funding-opportunities	Cutting-edge research to coordinating standards development
NIST – Resilience	https://www.nist.gov/oam/funding-opportunities	Focuses on the impact of hazards on buildings and communities and on post-disaster studies
Dept. of defense		
Defense advanced research projects agency	https://www.darpa.mil/work-with-us/opportunities	Explicitly reaches for transformational change instead of incremental advances
U.S. Army Medical Research & Materiel Command (USAMRA)	http://mrmc.amedd.army.mil/	Leading role in the advancement of military medicine
U.S. Army Research Laboratory (ARL)	https://www.arl.army.mil/www/default.cfm?page=8	Concentrates on scientific discovery, innovation, and transition of technological developments

Congressionally Directed Medical Research Programs (CDMRP)	http://cdmrp.army.mil/	Advancing paradigm shifting research, solutions that will lead to cures or improvements in patient care, or breakthrough technologies and resources for clinical benefit
U.S. Army Research Office (ARO)	https://www.arl.army.mil/www/default.cfm?page=29	The Army's principal extramural basic research agency in the engineering, physical, information and life sciences; developing and exploiting innovative advances to insure the Nation's technological superiority
U.S. Navy – Office of Naval Research (ONR)	https://www.onr.navy.mil/en/Contracts-Grants/Funding-Opportunities/Broad-Agency-Announcements	Innovative scientific and technological solutions to address current and future Navy and Marine Corps requirements
Defense Medical Research and Development (DMRDP)	http://cdmrp.army.mil/dmrdp/default	Manages and executes the Defense Health Program (DHP) Research, Development, Test, and Evaluation (RDT&E) appropriation
Agency for Healthcare Research & Quality (AHRQ)	https://www.ahrq.gov/funding/index.html	Federal agency charged with improving the safety and quality of America's health care system
National Board Medical Examiners – Stemmler Medical Education Research Fund	https://www.nbme.org/research/stemmler.html	Research or development of innovative assessment approaches that will enhance the evaluation of those preparing to, or continuing to, practice medicine
The Robert Wood Johnson Foundation	https://www.rwjf.org/en/how-we-work/grants-explorer/funding-opportunities.html	Funding of new and creative approaches to building a Culture of Health
The Josiah Macy Jr. Foundation	http://macyfoundation.org/grantees/c/society-for-simulation-in-healthcare	Improve the health of the public through innovative research and programs
National League of Nursing (NLN)	http://www.nln.org/professional-development-programs/grants-and-scholarships	Furthering the scholarship of teaching and promoting evidence-based nursing education
International Nursing Association for Clinical Simulation and Learning (INACSL)	https://www.inacsl.org/resources/inacsl-grant-opportunities/	Conduct investigations on issues relevant to nursing simulation. Designed to fund research that advances the science of simulation in healthcare and is related to at least one of the INACSL priorities
US Department of Health & Human Services, Health Resources & Services Administration	https://www.hrsa.gov/grants	
Canadian Institutes for Health Research	https://www.researchnet-recherchenet.ca/rnr16/LoginServlet?language=E	
Canarie Canada	https://www.canarie.ca/	
The virtual foundation	http://www.virtualfoundation.org/	
American academy of pediatrics	https://www.aap.org/en-us/about-the-aap/Sections/Section-on-Pediatric-Trainees/Pages/SOPT.aspx	
American heart association	https://professional.heart.org/professional/ResearchPrograms/ApplicationInformation/ScientistPrincipalinvestigators/UCM_316962_For-Scientists.jsp	
The association for surgical education	https://surgicaleducation.com/grants-awarded/	
Laerdal foundation	https://laerdalfoundation.org/history/	
Zoll foundation	http://zollfoundation.org/apply.html	
Grants.gov	https://www.grants.gov/web/grants/search-grants.html	Provide a common website for federal agencies to post discretionary funding opportunities and for grantees to find and apply to them
FedBizOpps	https://www.fbo.gov/	Government point-of-entry (GPE) for Federal government procurement opportunities over $25,000

Conclusion

Despite the current environment of dwindling national level funding within certain organizations, there are ways for an individual, team, and organization to increase their probability of obtaining funding outside of their respective organization to fund simulation and modeling research. Strategic planning, albeit sometimes painful, should significantly improve one's prioritization, where to look, completion of a submission, and probability of award.

Learn how to say "No" to funding opportunities that do not align well with your strategy and minimize taking a "shot-gun" type of approach. If all you are doing is playing the numbers game, than the 'shot-gun' approach may provide some funding. However, it will most likely will be scattered and it will be difficult for you and your organization to piece the funding together in order for you to produce subsequent projects based upon that specifically funded research.

Be strategic, have assistance in finding funding resources, know where your project resides within the research lifecycle, and do not allow the process to intimidate you.

Resources

AHRQ. Tips for grant applicants. Agency for healthcare research and quality, Rockville. http://www.ahrq.gov/funding/process/grant-app-basics/apptips.html. Accessed 18 Sept 2017.

Federal business opportunities. https://www.fbo.gov/.

Grants.gov. https://www.grants.gov/.

Institute of medicine: strategies to leverage research funding: guiding DOD's peer reviewed medical research programs. Chapter 2 sources of funding for biomedical research.

NIH. What to know before you start writing. National Institutes of health. https://grants.nih.gov/grants/how-to-apply-application-guide/format-and-write/write-your-application.htm. Accessed 18 Sept 2017.

U.S. Army Medical Research Acquisition Activity (USAMRAA). https://www.usamraa.army.mil/Pages/Main01.aspx.

https://www.nonprofitexpert.com/international-grants/

https://phpartners.org/grants.html

https://www.nap.edu/read/11177/chapter/8#45

https://www.rand.org/content/dam/rand/pubs/monograph_reports/MR1194/MR1194.appb.pdf

https://techcrunch.com/2017/06/01/the-meeting-that-showed-me-the-truth-about-vcs/

https://techcrunch.com/2016/09/16/venture-capital-is-a-hell-of-a-drug/

https://blog.grants.gov/2018/05/09/what-is-a-contract-5-differences-between-grants-and-contracts/#more-3709

Anatomy of a Successful Grant Proposal

37

Rosemarie Fernandez, Shawna J. Perry, and Mary D. Patterson

Overview

The mechanics of writing a research proposal are addressed elsewhere in this text (see Chap. 33). In this chapter we will address the factors that are relevant to the successful funding of a grant proposal. The scientific basis of a research proposal is paramount, but other factors, often termed 'grantsmanship', influence reviewers' perceptions and evaluations of a research proposal. This chapter uses the example of a United States based federal grant application, but this information is applicable to a variety of funding mechanisms.

The authors of this chapter have been successfully funded and served as grant reviewers for governmental agencies, private foundations and international agencies. The information presented here will assist the applicant in understanding the mechanics of grant review from the reviewer's perspective as well as methods to create enthusiasm for the application in the reviewers.

Practice Points

- No amount of grantsmanship can overcome a weak or poorly designed research plan.
- The person who reviews an application is not necessarily an expert in the field.
- The successful application will be clear and understandable even by someone without expertise in the domain.
- The applicant should make it easy for the reviewer to perform the review.
- The Specific Aims page is the most important section of an application.
- The reviewer wants to advocate for your proposal. It is your job to give them the necessary information to do so.
- It is always a good idea to have individuals that are not intimately familiar with the proposed work, review the application before submission.

Introduction

The ability to advocate for a research proposal is heavily influenced by the clarity of writing and ease of sensemaking for the reviewers. Despite this, the role and perspective of grant reviewers is often not a primary consideration during the writing process. Reviewers are seeking logical, concise, easily understood arguments and plans for meeting the specific aims proposed by the research team. The audience you are writing for are individuals of varying backgrounds and expertise who are anonymous to you. This chapter will offer some insights on the characteristics of competitive proposals from the viewpoint of the reviewer. It will highlight aspects of research proposals that support the reviewers' ability to evaluate, score and advocate on its behalf.

One widely held misconception is that every member of a grant review panel will be *the* world-renowned leading expert in *your* specific research area of interest, and as such, they will be exceedingly well versed in the subject your team wishes to study. This is very seldom the case. Review panels are composed of a number of highly

R. Fernandez
Department of Emergency Medicine and Center for Experiential Learning and Simulation, College of Medicine, University of Florida, Gainesville, FL, USA
e-mail: fernandez.r@ufl.edu

S. J. Perry
Department of Emergency Medicine, University of Florida College of Medicine – Jacksonville, Jacksonville, FL, USA

M. D. Patterson (✉)
Department of Emergency Medicine and Center for Experiential Learning and Simulation, College of Medicine, University of Florida, Gainesville, FL, USA
e-mail: m.patterson@ufl.edu

D. Nestel et al. (eds.), *Healthcare Simulation Research*, https://doi.org/10.1007/978-3-030-26837-4_37

educated individuals from a variety of domains. Reviewers are assigned proposals to critique and present to the larger group; however, reviewers are often assigned proposals that are outside of their primary area of expertise. Three reviewers are usually assigned to each application. For example, a proposal related to computer simulation of patient scheduling in a primary care clinic may be reviewed by an expert in computer simulation, another who is an expert in quality assurance in health care and a third who is a primary care clinician with expertise in public policy and access to care. Bearing this mind, the content and format of the grant should be written to persuade any reviewer to become an enthusiastic advocate of the project, thus allowing them to present it with ease to the larger review panel for discussion and scoring.

Timeline for Writing a Grant Proposal

Typically, the writing of a grant proposal should begin 4–6 months prior to the sponsor's deadline. While that may seem to be a long interval, there are myriad details that must be addressed during this time. If the applicant is associated with an academic organization, the required certifications and registrations with the sponsor are likely already in place. This may require more time in a healthcare organization that does not frequently submit proposals for funding. Most organizations also require some sort of additional internal review prior to submission to the funding organization and that must be accounted for. Not infrequently, funding organizations release requests for proposals (RFP) that have short submission deadlines, e.g., 6 weeks. Many successful applicants keep several partially written applications on hand that can be rapidly polished and submitted when a promising RFP is released.

The Research Question: Framing Your Research and Making an Argument for Reviewers

A successful research proposal should convince the reviewer that your work is (1) important, (2) feasible, and (3) aligned with the mission of the funding organization. The reviewer is likely to form an opinion on how well you've accomplished these goals within the first few pages of the proposal. Different funding sources may require different proposal layouts and formats. In general, the first few pages will provide you the opportunity to describe your overall research question and approach (Specific Aims), describe the current state of the science and explain the importance of your study (Significance and Rationale), and highlight why your study will advance the field (Innovation/Importance). We discuss each of these sections with a focus on how they are seen through a reviewer's lens.

Specific Aims

The Specific Aims section of your grant is the most important page you will write. In most federal grant review processes, it is the only page most of the study section members will have time to read due to the large number of grants being considered at each review session. This means that this section needs to communicate the knowledge gap you are trying to fill, the overall objective of the proposed work, your research questions and associated hypotheses, and the relevance of your outcomes to the funding institution or mechanism. Because the Specific Aims section must convey a great deal of information, the writing must be extremely focused and concise. Reviewers are not looking for in-depth details about your preliminary work, approach, or research team. If your team is extremely strong and uniquely positioned to do the proposed work, then a single sentence stating this might be warranted. However, details about the strengths of each investigator will take up valuable space and leave your reviewers wondering why you chose to discuss your team rather than provide clear information about what you are going to do, how you will do it, and what you hope to discover.

Recommendations

- Reviewers are looking to make sure your specific aims are independent. If the success of one aim depends on the success of another, reviewers will see this as a major threat to the feasibility of your proposal.
- Reviewers want to see that your aims clearly address hypotheses and have well-defined outcomes. Reviewers want to understand what you want to do, how you want to do it, and how it will be measured.
- Reviewers want to know exactly what your primary outcome(s) is.
- Reviewers want to see exactly how your proposal addresses the priority of the funding agency. If you are responding to a specific call for proposals, clearly state how your proposal is applicable.
- Reviewers are often not experts in your area and have not read your entire proposal. Limit the use of jargon to avoid unnecessary confusion.
- Reviewers want to be able to sell your proposal and be able to say how it will advance clinical, education, simulation, or safety science. In the last few sentences, be sure to state clearly to the reviewer how the successful completion of your study advances the field.

Significance/Background/Rationale

Different funding agencies will have different requirements, but ultimately they all want some background information that explains why your project is significant. The Background and Rationale section helps reviewers answer the question "So what?" In other words, why should we care about your research question? Is it because five million people are impacted by the disease every year? Is it because you are addressing a knowledge gap that prevents implementation of evidence-based medicine? It is important to remember that at least one or more of your reviewers will not be familiar with the clinical or educational question you seek to address. It is your job to orient and convince them that the problem or knowledge gap you are trying to address has significant impact and is important. By the end of your Background section, the reviewer should be able to clearly articulate the critical importance or your research problem or knowledge gap.

Some funding agencies also ask that you address the rationale for your approach as a separate section. A reviewer wants to understand why the investigator is choosing to answer this question with the techniques / research approach proposed. For instance, if you are proposing to use virtual reality-based training to develop lay-person CPR skills in high schools, a reviewer wants to know why virtual reality? Why CPR skills? Why this population of learners? How is this approach better than what is already done?

Recommendations

- Reviewers want to understand what knowledge gap you wish to address, why it is important, and why your approach makes sense. Your reviewer needs to be able to answer the "So what?" question. If your project is successful, so what?
- It is important to demonstrate that you have a thorough understanding of the existing science. Make sure the material you reference is up to date. If there is some disagreement in the literature around your topic, acknowledge it and provide a rational argument for your study.
- It is critical that this section clearly conveys understanding of the domain within which you will be doing research. Technical, jargon-filled language does not help your cause. If you use specialized terms, define them and be consistent with their use throughout. Define all abbreviations and ask yourself if it is really important that a term is abbreviated. You don't want to lose the reviewer's attention because s/he can't keep track of your abbreviations!

Innovation

This can be one of the toughest areas for new investigators to understand. There is a natural tendency to see an overlap between the Innovation section and the Significance section. These are, however, two very different content areas for reviewers. When reviewers look at your Significance section, they want to understand the importance of the problem you are addressing and why it deserves attention. In contrast, the Innovations section is expected to discuss why your approach and solutions to the problem are novel and advance the field. Not all grant applications require that you include this section. However, if your approach is novel, you want to emphasize why your project is innovative, especially if you are using a new method, technique, tool, perspective or technology in your Approach.

Recommendations

- Reviewers want to clearly understand what is innovative about your proposed work. You may be implementing your work in a novel population, or using a novel technique, or adapting a conceptual framework previously applied in a non-medical field. Whatever it is that makes your work innovative, make it clear for the reviewer.
- Avoid rehashing the content of your Background or Significance section with regard to innovation. This section is often short (less than one page), very direct, and to the point.
- Sometimes a proposal addresses a very important knowledge gap but does not necessarily meet the definition of "innovative". Reviewers understand this. If this is the case, you should use this as an opportunity to address how the overall project, with its proposed methodology, will result in a major leap forward for your field.

The Research Team: What Are Reviewers Looking For?

The composition of your research team is a critical component of your proposal. Reviewers will be specifically looking for information that demonstrates the team has the requisite expertise to execute the proposed work. This may sound simple, but the team composition is often an area that is heavily critiqued by reviewers. Successful proposals clearly identify the role and responsibility of each investigator, leaving no uncertainty about each individual's contribution to the project. This will begin with reviewers assessing the type of scientific expertise included on the research team. A study of informal clinical communication using smart phones between nurses and physicians in ICUs would be expected to include not only professionals from each discipline, but individuals with expertise in communication and perhaps sociology or human factors engineering. To this end, is also important that each team member's biosketch clearly demonstrate domain expertise that supports the work of the proposal and the budget justification delineates succinctly how that expertise will be expected to contribute and what will be the responsibilities of each team member.

Reviewers will also seek evidence that the level of research experience of the PI and team members is commensurate to the level of funding being requested. Specifically, does the team have experience in grants management necessary to execute the proposed project? For instance a team composed exclusively of junior investigators seeking several million dollars of funding would raise concern about the feasibility of completion of the project. This can be mitigated by including a more senior and seasoned investigator to the team, and clearly stating in the proposal and in his/her biosketch that grant management is one of their roles on the project.

Reviewers also want to see that each person on your team has enough financial support within the grant to "buy time" from their primary employer in order to execute the responsibilities to your project. This means that the amount of grant money allocated as salary support for each team member should accurately reflect his/her responsibilities and commitment to the project. Reviewers will be concerned if key personnel performing a number of critical roles within a 4-year project are only supported for a small percentage of their effort each year. This is particularly concerning if your team members are also involved in a number of other research projects. The question reviewers will be considering is whether or not each team member will have enough time to substantially contribute to the work of your proposal.

Your proposal must also show that the team can feasibly gather the data it seeks, i.e., recruit subjects and/or access databases. Reviewers are looking for evidence that you have not only local support, but support at all the sites where the project is being conducted. This frequently takes the form of letters of support from all entities participating in the project. The inclusion of co-investigators or consultants at each research site and descriptions of how the proposal will be supported, e.g., statements such as "the medical director for the clinic will assist with identifying potential subjects to include in the study", provide reassurance that the proposal has a good chance of being successful.

Finally, reviewers are also looking to understand how your collaborators are going to work together. It is common for investigators on a grant to come from multiple institutions, even if data are only collected at one site. This is somewhat expected, but does present challenges during the collaborative process. Reviewers want to know that you've considered this and have a plan to manage your distributed team. This may include virtual meeting software, budgeting for in-person meetings, or a successful track record of long distance collaboration.

Recommendations

- Reviewers want to see that your team has the expertise to complete the project.
- The involvement of professionals from domains outside healthcare is considered a significant positive. Depending upon the nature and focus of your project, including investigators from the social sciences, engineering, or humanities suggests to reviewers that your project is innovative and will make more than an incremental advance in the field.
- Reviewers like to see that investigators have a history of successful collaboration. While this is not always the case, be sure to highlight any shared projects you have with other investigators on your grant.
- All investigators are not seasoned scientists. Consider obtaining a letter of support for junior investigators from their mentors or direct supervisors that will ensure they have the support needed to fulfill their role.
- Reviewers know how difficult it can be to recruit and collect data at remote sites. Demonstrate you have the necessary support at each site.
- Inconsistencies within a proposal are very distracting and viewed negatively by reviewers. For instance, be sure personal statements of biosketches match role descriptions within the research proposal, budget justification, and letters of support. Reviewers notice when a biosketch reflects a previous project rather than the current proposal.
- Letters of support should not be identical; each letter of support should reflect the specific resources, responsibilities, and commitment of the individual or entity authoring the letter.

The Environment: Are You Set Up for Success?

Reviewers need to know that your institution and your study sites can support your work. For simulation, this may mean that you have the requisite simulation equipment as well as recording capability and video processing. If the simulation work to be done is significant, reviewers will want to see that you've budgeted for simulation faculty and staff time. If the institution is providing this as an "in-kind" contribution to the project, reviewers will be looking for a letter of support from the institution that clearly states what the level of support is.

While you may be submitting a proposal that centers on simulation, don't forget to describe other relevant components of the research environment. If you are recruiting nurses, the reviewers want to know that the clinical environment can support your recruitment plan and would like to see a letter of support from the nursing leadership ensuring that they will help you achieve the recruitment goals. You may also want to mention research infrastructure present at your institution, especially if these resources will be used in your project.

Recommendations

- Reviewers want to know that the study institution(s) has the resources needed for you to get your work done.
- It is important for reviewers to see evidence that your environment can support the recruitment plan you've outlined.
- Letters of support should clearly state how and what resources will be provided.

Methods/Approach

It could be argued that, in addition to the Specific Aims of an application, the Approach, or Methods, is the second most important section. Reviewers (including those not assigned to the application) will read the Approach after reading the Specific Aims. Chapter 33 provides detailed instructions on writing a research proposal; the focus here is on the preferred presentation and pitfalls to avoid.

The Approach should include enough background to enable the reviewer to grasp what is known and where the gaps in knowledge are. Previous work, especially by the applicants, may be included in the background or as part of the introduction to each proposed intervention. A description of related preliminary work performed by the team or team members engenders confidence in the reviewers.

Each experiment or intervention should be explicitly linked to a specific aim. The work should be feasible, and each intervention should be independent of other interventions, ie; each proposed intervention should not be dependent on the success of an earlier activity in the application. There are exceptions to this, but especially when a specific aim appears risky or less likely to be successful, the remaining specific aims should not rely on the successful completion of an aim that seems chancy. The work described should be feasible given the proposed effort, timeframe, and available resources. Reviewers are often skeptical of what they perceive to be overambitious projects. Members of the research team should have the skills to carry out all the proposed activities. Proposed methods that are not yet developed may hinder enthusiasm for the application.

The conceptual framework is a crucial aspect of the application; it is the foundation that enables reviewers to understand the theoretical construct supporting the proposed project. Paradoxically, it is often omitted, and this omission is often viewed as a fatal flaw by reviewers. Providing a well-referenced conceptual framework around which the study components (measures, outcomes, and analyses) are organized will help reviewers understand your work and believe that your work is well-grounded. The inclusion of a visual representation or diagram illustrating the key components of the conceptual framework is also helps with sensemaking by the reviewer of your proposal.

The study design is also critical. In general, the strongest design that can practically be carried out is desirable. A randomized controlled trial is not common in simulation research, but a stepped wedge design is a variation that is a considerably stronger design than a simple pre-and post study. Again, remember that not all reviewers are familiar with simulation or medical education research. Your work must be rigorous by standards of medical research overall. Careful understanding of the limitations and biases inherent in your work are a must and their inclusion benefits the reviewers understanding and ultimately advocacy for the proposal. Any step where you had to scale back for practical reasons should be acknowledged. Reviewers understand that study design is a careful balance of practicality and a desire for scientific rigor.

The choice of appropriate outcomes is key to a successful grant application. In simulation, there has been a tendency in the past to select weak outcomes that measure learner reactions to the simulation experience or the immediate change in knowledge or skill. Current successful grant applications are more likely linked to behavior change, clinical outcomes or a system/process measure. This does not mean that multilevel outcomes are not important, but rather outcomes should be supported by the conceptual model and match the rigor and funding level of the grant to which you are applying.

Finally, a Gantt chart or other type of timeline should be included to demonstrate the proposed interval for each grant activity. This should be followed by a section describing the limitations of the study. A paragraph or two on the limitation or alternative methods is often missing from grant applications, and reviewers are typically sensitive to this omission. They understand that all grants have limitations. However, the funding organization wants to know that the applicant has thought through the research process and has identified alternative methods that will result in meaningful contributions even if the primary intervention is not successful.

Recommendations

- Use clear, understandable language and avoid technical jargon.
- Make clear how the preliminary work supports the proposal.

- A well-organized figure that outlines each step in your study goes a long way to help with clarity!
- Be clear what your primary, secondary outcomes are. Be sure you state which outcome you are using for your sample size calculation.
- Have clear sections within the Approach that mimic a clinical manuscript: setting, subjects, intervention(s), outcomes, data collection, analyses, etc.
- Align the specific aims with each intervention.
- Use the strongest design and outcomes that can feasibly be accomplished.

Budget

The budget requirements are dictated by the funding organization and the resources that are required to complete the work. Reviewers want to ensure that proposed budget is sufficient to accomplish the work proposed, but they are also skeptical of anything that appears to be lavish or excessive. Budget instructions are typically quite detailed and should be followed without deviation. In the case of any ambiguity, the applicant should consult with the funding organization or agency on what is permissible. A business manager or someone associated with an organization's grants and development office is helpful in developing the grant budget. For most research grants, simulation equipment (simulators) is seen as an inappropriate expense, while simulation supplies would be expected expenditures. Be sure to include in the budget the cost of methods and tools necessary for collaboration across your team, e.g., travel for team meetings for data analysis at key intervals, teleconferencing, etc.

Key budget considerations in any grant proposal include:

- Total allowable budget over what time interval
- Including or excluding indirect costs
  - Many private foundations do not allow for indirect costs
- Modular budget or not (NIH uses modular budgets; many other organizations/agencies do not)
- Budget use for capital expenses (expensive equipment expected to last for several years, i.e., simulators.)
  - Many funding organizations limit capital expenses to a small percentage of the overall budget or don't allow for any capital expenses.
- Many funding organizations adopt the US federal government federal agency salary cap. All reimbursement for salaries is limited by the salary cap.
- Many foundations require some proportion of in-kind contribution from the applicant's organization. If included in the proposal, in-kind funds and resources should be outlined in a Letter of Support.

Human Subjects

See Also Chap. 34: Writing an Ethics Application

The protection of human subjects is mandated by funding organizations and agencies and is a required element of all grant applications. Often this aspect of the application is given short shrift by the applicant, typically being located at the end of the application. While a well done human subjects section will not necessarily gain any points with a reviewer, a sloppy or missing human subjects section may sink the grant application. Students, trainees, and healthcare professionals frequently serve as subjects in simulation research. These subjects are viewed as vulnerable research populations in light of their positions as students and/or employees of the healthcare institution. As such, they are entitled to additional protections and care needs to be exercised in terms of recruitment and de-identification of data. Any hint of coercion must be avoided. While the applicant may make the case that the proposed project is exempt from regulation as human subjects research, only an ethics board or Institutional Review Board (IRB) can make that determination. Ethics (or IRB) review is required in addition to the grant application. The ethics review is not necessarily completed by the time of the grant submission. However, timing of the ethics review should take into account the expected funding date, the interval required for ethics review in a particular institution and possibly the need for multiple organizations to perform an ethics review if multiple sites are participating.

Grantsmanship and Other Miscellaneous Points

Grantmanship is defined as "the art of obtaining grant funding" [1] and this section will focus on "the art". Being attentive to fine details, such as how the proposal 'looks' to the reviewer is important, not to mention spelling errors, grammar etc. More than last minute attention should be given to the page layout of the document, margins, line spacing, font size and figures or images as they affect the conveyance of ideas and comprehension for the reviewer. A proposal of 15 or more pages with narrow margins, single spaced with size 9 font can be off putting at first glance as it connotes a dense proposal that is full of information that will likely be difficult to follow or reference. This can also signal a proposal that has not been well thought out. Be sure to check with the grantor for submission specifications as some grantors will not accept proposals that do not meet their format and layout criteria. In the event there is a lack of specifications, a good rule of thumb is to not submit any proposal that you, a colleague, or family member would not want to read and evaluate.

Recommendations

- Be sure you read the call for proposals carefully (at least twice), making note of any and all requirements. These include:
- Formatting (font, spacing, margins, page limits)
- Required sections
- Key material that must be covered (consider bolding such information in your proposal to ensure it is not missed)
- Budget requirements or salary and effort requirements
- Project length
- Funding agency priorities
- The appropriate inclusion of flow diagrams or models to demonstrate important features of the proposal (e.g., relationship of specific aims to the research methodology, process of data collection, analysis, etc.) can often be helpful to the reviewer, especially if the project has more than one intervention arm or is complex.
  - Make figures readable and ensure they are necessary for the reviewer to understand your proposal
- Each funding entity will have specific minimum criteria it expects its reviewers to use for evaluation (e.g., responsiveness to request for applications, significance, methodology, inclusion of a specific population for study, etc.) These can often be found in the call for applications or on the grantor's website. As discussed earlier, the art of grantsmanship includes making the proposal understandable and easy to navigate. Specific criteria should be readily identified, as they can be easily overlooked in a poorly presented proposal, resulting in a non-competitive score.
- Make sure references are correct, current, and relevant to the subject matter. Reviewers do periodically check them to clarify their understanding of the proposal and overall validity
  - In citing references in the body of the application, use the author(s) and year in parentheses rather than a superscript. This uses slightly more space, but is very helpful to the reviewer.
  - Be certain to include classic or seminal references – if they are not cited, a reviewer will make note of it. One of those overlooked authors may also be a reviewer!
- Limit use of appendices to items that are *crucial* for making your case. An overabundance of appendices can be time consuming to review and often add little to the reviewers overall understanding of the proposed project. US federal granting agencies currently restrict the type of materials that can be placed in Appendices.
- The importance of the 'understandability factor' to reviewers cannot be emphasized enough. On occasion, proposals are not scored favorably despite being an innovative, potentially impactful project simply because it was difficult to understand (e.g., numerous complex equations with limited explanation of their relevance, run on sentences that contain too many ideas, etc.) Having the draft proposal read by several people (some of whom are not familiar with the subject matter) for clarity and comprehension can be an effective litmus test for understandability.
  - Avoid jargon
  - Use abbreviations sparingly and define them early. Avoid abbreviations in the Abstract or Specific Aims sections
  - Use the same term for the same concept throughout the proposal

When You Aren't Funded the First Time

After passing though Kubler-Ross's stages of grief [2] when your proposal receives a score that will not result in funding, it is important to critically analyze the proposal, the submission, and most importantly, the reviews. This would initially include deciphering the scoring system used by the grantor to determine how far your proposal is from a fundable score. This, along with a thorough vetting of the reviewers' comments, will assist in determining how much revision will be needed for a successful resubmission. The reviewers' comments will be provided in writing and will discuss strengths and weaknesses of the proposal. They will often include suggestions for refining and improving the proposal and project overall. These are offered as constructive criticism and are based on the reviewers' desire to advance scientific exploration. They are not personal in nature, although they may initially feel that way. Think of these comments as a roadmap to success when the proposal is resubmitted. If the application was triaged, meaning the preliminary reviewer scores were not high enough to require discussion by the entire review panel, the applicant will only receive the reviewers' written comments. If the application was discussed, the applicant will receive a summary of the discussion as well. In those cases, it is sometimes helpful to arrange a phone conversation with the science officer of the funding organization. The science officer may be able to provide more nuanced feedback concerning the reviewers' discussion.

If resubmitting to the same grantor, it is expected that there will be a cover letter that begins by thanking the reviewers for their review and explaining how the previous reviewers' comments were addressed in the new version of the proposal (or not addressed with an explanation). It can also be helpful and it is often required to highlight specific revisions made within the resubmitted proposal. This may be a point-by-point overview of the revisions made, as well as noting how the changes in the text can be identified (e.g., italics, highlighting, etc.) This can be helpful to the second review process, as on occasion, the same reviewers may be assigned to evaluate the re-submission. It is therefore very important to respond to each and every recommendation in your cover letter that you received from the first submission.

Closing

In general grant reviewers spend hours reviewing applications before the grant review meeting. Then they spend 2–3 days in windowless conference rooms discussing large numbers of grant applications. Each grant application is only discussed for 15–20 minutes. To top it off, all the reviewers assigned to your application may not be experts in your field. BUT, there is hope--Your best chance for success is to ensure that the reviewers assigned to your application are enthusiastic about and will strongly advocate for your grant application. It is our hope that this chapter will support you to that end.

Recommendations to Increase Your Chances of Success

- Tell a compelling, rational, exciting story in plain language.
- Ask several colleagues unfamiliar with your work to review application before submission to ensure it is easily understood by non-experts.
- Make it easy for reviewers to like your application and advocate for you.
- Use headings that match the review criteria- don't make reviewers search for it.
- Explicitly state how the application is responsive to the Request for Proposals.
- Use white space, figures, and diagrams to break up pages of print.
- Adhere to requirements for formatting, margins, and font.
- Identify strong and meaningful outcomes.

FAQS

What are the most common mistakes that reviewers see in Grant Applications?

- Not aligning the work with the funding organization's/RFP's stated priorities
- Research question or hypothesis that is not exciting/meaningful
- Too many specific aims for the timeframe of the grant
- Highly technical and incomprehensible language
- Specific aims that are interdependent
- Absence of a conceptual framework
- Absence of Limitations/Alternative Methods sections
- Not including specific expertise for the work proposed, especially for statistical analysis
- Promising too much for the time and effort allocated
- Weak outcomes
- Non-compliance with budget requirements
- Absent or inadequate human subjects section

References

1. Merriam Webster Online Dictonary. Available from: https://www.merriam-webster.com/dictionary/grantsmanship. Accessed 22 Jan 2018.
2. On Death and Dying: Elizabeth Kubler-Ross Foundation. Available from: http://www.ekrfoundation.org/five-stages-of-grief/. Accessed 22 Jan 2018.

Additional Resources

National Institute of Nursing Research. Writing a Successful Grant Application. Tully L. Available from: https://www.ninr.nih.gov/sites/www.ninr.nih.gov/files/Module3WritingaSuccessfulGrantApplication.pdf.Accessed 22 Jan 2018.

New Set of R01 Sample Applications: National Institute of Allergy and Infectious Disease. Updated January 11, 2017 New: https://www.niaid.nih.gov/grants-contracts/new-r01-sample-applications. Accessed 22 Jan 2018.

Write Your Application: NIH updated Jan 28, 2016. Available from: https://grants.nih.gov/grants/how-to-apply-application-guide/format-and-write/write-your-application.htm. Accessed 22 Jan 2018.

38 Establishing and Maintaining Multicenter Studies in Healthcare Simulation Research

Travis Whitfill, Isabel T. Gross, and Marc Auerbach

Overview
Multicenter research studies are a robust research tool that—if well-executed—offer a number of benefits over single center studies, including increased sample size, greater generalizability of findings, and shared resources. However, a successful multicenter research study takes significant preparation and execution strategies, many of which have unique considerations in simulation-based research. In this chapter, we offer a framework for designing and executing multicenter simulation-based studies: (a) pre-planning phase (defining the question, conducting pilot work, assembling the team); (b) planning phase (developing protocols, identifying and recruiting collaborators, executing paperwork, disseminating protocols, training sites to comply with protocols); (c) study execution (recruitment, enrollment, quality assurance, compliance); (d) study maintenance (communication, maintenance, consistency); and, (e) data analysis and dissemination (abstracts, social media, manuscripts). This chapter serves as a guide to conducting multicenter, simulation-based research studies with a focus on quantitative research questions.

Practice Points

- It's imperative to understand the problem, ensure that a multicenter approach is indicated, conduct pilot work, and plan your research and logistics before conducting a multicenter study
- Build a multidisciplinary team that complements each other and is an effective group
- Transparency and communication are critical to success
- Consider a "train-the-trainer" model and other ways to ensure quality of methods and data collection at all sites involved
- Beware of IRB/ethics review and plan well in advance

Introduction

Despite the advantages offered by multicenter research studies, a relatively small number of published simulation-based research studies involve a multicenter approach [1, 2]. Single-center studies comprise the majority of simulation-based research studies and are often limited in effect size, strength of findings, or generalizability [3]. Multicenter studies offer a number of important strengths, including larger samples, enhanced generalizability, opportunities for sharing of resources and ideas across institutions and disciplines, and linkage to other data repositories and registries [4–7]. In other fields of medical research, the majority of high-impact publications are the products of multicenter studies [5]. As simulation-based research studies often aim to link simulation to patient outcomes, particularly for rare events, a multicenter approach is needed. However, these larger scale projects involve additional resource requirements that create a variety of challenges related to standardization, funding, and scalability. Additionally, establishing collaborative simulation-based research without sufficient preparation will likely result in a lack of sustainability and wasted resources.

A clear benefit of a multicenter approach is that it often increases the number of study subjects available for recruitment and participation. Thus, a larger study can be conducted in a shorter amount of time, which is especially noteworthy when the expected effect size is small, a larger sample size is

T. Whitfill · I. T. Gross · M. Auerbach (✉)
Pediatrics and Emergency Medicine, Yale University School of Medicine, New Haven, CT, USA
e-mail: travis.whitfill@yale.edu; isabel.gross@yale.edu; marc.auerbach@yale.edu

D. Nestel et al. (eds.), *Healthcare Simulation Research*, https://doi.org/10.1007/978-3-030-26837-4_38

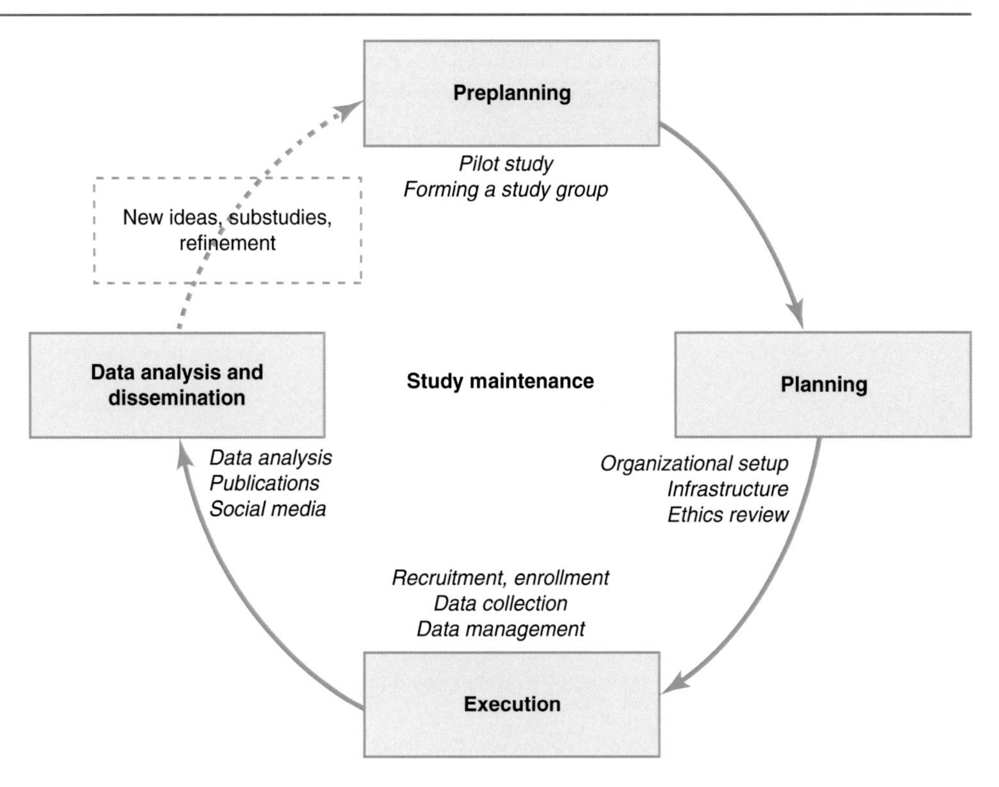

Fig. 38.1 Process of a multicenter simulation-based research study. A multicenter study involves the following phases: pre-planning, planning, study execution, study maintenance, and data analysis and dissemination. This process could be cyclical due to additional study questions, substudies, or new ideas resulting from a multicenter study

needed, or the studied outcome is rare [8]. It is important to note, however, that a larger number of study subjects does not necessarily translate to increased power of the study [9], as other factors such as study design, arms, measurement tools, and effect size heavily impact the power of the study. The design of a multicenter study is a primary determinant of sample size and power [10].

In general, multicenter studies increase the generalizability of the findings due to variations in practice or populations between sites [3, 11]. A well-conducted multicenter study helps increase the generalizability of its findings. This is particularly important in simulation-based studies, where extending the findings to patient outcomes is a goal. Significant findings from a multicenter study—despite inter-site variations in practice, implementation, or training—strengthen its generalizability.

Other secondary benefits come with multicenter studies. Sharing and pooling of resources and expertise across multiple institutions promotes capacity and productivity, while promoting interdisciplinary teamwork and mentorship between senior and junior investigators [2, 4, 12]. It can also spread the required funding responsibilities across institutions.

This chapter will offer a framework for designing and maintaining a multicenter study that is based on the framework described by Cheng et al. [2]:

I. Pre-planning phase (defining the question, conducting pilot work, assembling the team)
II. Planning phase (developing the protocol, identifying and recruiting collaborators, executing paperwork, disseminating protocol, training sites to comply with protocol)
III. Study execution (recruitment, enrollment, quality assurance, compliance)
IV. Study maintenance (communication, maintenance, consistency)
V. Data analysis and dissemination (abstracts, social media, manuscripts).

An overview of this framework is represented in Figs. 38.1 and 38.2 with granular details in Table 38.1.

Pre-planning: Key Considerations Before Choosing a Multicenter Approach

Forming a Study Question

The early process of forming a multicenter study is three-fold: (a) defining a study question; (b) collecting pilot data; and (c) forming a working group. Before embarking on a multicenter study, just as with a single center study, the

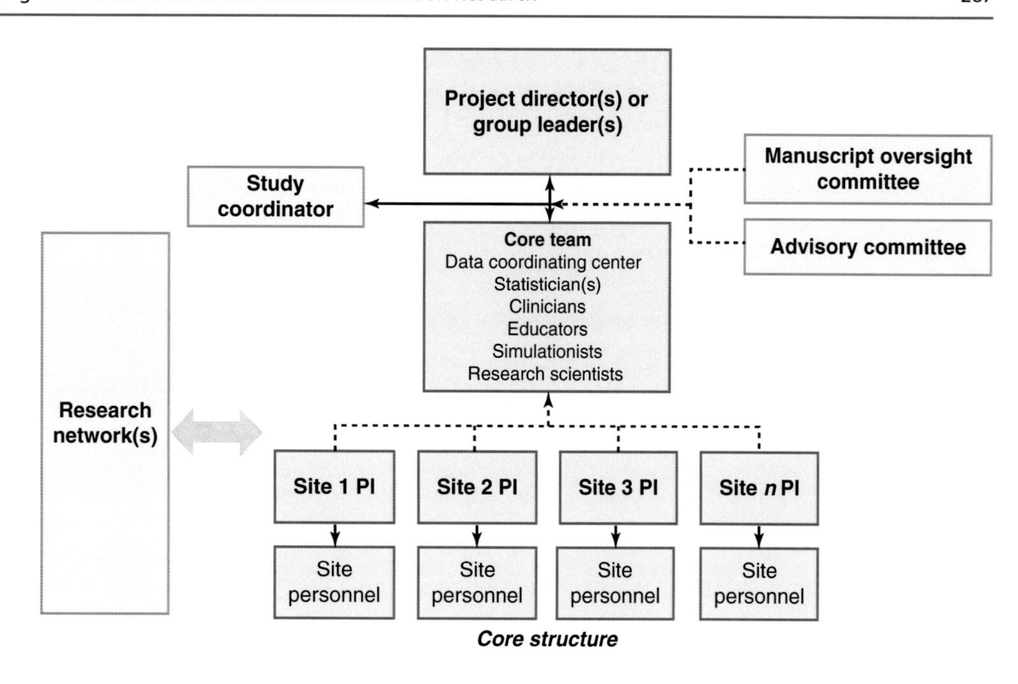

Fig. 38.2 Possible organizational structure of a multicenter study. The essential components of the structure (in blue) of a multicenter study include: The Project Director(s) or group leader(s) who guide study; the core research team, composed of team members who may be affiliated with various sites; and the sites with a PI and key personnel at each site. Non-essential—but beneficial—components of the organization of a multicenter study include: a manuscript (or authorship) oversight committee, advisory committee, coordinator, and possible relationship(s) with research networks

Table 38.1 Phases and activities of executing a simulation-based, multicenter study

Phase	Activities	Considerations
Preplanning	Study question formation Literature reviews Pilot study Study team assembly	Is the study question feasible, interesting, novel, ethical and relevant? How do the pilot data inform the design of a multicenter study?
Planning	Organizational setup Infrastructure Funding Protocol development IRB submissions Training of the study team	Are committees needed? (e.g., manuscript oversight committee) Are responsibilities for site PIs clearly defined before recruiting them? Is the study team multidisciplinary and contain the relevant expertise for the success of the study? Is authorship clearly defined in advance? Is a study coordinator needed? Have various funding sources been explored? How will the study team be trained?
Execution	Recruitment and enrollment Data management and abstraction	Is recruitment consistent across all sites, or are there differences? Who is handling the data—Data Coordinating Center? And how are the data being collected?
Maintenance	Quality assurance Communication Leveraging a multicenter collaborative	Is the core team ensuring data are being collected uniformly across all sites? What is the communication plan, and how is the PD/core team implementing or tracking it?
Analysis and dissemination of results	Data analysis Authorship Publications	Are the data analyses taking into account the unique attributes of multicenter studies? How will the results be disseminated beyond manuscripts and conferences?

study question formation is essential: is the research *feasible* to carry out; is it an *interesting* topic; is the idea *novel* and move the literature forward; is it *ethical* to study; and is the idea *relevant* and timely to the scientific or medical community [13, 14]? Does it impact learner or patient outcomes? Is simulation the appropriate study modality or subject to research the study question? A comprehensive review of the published literature is essential to answering these questions and determining a need for study to be conducted [2, 13], and a systematic literature review is a great first project for a multicenter group to establish a track record of collaboration. Then, outcome measures and study tools are carefully selected, with particular consideration for the validity evidence of tools used in context of the study question.

Pilot Data

In most cases, conducting a pilot study in a single institution is essential before proceeding to a multicenter study [15]. The benefits of this include:

I. Ensuring that the study question and research warrant a multicenter study
II. Collecting pilot data for later power analyses and sample size calculations
III. Refining the study protocol
IV. Guiding participant recruitment
V. Refining the analytical plan and data collection
VI. Assessing feasibility and projecting costs
VII. Estimating generalizability.

Outcomes and lessons learned in the pilot study should be integrated into the design of the multicenter study. Importantly, in conjunction with early buy-in from a statistician, the pilot study will inform the power calculation of the sample size of the multicenter study [14], and protocol deviations or recruitment difficulties seen in the pilot study will be need to be incorporated into this calculation [16].

Forming a Study Team

Second, in the early process of forming a multicenter study investigators will form a study team composed of individuals with expertise in the content area from within and outside of their own discipline. A study team can serve as the early basis for key collaborators on the multicenter study and will give early buy-in and feedback on the study question and study design. Many research networks exist that help facilitate working groups within a larger network. Simulation-based research networks in particular are an effective way to pool expertise and facilitate working groups and collaboration that form a breeding ground for multicenter studies. For example, the INSPIRE network (International Network for Simulation-Based Pediatric Innovation, Research and Education) is a pediatric simulation-focused research network with the goal of facilitating multicenter, collaborative, simulation-based research [17]. A number of multicenter studies have been generated from this network [18–22].

Forming a Study Timeline

The Project Director (PD) and study team should develop a timeline for each of the phases below. It is important to have deadlines for all these phases to help ensure study progression. Without clearly defined deadlines, many studies fail to progress through these phases.

Planning

Once a multicenter approach is chosen as the appropriate study type, extensive planning is required before beginning the study. Planning involves several key activities: (a) organizational setup; (b) infrastructure; (c) funding; (d) protocol development; (e) ethics approval; and (f) training of the study team.

Organizational Setup

Site selection and team composition are vital to a successful multicenter study. Team members of a multicenter study could include clinical investigators, educators, simulation specialists, statisticians, and non-clinical research scientists. A single individual PD or a small group should be clearly identified as the team leader(s). Sharing leadership between a more junior team leader and a senior mentor may also be an effective strategy. An advisory committee, such as a manuscript oversight committee or steering committee structure, can also be used. A proposed organizational setup is laid out in Fig. 38.1.

The PD is responsible for bringing the team together in the planning phase. He or she often has a clinical background and facilitates engagement of the other team members. The educators, statisticians and non-clinical team members may be affiliated with the PD's institution or a different study center. At each participating center, the PD must identify a site leader, or site Principal Investigator (PI).

It is important to articulate clear expectations to potential sites. This includes site PI's roles, responsibilities, timeline, and time commitment. An explicit description of the study protocol must be balanced with efforts to engage collaborators on the research team. We recommend disseminating the protocol with collaborators to request feedback early.

When inviting collaborators, we suggest being clear about why you are inviting them and their role in the project. Authorship rules and roles should be established early and should be clearly defined. This is in addition to discussions about author inclusion where guidelines from the International Committee of Medical Journal Editors (ICMJE) should be followed [23]. We propose identifying the primary authors (first, second and last) of the proposed academic output as early as possible. The other collaborators should be given clear information as to the expectations that they must meet to achieve authorship status. This can include attendance at a set number of study meetings, enrollment of a minimum number of teams or other explicit contributions. A manuscript oversight committee within the organizational setup is very helpful to ensure ethical authorship standards are enforced throughout the research study process—especially if multiple academic outputs are produced from the research.

It is also very important to involve experts from other disciplines. For example engage a statistician early in the process to provide statistical support and an analytical plan a priori to maximize likelihood of success in a multicenter study, particularly given the statistical complications of a multicenter versus single center study. Other examples could include psychometricians, educators, human factors engineers, simulation technicians, and psychologists. Increasingly, patients and lay stakeholders are participating in research processes as active contributors to what should be investigated.

Infrastructure

Infrastructure is key to scaling a multicenter study. The primary site's and auxiliary sites' capabilities, simulation facility, equipment, staffing and administrative support, and data collection platforms may serve as an initial foundation for the infrastructure. This will enhance the consistency of the study and strengthen collaboration with simulation technicians.

The coordination or administration of a multicenter study will require a large time investment. This work should be accomplished by the PD with the support of a study coordinator. This often will require funding—although an alternative strategy is the opportunity for authorship for this individual, which may be enough to recruit someone to this role when funds are not available to pay him or her. The coordination involves developing a project plan, a contact list, roles and responsibilities, timelines, and the Institutional Review Board (IRB) or other human research ethics process.

Funding

Funding for a multicenter study is necessary to support the infrastructure, including administrative needs, communication platforms, and meetings and travel [24]. Funding sources and agencies may vary and can include federal sources, internal sources, or non-profit organizations. An often overlooked benefit of preparing a grant submission is forcing investigators to carefully detail the methodology and meticulously plan out the research strategy, which—even if the project does not get funded immediately—often strengthens the overall research plan.

Protocol Development

Thorough protocols serve as the backbone of the study and ensure consistency across multiple sites. Protocol development needs to be done carefully to avoid methodological mistakes before the study begins. The involvement of a diverse team will help optimize the study design and avoid mistakes in the data collection before the start of the study. Lessons from non-simulation-based, multicenter studies have revealed that study design, data management, and data analysis need to be planned and considered carefully to avoid biases and to facilitate generalizability [14]. Additionally, one may consider publishing protocols to ensure thorough review and standardization of all facets of the research.

As part of the protocol development process, authorship and institutional credit for the multicenter involvement should continue to be discussed during this process. An ongoing commitment and willingness to collaborate seem to be two of the most important factors to facilitate success and productivity of multicenter studies [25].

In multicenter simulation-based research, standardization of the simulation environment and the specific intervention(s) must be well defined and described in the protocol [2]. In addition, the protocols for data acquisition are important to lay out in advance. A manual of operations and statistical analysis plan must be included in the protocol to provide operational definitions for data elements, expectations for data quality, data management, and data analysis.

Institutional Review Board or Ethics Approval

After developing the final protocol, the study team must submit for IRB or human research ethics review. While there are a number of central IRBs or human research ethics boards that are being used in clinical research [26], we find that the best strategy for a simulation-based study is to have the primary site complete the review first and each individual site submit for review subsequent to the primary site's approval. Depending on the nature of the simulation study, the review process can vary from an exempt protocol that relates to standard educational practices to a full board review when patients with protected health information are involved. We strongly encourage that the site investigators have a discussion with their local board prior to submission and that the PD provide a timeline and his or her approved protocol for this discussion. The PD should recognize that this process is variable and provide ample time for completion of approval at each site.

If the authors are conducting a randomized trial, many journals and funding agencies require registration of the study protocol on a site such as clinicaltrials.gov or publication of the protocol in a peer-reviewed journal. This must be done in advance of beginning the study.

Training of the Study Team

Training the study team is key to maintaining consistency in a multicenter, simulation-based study and should be conducted before the study begins. Training of each study site PI and team through in-person or remote train-the-trainer sessions may help ensure compliance. This can include orientation to the simulation methods, the data entry approach, and expectations for ancillary factors. Explicit details on the environment of the simulation event including the simulator, environment, and actors can be augmented by using prescripted and preprogrammed scenarios. An example of training for confederates in a simulation-based study is described in Adler et al. [27].

Compliance with the protocol must be audited over time. One strategy is to have a research coordinator or investigator travel to each site to observe the sessions. Recent simulation reporting guidelines describe important elements for standardization, including participant orientation, simulator type, simulation environment, simulation event and scenario, instructional design, and debriefing [27].

Data Management and Abstraction

If working with quantitative data in a multicenter study, data management may be challenging but essential to the findings of the study. A variety of methods can be used for data acquisition. We recommend a centralized approach to data management, where the lead institution—or a single distinct institution—collects, stores and manages all data and serves as the Data Coordinating Center (DCC). The DCC assumes responsibility for collecting data from all participating sites, ensures quality of the data, manages data collected locally, monitors study sites, and performs statistical analyses. Several practical data management systems exist, but a commonly used research tool is REDCap (Research Electronic Data Capture) [28].

Rater orientation and calibration is needed when using performance checklists locally. When possible, centralized rating can allow for blinding and improved reliability. Inter-rater reliability calculations are crucial when multiple raters are used, and should be calculated for a subset of scores.

Study Execution

After extensive pre-planning and planning, the study execution phase is an opportunity to begin the actual multicenter research. Many components and moving parts comprise this phase, but extensive planning will help execute the study appropriately.

Recruitment and Enrollment

The strategies for recruitment and enrollment will depend on the study protocol. Some protocols involve each site enrolling a set number of subjects or teams, while other protocols will have a minimum number of subjects or teams to participate. The strategy towards recruitment and enrollment will have to be outlined in the ethcs application. It is also important to describe a deadline or timeline for recruitment and what will happen if a site cannot achieve this (e.g., will not be included in the study). Some studies will recruit backup sites that will only get involved should another site not be able to complete their enrollment.

Selection bias may result from certain participants self-volunteering in the study [8]. Clear inclusion and exclusion criteria are needed to ensure similar participants across study sites. In addition, randomization can ensure participant characteristics are similar between study groups. Block randomization can ensure equal allocation of study groups within sites [2].

Study Maintenance

A commonly overlooked aspect of multicenter studies is their maintenance, which may take place over the course of many months or years, and require active management of and engagement with all study sties. Maintenance requires significant time and resources, but is critical to the success and quality of the study.

Quality Assurance

Quality assurance helps the investigators to prevent problems and is an ongoing process for the duration of the study. While local empowerment and ownership of sites that are part of multicenter studies is vital for their success. It is very important to ensure the quality and comparability of the collected data and adherence to the research protocol across institutions. There are three key components for quality assurance [29]. The first component is prevention and can be achieved by a well-written protocol, investigator training, and site visits. Secondly, central data monitoring, data review and statistical investigations help detect quality concerns throughout the study. Lastly, actions need to be taken to correct errors and protocol violations need to be reported in the publication. Centralized monitoring, analysis and storage of data and video documentation of the local simulation study components are helpful [2].

Communication

Communication is critical to the success of a multicenter study, and involves deliberate, planned interactions between the PD, core team, and study sites. Face-to-face interaction is also highly encouraged, although sometimes it may not be feasible. Often times study investigators can convene at an academic conference, which can serve as an opportunity to meet about the study project. Web-conferencing platforms and group conference lines can provide for ongoing interactions among the entire study team or also serve to facilitate smaller working groups within the broader study team. Concurrently, the PD or project coordinator should make efforts to have regular communication with individual investigators (through email or phone). Additionally, regular updates on the study progress for the entire team and site-specific measures are helpful through an email or newsletter. Collaborative work sites and file sharing sites are frequently used to create a virtual work environment for team members to interact asynchronously—although data must be protected if stored on a shared server.

Analysis and Dissemination of Results

Data Analysis

A key aspect of data analysis unique to multicenter studies is correlation of the data. This is because study participants from the same center are more similar to each other than are those from different centers [10]. If teams participated in the study within each site, this adds another layer of data clustering. These can all be addressed statistically, e.g., generalized estimating equations or mixed-effects models that can both account for data clustering. Methods for analyzing data from multicenter studies have been well described in the literature [30].

Missing data are common in simulation-based studies. For example, some sites or teams may experience technical difficulties and may not be able to videotape the simulated session. Other times, study participants may not consent to videotaping. All of these should be thoroughly addressed, and the data should be assessed if the missing values are missing at random (MAR) (or completely at random [MCAR]) or not [31]. Data imputation can be used for MAR or MCAR values, which can include deletion or imputation [32]. Imputation can include simple replacement with central tendency (i.e. mean or median) or with regression or multiple imputations. If the data are not missing at random, deletion or imputation is not recommended, as this may bias the study results. In all of these cases, the statistician will be able to guide the team through the nuances and issues of data analysis.

Publications and Authorship

A strategic dissemination plan is needed for any research project. Beyond conference abstracts and peer-reviewed manuscripts, websites and social media may help increase dissemination and promote idea sharing [33]. As we mentioned earlier, authorship must be considered as early as possible and discussed on an ongoing basis. Credit and accountability for all authors and investigators is a cornerstone of successful collaborations. If the research group has a name, it may be acknowledged on the byline of the publications.

Multiple work products may be generated from a multicenter study. With the large number of investigators, teams should consider dividing the authorship of additional work products amongst the team. Additionally, the team should consider ongoing adaptations to authorship based on individuals who could not complete their assigned role (e.g., if lead author of a paper does not complete draft by a deadline the second author could become the lead). These issues are often overseen or resolved by the Manuscript Oversight Committee. We suggest that these ancillary manuscripts and abstracts are submitted after the primary work has been completed and published.

Leveraging a Multicenter Collaborative

Once the study is complete, a consideration for the team is what to do next. This team has an infrastructure that supported a project and can be used for additional projects. As such, the team can consider shifting the leadership structure for subsequent projects and may want to expand or contract the number of sites based on the initial project experience. The group can also consider leveraging this process into the formation of a research network (this is the origin of the INSPIRE network [34]).

Challenges in Multicenter, Simulation-Based Studies

Scalability, sustainability, and feasibility of multicenter studies and programs are common challenges . The PD and team do noy need to be physically present to conduct the study in the different centers [35]. Close collaboration with the PD will provide support while multi-directional collaboration and networking between centers can help improve the quality of the study and support local solutions and ideas [36]. Standardized train-the-trainer programs can help ensure compliance with the protocol at each site. While local cultural variations and local empowerment are key, quality assurance is important to prevent heterogeneity in the collected data as well as diluted assessor orientation to simulation research tools.

Tele-simulation can be a cost-effective and convenient tool to both initiate train-the-trainer efforts as well as to support sustainability and quality assurance of multicenter simulation research studies. This is particularly important in resource-limited settings including regions in resource limited countries or smaller centers in the United States that have limited resources for simulation-based research. Tele-simulation can be conducted with different levels of involvement of the trainer: The trainer can run the local simulator remotely via software and function as a facilitator during the simulation and debrief via audio and video transmission. The trainer can also observe and assist the learner during facilitation and the debrief or even observe the tele-simulation and give the learner feedback in form of a meta debrief at the end of the simulation [37]. Although tele-simulation can help decrease resources and in-person time provided by experts, it can cause challenges including technical difficulties as well as a distance to the local learner [38].

Conclusions

Multicenter studies require significant resources and time for pre-planning, planning, execution, maintenance, and dissemination. Multicenter studies offer significant advantages over single center studies including larger samples, enhanced generalizability, opportunities for sharing of resources and ideas across institutions and disciplines, and linkage to other data repositories and registries. Multicenter studies are necessary for simulation-based research as we work to advance our field and leverage simulation to improve patient outcomes.

References

1. Cook DA, Hatala R, Brydges R, Zendejas B, Szostek JH, Wang AT, et al. Technology-enhanced simulation for health professions education: a systematic review and meta-analysis. JAMA. 2011;306(9):978–88.
2. Cheng A, Kessler D, Mackinnon R, Chang TP, Nadkarni VM, Hunt EA, et al. Conducting multicenter research in healthcare simulation: Lessons learned from the INSPIRE network. Adv Simul (London, England). 2017;2:6.
3. Bellomo R, Warrillow SJ, Reade MC. Why we should be wary of single-center trials. Crit Care Med. 2009;37(12):3114–9.
4. Schwartz A, Young R, Hicks PJ. Medical education practice-based research networks: facilitating collaborative research. Med Teach. 2016;38(1):64–74.
5. Payne S, Seymour J, Molassiotis A, Froggatt K, Grande G, Lloyd-Williams M, et al. Benefits and challenges of collaborative research: lessons from supportive and palliative care. BMJ Support Palliat Care. 2011;1(1):5–11.
6. O'Sullivan PS, Stoddard HA, Kalishman S. Collaborative research in medical education: a discussion of theory and practice. Med Educ. 2010;44(12):1175–84.
7. Huggett KN, Gusic ME, Greenberg R, Ketterer JM. Twelve tips for conducting collaborative research in medical education. Med Teach. 2011;33(9):713–8.
8. Cheng A, Auerbach M, Hunt EA, Chang TP, Pusic M, Nadkarni V, et al. Designing and conducting simulation-based research. Pediatrics. 2014;133(6):1091.
9. Vierron E, Giraudeau B. Design effect in multicenter studies: gain or loss of power? BMC Med Res Methodol. 2009;9:39.
10. Localio AR, Berlin JA, Ten Have TR, Kimmel SE. Adjustments for center in multicenter studies: an overview. Ann Intern Med. 2001;135(2):112–23.
11. Sprague S, Matta JM, Bhandari M, Dodgin D, Clark CR, Kregor P, et al. Multicenter collaboration in observational research: improving generalizability and efficiency. J Bone Joint Surg Am. 2009;91(Suppl 3):80–6.
12. Payne S, Seymour J, Molassiotis A, Froggatt K, Grande G, Lloyd-Williams M, et al. Benefits and challenges of collaborative research: lessons from supportive and palliative care. BMJ Support Palliat Care. 2011;1(1):5.
13. Hulley SBCS, Browner WS. Designing clinical research: an epidemiologic approach. 2nd ed. Philadelphia: Lippincott Williams and Wilkins; 2001.
14. Chung KC, Song JW. A guide to organizing a multicenter clinical trial. Plast Reconstr Surg. 2010;126(2):515–23.
15. van Teijlingen E, Hundley V. The importance of pilot studies. Nurs Stand. (R Coll Nurs (Great Britain): 1987). 2002;16(40):33–6.
16. Lancaster GA, Dodd S, Williamson PR. Design and analysis of pilot studies: recommendations for good practice. J Eval Clin Pract. 2004;10(2):307–12.
17. Cheng A, Auerbach M, Calhoun A, Mackinnon R, Chang TP, Nadkarni V, et al. Building a community of practice for researchers: the international network for simulation-based pediatric innovation, research and education. Simul Healthc J Soc Simul Healthc. 2017;13:S28–34.
18. Kessler D, Pusic M, Chang TP, Fein DM, Grossman D, Mehta R, et al. Impact of just-in-time and just-in-place simulation on intern success with infant lumbar puncture. Pediatrics. 2015;135(5):e1237–46.
19. Kessler DO, Walsh B, Whitfill T, Dudas RA, Gangadharan S, Gawel M, et al. Disparities in adherence to pediatric sepsis guidelines across a spectrum of emergency departments: a multicenter, cross-sectional observational in situ simulation study. J Emerg Med. 2016;50(3):403–15. e1-3
20. Auerbach M, Whitfill T, Gawel M, Kessler D, Walsh B, Gangadharan S, et al. Differences in the quality of pediatric resuscitative care across a spectrum of emergency departments. JAMA Pediatr. 2016;170(10):987–94.
21. Kessler DO, Peterson DT, Bragg A, Lin Y, Zhong J, Duff J, et al. Causes for pauses during simulated pediatric cardiac arrest. Pediatr Crit Care Med J Soc Crit Care Med World Fed Pediatr Intensive Crit Care Soc. 2017;18(8):e311–e7.
22. Cheng A, Brown LL, Duff JP, Davidson J, Overly F, Tofil NM, et al. Improving cardiopulmonary resuscitation with a CPR feedback device and refresher simulations (CPR CARES Study): a randomized clinical trial. JAMA Pediatr. 2015;169(2):137–44.
23. ICMJE. Defining the Role of Authors and Contributors [Available from: http://www.icmje.org/recommendations/browse/roles-and-responsibilities/defining-the-role-of-authors-and-contributors.html Accessed 2019.
24. Schwartz A, Young R, Hicks PJ, Appd Learn F. Medical education practice-based research networks: facilitating collaborative research. Med Teach. 2016;38(1):64–74.
25. Hogg RJ. Trials and tribulations of multicenter studies. Lessons Learn Experiences Southwest Pediatr Nephrol Study Group (SPNSG) Pediatr Nephrol. (Berlin, Germany. 1991;5(3):348–51.
26. Mascette AM, Bernard GR, Dimichele D, Goldner JA, Harrington R, Harris PA, et al. Are central institutional review boards the solution? The national heart, lung, and blood institute working group's report on optimizing the IRB process. Acad Med J Assoc Am Med Coll. 2012;87(12):1710–4.

27. Adler MD, Overly FL, Nadkarni VM, Davidson J, Gottesman R, Bank I, et al. An approach to confederate training within the context of simulation-based research. Simul Healthcare J Soc Simul Healthc. 2016;11(5):357–62.
28. Harris PA, Taylor R, Thielke R, Payne J, Gonzalez N, Conde JG. Research electronic data capture (REDCap)—a metadata-driven methodology and workflow process for providing translational research informatics support. J Biomed Inform. 2009;42(2):377–81.
29. Knatterud GL, Rockhold FW, George SL, Barton FB, Davis CE, Fairweather WR, et al. Guidelines for quality assurance in multicenter trials: a position paper. Control Clin Trials. 1998;19(5):477–93.
30. Austin PC. A comparison of the statistical power of different methods for the analysis of cluster randomization trials with binary outcomes. Stat Med. 2007;26(19):3550–65.
31. Bhaskaran K, Smeeth L. What is the difference between missing completely at random and missing at random? Int J Epidemiol. 2014;43(4):1336–9.
32. Taljaard M, Donner A, Klar N. Imputation strategies for missing continuous outcomes in cluster randomized trials. Biom J Biometrische Zeitschrift. 2008;50(3):329–45.
33. Rolls K, Hansen M, Jackson D, Elliott D. How health care professionals use social media to create virtual communities: an integrative review. J Med Internet Res. 2016;18(6):e166.
34. Cheng A, Auerbach M, Calhoun A, Mackinnon R, Chang TP, Nadkarni V, et al. Building a community of practice for researchers: the international network for simulation-based pediatric innovation, research and education. Simul Healthc J Soc Simul Healthc. 2018;13(3S Suppl 1):S28–s34.
35. Yarber L, Brownson CA, Jacob RR, Baker EA, Jones E, Baumann C, et al. Evaluating a train-the-trainer approach for improving capacity for evidence-based decision making in public health. BMC Health Serv Res. 2015;15(1):547.
36. Madah-Amiri D, Clausen T, Lobmaier P. Utilizing a train-the-trainer model for multi-site naloxone distribution programs. Drug Alcohol Depend. 2016;163:153–6.
37. Hayden EM, Navedo DD, Gordon JA. Web-conferenced simulation sessions: a satisfaction survey of clinical simulation encounters via remote supervision. Telemed J E-health Offic J Am Telemed Assoc. 2012;18(7):525–9.
38. Hayden EM, Khatri A, Kelly HR, Yager PH, Salazar GM. Mannequin-based telesimulation: increasing access to simulation-based education. Acad Emerg Med. 2018;25(2):144–7.

Supervision in Healthcare Simulation Research: Creating Rich Experiences

39

Debra Nestel, Andree Gamble, Grainne Kearney, and Gerard J. Gormley

Overview
Supervision is an important facet of research. Highly relational in nature, it has the potential to provide rich experiences for all involved. The supervisory relationship has an important role in stewarding standards of research practice and achieving shared goals. In this chapter, we focus on formal supervisory relationships between higher degree research students and their supervisors. Although there are variations in these practices globally, we believe that a focus on this relationship may have universal relevance. We draw on our experiences in healthcare simulation research sharing perspectives from supervisors and their graduate students.

Practice Points

- Supervision in healthcare simulation research is an important professional activity for scholars in this field.
- There is wide variation in the nature of the supervisory role and standards of practice expected of supervisors and student researchers.
- Supervisors are encouraged to undertake professional development activities to best prepare them for their role.
- Published approaches to supervision have shifted in their locus of control with students rather than supervisors increasingly positioned as 'driving' their research.
- The quality of the supervisory relationship has a profound impact on shaping future practices. This relationship is a key driver through which ethical research and writing practices are developed.

D. Nestel (✉)
Monash Institute for Health and Clinical Education, Monash University, Clayton, VIC, Australia

Austin Hospital, Department of Surgery, Melbourne Medical School, Faculty of Medicine, Dentistry & Health Sciences, University of Melbourne, Heidelberg, VIC, Australia
e-mail: debra.nestel@monash.edu; dnestel@unimelb.edu.au

A. Gamble
Nursing Department, Holmesglen Institute, Southbank, VIC, Australia

Faculty of Medicine, Nursing & Health Sciences, Monash University, Clayton, VIC, Australia
e-mail: Andree.gamble@holmesglen.edu.au; andree.gamble@monash.edu

G. Kearney · G. J. Gormley
Centre for Medical Education, Queen's University Belfast, Belfast, UK
e-mail: g.kearney@qub.ac.uk; gkearney03@qub.ac.uk; g.gormley@qub.ac.uk

Introduction

> A quality supervisory relationship goes beyond knowledge transfer and institutional protocols to foster norms and expectations that enable supportive processes of knowledge production. [1]

Our focus in this chapter is on the supervision of researchers undertaking higher degrees. Although there is much published about supervision, this work often focuses on the functional, practical elements of this important role [2–6]. There seems to be an ever-increasing list of supervisory tasks and checklists to complete as universities tighten governance and other procedures, including funding arrangements associated with research-based higher degrees. However, there are parallel changes with the student's role also now more clearly articulated; with them positioned as 'driving' their research. As the quote above implies, supervision is about a working-relationship and much more than the production of a thesis – it may also have a profound influence on future research practices and students' career choices. Although successful completion of a thesis is an obvious purpose of the supervisory relationship, there are many elements of research processes that are also acquired during this period.

The supervisory relationship is also situated in a socio-political environment, which influences variously the

D. Nestel et al. (eds.), *Healthcare Simulation Research*, https://doi.org/10.1007/978-3-030-26837-4_39

progression (or not) of the research and of the student researcher. Supervision has been identified as influencing degree completion, length of candidacy, student well-being and satisfaction with the overall doctoral experience as well as competencies developed while studying [7]. While this chapter sits in a book on healthcare simulation research, the content likely has relevance in other settings as well. However, local conditions and contexts will influence relevance.

First, we share a little about our own experiences before considering the broader topic of the research supervision. The authors have experienced a range of supervisory relationships and in different institutions. DN completed her doctorate in Hong Kong – a mixed methods study on simulated patient methodology. DN is currently co-supervising AG in her PhD exploring ethical issues associated with children and adolescents as simulated patients. GG completed his doctorate in Belfast – exploring the use of simulation to enhance general practitioner abilities to perform joint and soft tissue injections. GG is currently co-supervising GK in her PhD an inquiry of Objective Structured Clinical Examinations (OSCEs) using an approach called Institutional Ethnography. Currently DN is supervising 12 and GG 7 graduate research students. The content of these research projects is mainly health professions education, using qualitative or mixed methods research designs. While drawing on current literature, we also share our personal experiences to make meaning and sense of research findings. It is also important to note that supervisors' personal experiences of their doctorate supervision can have a profound impact on the way in which they supervise [4].

We begin by reporting ways in which supervisory relationships are characterised – approaches and a taxonomy – and consider the important roles of forming supervisory relationships and co-supervision. We report on quality supervisory practices from supervisor and student perspectives. We list steps to supervision from getting started, through the supervision and on completion before focusing on two key issues in supervisory practices – learning about ethics and writing.

Supervisory Relationships

The research supervisory relationship has been extensively studied. There are classifications of approaches to supervision such as that by Lee [4, 8]. These approaches are likely to have different weightings across a research project. Awareness of these approaches may be helpful for supervisors and students to reflect on their current relationship.

1. "Functional – where students' projects are managed
2. Enculturation – where students are encouraged to become members of the disciplinary community
3. Critical thinking – where students are encouraged to question and analyse their work
4. Emancipation – where students are encouraged to question and develop themselves
5. Developing a quality relationship – where students are enthused, inspired and cared for" [5]

Box 39.1 offers DN and AG's descriptive reflections of their supervisory relationship relevant to these approaches [5].

Box 39.1 Descriptive reflections on approaches to supervision by supervisor (DN) and student researcher (AG) at the midway point of a PhD

Approach to supervision	Descriptive reflections
Functional	The doctoral programme in the Faculty of Medicine, Nursing and Health Sciences (FMNHS) at Monash University positions the student as manager of most aspects of their progression including the arrangement of milestone activities. FMNHS Graduate School administration processes automatically generate emails to supervisors and students. These emails outline the requirements of milestones (oral and written reports) and provide a portal through which milestone reports are submitted to the panel after supervisor review. Although assessment responsibilities sit with the supervisors (e.g. identification of panellists) the logistics of such activities (liaising with panellists, logistical arrangements) sit with the student while the Graduate School monitors overall governance. AG is also responsible for completing Human Research Ethics Committee (Institutional Review Board) annual reports, coordinating all monthly supervisory meetings, setting meeting agendas, and writing notes for sharing with supervisors
Enculturation	DN has facilitated AG into the academic and practitioner community of simulated patients (SP)methodology – nationally (Simulation Australasia annual congress, especially participation in their SP special interest group; Enabling connections with colleagues to facilitate recruitment of study participants; Enabling participation in the development of resources for a national faculty development programme for simulation educators – NHET-Sim – www.nhet-sim.edu.au); and, internationally (through personal introductions to colleagues). DN has fully supported AG's initiatives to participate in local activities (Victorian Simulation Alliance; SP Network – running workshops, membership, attending activities). DN has also encouraged AG to attend and/or present at both national and international conferences in order to disseminate research and develop additional collegial relationships

Approach to supervision	Descriptive reflections
Critical thinking	From the outset, AG has been encouraged by her co-supervisors to think critically about all facets of her research. Key areas have included analysis of current literature in preparation for a systematic review, selection of suitable theoretical frameworks (on 'power') and methodology (phenomenology). AG has been directed to various readings and often tangential ones in an effort to justify research decisions. Given the content of power and ethics in the subject of the thesis, this has also been an area for critical thinking. The development of critical thinking has also ensued as a result of co-supervisors differing perspectives with final decisions often resting with AG after a reflective evaluation of both inputs. The provision of critical feedback also provides opportunity for analysis and deep thinking designed to improve the quality of both research and writing outputs
Emancipation	Having successfully passed mid-candidature, published one peer reviewed paper (systematic review), undertaken data collection for a major part of the thesis, analysed data for two manuscripts, AG is now encouraged to offer and justify directions in her research. Because there is co-supervision and both supervisors are committed to the research project, both supervisors make contributions to the shaping of the manuscripts. However, the framing of the manuscripts as a body of work comprising the thesis is the 'total' responsibility of AG. This level of emancipation, and subsequent increase in responsibility for significant decisions, has been challenging for AG. The ability of supervisors to recognize varying levels of confidence, and offer continued support and guidance whilst also encouraging independence, is a crucial part of the emancipatory process
Developing a quality relationship	The research team aims to meet monthly and over almost 4.5 years have developed a warm collegial and supportive relationship. This encourages AG to further develop with the knowledge that both supervisors are available and committed to the project. The relationship also offers opportunities to consider career development beyond the PhD although our priority is currently with important outputs of the research

Similarly, a taxonomy of five supervisory practices has been proposed by Halse and Malfroy [3].

1. A learning alliance – Agreement between supervisor and student on a common goal
2. Habits of mind – Capacity to learn, reflect and make decisions in an ethically appropriate way
3. Scholarly expertise – Theoretical knowledge of the discipline of study
4. Technê – Creative and productive use of expert knowledge and skill
5. Contextual expertise – 'Know how' of the discipline and setting

In Box 39.2, GG and GK share their experiences of their supervisory relationship through the lens of this taxonomy [3].

Box 39.2 Descriptive reflections on a taxonomy of supervisory practices by supervisor (GG) and student researcher (GK)

Approach to supervision	Descriptive reflections
A learning alliance	GG and GK' s learning alliance started early with informal and open chats prior to the official commencement of her doctoral work, encouraging an atmosphere of shared engagement from the outset. In time, they set more a formal agreement towards achieving the common goal of producing a doctoral thesis they would both be proud of whilst recognising that the development of both of their research interests and abilities was of equal importance. They agreed where the various responsibilities lay within the alliance for example it is GK's responsibility to organise the times for their meeting, write up the summary and action points and to circulate these to the entire supervisory team. This alliance, founded on the mutual respect which has grown as GG and GK have worked together, is also necessarily flexible at times depending on commitments both inside and outside work, allowing for more informal discussions when this is what's required. Goals can and have changed through this process, GG and GK have reappraised when required and adapted to these changes in order to promote and protect each other's interests
Habits of mind	As an experienced researcher, GG knows only too well that research projects generally never follow a linear smooth course. Challenges, hurdles and curve balls have a habit of unexpectedly presenting themselves in the pathway of research and PhDs!. From difficultly in recruiting research participants to 'why am I doing this PhD?' – we've all been there! GG draws on his varied experiences as a researcher, supervisor, mentor and even clinician in order to help develop GK's ability to reflect and engage with her research and to navigate these trials and tribulations. Though study, discussion and feedback, GG has supported GK to see these moments as learning opportunities. GG encourages GK equally to draw on her "lived knowledge" in her previous academic and clinical work, in order to help take ownership of, plan and deliver this research project. For both GG and GK, this support is underpinned with values (e.g. research ethics) and standards (e.g. research governance) – both in actions and words, advocating for social justice through research for the benefit of learners, educators and of course, patients

Approach to supervision	Descriptive reflections
Scholarly expertise	GG has long been involved in the topic of OSCEs through active engagement in programmatic research, joining important conversations, numerous publications and on the ground convening. This has afforded him a deep, substantive knowledge in this area and an ongoing desire to advance this expertise. GG's enthusiasm for this area has inspired GK to broaden her knowledge base on OSCEs, and their process of continuous dialogue and reflection has encouraged her to think independently and critically. As they look at the delivery of OSCEs both on the ground and how they are organised from above, GG and GK aim to make the familiar world of OSCEs strange. As a research team, GK and GG have successfully published two articles in peer reviewed journals stemming from this research
Technê	Those new to qualitative research are often challenged by the many new terms and concepts. Listening to qualitative researches discuss their work can at times seem like a foreign language. From epistemology to ontology; from reflexivity to positioning – research students (GK included) often struggle when grappling with these new terms. As GK's knowledge and confidence grew, GG encouraged her to develop adaptability in presenting her work, depending on her audience which varied from those leaning more towards positivist, non-qualitative research through to those versed in Institutional Ethnography and other critical approaches. This was a useful exercise as it supported GK towards future dissemination of their research to participants, policy makers, practitioners, researchers and academics
Contextual expertise	With regards to the knowledge and understanding of the institutional and disciplinary context of a doctoral study, GG having supervised many postgraduate students both within the institution where he works and outside, has developed knowledge of the regulations and requirements of these institution; in addition to access to resources and administrative support when required. With 18 years of research experience in medical education, he has an understanding of the dynamic disciplinary requisites and expectations through engagement with this community at conferences and his various roles in peer reviewed journals. In addition, GG's connections in research networks facilitated many interesting introductions for GK! Moreover, the various facets of contextual expertise allow GG to see value in the work that he and GK are involved in, promoting a collective sense of pride and ambition

Getting Connected

Having made a decision to undertake a research-based higher degree, identifying a supervisor is a key step in getting started. Some might argue it is the most important step! The process of finding a supervisor (or of finding a student) will depend on the setting in which the research is undertaken and may depend on local regulatory issues. Sometimes there may be little choice while in others, there can be an extensive process of selection. Supervisors and students may be connected through expertise in content or methods, through workplace or other institutional contexts. Whatever the process of connecting, it is important to acknowledge the potential power dynamics within the relationship. The power base may reside with either the supervisor or the student. For example, the Simulation Centre Director is a clinician who has shifted towards an important educational role in her career. One of the simulation centre faculty members has a PhD in education and a faculty appointment at the University where the student (Centre Director) is enrolled in their degree. Acknowledging these relationships and potential complexities at the outset of the relationship and regularly checking-in may identify emerging problems early. Periodically checking-in to ensure adherence to university policies is also a crucial element of the supervisory relationship. Ensuring both supervisors and students are aware of their responsibilities and intermittent reporting requirements helps to ensure a more seamless PhD journey. It is also important to declare potential conflicts of interest. These are usually clearly stated in university regulations and often times can be managed once declared and procedures followed. Although described for a different setting, Schmutz et al. [9] describe approaches to team reflexivity. It may be worth considering their application in supervisory relationships [9].

Co-supervision

At the doctoral level, it is common to have two or more supervisors. Co-supervision has benefits but can also come with challenges. Students may struggle navigating both the different perspectives of their supervisors and in some instances, the relationship between the supervisors too. It is important that supervisors check in with each other about how 'co-supervision' is progressing from supervisors and student perspectives. One of the authors (DN), reflects on the enormous benefits of co-supervision relative to tapping into the expertise of her co-supervisors, of challenging her own thinking and practice about the content and methods of the research and of approaches to supervision. Such diversity in supervisors can enrich the students' experiences and their research. However, it is a dynamic process that needs all involved to be continually reflexive and share the common goal of transforming the student and their work. This experience is key to the title of the chapter as 'creating rich experiences'. Finally, connecting experienced and less experienced researchers as supervisors provides a means of professional development for new supervisors.

Quality Supervisory Practices

The purpose of the supervisory process has been outlined above but in summary comprises guidance in intellectual content, research process, auditing research practice (i.e. research carried out as intended) and assisting the student to work to standards. Here we consider quality supervisory practices to address the purposes as documented in literature. Twenty years ago, James and Baldwin (1999) reported eleven practices of effective postgraduate supervisors. [10] (Box 39.3) They divide the practices into foundations (1–4), momentum (5–9) and final stages (10–11). The authors frame these practices with underpinning principles seeing supervision as an intensive form of 'teaching' but in a much broader sense than simply transferring information. This is reflected in some of the approaches described above as supervisors facilitate students' progression into new communities of practice where students learn elements of research practices as 'lived experiences' or by being 'situated' in a research community. Other principles of effective teaching are also acknowledged such as being organized, enthusiastic, knowledgeable and providing timely feedback on student progress. They position the supervisor as a role model. The practices all seem important; however, they are all from the perspective of supervisor with less acknowledgement of the students' practices. In our experience, contemporary approaches to supervision have shifted more of the locus of control to the student.

Box 39.3 Eleven practices of effective postgraduate supervisors (James and Baldwin, Ref 10)

Effective supervisors:

1. Ensure the partnership is right for the project
2. Get to know students and carefully assess their needs
3. Establish reasonable, agreed expectations
4. Work with students to establish a strong conceptual structure and research plan
5. Encourage students to write early and often
6. Initiate regular contact and provide high quality feedback
7. Get students involved in the life of the department
8. Inspire and motivate
9. Help if academic and personal crises crop up
10. Take an active interest in students' future careers
11. Carefully monitor the final production and presentation of the research

More recently, van Schalkwyk et al. summarised supervisory practices for doctoral studies in health professions education research [11]. These are listed in Box 39.4 and are a mix of supervisor and student activities. Although obviously limited by reducing the complexity of supervision to 'twelve tips' they have been developed by experienced supervisors and offer a quick summary of considerations. Each tip would likely add value and make conversation starters but again the list is largely positioned from a supervisor perspective.

It is clear from Boxes 39.3 and 39.4 that supervisors are likely to require orientation to these complex supervisory relationships. Most universities offer development opportunities for supervisors that cover diverse content and includes local regulatory higher degree requirements and revisiting policy documents such as codes of conduct for responsible research practice. At doctoral level, supervision qualifications are usually mandated.

If we shift the focus to student practices or responsibilities in the supervisory relationship, then we are likely to see complementarity. However, as indicated above, in our contemporary practices, the locus of control sits more firmly with the student than those set out in Boxes 39.3 and 39.4. Students commonly are required to initiate all aspects of the research process while being 'supported' by the supervisor/s. Of course, this will also depend on the nature of the research project and the supervisor's position within it. That is, if the project is one that the student conceived of and is largely conducting independently or if it is part of a broader research project involving other researchers (and perhaps other students) will influence who initiates and maintains activities. Franke & Arvidsson (2011) distinguish these types of supervision as "research-practice oriented supervision" in which the supervisor and researcher are involved in the research and "research-relation oriented supervision" in which the supervisor and student lack a common research practice. Both approaches can work effectively although this will likely depend on awareness of this positioning [6].

Box 39.4 Twelve tips for supervising (van Schalkwyk SC et al. Ref 11)

1. Clarify purpose of supervision and gain some experience
2. Get to know the student
3. Co-supervise
4. Choose the right approach
5. Be realistic about the project
6. Time management
7. Negotiate a research plan with measurable tasks
8. Complete ethical requirements
9. Discuss the level of support required
10. Agree on a communication plan
11. Accept frustrations as they arise
12. Find avenues for dissemination of research

Processes for Supervision

The following guidelines are dependent on the nature of the degree and local requirements. Below we share ideas for consideration in getting started, during the supervision and following successful completion of the research degree.

Getting started in the supervision

- At the outset, ensure as far as possible that the project topic, the student, the supervisory team and the timescale are a good fit.
- Establish 'ground rules' for all including remits, roles and processes.
- If there is co-supervision – each supervisor needs to know their remit and also how they work together; competent supervisors may not make a competent supervisory team.
- Articulate goals for each member of the research project team. (Tease out similarities and differences in goals for each member. Although knowledge discovery will be important for all team members, for the student, successful completion of their degree and for the supervisors, extending their supervision repertoire, peer reviewed publications and other academic outputs are likely to be important.)
- Share examples of the standard of the 'end product' so that the student is aware of the expected output.
- Define boundaries of the supervisory relationship.
- Agree to mention 'issues' early.
- Agree to attend to well-being (both supervisors and students).
- Agree to a schedule of regular meetings (additional as required) and how they are structured – student-led agenda.
- Meet as agreed to exchange ideas, check progress and assist the student to develop the graduate attributes for their qualification.
- Agree approach to human research ethics approvals.
- Share research standards and guidelines for reporting research.

During the supervision

- Maintain active interaction while acknowledging for all research team members that other activities will at times take precedence – expected or unexpected.
- Continue regular formal research meetings.
- Enable impromptu informal conversations.
- Checking in often with alignment to objectives.
- Continually check that boundaries, initially set between supervisor and student – are being maintained through the period of supervision.
- Conduct 'major' reviews of student progress before milestones.
- Revisit format of thesis and plan for production.
- Within reason, student to submit written work in advance of scheduled meetings and supervisor to read as preparation.
- Supervisor to provide regular feedback on the student's work and the student to respond to feedback – if progress is not made then check in as to why. This is commonly identified in feedback on written submissions.
- Supervisors to encourage the student to disseminate/publish their work and to participate in professional communities.
- Promote participation in student activities (e.g. writing retreats)
- Students to initiate opportunities for participation in professional communities.
- Supervisors to facilitate access to resources at institutional level (e.g. graduate school, library etc.) and students to share resources they identify that supervisors may not know (e.g. social media, graduate student informal activities etc.).
- Draw on professional (clinical) background for skills such as flexibility, responsiveness to changing needs.
- Tailor development to individual students (e.g. writing skills workshop).
- Complete human research ethics reporting requirements.

Completing the supervision

- Review final submission of thesis.
- Supervisors may play an important role in the professional development of the student beyond the thesis and in doctoral studies, it can be important to consider career planning beyond the PhD.
- Plan to complete or continue dissemination of research outputs that fall outside of the thesis submission.

One of the authors (DN) convenes a research-based higher degree. Students participate in an activity in which they identify responsibilities of students. Box 39.5 summarizes responses from several years of running the activity but included here is solely the list of students' responsibilities. Although they complete the same task relative to expectations of supervisors, these are not included here and are largely complementary. They differ to the above in terms of how the student 'drives' the research.

Box 39.5 Summary of students' responses to being asked about students' responsibilities during a research-based masters' programme

What are the student's responsibilities in the supervisory relationship?

- First, be aware of their responsibilities
- Be proactive in all facets of the research process – 'drive' the thesis
- Have an early discussion of supervision logistics
- Set up meetings with supervisor/s
  - Propose fortnightly or monthly appointments – 30 to 60 minutes or as necessary.

- Document meetings and send to supervisor for acknowledgement/additions
  - Maintain a running file of meetings with recent entries at the top as a continuous record of activity
- Devote time to study
  - Gain knowledge of content and methods – reading, participating in educational research activities
- Be organized with a file management system – use a systematic approach to naming of files, include date or some other timing sequence
  - Consider a shared drive for information sharing with supervisor
  - Set up folders such as meetings, reference management system, ethics, proposal, conferences, thesis folder with sub files – introduction, methods, results, discussion, appendices etc.
- Keep a journal – be aware of reflexivity (articulate assumptions/biases)
- Be open-minded, flexible in thinking, in planning and be willing to adapt to change
- Be responsive to feedback
- Be aware of strengths and of limitations of their skills in research content and process
- Be aware of complexities in supervisory relationship

Learning About Ethics and Writing Through the Supervisory Relationship

Now we focus on two important and sometimes related aspects of the supervisory process, ethical practices and writing. Research ethics are often learned through the supervision process [7]. Although this occurs directly through supervision it is also as membership of an academic department [7]. The supervisor may play an important role in facilitating this process of integration into an academic department. Through membership students are likely to learn about ethical issues associated with research processes (e.g. recruitment of study participants, managing de-identification of participants, storage of data, data retention, reporting to funding bodies, authorship processes etc.) [7].

Writing is a significant component of thesis work and is often reported to be a highly challenging area for supervisors and students alike [5, 12]. Questions related to the amount of feedback, when and how it is offered are important. The supervisory approaches outlined at the beginning of the chapter are likely to influence the answers to these questions. The supervisor can be an expert reviewer, active contributor, proof reader or all of these. It is important that regular conversations are held as research progresses to ensure that expectations of supervisors and students are met. Supervisors and students should work to codes of conduct in writing for publication such as the standards set by the International Committee of Medical Journal Editors (http://www.icmje.org/recommendations/browse/roles-and-responsibilities/defining-the-role-of-authors-and-contributors.html). Each discipline has particular ways of dealing with authorship and supervisors ought to direct students to them. It is also important to acknowledge different approaches to writing for different audiences as part of the research process.

Closing

In this chapter, we have shared experiences of research supervision. The relationship is complex and usually sustained and occurs during key professional development stages with significant identity shifts for students. The supervisory relationship potentially offers opportunities for rich learning for all involved, for participation in research that advances the field that either may not have the chance without each other. It is also an important and exciting professional activity that usually results in building new knowledge and scholars of the future. Supervision is a specialised practice and requires targeted professional development.

References

1. Halbert K. Students' perceptions of a 'quality' advisory relationship. Qual High Educ. 2015;21(1):26–37.
2. Murnan J, Cottrell R, Rojas-Guyler L. Survey of practices in health promotion and education supervision of theses and dissertations. Heath Educ. 2009;41(1):11–8.
3. Halse C, Malfroy J. Retheorizing doctoral supervision as professional work. Stud High Educ. 2010;35(1):79–92.
4. Lee A. How are doctoral students supervised? Concepts of doctoral research supervision. Stud High Educ. 2008;33:267–81.
5. Lee A, Murray R. Supervising writing: helping postgraduate students develop as researchers. Innov Educ Teach Int. 2015;52(5):558–70.
6. Franke A, Arvidsson B. Research supervisors' different ways of experiencing supervision of doctoral students. Stud High Educ. 2011;36(1):7–19.
7. Lofstrom E, Pyhalto K. Ethics in the supervisory relationship: supervisors' and dcotral students' dilemmas in the natural and behavioural sciences. Stud High Educ. 2017;42(2):232–47.
8. Lee A. Successful research supervision. Abingdon: Routledge; 2012.
9. Schmutz J, Kolbe M, Eppich W. Twelve tips for integrating team reflexivity into your simulation-based team training. Med Teach. 2018;40:721–7.
10. James, R. and G. Baldwin, Eleven Practices of Effective Postgraduate Supervisors. The University of Melbourne, Centre for the Study of Higher Education and The School of Graduate Studies,

Editor. Parkville: Centre for the Study of Higher Education and The School of Graduate Studies; 1999.
11. van Schalkwyk SC, et al. The supervisor's toolkit: a framework for doctoral supervision in health professions education: AMEE Guide No. 104. Med Teach. 2016;38(5):429–42.
12. McAlpine L, McKinnon M. Supervision – the most variable of variables: student perspectives. Stud Contin Educ. 2013;35:265–80.

Additional Resources

For example, National Health and Medical Research Council, Australian Research Council, & Universities Australia. (2018). Code for Responsible Conduct of Research: Australian Government.
See national and international guidelines for ethical research.

Social Media

@researchwhisper
@PHDComics
@PhDForum
@thesiswhisperer

Project Management in Healthcare Simulation Research

40

Cylie M. Williams and Felicity Blackstock

Overview

There are many frameworks available that can guide the researcher through the process of project management for simulation research. Projects are not routine business, but are a unique specified set of operations designed to accomplish specified goals. In the context of simulation research, the overarching goal is usually to answer a discrete research question. Projects therefore frequently bring together a team of people and stakeholders, who do not usually work together, and are often from different organizations, different career backgrounds, and across multiple geographical sites. This chapter explores the components of common key frameworks. Sequentially to a project management framework is the need for risk mitigation strategies to ensure smooth running of your project. Piloting your project can help identify many of the challenges you may face during the project. There are also many risks that are specific to simulation research. Many of these risks can be avoided with careful planning, attempting to manage the unmanageable, and the development of detailed timelines and budgets that are specific to research projects.

Practice Points

- If you only take away one thing from this chapter, let it be that piloting your intervention is one of the keys to success.
- There are a number of project management frameworks that can be considered during the planning phase of the project. Each has its own strengths and limitations.
- Regardless of the framework chosen to guide the research, there are key sections in all project management knowledge areas that will help researchers through the stages.
- Managing the unmanageable is key to risk mitigation, and many of these key components are external to the project.
- Working in teams and identifying key players external to your team is your other key to success.

Introduction

Regardless of whether a research project evaluating the impact of simulation has a sample size of 20 with no formal funding, or is a large multicentre randomized controlled trial that is funded for millions of dollars, the project will benefit from a structured approach to operationalization. Over this chapter, we will explore the different project management frameworks, systems and resources available to a project team, and contextualize them to simulation research. Risks associated with conducting projects will be outlined, with strategies to avoid adverse outcomes recommended.

Project Management Frameworks

A project is a distinct temporary endeavor undertaken to create a unique product, service, or result, within the bounds of specified funding [1]. As such, a project has a defined beginning and end in time with specified scope and resources. Projects are not routine business, but are a unique specified set of operations designed to accomplish specified goals [2].

C. M. Williams
School of Primary and Allied Health Care, Monash University, Frankston, VIC, Australia
e-mail: cylie.williams@monash.edu

F. Blackstock (✉)
School of Science and Health, Western Sydney University, Campbelltown, NSW, Australia
e-mail: F.Blackstock@westernsydney.edu.au

D. Nestel et al. (eds.), *Healthcare Simulation Research*, https://doi.org/10.1007/978-3-030-26837-4_40

In the context of simulation research, this is usually to answer a discrete research question. Projects therefore frequently bring together a team of people and stakeholders, who do not usually work together, and are often from different organizations, different career backgrounds, and across multiple geographical sites. As a research question is being explored, the people involved may also be using techniques they have not trialed in a research context or as a component of their usual work routines. Projects can therefore lead to complex interactions between team members, which, if not addressed, can lead to poor outcomes for the team members and the project goals being not achieved.

Project management is the application of knowledge, skills, tools, strategies, and techniques to project activities to ensure the project meets the requirements and achieves the goals in a timely manner [3]. Many simulation research projects practice project management informally, with project leaders formulating the project plans and actions in a non-structured manner. This can lead to a chaotic approach to completion of stages and tasks in the project, which in turn can lead to project goals not being achieved or achieved inefficiently. To address this, experts in project management have developed frameworks that can be used to support project leaders in simulation research. Use of such frameworks supports project implementation by:

1. Developing, and then replication of, accepted practice within the project
2. Establishing clear communication within the team, and across stakeholders, through the use of a common language
3. Streamlining the use of tools, templates, software, processes, and systems
4. Affirming consistent approaches across the project for all team members and stakeholders
5. Providing focus across all stages of the project, particularly in early stages of the project lifecycle

Project management therefore provides an organization with tools that improve planning, implementation, and control over the activities within a project, as well as ways to engage people and channel resources appropriately [4].

A vast array of project management frameworks exist to support researchers designing and implementing projects with an equally vast array of aims and objectives. Many of the frameworks have been designed to support software development teams, producing rapid updates to software products to meet market demand. To date, no specific simulation research project framework has been identified or evaluated for use. Therefore, where a simulation research project requires a framework for implementation, the generic frameworks currently published for broad application are most appropriate for use. Selection of framework will depend on the scope of the project. Larger more dynamic projects, with complex curriculum, multiple organizations, a large human resource team involved in implementation and a significant budget to monitor are likely to have different framework requirements to smaller projects which are single sites with only a small staffing base, brief learning activities, and one stakeholder. Some organizations also have policies related to selection of project management frameworks, which may also be used to guide the research team.

The aim of all frameworks and methodologies in project management is to provide structure to the project lifecycle including design, implementation, and project close off [3]. To support the project team, most frameworks also have complementary templates, tools, and software that accompany the framework and can be readily used in simulation research. Table 40.1 summarizes commonly used project management frameworks and methodologies, providing an outline on the features of each framework.

It is important to note that subprojects within a larger project may have different methods or frameworks to the overarching project. Selection must be based on the aims and objectives of each component of the project.

In addition to the aforementioned methodologies, the Project Management Institute has developed *A Guide to the Project Management Body of Knowledge (PMBOK® Guide)* [7]. The PMBOK® Guide outlines that project management processes fall into five groups: (i) initiating, (ii) planning, (iii) executing, (iv) monitoring and controlling, and (v) closing, while project management knowledge is drawn from ten different areas (Box 40.1) [7].

Box 40.1 Project Management Key Knowledge Areas

1. Integration
2. Scope
3. Time management
4. Cost
5. Quality
6. Procurement
7. Human resources
8. Communications
9. Risk management
10. Stakeholder management

All management is concerned with these topic areas; the difference is that project management contextualizes these areas to the specific goals and operations of the project. Key attributes of a project manager are therefore an ability to develop and maintain relationships within the team and with stakeholders, communication skills across all media, time management, and an ability to support others in time management. In light of the extensive knowledge base and diverse areas of expertise, for complex and dynamic projects, a formally trained project manager is recommended to be

Table 40.1 Examples of project management frameworks and methodologies

Framework	Philosophy and principles	Strengths	Limitations
Waterfall method [5, 6]	Linear sequential design approach, where the project flows steadily through the following stages: 1. Conception and feasibility 2. Plan and analysis 3. Design 4. Construction 5. Testing 6. Deployment 7. Support and maintenance	Potential issues can be identified during the development stage, and proactive solutions can be implemented early Emphasis on documentation at all stages As linear, may be easier to understand by less experienced teams	In curriculum design, may not know exactly what features are needed or may need to change as research progresses; the waterfall method does not easily accommodate change The process is very fixed and not suitable where there is risk of project modifications or changes
Scope statements [7]	A structured template that provides a baseline understanding of the scope of a project, including deliverables, tasks, and timelines, that ensures a common understanding of the project's scope. The statement defines: 1. Purpose and justification 2. The project owner, sponsors, and stakeholders 3. The project goals and objectives 4. The project requirements 5. Project scope management plan 6. Project deliverables 7. Acceptance criteria 8. Project constraints, including non-goals and what is out of scope 9. Milestones 10. Cost estimates and benefits	A structured template that is completed at the commencement of the project, with proactive planning and highly organized project stages Conflict resolution is facilitated by the pre-project planning documentation; when disputes arise, teams can review the agreed project plans Easier implementation for inexperienced teams and managers	Time-consuming process at the commencement of a project May be created and then never used, which could be a waste of time Flexibility to change direction is limited and requires revision of the scope statement
Agile method [8, 9]	Incremental and iterative approach to project management, allowing for concurrent stages being implemented simultaneously using cross-functional team interactions. The goal of each iteration stage of the project is to produce a product which can then receive feedback and adapt for the next iteration of the project. Efficient face-to-face communications and short feedback loops are a feature	Each iteration of a product is quicker, and successive iterations can be delivered frequently Rapid changes in direction can be accommodated easily and quickly Close collaboration between stakeholders as feedback frequently sort Explicit opportunities for continuous improvement Highly transparent as teams connect for feedback frequently	Method is harder to understand and implement effectively, particularly with inexperienced teams and/or managers Documentation can sometimes be neglected with a face-to-face focus for team communications When implemented poorly, can lead to inefficiencies
Scrum method [8]	Based on the principles of Agile methodology, with an emphasis on decision-making from real-world results rather than speculation. Time is divided into short work periods, referred to as sprints (of a week or two). At the end of each sprint, stakeholders and team members meet to see the product and plan the next steps	Large projects are divided into easily manageable sprints Works very well for fast-moving projects Transparent communications in "scrum meetings" Adopts feedback from all stakeholders and rapid changes in project direction and product easily accommodated	Often leads to scope creep due to lack of defined end date Chances of project failure are high if team members are not committed or cooperative Use of the framework with large teams is challenging Experience necessary for successful application Staff turnover can negatively impact on success of project as knowledge of project development by team members essential
PRojects IN Controlled Environments (PRINCE2) [10]	The framework is process orientated, where projects are divided into stages and each stage has its own plan and operations to follow. Documentation of the project underpins the framework	For accurate implementation, the project manager should be accredited Documentation mitigates risks Templates to support documentation and guide teams	Complex methodology, which requires accredited project manager for implementation Rapid change in project direction or product is difficult to achieve as the processes are cumbersome and substantial documentation requires amendments

employed. When recruiting a project manager to lead simulation research, these knowledge areas outlined by the Project Management Institute could be used to provide guidance on selection criteria and required expertise for successful project managers.

Key Strategies for Successful Project Management in Simulation Research

Planning your research has many similarities to planning the curriculum within in an education unit or even a family holiday. Set the aims and what outcomes you hope to achieve to help gather knowledge of success.

Regardless of which framework or process is implemented, there are strategies particular to simulation research that should be considered. The chosen framework should help give you a road map to completion, with key factors to include for planning and risk mitigation strategies. Additional factors to consider in project management are planning to manage the unknown. Reducing the unknown is one of the key things to ensure success. The following sections explore a number of strategies that can assist in reducing the unknown and mitigate the risks encountered during project management.

Managing the Unmanageable

Key to risk mitigation is awareness of the components you have little control over and having strategies in place to ensure these have minimal project impact. These components should be highlighted in your timeline. If you have an established team in place, and have navigated these components with previous projects, you have a great advantage. If you are new to the organization and recently employed to manage a project or a student researcher, finding out the key players, committees, and departments across different organizations, together with their processes, can take time, and it is imperative that they must be planned for.

Some of key considerations essential to your research, but operating within their own timelines can include:

- Approval and then advertising for employment of appropriate staff (this may include auditioning for actors or trials for technicians for particular equipment or IT)
- Allocation of a cost center manager or business unit for financial accountability
- Purchasing departments for larger equipment that is being ordered from another country and processes for purchasing of smaller consumables
- Institutional review boards or human research ethics committee approvals
- Legal or contract departments of both yours and any partner organizations

Understanding individual committees and unit processes is essential to effective project management. Sometimes the smallest thing can hold up a project. Many a project has suffered delays from something as simple as an essential form with an approval signature missing during an application to a review board.

Piloting Your Intervention

It sounds too simple, but a quality pilot of your intervention will often identify elements that have potential to cause havoc in your project. Many people consider a run through of the research scenario enough. However, this does not let you truly test what will happen in a timed, and at times, high-pressured situation. There are many factors that have potential to hold up your research. Research participants may be unpredictable in their responses to the situation, or equipment may fail or be temperamental, or the timing blows out. A pilot intervention with any actors who are playing simulated patients, or a small group of students in peer simulation or high-fidelity simulation equipment, will enable you to find these pressure points. This also enables you to develop strategies to manage or even change and remove components where needed, to ensure each future simulation scenario runs smoothly. Lastly, this pilot testing will enable you to see if you have the right skill mix in your team or if there is additional training or practice is needed among your team members. This is particularly key when leaders of the project or components may be inexperienced or students themselves.

Identifying and Managing Key Players

Researchers commonly identify the key research personnel or investigators during the grant writing or protocol process. What may not be considered at this point are any additional key players external to your research team but essential to your research project. This may include researchers with additional skill sets such as a statistician or health economist. Your research setting will also often determine these key players with some overlap between settings but some having distinct people for certain roles. This is particular to research being undertaken between universities and healthcare organizations.

While many of these were identified in managing the unmanageable section, it is also worth identifying any particular people you will need to liaise with early in your project. These will include people who are responsible for:

- Room bookings and access to equipment
- Funding or cost center management
- Approving any purchasing or petty cash reimbursement
- Contract negotiation
- Consumer engagement

Authentic consumer engagement at the beginning of your project can also ensure that target language and strategy of your intervention is right. This will enable greater translational impact [11]. Development of a project management group of student representatives (university setting) or staff, managers, and community consumers (healthcare setting) has a multitude of benefits for applicability and translation.

A well-functioning team with complementary skills is paramount to a successful research project. Early identification of skill mix is important, as well as understanding who are the leaders and who are the managers. While many roles are intertwined, clear identification of responsibilities for overall and day-to-day project management is essential. Knowledge of acting roles or equipment should be identified and skills of any employed research assistants or students you embed into the project should be complimentary. It is important to consider who will also provide day-to-day management of these staff members in accordance to employment policies.

Authorship plans [12] and intellectual properties agreements [13] are not unique to simulation research project management. These must be considered during formation of protocol development or funding applications. Student projects are governed by an institutional agreement of intellectual property and authorship on subsequent publications. Research collaborators should also discuss this early in the project. Consider developing a standing agenda item for authorship plan for any project committee meetings and regularly revisit this during the implementation.

Setting Your Timelines and Budget

Setting realistic timelines is intertwined with your budget and accountabilities to stakeholders. It can be difficult without an experienced project manager or researcher in your team. There are many factors considered to minimize the risk of not meeting either the timeframe or budget. If you are bound by a funder's timeline, set this and work backwards and identify each element where a risk to your timeline or budget may present.

Considerations or external factors affecting the timeline and budgets could be anything from the funder's rules around costing through to timetables of the students you are looking to engage in the research. Additionally, competing demands nearing the end of your project, budget over or underspend, and staffing mix while moving into the next component of research is often what impacts the translation or publication of findings. Figure 40.1 presents a brief summary of key pressure points at each component of the research which may impact the timeline and budget. This list is by no means exhaustive, but presents areas that commonly impact projects significantly.

Gantt charts are the most common visual planning tool to set and identify timelines but there are a many other project management tools that you also may consider. There are a number of free online project management software suites that may suit the needs of your team or for smaller projects, and the good old whiteboard may also suffice.

The Final Product and Concluding Your Project

Just when you think the end is coming, there is still a lot of work to be done. At this time, staff may have moved on, and the team are ready to move into the next project. This is often where some of the essential components can fall by the wayside, including:

- Ethics final reports
- Data archiving
- Curriculum archiving
- Dealing with requests from other institutions or researchers
- Presenting findings at conferences or in publications

Many of these essential components may be built into the next project you have moved to, however need to be planned for within the project's timelines. For translation, it is essential that you consider the different dissemination strategies and which conferences will be best to get your message out. Consider engaging with your institutional media liaisons to ensure your outcomes are reported on websites and within social medial.

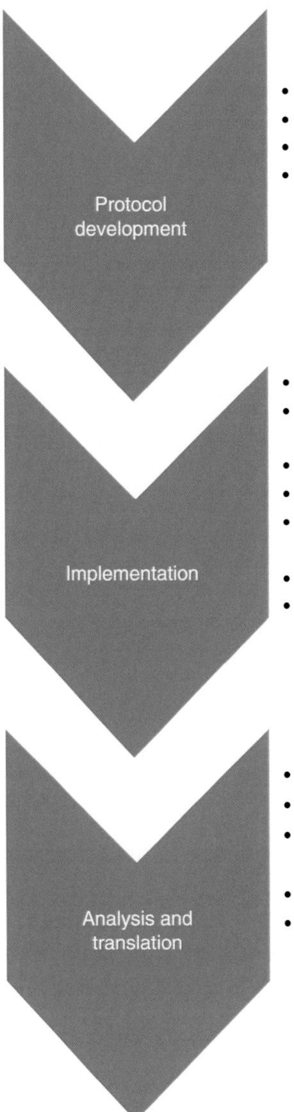

- Set up of fund management structures
- Approval to, and purchase of, equipment
- Institutional review board or human research ethics committee approvals
- Training of team members

- Advertising and recruitment of staff to undertake research
- Staffing oncosts including department costs, tax, medical insurance, superannuation etc
- Realistic recruitment of participants
- Availability of research space/simulation environments
- Staffing changes or absences due to illness, leaving for other emplyoment or family carer responsibilities
- Student timetables or organisational rosters or rotations of staff
- Budget reports

- Data extraction, coding, trasncription, cleaning
- Statistical analysis
- Dedicated time for results documentation and format for funders or educational requirements
- Funding reports and management of over or understand
- Data archiving

Fig. 40.1 Key conditions to consider during timelines and budget setting

Conclusion

Project management is needed for all projects, but not everyone has the expertise to lead. However, with the right framework and risk mitigation strategies in place, project management and tasks should be less daunting. Many institutions will have learning packages or leaders in this area that may mentor those new to project management. What is common to all successful large funded projects is careful planning and management at each stage. Simulation research has a number of unique factors in equipment and staffing, but many of the frameworks are applicable to simulation research.

References

1. Kerzner HP. Project management: a systems approach to planning, scheduling and controlling. 12th ed. Hoboken: Wiley; 2017.
2. Rosenau MD, Githen GD. Successful project management: a step-by-step approach with practical examples. 4th ed. Hoboken: Wiley; 2005.
3. Project Management Institute. A guide to the project management body of knowledge. 6th ed. Newton Square: Project Management Institute; 2017.
4. Meredith JK, Mantel SJ. Project management: a managerial approach. 8th ed. Wiley: Hoboken; 2012.
5. Royce W. Managing the development of large software systems. Technical Papers of Western Electronic Show and Convention. Los Angeles: WESCON; 1970.

6. Collyer S, Warren CMJ. Project management approaches for dynamic environments. Int J Proj Manag. 2009;27(4):355–64.
7. Snyder CS. A guide to the project management body of knowledge: PMBOK (®) guide. Newtown Square: Project Management Institute; 2014.
8. Cervone HF. Understanding agile project management methods using Scrum. OCLC Syst Servi Int Digit Library Perspect. 2011;27(1):18–22.
9. Highsmith J. Agile project management: creating innovative products. 2nd ed. Boston: Pearson Education; 2010.
10. Sargeant R, Hatcher C, Trigunarsyash B, Coffey V, Kraatz JA. Creating Value in Project Management using PRINCE2. Office of Government Commerce: London. 2010. Available from: https://eprints.qut.edu.au/52853/1/Final_Report__v1.0e%5B1%5D.pdf
11. Leape L, Berwick D, Clancy C, Conway J, Gluck P, Guest J, et al. Transforming healthcare: a safety imperative. BMJ Qual Saf. 2009;18(6):424–8.
12. Newman A, Jones R. Authorship of research papers: ethical and professional issues for short-term researchers. J Med Ethics. 2006;32(7):420–3.
13. Hertzfeld HR, Link AN, Vonortas NS. Intellectual property protection mechanisms in research partnerships. Res Policy. 2006;35(6):825–38.

41 Disseminating Healthcare Simulation Research

Adam Cheng, Brent Thoma,
and Michael J. Meguerdichian

Overview
Effective dissemination of research ensures that relevant end-users can utilize results to inform education, clinical care, and/or the future trajectory of the field. In this chapter, we describe the various methods that simulation-based researchers can use to share their work, including conferences, publication, courses, massive online open courses, social media, blogging, and podcasts. We also share ways of measuring the dissemination of research, namely: journal impact factor, article downloads, Altmetrics, and open educational resource metrics. These hold particular importance for researchers as it relates to promotion, funders' requirements, personal network and individuals' choices about target audience.

Practice Points

- Many options for disseminating research exist.
- Researchers should thoughtfully consider the pros and cons of the many different vehicles for disseminating research as part of the research process.
- Metrics for measuring the dissemination of research may play a role in supporting academic promotion in the future.

A. Cheng (✉)
Department of Pediatrics & Emergency Medicine, Albert Children's Hospital, University of Calgary, Calgary, AB, Canada

B. Thoma
Department of Emergency Medicine, University of Saskatchewan, Saskatoon, SK, Canada

M. J. Meguerdichian
Department of Emergency Medicine, Harlem Hospital Center, New York, NY, USA

Introduction

Simulation-based research involves studying the value of simulation as an educational modality, using the simulation as investigative methodology, or applying simulation as a vehicle to explore healthcare teams and systems [1]. Regardless of its application, healthcare simulation research typically has one ultimate goal: to improve healthcare outcomes. Critical in the mission to achieve this goal is the effective dissemination of research results to the appropriate end users (e.g. healthcare providers, educators, administrators, researchers etc.). Effective dissemination improves the likelihood that completed research informs the future trajectory of education, clinical care and/or inquiry in a particular field [2]. In this chapter, we explore the various means of disseminating research and discuss the measurement of research dissemination.

Disseminating Results of Simulation Research

Conferences

Conferences offer an ideal venue for disseminating results of simulation-based research. Presenting ongoing or completed research at conferences provides the advantage of human interaction – giving researchers a chance to communicate their message in person and to seek input, advice and feedback from other researchers in the field. Interactions between researchers at conferences can lead to important revisions to study protocols and foster the development of collaborations between individuals or groups with shared research interests.

Simulation research can be presented in various different formats at conferences. Most conferences allow for submission of research abstracts, which, if accepted, the presentations are in either oral format or as a poster. While poster sessions may be perceived as less prestigious than an invitation to give an oral presentation, they provide opportunity

D. Nestel et al. (eds.), *Healthcare Simulation Research*, https://doi.org/10.1007/978-3-030-26837-4_41

for investigators to receive feedback from experts in the field [3]. Posters offer a short, succinct summary of the ongoing or completed research project in visual form, supported by details provided by the investigator standing next to the poster [3]. Conference attendees have the chance to interact with investigators during poster sessions, as well as have any questions answered related to the research project [4]. As only a minority of delegates attending international conferences actually stop to view posters, investigators should design posters to maximize visual impact by presenting data as figures or graphs [5]. Some conferences provide opportunity to present posters in electronic format, known as e-posters, which involve displaying the poster on a screen as opposed to a printed version. The rate of publication of abstracts presented at international simulation conferences is relatively low in comparison to other types of healthcare conferences [6]. "*Professor rounds*" coupled to poster presentations can provide a forum for experts to guide investigators on writing manuscripts for publication while fostering the development of new ideas and innovations. Other formats for presenting scholarly work include: (a) pre-conference courses and workshops, where research findings are offered as practical strategies to help educators and/or researchers improve their practice; (b) expert panels, where a group of thought leaders share their expertise in a structured format; or (c) podium presentations. The latter is where investigators have an opportunity to present a project or program of research in oral form.

Simulation researchers have a variety of options when it comes to presenting at conferences that include simulation-based research. Many healthcare simulation conferences exist, at the local, national and/or international level – most of which allow work to be presented even if it has already been presented at a prior conference. We recommend investigators review abstract submission requirements closely to ensure they are compliant with the rules for individual conferences. Key international conferences in healthcare simulation are offered by leading simulation societies and include: (a) the International Meeting for Simulation in Healthcare (IMSH) offered by the Society for Simulation in Healthcare; (b) the International Nursing Association for Clinical Simulation (INACSL) annual conference; (c) the Society in Europe for Simulation Applied to Medicine (SESAM) annual meeting; (d) the SimCongress conference offered by Simulation Australasia; and (e) the Royal College Simulation Summit hosted by the Royal College of Physicians and Surgeons of Canada. A list of key healthcare simulation conferences is provided in Table 41.1.

Besides healthcare simulation conferences, researchers may also look to specialty or profession-specific conferences, patient safety conferences, and/or medical education conferences as possible venues to share and disseminate their work. Researchers should review categories of abstracts and/or abstracts accepted to previous conferences to help gauge the level of interest in simulation-based research for specific conferences.

Publication

Publication of completed research in a peer-reviewed journal remains the gold standard for disseminating results. A number of avenues exist for publishing completed projects: traditional peer-reviewed journals, open access journals, and web-based resources. Traditional, peer-reviewed jour-

Table 41.1 Healthcare simulation conferences

Conference name	Organizing group and website	Host region
The International Meeting for Simulation in Healthcare	Society for Simulation in Healthcare https://www.ssih.org	United States
The International Nursing Association for Clinical Simulation Annual Conference	International Nursing Association for Clinical Simulation https://www.inacsl.org	United States
The Society in Europe for Simulation Applied to Medicine Annual Meeting	Society in Europe for Simulation Applied to Medicine https://www.sesam-web.org	Europe
The Royal College of Physicians and Surgeons of Canada Simulation Summit	Royal College of Physicians and Surgeons of Canada http://www.royalcollege.ca	Canada
Simulation Congress	Simulation Australasia http://www.simulationaustralasia.com	Australia
The International Pediatric Simulation Symposium and Workshops	International Pediatric Simulation Society https://www.ipssglobal.org	Global
Pan Asia Simulation Society in Healthcare Conference	Pan Asia Simulation Society in Healthcare http://passh.org/events.html	Asia
The Association for Simulated Practice in Healthcare Annual Conference	Association for Simulated Practice in Healthcare https://aspih.org.uk	United Kingdom
Saudi Health Simulation Conference	Saudi Society for Simulation in Healthcare http://shscmoh.com	Middle East

nals require submission of the manuscript for peer review, with no charge to the author if the manuscript is ultimately accepted. A number of traditional, peer-reviewed journals with a focus on healthcare simulation exist, including *Simulation in Healthcare*, *BMJ Simulation and Technology Enhanced Learning* and *Clinical Simulation in Nursing*. These journals usually also offer *Open Access* for a fee. Researchers also have the option of exploring publication in specialty or profession-specific journals. For example, an investigator completing a simulation-based research project related to cardiac arrest care might explore publication in an emergency medicine journal such as *Resuscitation*, *Academic Emergency Medicine*, or *Annals of Emergency Medicine*. Medical education journals (e.g. *Academic Medicine*, *Medical Education*), patient safety journals (e.g. *BMJ Quality and Safety*), and/or modeling and gaming journals (e.g. *Simulation and Gaming*) are other alternative targets for publication.

Open access journals are an attractive alternative for many researchers because when the manuscript is published, it is made available to the general-public free of charge. This is in contrast to most traditional peer-reviewed journals, whose articles are accessible by paying a fee or through institutional access (e.g. university library). The benefit of open access journals is that articles are downloaded and shared easily, making dissemination of results quick and efficient [7]. Some funding bodies make open access publishing mandatory. There is conflicting evidence suggesting that open access articles are cited more frequently [7–9]. Most open access journals require the researcher to pay a publication fee once the manuscript is accepted; although in some cases this may be covered if the researcher is affiliated with an academic institution that already pre-paid this fee. *Advances in Simulation* is an open-access healthcare simulation journal that publishes its articles exclusively online, includes peer review of submissions, but requests authors pay a publication fee. *Cureus* is a relatively new web-based open access journal that offers quick peer review and extremely short timelines to publication. Other specialty and/or profession-specific open access journals exist that may publish simulation-based research. Lastly, some web-based resources such as *MedEdPortal* offer opportunities for publication of simulation scenarios and/or curriculum; advantages include peer review and the ability to cite the work once it is published online.

When selecting a potential target journal, researchers should consider various factors, including (a) target audience and readership of the journal; (b) quality of the work relative to the impact factor of the journal; (c) desire to have the publication as open access, and if so, ability to pay publication fee; (d) research funder conditions; and (e) prior track record of journal in publishing simulation-based research, which can be gauged by reviewing past issues of the journal for content type. Researchers new to publishing simulation-based research might consider seeking advice from well-published colleagues who can provide suggestions for the ideal target journal.

Simulation Courses

Simulation courses are another avenue for sharing research results. Multiple instructor courses exist at various institutions with a focus on best practices. Often, research is cited for the purposes of making compelling arguments to, for example, create a curriculum or design a scenario in a particular way. In many of the training programs, those individuals contribute and shaping the literature are additionally the ones offering the courses, lending more opportunities for dissemination.

Similarly, fellowships train future educators and use literature and research as the foundation of their curriculum. In a recent survey evaluating simulation fellowships internationally, 84% of programs use tools like journal club, a meeting where research is discussed and appraised, as a development strategy [10]. Fellowship programs also promote attendance to the aforementioned conferences where attendees are exposed to abstracts, posters and presentations on current simulation-based research [11].

Massive Online Open Courses

Massive Online Open Courses, also known as MOOCs, are another knowledge sharing method by which simulation research can be disseminated [12]. MOOCs curate to an international community and, by design, deliver the research to thousands of learners simultaneously. Platforms like *Coursera*, *Edx*, *Khan Academy*, and *Udacity* are the most prevalent platforms for hosting MOOCs. Other sources for MOOCs can be found through academic campuses. Similar to the real classroom, most MOOCs are structured as a combination of lecture and group interaction. One benefit of this virtual classroom is that it focuses groups on shared interests [13]. The social network that is created by MOOCs allows for cases studies that promote group interaction and feedback; an opportunity to share literature to support arguments. There are few notable limitations to MOOCs in terms of research dissemination. Most of these on-line classrooms are facilitated in English, thus limiting its reach to certain international communities. Another criticism of these forums is that they are too loosely structured and are highly dependent on both the developer of the program as well as very vocal individuals to sustain the sharing of information [12].

Social Media

Social media is arguably the newest means for research dissemination and continues to experience a surge as a platform over the past two decades [14]. It is especially important to harness this modality's potential to broadcast advances in simulation research and literature. What distinguishes social media from other means of research dissemination is that the end user is no longer pulling the information from the literature base. Instead, the user pushes the relevant information to other users within his/her social network [15]. Sites that are particularly popular include *Facebook®*, *Linkedin®*, *Twitter™*, *Academia.edu*, *ResearchGate*, and *Mendeley* (Table 41.2). Understanding the advantages of each social media modality allows the user to decide which platform to explore and consider sharing their contribution to the simulation community of practice.

Facebook is traditionally used more for informal relationships and sharing of personal experiences. *Facebook*, like other social media sites, offers the opportunity to create thematic groups where individuals can share research and discuss findings [16, 17]. Posts can share abstracts and references to research where others can comment and contribute. *Linkedin* is a similar platform that is known for the opportunity to network and search for jobs. The platform makes a particular effort to allow the user to post and share their publications alongside their resume or curriculum vitae through their profile. It helps foster relationship by offering the opportunity to create groups and connect with other like-minded people. *Twitter*, another popular social media platform, allows the user to send a "tweet" limited to 280 characters paired with an option to pair with a picture, animations such as GIF's, or polls. *Twitter* has been cited as the most often used resource for the purposes of professional development [18–20]. The platform has been demonstrated to draw significant attention to scholarly publications and has been suggested to increase the potential for citation [19]. Another advantage to using *Twitter* is that it offers metrics that aids the user to gauge the visibility of the post sharing their publication. There is a developing virtual community of practice on *Twitter*, which can be found by searching the #FOAMsim hashtag [21].

Table 41.2 Opportunities for dissemination on social media platforms

Platform	Dissemination opportunities	Source
Facebook®	Create posts to share articles Create thematic interest groups	www.facebook.com
LinkedIn®	Job networking with CV/resume sharing Create posts to share articles	www.linkedin.com
Twitter™	Post scholarly work and link to similar themes using hashtags Pre-existing and growing virtual simulation community of practice Monitor with metrics to gauge visibility	www.twitter.com
ResearchGate	Networking and sharing PDFs with other scholars Monitor with metrics to gauge visibility and downloads Follow other researchers of interest	www.researchgate.net
Mendeley	Develop reference library to help identify link like-minded readers Create thematic interest groups	www.mendeley.com

Academia.edu is a site that allows for the sharing of the user's research articles and through focus groups helps target individuals sharing mutual interest. It too offers metrics to monitor the number of views, downloads, and then generates an impact score for the individual article. For a fee, the service will offer advanced metrics and indicate who is reading and citing the user's research. *ResearchGate* offers a similar experience, as it monitors the number of publication reads [22]. It has a stronger social networking emphasis by linking individuals through co-authorship. The platform allows for "following" individuals. This following reveals the user to a researcher's interests as well as current and past works. *ResearchGate* also promotes the sharing and requesting of full articles from colleagues. Some research suggests that the sharing of an article on *ResearchGate* increases the likelihood of citation by other authors [23].

Recently reference managers, like *Mendeley*, have developed a social networking component to their repertoire. *Mendeley* uses your reference library to push topic-specific articles to the user that may be of interest to them by email. Increased number of downloads of an article to a *Mendeley* library increases the likelihood of readership as well as citation [24]. Like other social media resources, groups can be created to better network and share published works.

Infographics and visual abstracts are visualizations used for effective communication of findings and concepts of literature. Visual abstracts present key findings/concepts related to data through visual media so that the audience can easily process the findings [25, 26]. Infographics are used to help guide the audience or draw conclusions about the represented literature. These representations applied on social media like Twitter and facebook promote journal readership as the reader gets to preview the topic through clear messaging prior to accessing the entire article. Paired with social media, on platforms like Twitter, visual abstracts were linked to increased sharing as well as review of the full article [25, 26]. Results for infographics collaborating with social media demonstrate increased abstract views [14].

Social media is powerful because these platforms can also communicate with each other. The use of hashtags, helps link and market research to particular audiences. For example, #debriefing, will pool all posts with that hashtag to a common repository with related posts around debriefing. Social media has very little policing and has become increasingly utilized as free open access medical education has risen in popularity. With little control over what is communicated through social media, a user may take pause when considering sharing information through these platforms. It must be emphasized that social media does not serve as a substitute for reading the entirety of an article [27]. The reliability and accuracy of content on social media must be considered to avoid miscommunication of literature content. Recent efforts have been made to develop instruments to gauge the quality of blogs and podcasts so that the content reliably reflects the evidence [28–31].

To best harness the power of social media as a means of knowledge sharing, trust building must be considered due to the uncontrolled nature of on-line communication. Relationships are typically built on previous personal interaction, peer recommendation, the way information is shared with an eye toward relevance and framing of content, professional standing and consistency of communication through the online modality [32, 33].

Blogging and Podcasts

Blogging is another useful approach to disseminate scholarly works. Blogs tend to encapsulate concepts in a succinct manner using informal language to translate them for the end-user. Blogs are typically arranged around central themes, such as 'debriefing', 'ultrasound' or 'game-based learning'. Blog posts can promote a single article or may encompass multiple article citations around a related topic [21]. They also can be structured to incorporate an infographic or visual abstract to communicate data. The social media release of an article through a blog will likely increase views and downloads of an article [34].

Podcasts, another medium of distribution, are digital audio recordings that are downloaded and streamed by the user. Podcasts are being incorporated into many asynchronous curricula due to their accessibility in a mobile fashion [35, 36]. There are several podcasts published on simulation education that encapsulate ideas and promote research through interviews and discussions around a specific topic [21]. Thoma et al. recognized that podcasts increase abstract readership but does not necessarily influence full-text article views [14]. More research needs to be conducted explicitly around simulation literature to better inform the impact of social media on readership.

Measuring the Dissemination of Research

Historically, the dissemination of a research paper was estimated largely by the number of subscribers to its publishing journal. As we fully enter the digital age, journal articles are frequently accessed and universally published online. It is now much more common to quantify an article's dissemination using online metrics. These metrics are also increasingly available for less traditional forms of scholarship such as blogs and podcasts.

Journal Impact Factors

The Journal Impact Factor (JIF) was first described by Garfield [37, 38] and is now calculated annually by Thomson Reuters (New York, NY) for publication in their Journal Citation Report. It uses citations from scientific articles to determine the impact of a journal. The JIF is equal to the number of citations in a given year to articles published in the previous two years divided by the number of citable articles that have been published during that year [39]. While it is widely used to quantify impact, it has been criticized for being open to gaming [40], secretive [41], and narrowly focused [42]. In response to these criticisms other metrics have been developed (e.g. Eigenfactor metrics), but they have either failed to address these concerns or created new ones [43, 44].

Journal-level metrics such as the JIF gain their importance based upon the assumption that higher impact journals are read more broadly and that the articles published within them are therefore more impactful. However, especially in the digital age when dissemination occurs more frequently at the level of the article, impactful journals may publish lower impact articles that generate little interest and while a lower impact journal may end up publishing work that generates immense enthusiasm. Fortunately, it has become increasingly possible to quantify the impact of an article at the level of the article, rather than the journal.

Article Downloads

While it is not possible to determine the number of times that an article has been read, it is possible to track the number of times an article has been downloaded and/or viewed in full-text online. Most publishers track these metrics internally, some (e.g. Elsevier) disseminate this data to the articles authors [45] while others (e.g. Cambridge University Press) publish them directly on their website. Other article repositories track similar statistics. As mentioned earlier, *ResearchGate* and *Academia.edu* are academic social networks, which allow researchers to upload PDF's of their publications and track their views and downloads [22]. Finally, institutional repositories (e.g. eScholarship) may allow authors to upload their publications to their institutional repository, which also tracks views and downloads [46].

Unfortunately, given the numerous websites and online repositories that are available, it is difficult to get a complete picture of the number of times that an article has been accessed. Even if these sources could be accounted for, such metrics would not account for dissemination through direct sharing of documents via authors, printed articles, journal clubs, research groups, or underground dissemination websites such as Sci-Hub – a website which provides untracked access to nearly all scholarly literature [47].

Altmetric Score

Also known as "alternative metrics" or "article-level metrics," Altmetrics measure the dissemination of an article using social media metrics to reflect the overall interest in an article online [48]. The Altmetric Score is a proprietary service that tracks, among other things, citations in Wikipedia, blog posts, and news articles; mentions on Twitter and facebook; and the number of users that have the article in their Mendeley library [49].

Potential advantages to the use of the Altmetric Score include its speed (it is updated almost instantaneously), article-level specificity, and public accessibility. Drawbacks of the Altmetric Score include:

(a) high scores received by articles that are controversial (i.e. may receive significant attention but be flawed or incorrect);
(b) its proprietary nature (i.e. the exact calculation used to determine the Altmetric Score has not been disclosed); and.
(c) a relative advantage of newer (e.g. benefit from social network growth) over older publications [48]. Ultimately, Altmetrics may allow for the identification of articles that are shared and used widely but not be frequently cited.

Articles with high 'disseminative impact' may be of significant scholarly value despite a lack of more traditional 'scholarly impact' quantified by citations [48].

Open Educational Resource Metrics

Work traditionally considered 'grey literature' (i.e. not published in a traditional journal) is increasingly being viewed by scholars. Open educational resources (OERs) are available on websites or in the form of blogs and podcasts, which can broadly disseminate new ideas [50]. Their creators can have substantial influence on a field without publishing traditional research articles [27]. We are unaware of any broadly accepted parameters for quantifying the impact of these resources; however, several are described in the literature. The impact of websites can be tracked by their overall number of pageviews (e.g. using Google Analytics) as well as proprietary tools such as the Alexa Score, which ranks websites based on traffic and inbound links to its content [51]. The number of times that a podcast is downloaded can also be readily tracked. As institutions increasingly recognize these less traditional forms of scholarship, we anticipate that reporting standards will develop to compare their relative impact.

Using Metrics to Support Academic Advancement

The value of impactful metrics for supporting academic promotion is increasingly being recognized in the literature and by institutions. Sherbino et al. have defined criteria for social media-based scholarship in health professions education, which could guide the recognition of novel forms of scholarship, which are impactful, but do not fit within the traditional paradigms of academic advancement [52]. Cabrera et al. described the use of a social media portfolio to quantify and demonstrate the impact of these novel forms of scholarship while calling for increased recognition of alternative metrics for decisions regarding promotion and tenure [53, 54]. The Mayo Clinic is an example of an institution that has successfully embraced the use and importance of alternative metrics and social media to support its mission [54]. Notably, the majority of the literature on this topic has been published recently and it is anticipated that it will evolve dramatically over the next decade.

Closing

In summary, research dissemination is a critical step in the process of research. Researchers should plan a dissemination strategy from the outset, with room for flexibility as the project unfolds. Effective dissemination of research ensures that results can help to inform and shape the future of the field – changes that may ultimately improve patient care. Many metrics exist to measure the dissemination of research. Researchers should consider accessing and/or tracking these measures to support academic advancement and/or promotion.

References

1. Cheng A, Auerbach M, Hunt EA, Chang TP, Pusic M, Nadkarni V, et al. Designing and conducting simulation-based research. Pediatrics. 2014;133(6):1091–101.
2. Wilson PM, Petticrew M, Calnan MW, Nazareth I. Disseminating research findings: what should researchers do? A systematic scoping review of conceptual frameworks. Implement Sci. 2010;5:91.

3. Rowe N, Ilic D. Poster Presentation a visual medium for academic and scientific meetings. Paediatr Respir Rev. 2011;12:208–13.
4. Sherbinski LA, Stroup DR. Developing a poster for disseminating research findings. J Am Assoc Nurse Anesth. 1992;60(6):567–72.
5. Goodhand JR, Giles CL, Wahed M, Irving PM, Langmead L, Rampton DS. Poster presentations at medical conferences: an effective way of disseminating research? Clin Med. 2011;11(2):138–41.
6. Cheng A, Lin Y, Nadkarni V, Wan B, Duff J, Brown L, et al. The effect of step stool use and provider height on CPR quality during pediatric cardiac arrest: a simulation-based multicentre study. CJEM. 2017;30:1–9.
7. Davis PM, Lewenstein BV, Simon DH, Booth JG, Connolly MJL. Open access publishing, article downloads, and citations: randomised controlled trial. BMJ. 2008;337:a568.
8. Eysenbach G. Citation advantage of open access articles. PLoS Biol. 2006;4(5):e157.
9. Craig I, Plume A, McVeigh M, Pringle J, Amin M. Do open access articles have greater citation impact? A critical review of the literature. J Informet. 2007;1(3):239–48.
10. Natal B, Szyld D, Pasichow S, Bismilla Z, Pirie J, Cheng A, et al. Simulation fellowship programs: an international survey of program directors. Acad Med. 2017;92(8):1204–11.
11. Frallicciardi A, Vora S, Bentley S, Nadir NA, Cassara M, Hart D, et al. Development of an emergency medicine simulation fellowship consensus curriculum: initiative of the society for academic emergency medicine simulation academy. Acad Emerg Med. 2016;23(9):1054–60.
12. Gillani N, Yasseri T, Eynon R, Hjorth I. Structural limitations of learning in a crowd: communication vulnerability and information diffusion in MOOCs. Sci Rep. 2014;4:6447.
13. Goldberg LR, Crocombe LA. Advances in medical education and practice: role of massive open online courses. Adv Med Educ Pract. 2017;8:603–9.
14. Thoma B, Murray H, Huang SYM, Milne WK, Martin LJ, Bond CM, et al. The impact of social media promotion with infographics and podcasts on research dissemination and readership. CJEM. 2018;20:300–6.
15. Allen HG, Stanton TR, Di Pietro F, Moseley GL. Social media release increases dissemination of original articles in the clinical pain sciences. PLoS One. 2013;8(7):e68914.
16. Tripathy JP, Bhatnagar A, Shewade HD, Kumar AMV, Zachariah R, Harries AD. Ten tips to improve the visibility and dissemination of research for policy makers and practitioners. Public Health Action. 2017;7(1):10–4.
17. Seidel RL, Jalilvand A, Kunjummen J, Gilliland L, Duszak R Jr. Radiologists and social media: do not forget about facebook. J Am Coll Radiol. 2017;15:224–8.
18. Alsobayel H. Use of social media for professional development by health care professionals: a cross-sectional web-based survey. JMIR Med Edu. 2016;2(2):e15.
19. Schnitzler K, Davies N, Ross F, Harris R. Using Twitter to drive research impact: a discussion of strategies, opportunities and challenges. Int J Nurs Stud. 2016;59:15–26.
20. Choo EK, Ranney ML, Chan TM, Trueger NS, Walsh AE, Tegtmeyer K, et al. Twitter as a tool for communication and knowledge exchange in academic medicine: a guide for skeptics and novices. Med Teach. 2015;37(5):411–6.
21. Thoma B, Brazil V, Spurr J, Palaganas J, Eppich W, Grant V, et al. Establishing a virtual community of practice in simulation: the value of social media. Simul Healthc. 2018;13:124–30.
22. Thelwall M, Kousha K. ResearchGate: disseminating, communicating, and measuring scholarship? J Assoc Inf Sci Technol. 2015;66(5):876–89.
23. Batooli Z, Ravandi SN, Bidgoli MS. Evaluation of scientific outputs of Kashan University of Medical Sciences in Scopus Citation Database based on scopus, researchgate, and mendeley scientometric measures. Electron Physician. 2016;8(2):2048–56.
24. Kudlow P, Cockerill M, Toccalino D, Dziadyk DB, Rutledge A, Shachak A, et al. Online distribution channel increases article usage on Mendeley: a randomized controlled trial. Scientometrics. 2017;112(3):1537–56.
25. Ibrahim AM. Seeing is believing: using visual abstracts to disseminate scientific research. Am J Gastroenterol. 2017;266:e46–8.
26. Ibrahim AM, Bradley SM. Adoption of visual abstracts at circulation CQO: why and how we're doing it. Circ Cardiovasc Qual Outcomes. 2017;10(3). https://doi.org/10.1161/CIRCOUTCOMES.117.003684
27. Chan T, Seth Trueger N, Roland D, Thoma B. Evidence-based medicine in the era of social media: scholarly engagement through participation and online interaction. CJEM. 2017;0(0):1–6.
28. Thoma B, Chan TM, Paterson QS, Milne WK, Sanders JL, Lin M. Emergency medicine and critical care blogs and podcasts: establishing an international consensus on quality. Ann Emerg Med. 2015;66(4):396–402 e4.
29. Chan TM, Thoma B, Krishnan K, Lin M, Carpenter CR, Astin M, et al. The derivation of two simplified critical appraisal scores for use by trainees to evaluate online educational resources: a METRIQ study. West J Emerg Med. 2016;17(5):574–84.
30. Chan TM, Grock A, Paddock M, Kulasegaram K, Yarris LM, Lin M. Examining reliability and validity of an online score (ALiEM AIR) for rating free open access medical education resources. Ann Emerg Med. 2016;68(6):729–35.
31. Lin M, Thoma B, Trueger NS, Ankel F, Sherbino J, Chan T. Quality indicators for blogs and podcasts used in medical education: modified Delphi consensus recommendations by an international cohort of health professions educators. Postgrad Med J. 2015;91(1080):546–50.
32. Panahi S, Watson J, Partridge H. Fostering interpersonal trust on social media: physicians' perspectives and experiences. Postgrad Med J. 2016;92(1084):70–3.
33. Panahi S, Watson J, Partridge H. Social media and physicians: exploring the benefits and challenges. Health Informatics J. 2016;22(2):99–112.
34. Buckarma EH, Thiels CA, Gas BL, Cabrera D, Bingener-Casey J, Farley DR. Influence of social media on the dissemination of a traditional surgical research article. J Surg Educ. 2017;74(1):79–83.
35. Vogt M, Schaffner B, Ribar A, Chavez R. The impact of podcasting on the learning and satisfaction of undergraduate nursing students. Nurse Educ Pract. 2010;10(1):38–42.
36. Nwosu AC, Monnery D, Reid VL, Chapman L. Use of podcast technology to facilitate education, communication and dissemination in palliative care: the development of the AmiPal podcast. BMJ Support Palliat Care. 2017;7(2):212–7.
37. Garfield E. The history and meaning of the journal impact factor. J Am Med Assoc. 2006;295(1):90–3.
38. Garfield E. Citation analysis as a tool in journal evaluation. Science. 1972;178(4060):471–9.
39. Thomson-Reuters C. 2017 Journal Citation Reports® Science Edition.
40. Arnold DN, Fowler KK. Nefarious numbers. Noti AMS. 2011;58(3):434–7.
41. Rossner M, Van Epps H, Hill E. Show me the data. J Cell Biol. 2007;179(6):1091–2.
42. Brody T, Harnad S, Carr L. Earlier web usage statistics as predictors of later citation impact. J Assoc Inf Sci Technol. 2006;57(8):1060–72.
43. West JD, Bergstrom TC, Carl T. The eigenfactor metrics: a network approach to assessing scholarly journals. Coll Res Libr. 2010;71:236–44.

44. Rizkallah J, Sin DD. Integrative approach to quality assessment of medical journals using impact factor, eigenfactor, and article influence scores. PLoS One. 2010;5(4):e10204-e.
45. Willems L. Article Usage Reports enable authors to track downloads and views: personalized dashboards provide metrics to gauge an article's impact. 2013.
46. Palmer LA, Gore SA. Taking Flight to Disseminate Translational Research: A Partnership between the UMass Center for Clinical and Translational Science and the Library's Institutional Repository. Library Publications and Presentations. 2016:Paper 195-Paper.
47. Himmelstein DS, Romero AR, McLaughlin SR, Tzovaras BG, Greene CS. Sci-Hub provides access to nearly all scholarly literature. PeerJ PrePrints. 2017;5:e3100v2.
48. Trueger NS, Thoma B, Hsu CH, Sullivan D, Peters L, Lin M. The altmetric score: a new measure for article-level dissemination and impact. Ann Emerg Med. 2015;66(5):549–53.
49. Zaugg H, West RE, Tateishi I, Randall DL. Mendeley: creating communities of scholarly inquiry through research collaboration. TechTrends. 2011;55(1):32–6.
50. Cadogan M, Thoma B, Chan TM, Lin M. Free open access meducation (FOAM): the rise of emergency medicine and critical care blogs and podcasts (2002–2013). Emerg Med J. 2014;31(e1):e76–7.
51. Thoma B, Sanders JL, Lin M, Paterson QS, Steeg J, Chan TM. The social media index: measuring the impact of emergency medicine and critical care websites. West J Emerg Med. 2015;16(2):242–9.
52. Sherbino J, Arora VM, Van Melle E, Rogers R, Frank JR, Holmboe ES. Criteria for Social Media-based Scholarship in Health Professions Education. Postgrad Med J. 2015;91:551–555.
53. Cabrera D, Roy D, Chisolm MS. Social Media Scholarship and Alternative Metrics for Academic Promotion and Tenure. J Am Coll Radiol. 2018;15:135–141.
54. Cabrera D, Vartabedian BS, Spinner RJ, Jordan BL, Aase LA, Timimi FK. More than Like and Tweets: Creating Social Media Portfolios for Academic Promotion and Tenure. J Grad Med Educ. 2017;9(4):421–425.

42 Writing for Publication

William C. McGaghie

Overview
This chapter addresses writing for publication from eight perspectives: (a) motivation to write, (b) types of publications, (c) International Committee on Medical Journal Editors (ICMJE) recommendations, (d) study design and reporting conventions, (e) peer review, (f) words of wisdom, and (g) the craft of writing. Writing for professional publication is hard work that can improve with practice and feedback. Newcomers to professional writing are encouraged to seek advice from experienced colleagues and from many other published resources. The future of healthcare simulation research will depend, in part, on scholars and writers who advance this craft.

Practice Points

- Writing for publication is hard work.
- Research study designs align with research reporting conventions.
- Research reporting conventions can simplify manuscript preparation.
- Read and follow journal instructions to authors.

Writing for publication is hard work. Academic healthcare professionals in nursing, pharmacy, physical therapy, clinical psychology, medicine, and many other specialties know that publications are difficult to produce. These healthcare professionals also know that publications are often a key to career advancement yet few are educated about the goals and skills of professional writing. The strategy and tactics of scholarly expression, its variety of forms and outlets, reporting conventions that govern scholarly work, and how professional writing is submitted and judged are obscure, opaque, and murky. These are some of the mysteries of professional writing that amplify its labor intensity.

The pathway to scholarship and publication in the health professions is often a rocky road, especially for novice clinicians. I wrote in an earlier monograph that, "Newcomers enter the realm of scholarship, publication, and career advancement in health professions education as if going into an alien culture. This alien culture has a language, code of conduct, transaction patterns, and rules of engagement that express core ideas that are different from ideas usually found in clinics and classrooms. Newcomers and established scholars alike must understand, accept, work on, and extend the field's core ideas. Several core ideas (with examples) expressed in health professions scholarship and publication include:

- **Values** – primacy of advancing knowledge and professional practice; conceptual thinking and theory building; clear and simple writing.
- **Aspirations** – conduct 'cutting edge' biomedical, clinical, and behavioral research; publish research reports in peer-reviewed journals; express scholarship in teaching, program development and administration, community service, and many other ways; improve education via research; personal and career development.
- **Key practices** – individual and team science; collegial disputation; reading; writing.
- **Diverse forms of activity** – writing journal articles and other publications; preparing grant applications;

W. C. McGaghie (✉)
Feinberg School of Medicine, Department of Medical Education and Northwestern Simulation, Northwestern University, Chicago, IL, USA
e-mail: wcmc@northwestern.edu

D. Nestel et al. (eds.), *Healthcare Simulation Research*, https://doi.org/10.1007/978-3-030-26837-4_42

teaching; attending and participating in scientific and professional meetings; evaluating papers and grant applications written by peers; professional portfolio management.
- **Judgment criteria** – importance and publishability of written work; methodological rigor of research studies; clear goals, scholar preparation, proper methods, significant results, effective presentation, and reflective critique of scholarly products; quality of writing.
- **Quality standards** – uniformly high, competitive standards for submitted papers; peer review of scholarship; acknowledge the utility of 'connoisseurship' as needed.
- **Recurring conflicts and tensions** – judging scholarly quality and quantity; annual journal page limits; tension about authorship credit; unclear rules about professional advancement and promotion; potential for bias due to financial support or sponsorship" [1].

These core ideas about scholarship and publication in the healthcare professions are frequently tacit, unspoken, and learned from workplace experience. There are few everyday opportunities for health professionals to acquire and refine writing savvy and skills that lead to professional publications. This is despite the worldwide increase in advanced degree programs designed to equip health professionals with scholarly skills and other leadership qualities [2].

The purpose of this chapter is to clarify and simplify some of the inside knowledge needed to boost one's chances of success at writing for publication. The intent is to sweep away the mystery that inexperienced health professions scholars perceive to affect the publication process. The chapter has eight sections: (a) motivation to write, (b) types of publications, (c) International Committee of Medical Journal Editors (ICMJE) recommendations, (d) study design and reporting conventions, (e) peer review, (f) words of wisdom, and (g) the craft of writing.

The chapter does not contain a primer on basic writing styles and skills that are taught and learned in beginning English composition courses. These styles and skills include outlining a manuscript, sentence and paragraph structure, nouns and pronouns, noun-verb agreement, rhetorical devices, developing an argument, writing a conclusion, and many other writing matters. Information and advice about such writing essentials are available from other sources [3, 4].

Motivation to Write

The motivation to write for professional publication stems from external and internal sources. External motives come from one's professional workplace and one's larger national and international community of practice. Internal motives come from an individual's personal hard work and ambition. The two sources of motivation to write professionally are complementary, not separate. External and internal motives to write compel authors to produce good work and meet deadlines.

External motivation for professional writing and publication is especially strong for academic health professionals. Writing that produces professional publications is the "coin of the realm" in most college, university, and teaching hospital academic settings. Individual career advancement is often linked directly to publication productivity, especially from publications in journals with high impact factors. Professional writing is prompted by academic pressure and other external motives including practice improvement, commitment to patient care, promotion of new patient care and educational technologies, and a desire to advance one's field or discipline.

Internal motives to write professionally stem from one's achievement impulses, career advancement and status improvement ambitions, a need for collegial approval, a response to challenge, the desire to make creative contributions, and an aspiration to achieve visibility and recognition. Writing is a form of personal expression. Writing and the visibility it confers establishes an individual's professional image or "brand." Professional writing and publication is also a source of personal satisfaction and builds self-efficacy from serial accomplishments.

External and internal motives to write coalesce to make healthcare education scholars productive and fulfilled.

Types of Publications

There are many types of publications, all of which contribute to the body of skill and knowledge that informs education and practice in the healthcare professions. The journal *Simulation in Healthcare* (SIH) lists ten publication categories that account for its set of eligible reports. A neighbor journal, *Advances in Simulation,* has a similar, but not identical, set of publication categories. The SIH list of ten publication categories is:

1. **Empirical investigations**, divided into qualitative reports (see Sect. 3, Chaps. 9–21 of this volume), quantitative reports (see Sect. 4, Chaps. 22–31 of this volume), and mixed method reports (see Sect. 5, Chaps. 32–33 of this volume) [5].

2. **Technical reports**, which include descriptions of new simulation technologies, novel research methods, and simulation engineering refinements.
3. **Concepts and commentaries**, statements about new directions in healthcare simulation and debates about best healthcare simulation practices
4. **Case reports and simulation scenarios**, informative descriptions of clinical cases and simulation scenarios that have demonstrated evidence-based educational effectiveness among health professions learners
5. **Economic or health policy articles**, reflective reports about such issues as return on investment (ROI) from simulation-based educational interventions and the science of implementing healthcare simulation.
6. **Review articles**, integrative scholarship that addresses at least five research review traditions: narrative, systematic, scoping, critical realist, and open peer commentary [6]. These are covered in Sect. 2, Chap. 7–8 of this volume. Review articles synthesize existing knowledge and point out directions for new investigations.
7. **Special articles**, invited and voluntary written contributions that treat new ideas, shed light on recurring problems or issues, or simply salt the editor's taste
8. **Correspondence**, letters to the editor; scholarly reactions to published reports; expressions of opinion about healthcare research, its consequences, and sequelae
9. **Meeting reports**, minutes and summaries of healthcare policy and research meetings of interest to the simulation community, and
10. **Other items**, a blanket category of reportable simulation research writings that defy placement into the nine preceding buckets.

The point in presenting this list of publication types from one of the leading journals in healthcare simulation is to show the wide variety of writing opportunities that are available to scholars. Healthcare simulation research journals are eager to receive substantive and well-written manuscripts that address topics from many professional angles. Qualitative, quantitative, and mixed-methods empirical reports, the customary routes of journal publication, are only a few categories of scholarly writing sought by prominent journals.

International Committee of Medical Journal Editors (ICMJE) Recommendations

The International Committee of Medical Journal Editors (ICMJE) "is a small working group of general medicine journal editors whose participants meet annually and fund their own work on the Recommendations for the Conduct, Reporting, Editing, and Publication of scholarly work in medical journals." [7] The organization states the purpose of the Recommendations: "ICMJE developed the recommendations to review best practice and ethical standards in the conduct and reporting of research and other material published in medical journals, and to help authors, editors, and others involved in peer review and biomedical publishing create and distribute accurate, clear, reproducible, unbiased medical journal articles." [7]

The ICMJE recommendations have become a de facto code of conduct for journal publication in a wide variety of health science disciplines, including healthcare simulation. The Recommendations are divided into four parts.

1. About the recommendations
2. Roles and responsibilities of authors, contributors, reviewers, editors, publishers, and others
3. Publishing and editorial issues related to publication in medical journals
4. Manuscript preparation and submission

Authors of healthcare simulation research manuscripts can go to the ICMJE Recommendations for advice about many issues involved in journal article publication and other scholarly works such as book chapters. In particular, the ICMJE Recommendations give practical advice about authorship credit. The document states, "The ICMJE recommends that authorship be based on the following 4 criteria.

1. Substantial contributions to the conception or design of the work; or the acquisition, analysis, or interpretation of data for the work; AND
2. Drafting the work or revising it critically for important intellectual content; AND
3. Final approval of the version to be published; AND
4. Agreement to be accountable for all aspects of the work in ensuring that questions related to the accuracy or integrity of any part of the work are appropriately investigated and resolved." [7]

These four authorship criteria have great weight in the academic health professions community. Many journals now insist that all authors listed on a manuscript submitted for publication must affirm in writing that they fulfill the four criteria. The Recommendations continue, "It is the collective responsibility of the authors, not the journal to which the work is submitted, to determine that all people named as authors meet all four criteria; it is not the role of journal editors to determine who qualifies or does not qualify for authorship or to arbitrate authorship conflicts." [7] The goal, of course, is to ensure that authorship credit is genuine, earned, and not a result of professional courtesy or administrative fiat.

Health professions scholars can also consult an on-line report published by the Committee on Publication Ethics

(COPE) titled, “Promoting integrity in research publication,” for advice about academic etiquette and codes of professional conduct [8].

Study Design and Reporting Conventions

Effective scientific and professional scholarship not only relies on clear and simple writing but also on manuscripts that are structured uniformly. The simplest manuscript structure for an empirical research report is the common Introduction, Methods, Results, and Discussion (IMRaD) format. The IMRaD format has been used in research reporting for decades and will likely remain a research reporting mainstay in the future.

Scholars in healthcare simulation research and many other disciplines are mindful that not all research reports are suited to the simple IMRaD structure. Studies that feature new and sophisticated research designs, quasi-experiments, health care simulation, meta-analyses, mastery learning, qualitative investigations, and many other variations need to be published using reporting conventions that go beyond IMRaD rules.

Table 42.1 presents 12 study designs whose results cannot be reported effectively using the IMRaD format. For each study design, the table also presents one or more citations to published guidelines that address the structure and order of separate research report sections. Many of the reporting convention guidelines shown in Table 42.1 also include a checklist for authors so that key methodological and reporting items are not omitted. Many journals now require that research studies grounded in the 12 designs must be written using the correct reporting conventions. Newcomers and experienced investigators alike will find that following the reporting conventions given in Table 42.1 will make outlining and writing manuscripts for publication rule governed and standardized. An added bonus is that use of reporting conventions and their checklists during research planning will boost the rigor and integrity of a proposed study.

Information about research study designs and reporting conventions is also available from the EQUATOR Network report, “Enhancing the quality and transparency of health research” [24].

Table 42.1 Study design and reporting conventions

Study design	Reporting convention
1. Randomized controlled trials and quantitative reports	CONSORT Statement [9] and APA reporting standards for quantitative research [10]
2. Observational studies (cohort, case-control, observational)	STROBE Statement [11]
3. Non-randomized comparative studies	TREND Statement [12]
4. Health care simulation research	Extension of the CONSORT and STROBE Statements [13]
5. Meta-analyses: randomized controlled trials	QUOROM Statement [14]
6. Meta-analyses: observational studies	MOOSE Statement [15]
7. Mastery learning reports	ReMERM Statement [16]
8. Qualitative research reports	COREQ Guidelines [17]; SRQR Statement [18]; and APA reporting standards for qualitative research [19]
9. Realist synthesis	RAMSES Standards [20]
10. Network analysis	Network structure [21]
11. Diagnostic accuracy	STARD Statement [22]
12. Health care improvement	SQUIRE Statement [23]

Peer Review

Peer review of a manuscript presenting a healthcare simulation research report is based on an academic tradition which holds that the value of scholarly work is best judged by a community of scholars. A journal editor, herself an accomplished scholar and writer, selects a small number of reviewers (usually between three and five) from a pool of people knowledgeable in the field or the subject concerning a manuscript. Reviewers may be members of the journal’s editorial board, distinguished professionals, scholars who publish frequently in the journal, or persons who have volunteered to review manuscripts dealing with certain topics.

Peer review subjects an author’s or a research team’s scholarly papers or ideas to the scrutiny of other experts in the same field. This is done before a manuscript is published in a journal, conference proceedings, or in a monograph or book. Peer review informs a journal editor or editorial board if a manuscript should be accepted for publication, reconsidered after review and revision, or rejected. Peer review is not flawless but is still the state-of-the-art procedure for judging scholarly work [25].

Manuscripts are submitted to journals for critical review and publication consideration via the journal’s website. The first review that a manuscript receives in the journal office is performed by the editor or editorial staff to determine that basic submission requirements have been met. Basic manuscript submission requirements usually include a cover letter describing the paper with assurance that it has not been sent to another journal, a conflict of interest guarantee, and acknowledgement of authorship credit. Next, the manuscript will be checked to be sure it conforms with the journal’s instructions to authors regarding the “fit” of the paper’s topic with the journal’s intent and scope and its length, format, tables, figures, and references. Failure to follow these detailed instructions usually means a submitted manuscript will be returned or discarded.

Following initial screening, the manuscript is sent out for peer review along with the specific criteria that the judges use to evaluate the paper. The review criteria usually include scientific accuracy, coherence and clarity of writing, and propriety for publication in the receiving journal. Manuscripts submitted to *Simulation in Healthcare* and *Advances in Simulation* are evaluated using a combination of quantitative and qualitative criteria [13]. Other reports provide detailed descriptions of the methods and criteria that are used to review manuscripts submitted to professional and scientific journals [26–28].

The journal editor or editorial board make one of three decisions about each submitted manuscript after the peer reviews have been returned: (a) accept, (b) reject, or (c) revise and resubmit (R & R). Accept and reject decisions are straightforward and final. A R & R decision means the editor believes the paper has promise for the journal but needs improvement. Authors should take comfort from a R & R decision because a revised paper that incorporates reviewers' suggestions has a good chance for journal acceptance. Most journals in the healthcare professions, including those that publish healthcare simulation research, ultimately publish about 10% of submitted manuscripts. Thus an accept or R & R editorial decision is a sign of encouragement.

Several studies have reported the most common reasons for rejection of research manuscripts submitted to journals in the healthcare sciences. Consistent findings from these investigations include failure to identify a clear research question, flawed or weak research design, inappropriate in incomplete data analysis, small or biased research sample, and failure to address a research topic aligned with the mission of the journal. Other issues include an incomplete, inaccurate, or dated literature review, poor writing, and questionable scientific conduct [29–31]. Authors who attend to these and other matters during research planning and research report writing will increase the probability that their work will be published.

Words of Wisdom

New and experienced healthcare simulation scholars need not work alone. Specific advice about how writing can be accepted for publication in professional journals and books is available from several sources. Two sources are highlighted here. The first is *AMEE Guide No. 43: Scholarship, Publication, and Career Advancement in Health Professions Education* [1]. The second is a journal article authored by psychologist Robert J. Sternberg titled, "How to win acceptances by psychology journals: 21 tips for better writing" [32]. Here, with sampling and in abbreviated form, is what the two sources have to say.

AMEE Guide No. 43 [1]

1. Know local rules at your university, college, or hospital and tailor your publication strategy to meet the rules.
2. Set writing goals and keep your purposes clear. Know exactly the products you aim to write and prepare a timetable for each.
3. Plan and organize your writing. Set a theme for your work and stick to it.
4. Read widely and in depth. Read and learn within and beyond your primary discipline.
5. Acknowledge publication competition and quotas. Journals have limited page space and only the "cream of the crop" will appear in print.
6. Set high standards for yourself and all scholarly collaborators.
7. Never plagiarize.

Sternberg's Advice [32]

1. **Attend carefully to what you say:** start strong with declarative statements; tell readers why they should be interested; describe methods and results clearly and simply; consider alternative interpretations of the data; end strong with a clear take-home message
2. **Carefully craft how you say it:** write clear and concise sentences; write for logical flow and organization; give concrete examples; write to be interesting and tell a story; write for a broad and technically less skilled audience; avoid autobiography
3. **What to do with what you say:** proofread your work; check for fit with journal guidelines and subject matter; get feedback from local colleagues before submitting a manuscript to a journal
4. **What to do with what others say:** take journal reviews seriously but acknowledge that reviewers are not gods; don't take reviewers' comments personally; perseverance usually pays off

Attention to this advice will not guarantee success in writing and professional publication. However, the statements suggest in unison that good professional writing is planful, focused, deliberate, and takes time.

Craft of Writing

Professional writing expertise is challenging work that originates from verbal ability, broad and deep reading, and sustained deliberate practice [33]. There are many useful resources now available to help healthcare simulation scholars to sharpen their writing skills [1, 3, 4, 34, 35], especially for authors who have difficulty writing in English [1, 35]. Young writers can also learn about writing from seasoned colleagues in their home settings, from skill development workshops at professional meetings such as the annual International Meeting on Simulation in Healthcare (IMSH), and in the Writer's Craft series of articles in the journal, *Perspectives on Medical Education.* Writing for publication, like other professional skills, can improve from practice with feedback. New authors, in particular, are encouraged to share their data and ideas in the professional literature by writing with clarity and conviction.

References

1. McGaghie WC. Scholarship, publication, and career advancement in health professions education. AMEE Guide No. 43. Med Teach. 2009;31(7):574–90.
2. Tekian A, Harris I. Preparing health professions leaders worldwide: a description of masters-level programs. Med Teach. 2012;34(1):52–8.
3. Strunk W, White EB. The elements of style. 4th ed. New York: Longman; 2000.
4. Singh AA, Lukkarila L. Successful academic writing: a complete guide for social and behavioral scientists. New York: Guilford Press; 2017.
5. Creswell JW, Plano Clark VL. Designing and conducting mixed methods research. 3rd ed. Los Angeles: Sage Publications; 2018.
6. McGaghie WC. Varieties of integrative scholarship: why rules of evidence, criteria, and standards matter. Acad Med. 2015;90(3):294–302.
7. International Committee of Medical Journal Editors (ICMJE). Recommendations for the Conduct, Reporting, Editing, and Publication of Scholarly Work in Medical Journals. Available at: http://www.icmje.org
8. Committee on Publication Ethics (COPE). Promoting integrity in research publication. Available at: https://publicationethics.org/. Accessed 2019
9. Schulz KF, Altman DG, Moher D, et al. CONSORT 2010 Statement: updated guidelines for reporting parallel group randomized trials. Ann Intern Med. 2010;152(11):1–7.
10. Appelbaum M, Cooper H, Kline RB, Mayo-Wilson E, Nezv AM, Rao SM. Journal article reporting standards for quantitative research in psychology: the APA publications and communications board task force report. Am Psychol. 2018;73(1):3–25.
11. Vandenbroucke JP, von Elm E, Altman DG, et al. Strengthening the reporting of observational studies in epidemiology (STROBE): explanation and elaboration. PLoS Med. 2007;4(10):e297. https://doi.org/10.1371/journal.pmed.0040297.
12. Des Jarlais DC, Lyles C, Crepaz N, et al. Improving the reporting quality of nonrandomized evaluations of behavioral and public health interventions: the TREND Statement. Am J Pub Health. 2004;94(3):361–6.
13. Cheng A, Kessler D, Mackinnon R, et al. Reporting guidelines for health care simulation research: extensions to the CONSORT and STROBE statements. Simul Healthc. 2016;11:238–48.
14. Moher D, Cook DJ, Eastwood S, et al. Improving the quality of reports of meta-analyses of randomized controlled trials: the QUOROM statement. Lancet. 1999;354:1896–900.
15. Stroup DF, Berlin JA, Morton SC, et al. Meta-analysis of observational studies in epidemiology: a proposal for reporting. JAMA. 2000;283(15):2008–12.
16. Cohen ER, McGaghie WC, Wayne DB, et al. Recommendations for reporting mastery education research in medicine (ReMERM). Acad Med. 2015;90(11):1509–14.
17. Tong A, Sainsbury P, Craig J. Consolidated criteria for reporting qualitative research (COREQ): a 32-item checklist for interviews and focus groups. Int J Qual Health Care. 2007;19(6):349–57.
18. O'Brien BC, Harris IB, Beckman TJ, et al. Standards for reporting qualitative research: a synthesis of recommendations. Acad Med. 2014;89:1245–51.
19. Levitt HM, Bamberg M, Creswell JW, Frost DM, Josselson R, Suárez-Orozco C. Journal article reporting standards for qualitative primary, qualitative meta-analytic, and mixed methods research in psychology: the APA publications and communications board task force report. Am Psychol. 2018;73(1):26–46.
20. Wong G, Greenhalgh T, Westhorp G, et al. RAMSES publication standards: realist syntheses. BMC Med. 2013;11:21.
21. van de Wijngaert L, Bouwman H, Contractor N. A network approach toward literature review. Qual Quant. 2014;48:623–43.
22. Bossuyt PM, Reitsma JB, Bruns DE, et al. The STARD statement for reporting studies of diagnostic accuracy: explanation and elaboration. Ann Intern Med. 2003;138:w1–w12.
23. Ogrinc G, Mooney SE, Estrada C, et al. The SQUIRE (Standards for Quality Improvement Reporting Excellence) guidelines for quality improvement reporting: explanation and elaboration. Qual Saf Health Care. 2008;17(Suppl. 1):i13–32.
24. EQUATOR Network. Enhancing the quality and transparency of health research. Available at: http://www.equator-network.org/. Accessed 2019.
25. Cicchetti DV. The reliability of peer review for manuscript and grant submissions: a cross-disciplinary investigation. Beh Brain Sci. 1991;14:119–86.
26. Sharts-Hopko NC. How does a peer review scholarship? J Assn Nurses Aids Care. 2001;12(6):91–3.
27. Hancock GR. In: Mueller RO, editor. The reviewer's guide to quantitative methods in the social sciences. New York: Routledge; 2010.
28. Durning SJ. In: Carline JD, editor. Review criteria for research manuscripts. 2nd ed. Washington DC: Association of American Medical Colleges; 2015.
29. Bordage G. Reasons reviewers reject and accept manuscripts: the strengths and weaknesses in medical education reports. Acad Med. 2001;76(9):889–96.
30. Cook DA, Beckman TJ, Bordage G. Quality of reporting of experimental studies in medical education: a systematic review. Med Educ. 2007;41:737–45.
31. Meyer HS, Durning SJ, Sklar D, Maggio LA. Making the first cut: an analysis of *Academic Medicine* editors' reasons for not sending manuscripts out for external peer review. Acad Med. 2018;93(3):464–70.
32. Sternberg RJ. How to win acceptances by psychology journals: 21 tips for better writing. Pan-Pacific Mgmt Rev. 2008;11(1):51–9.
33. Kellogg RT. Professional writing expertise. In: Ericsson KA, Charness N, Feltovich PJ, Hoffman RR, editors. The Cambridge handbook of expertise and expert performance. New York: Cambridge University Press; 2006. p. 389–402.
34. Belcher WL. Writing your journal article in 12 weeks: a guide to academic publishing success. Los Angeles: Sage Publications; 2009.
35. Rocco TS, Hatcher T, editors. The handbook of scholarly writing and publishing. San Francisco: Jossey-Bass; 2011.

43 Peer Review for Publications: A Guide for Reviewers

Debra Nestel, Kevin Kunkler, and Mark W. Scerbo

Overview
In this chapter, we offer guidance on the processes of peer review for scholarly publications. We explore the purpose of peer review and summarise approaches from healthcare simulation journals. We position the role of reviewer as one of privilege and responsibility. We illustrate reviewer reports and set expectations for approaches to author responses. Although the chapter is written for reviewers it has relevance for anyone involved in the reviewing process.

Practice Points

- Peer review is an important professional responsibility.
- Peer review takes many forms and our focus is on manuscripts for publication.
- There are different types of peer review such as open, single-blind and double-blind.
- There are many considerations in peer review processes which may not be immediately obvious to reviewers when they accept invitations to review.

D. Nestel (✉)
Monash Institute for Health and Clinical Education, Monash University, Clayton, VIC, Australia

Austin Hospital, Department of Surgery, Melbourne Medical School, Faculty of Medicine, Dentistry & Health Sciences, University of Melbourne, Heidelberg, VIC, Australia
e-mail: debra.nestel@monash.edu; dnestel@unimelb.edu.au

K. Kunkler
School of Medicine – Medical Education, Texas Christian University and University of North Texas Health Science Center, Fort Worth, TX, USA
e-mail: k.kunkler@tcu.edu

M. W. Scerbo
Department of Psychology, Old Dominion University, Norfolk, VA, USA
e-mail: mscerbo@odu.edu

Introduction

In this chapter, we explore different types of peer review with a focus on scholarly publication in journals. *Peer review* is defined as "a process of subjecting an author's scholarly work, research or ideas to the scrutiny of others who are experts in the same field." [1] Manuscripts are assessed for publication suitability based on specified criteria that usually consider quality (methodologically sound), originality, significance, and coherence. These reviews inform Editors' decisions to publish manuscripts. In this chapter, our intended audience is anyone undertaking reviews of manuscripts for publication in journals although it has relevance for anyone involved in the process including authors. We start by sharing the purposes of peer review, different types of peer review, and then describe considerations during peer review including peer preview processes, peer review reports, conflicts of interest (and what can be done about them), responses to reviews and finally, ways to become a reviewer.

Purpose of Peer Review

The purpose of peer review in scholarly journals is to ensure standards of research are maintained or advanced through independent scrutiny of manuscripts by experts in the field. Peer review is also intended to improve the manuscript under review, although this is a secondary purpose. Most journals have a minimum of two reviewers per manuscript. Peer review informs the decision to publish a manuscript. This decision is made by an Editorial Board member (Associate Editor or Editor) and usually endorsed by an Editor-in-Chief. The Editor mediates exchanges between authors and reviewers and has a responsibility to ensure that the peer review is fair, respectful and timely. The Editor-in-Chief also makes the final decision on manuscripts. In most instances, the Editor-in-Chief accepts the decision of the Associate Editor or Editor. Sometimes,

D. Nestel et al. (eds.), *Healthcare Simulation Research*, https://doi.org/10.1007/978-3-030-26837-4_43

Table 43.1 Journals focused on publishing healthcare simulation, type of peer review and links to websites

Title	Type of peer review	
Advances in Simulation	Single-blind peer review	https://advancesinsimulation.biomedcentral.com/
BMJ Simulation and Technology Enhanced Learning	Single-blind peer review	https://stel.bmj.com/
Clinical Simulation in Nursing	Double-blind peer review	https://www.nursingsimulation.org/
Journal of Surgical Simulation	Double-blind peer review	http://www.journalsurgicalsimulation.com/
Simulation in Healthcare	Single-blind peer review	https://journals.lww.com/simulationinhealthcare/pages/default.aspx

however, the Editor-in-Chief must make a final decision when there are mixed opinions about a manuscript or even overrule decisions or recommendations made by the Associate Editor or Editor if there are compelling circumstances.

Reviews are undertaken independently by two or more reviewers deemed to have relevant expertise. Reviewers are usually not paid. It is important to note that healthcare simulation research is published in a diverse range of clinical, education, engineering and other journals and that they will have similar reviewer processes. Table 43.1 lists examples of journals focused on healthcare simulation.

Different Types of Peer Review

There are different types of peer review for journal publication. They have benefits and challenges depending on your perspective. In open-peer review (e.g., *BMJ Open* and *BMC Medical Education*), reviewers' identities are shared with authors and readers (if the manuscript is published). Reviewer reports are also published in some open approaches. See Box 43.1 for examples of articles and their history of submissions and exchanged among the authors, reviewers, and the editor.

Box 43.1 Two examples of open review processes

Example 1:
- Nestel, D., Bearman, M., Brooks, P., et al. A national training program for simulation educators and technicians: Evaluation strategy and outcomes. *BMC Medical Education*, 2016;16(25). https://doi.org/10.1186/s12909016-0548-x. The full reviewing history is available at:https://bmcmededuc.biomedcentral.com/articles/10.1186/s12909-016-0548-x/open-peer-review

Example 2:
- Huddy, J.R, Weldon, S., Ralhan, S., et al. Sequential simulation (SqS) of clinical pathways: a tool for public and patient engagement in point-of-care diagnostics. *BMJ Open* 2016;**6:**e011043. https://doi.org/10.1136/bmjopen-2016-011043. The full reviewing history is available at: https://bmjopen.bmj.com/content/bmjopen/6/9/e011043.reviewer-comments.pdf

In single-blind peer review, the reviewers are aware of the names and affiliations of the authors, but the reviewers' reports are anonymous. However, sometimes reviewers will request to add their names to their report. *Advances in Simulation, Simulation in Healthcare,* and *BMJ Simulation and Technology Enhanced Learning* use a single-blind peer review system. Double-blind peer review is where neither authors' or reviewers' names are disclosed to each other during the review process. Manuscripts in this review process need to be de-identified by the authors at submission and only after acceptance do the authors add their names. *Clinical Simulation in Nursing* and *Journal of Surgical Simulation* use this process.

The Peer Review Process

Although every journal has its own process for handling peer review, most follow these basic steps. (1) An Editor-in-Chief reviews the submitted manuscript and makes an initial decision about whether to send the manuscript for review. An Editor-in-Chief can reject a paper at this stage for several reasons (e.g., the topic lies outside the scope of the journal or the work is underdeveloped). More often, the Editor-in-Chief assigns the manuscript to an Associate Editor or Editor, who can also recommend rejection prior to review, but typically invites a set of reviewers based on the expertise required to evaluate the work.

The reviewers accept the invitation and are asked to provide a report by a certain date. When the Associate Editor receives the reports from the reviewers, he or she prepares a synthesis of the comments for the authors and makes a publication recommendation. The Editor-in-Chief then reviews all of the comments and makes the final decision to accept or reject the manuscript, or asks the authors to revise and resubmit the manuscript. If the authors accept the invitation to revise and resubmit, the process described here is repeated until a final decision is made to accept or reject the manuscript

Considerations During Peer Review

Before accepting an invitation, reviewers make a judgement (usually based on the abstract) as to whether they have relevant knowledge to serve as a reviewer. Although the selection process of reviewers is designed to identify 'peer' reviewers, from an author's perspective they may not always appear to align. Depending on the process (e.g., single- or double-blind, etc.) as well as the organization, the expertise of the reviewer may be more or less relevant for the manuscript. Some organizations do allow the authors to nominate potential reviewers as well as provide a list of reviewers who might have a conflict. Invitations may also be sent on the basis of keywords a reviewer has selected in a journal reviewer database, of references cited in the manuscript, or of the professional networks of the Editor/s. Also, individuals with the requisite knowledge and expertise may not accept the invitation to review. These strategies may not always result in the most appropriate match between manuscript and reviewer.

When invited, a time frame is usually provided in which to accept the invitation and to complete the review. If the reviewer is unable to accept, it is helpful to decline quickly so as not to slow down the review process. If the reviewer thinks they will be unable to complete the review in the time-period outlined then they should decline or request a different due date that can be met. If the reviewer regularly undertakes reviews for a journal then the editorial management system may provide an option to indicate availability. On declining, it can also be helpful for the individual to provide the names of other potential reviewers who may be suited to evaluate the manuscript. On acceptance, check the journal guidelines for manuscripts, format for review and it can be helpful to check the Equator network for reporting standards (https://www.equator-network.org/reporting-guidelines/).

During the review process, it is important that reviewers stay focused on the author/s work. It can be helpful for the reviewer to read the manuscript in full and with an open mind before starting to write the review. The author/s of the manuscript may take a different perspective to the reviewer's and of course this does not mean that it is wrong, but just different. Reviewers are welcome, indeed requested to offer honest critique – positive and negative. Although it can sometimes be very frustrating, the reviewer ought to try to hold the author/s in high regard. Comments should be phrased respectfully. Reviews can have a powerful impact on author/s. Reviewers should try to write the kind of review they would want to receive on their own work.

As an Editor-in-Chief, DN will not change a reviewer's comments but may communicate with the reviewer to adjust the tone and sometimes the content of the review if the review 'feels' patronising, unhelpful or uninformed. Reviewers have always responded positively to these rare requests. Ocassionally, Editor-in-Chief MS has made a decision to delete inappropriate comments, and these deletions are communicated accordingly to the reviewer. As an author, DN has received reviews that have frankly been outrageous (personally insulting and uninformed) which should never have permitted to be sent to an author/s. This is not a censoring process but mediating respectful relationships between scholars. Although it takes skills to be constructive, honest and inspiring when a manuscript is of poor quality or challenges the reviewer's own work, a thoughtful review is a gift to the authors. Authors sometimes have poor (or absent) mentors. The reviewer's role is not to step into this mentoring role but to offer an independent judgement on the manuscript in front of them. Reviewers are expected to offer some justification for their judgements. The level of detail of this justification can be challenging to navigate. While detailed guidance can be extremely valuable to author/s it is not the role of the reviewer to rewrite the manuscript. Furthermore, the reviewer is not expected to be a copy editor. Although pointing out some edits can be helpful, it may be more valuable over the duration to the author/s if the review simply offers some examples and requests the author/s reconsider their own work for similar examples.

Occasionally, reviews can seem more about the reviewers than the manuscript they have been invited to review. It is easy for this to happen because the reviewers have likely been invited because of their expertise and enthusiasm for the topic. However, reviewers should take some time to focus and make the review about the manuscript. If the reviewer is aware that a manuscript has evoked a particularly emotional response, then they should pause and reread the review before it is submitted. On re-review, perhaps a day later, it may be easier to alter the tone to be respectful and will benefit the author/s.

Manuscripts are confidential. They should not be forwarded to anyone (even a member of your research team without first seeking approval from the Editor). Requesting another person to undertake a review on your behalf is unacceptable and to some may be considered unethical.

If the reviewer has any serious concerns about a manuscript, then review processes usually enable the reviewer to provide confidential comments to the Editor. The most common concerns are associated with plagiarism of others' work, self-plagiarism, publishing the same or very similar content in multiple venues, data manipulation, or ethical issues in some phase of the conduct of the research. Not identifying the funding source(s) for the research may also be a concern as there could be a perceived conflict of interest of the author with the particular manuscript.

DN discourages reviewers from asking author/s to undertake more data collection. If this is thought to be necessary, then the reviewer should recommend rejecting the manuscript. Other Editors (MS, KK) however have asked authors to consider additional data and/or information that they may have on hand and/or consider data and information from other published manuscripts that might support their own manuscript. Offering feedback that the current dataset is insufficient (and perhaps why) is helpful. An Editor may invite resubmission if data collection seems as though it may be feasible. However, when it is obvious that gathering additional data is not possible the reviewer should reject the manuscript.

Reviewers may not feel comfortable responding to all aspects of a manuscript. This can happen when authors cite unfamiliar theories or clinical procedures or use unfamiliar methods or statistical approaches. If the reviewer does not feel confident of their expertise on certain areas, then it is important to just let the Editor know.

Peer Review Reports

Reviewer reports take different forms and may include online tick box judgements or checklists for structural and content components of the article (e.g. research question/hypothesis clearly stated; statement of human research ethics etc.). An example of such checklists can be found in a paper by Cheng et al. (2016) published jointly by *Advances in Simulation, BMJ Simulation & Technology Enhanced Learning, Clinical Simulation in Nursing,* and *Simulation in Healthcare* [2]. The authors modified reporting standards for randomized control trials (CONSORT, Consolidating Standards of Reporting Trials) and observational studies (STROBE, Strengthening the Reporting of Observational Studies in Epidemiology). These checklists are based on a set of guidelines intended to reduce inconsistencies found in the empirical research literature. They provide a standardized list of details that should be included in reports describing specific research designs. The checklists published by Cheng et al. were developed specifically for simulation-based research in healthcare.

There may also be options for recommendations to accept or reject the manuscript and variations in between. The Editor may have several options to select from in their judgement (e.g. minor revision; major revision; revise before peer review; reject and transfer; reject before peer review; reject after peer review and invite resubmission; reject after peer review etc.). It is reasonable to expect some explanation of the rationale for the decision. Box 43.2 contains guidance for peer reviewers of manuscripts submitted to *Advances in Simulation* and to *Simulation in Healthcare*.

Box 43.2 Considerations for reviewers during the peer review process

1. Only accept the invitation to review if it is within your scope of knowledge and practice and the deadline can be met. Let the Editor know if there are aspects of the manuscript that you do not feel qualified to consider.
2. Undertake the review in a timely fashion.
3. Remain open-minded while reading the manuscript and throughout the review process.
4. Remain focused on the submitted manuscript.
5. Do not share the manuscript with others.
6. Use the editorial management system for all communication about a manuscript with the Editor.
7. Do not contact the author/s directly during the review process. If you feel you need additional information or clarity from the authors, ask for that in your review (or contact the Editor).
8. Read the journal guidelines and seek advice from the Editor if you have any questions.
9. Use the function 'confidential notes to the Editor' (or equivalent) if you have concerns about the manuscript that you do not want to share with the author/s.
10. Structure your review – use headings and subheadings, reference the manuscript line and page numbers and number each point. These structures can facilitate systematic and complete responses from author/s.
11. Respond to requests from the Editor that may include clarification of specific points in your review.
12. If the authors are offered the opportunity to revise their manuscript, be prepared to be re-invited by the Editor to review the revised manuscript and consider whether you think the author/s have attended satisfactorily to your points.

Declaring Potential Conflicts

There are several potential conflicts of interest (COI), which may arise and affect an individual's ability to review. The most obvious COI is that reviewers usually do not review the work of their immediate colleagues, those in their same or recent employment or mentorship. Additionally, reviewing manuscripts of family members or close relatives is deemed a COI. If this situation arises, inform the Editor-in-Chief and the review may be reassigned. Under double-blind peer review, the reviewer will not know the identity of author/s. In this case it is not possible to declare conflicts. Another important COI concerns potential financial incentives. Reviewers need to be acutely cognisant of financial conflicts, such as receipt of funds from the members of the authoring team or their respective organization. If at any time in the review process a conflict may arise (since the full manuscript is only sent after acceptance) simply inform the Editor-in-Chief.

Examples of Responses to Peer Review

When authors are given an opportunity to revise and resubmit their manuscript, it is customary to respond to the reviewers. There are several ways to approach responses to reviewer comments. It is important to follow guidelines offered by the Editor-in-Chief. Authors should be systematic in responding to reviews and offer a point-by-point commentary to avoid making the reviewer search for the changes that have been made. Comments should be addressed in order with an indication to where in the manuscript the changes have been made. It is acceptable for authors to disagree with reviewers' comments and suggestions. They should simply provide reasons for rejecting the reviewer's idea/s with full, thoughtful and respectful responses. It may be appropriate for authors to acknowledge the effort that reviewers have made, especially when they have offered guidance and detailed justification for their feedback. An *author response table* is often used to address reviewer feedback. In an author response table, the author copies all reviewers' comments into the first column. Each reviewer comment has its own row. In the second column, the authors provide their detailed responses to each point *(whether positive and negative)* in each row. A third column may be used to indicate the page and line numbers where changes have been made. Sometimes, authors are requested to indicate in coloured text the changes they have made in their manuscript.

Becoming a Reviewer

Most journals welcome new reviewers. With experience of research and associated publications, it is likely that journals would consider adding new reviewers to their database. Some journals offer online reviewer training while others, especially those associated with professional associations may provide reviewer training at their conferences. Some journals also offer mentoring with reviews, which is an excellent way to transition to the role.

Closing

In closing, the role of reviewer is integral to advancing our professional community. It is important to undertake the role with an open mind. It is a privilege to review research that offers new ideas and is undertaken ethically and with integrity even though it may supersede or shift thinking and practice around the reviewer's own specific interest away from their work. Reviewing manuscripts often helps to refine one's own writing skills. Being a reviewer has developed our knowledge and practice of research. It is an exciting professional contribution.

References

1. Kelly J, Sadeghieh T, Adeli K. Peer review in scientific publications: benefits, critiques, and a survival guide. EJIFCC. 2014;25(3):227–43.
2. Cheng A, et al. Reporting guidelines for health care simulation research: extensions to the CONSORT and STROBE statements. Adv Simul. (1) 2016; Simul Healthc. 11:238–48.

Additional Resources

Bordage G. Reasons reviewers reject and accept manuscripts: The strengths and weaknesses in medical education reports. Acad Med. 2001;76:889–96.

JPGM Gold Con: 50 yrs of medical writing (huge web-oriented bibliography) www.jpgmonline.com/goldcon.asp Accessed 2004.

McGaghie WC. Scholarship, publication and career advancement in health professions education. In: Association for Medical Education in Europe. Dundee: AMEE; 2010.

Part VII

Getting Started in Healthcare Simulation Research: Tips and Case Studies

44 Unpacking the Social Dimensions of Research: How to Get Started in Healthcare Simulation Research

Margaret Bearman, Adam Cheng, Vinay M. Nadkarni, and Debra Nestel

Overview
Research makes a critical contribution to healthcare simulation. Many simulation educators and practitioners wish to start researching but are not sure where to begin. This book has provided many approaches. In this chapter, we introduce practical strategies for individuals new to healthcare simulation research to help navigate the social dimensions of research processes. We offer ten tips range across the lifecycle of a research project: from seeking guidance at the start to publication at the end.

Practice Points

- Conceptualising, conducting and publishing research are underpinned by a range of social practices.
- There are helpful, but often tacit, strategies to guide novices through the simulation healthcare research landscape.
- Consider how you can engage with others to develop as a researcher.
- Think about the audience for your research from the outset of your study – while acknowledging it may change!

M. Bearman
Centre for Research in Assessment and Digital Learning (CRADLE), Deakin University, Docklands, VIC, Australia
e-mail: margaret.bearman@deakin.edu.au

A. Cheng
Department of Pediatrics & Emergency Medicine, Albert Children's Hospital, University of Calgary, Calgary, AB, Canada

V. M. Nadkarni
Department of Anesthesiology, Critical Care and Pediatrics, The Children's Hospital of Philadelphia, Perelman School of Medicine, University of Pennsylvania, Philadelphia, PA, USA
e-mail: Nadkarni@email.chop.edu

D. Nestel (✉)
Monash Institute for Health and Clinical Education, Monash University, Clayton, VIC, Australia

Austin Hospital, Department of Surgery, Melbourne Medical School, Faculty of Medicine, Dentistry & Health Sciences, University of Melbourne, Heidelberg, VIC, Australia
e-mail: debra.nestel@monash.edu; dnestel@unimelb.edu.au

Introduction

Research exploring design, implementation, and impact of healthcare simulation provides critical information [1, 2]. Healthcare simulation research is diverse. Some research studies simulation as the subject (e.g. 'educational intervention') and other research utilizes simulation as an investigative methodology (e.g. probe of process)[3]. Research methods encompass both in-depth explorations of how and why simulation can enhance learning and practice (Sect. 2), and experimental approaches that compare different interventions (Sect. 4). For those who are new to research but curious about what, how and why simulation might affect healthcare process of care and outcomes, this diversity can be both exhilarating and confusing.

Developing healthcare simulation researchers is essential. Many healthcare simulation educators and practitioners lack specific research skills. We find this is particularly true for those without formal research training, but it also applies to clinical research scientists who may be less familiar with educational or social science research methods. Fortunately, there is a plethora of introductory literature on how to conduct educational and social research [4–8] (See Box 44.1). However, as active researchers who mentor others, we see a need to elaborate some tactics for navigating the research landscape. The social interactions that provide an entree into a research community can be hidden beneath the overt and often demanding tasks associated with undertaking research. For graduate students, this social know-how is most often tacitly role-modelled by their supervisors and colleagues (Chap. 39). For those outside of formal research structures these notions can be more difficult to come by. We sum-

D. Nestel et al. (eds.), *Healthcare Simulation Research*, https://doi.org/10.1007/978-3-030-26837-4_44

Box 44.1: Introductions to educational and social research

Tavakol, M., & Sandars, J. (2014). Quantitative and qualitative methods in medical education research: AMEE Guide No 90: Part I. *Medical Teacher*, 36(9), 746-756.

Tavakol, M., & Sandars, J. (2014). Quantitative and qualitative methods in medical education research: AMEE Guide No 90: Part II. *Medical Teacher*, 36(10), 838-848.

Ringsted, C., Hodges, B., & Scherpbier, A. (2011). 'The research compass': An introduction to research in medical education: AMEE Guide No. 56. *Medical Teacher*, 33(9), 695-709.

Cleland, J., & Durning, S. (Eds.). (2015). Researching Medical Education. West Sussex: Wiley Blackwell.

Creswell, J. (2013) *Research design: Qualitative, quantitative, and mixed methods approaches.* Sage Publications.

Mertens, D. M. (2014). *Research and evaluation in education and psychology: Integrating diversity with quantitative, qualitative, and mixed methods.* Sage Publications.

Simulcast: Getting Started in Simulation Research An interview with Margaret Bearman

https://simulationpodcast.podbean.com/e/ep-3-getting-started-in-simulation-research/

marise some of the tacit knowledge that we have collectively gathered over many years, often through trial and error.

The purpose of this chapter is to guide individuals new to healthcare simulation research through the social dimension of conducting research. This complements the extensive 'how to' provided by other resources. In particular, we offer condensed and practical tips, outlining the interactions that might take a 'novice healthcare simulation researcher' from the spark of an idea to publication (Box 44.2).

Box 44.2: Tips for conducting healthcare simulation research

1. Invest time at start by finding a research network
2. Know the literature to find the conversation you want to join
3. Research questions take time and input from many people
4. Think about ethics in a different way – beyond clinical research
5. Start with your network, but be prepared to move beyond
6. Learn what quality looks like
7. You don't have to do everything yourself
8. Don't be shy. Present at conferences!
9. Think of publication as a persuasion
10. Always begin with the end in sight: journals, reviewers and all that jazz

Our advice is particularly aimed at simulation practitioners who are health professional educators. We provide tips that are drawn from our experiences interacting with individuals with educational expertise who are seeking to expand their scholarly activities. For example, a team who have put together an innovative program for teaching leadership and followership in interprofessional settings might want to know if their work makes a difference. There are many paths this team could follow and still undertake scholarly work. If they are interested in finding out if their particular program helped people work in order to find out how to do it better, then this is program improvement and may be considered *scholarship of teaching* [9]. On the other hand, if the team is interested in understanding the phenomena of leadership and interprofessional education more generally, then this is what Boyer [9] calls the *scholarship of discovery*. This chapter is focussed on this latter type of scholarship, which is often termed "research". We present ten tips for those interested in undertaking research.

TIP #1 Start by Finding Guidance

Collegiate interactions lie at the heart of the research process. Without these interactions, it is challenging to calibrate work against the explicit and implicit standards of the broader research world. Novices without guidance can spend a long time in unproductive – or even counterproductive – tasks. When we suggest 'guidance', this can take several forms. Guidance can come from a formal mentor, research collaborators who are experienced, a stable research team, or even 'critical friends', who are more experienced but not able to invest in a more significant mentoring relationship.

It's not always easy to find a mentor, collaborators or even sympathetic colleagues, but it is a vital first step. Networking can be difficult – you can't be shy and nor should you be intrusive. Sometimes, a mutual interest in improving simulation and healthcare provides a useful platform for a connection to others. It's helpful to think about what you bring to any conversation, collaboration or partnership – it may not be research skills, but it may be access to a research population, a willingness to write grants and/or recruit participants, or simply enthusiasm and passion. Another way of finding a research community is to undertake courses in research methods, in simulation, or to attend events run by your local simulation society (see Box 44.3).

Box 44.3: Examples of simulation societies
ASPE – Association of Standardized Patient Educators
ASPiH – Association for Simulated Practice in Healthcare (UK based)
ASSH – Australian Society for Simulation in Healthcare
PASSH – Pan Asian Society for Simulation in Healthcare
SESAM – Society in Europe for Simulation Applied to Medicine
SSH – Society of Simulation of Healthcare (US based)

Ideally, you will find a mentor – for graduate students this is likely your supervisor (See Chap. 39). Unfortunately, there is no recipe for establishing an effective mentoring relationship but there are general principles that can assist. As mentors of emerging researchers, some of the aspects of successful mentoring we reckon are:

- Strong enthusiasm for the topic from mentor and mentee.
- A mentor who empowers the mentee, and a mentee who is willing to embrace responsibility. This entails diligence and proactivity on behalf of the mentee who must be patient with the mentor's often competing priorities but simultaneously politely drive the joint work. It also requires sufficient time and capability on behalf of the mentor.
- Mutual respect and a capacity to compromise; both parties should have some recognition of their own strengths and weaknesses.
- Reciprocal benefits and no exploitation (i.e. the mentee doesn't just pump mentor for resources/information and mentor doesn't ask the mentee to do tasks which only benefit the mentor).

TIP #2: Know the Literature to Find the Conversation You Want to Join

Understanding what's already been researched helps you develop expertise and refine your research question. Conducting a thorough search of existing literature to identify relevant studies also ensures you are not replicating prior work that has already answered your question in a satisfactory fashion. Most importantly, knowledge of the literature allows you to join the existing 'conversation', ensuring that your work builds on existing dialogue within a certain community of practice and advances the field [10]. This will ensure that your work will add to understanding of the topic being researched (See Chaps. 4, 7, 9, 10, and 45).

It can be valuable to find published examples of research that you'd like to do, whether they are in simulation-based education or not. Reviewing publications enables novice researchers to conceptualize what research outputs looks like [8]. Some of the best places to start are dedicated simulation journals (Box 44.4). However, you will also find healthcare simulation research in broader health professions education and a range of clinical journals.

At this point, you may choose to join the conversation by reaching out to researchers who have published in the field. Sometimes these may be people you know, or perhaps friends of your colleagues. Other times, reaching out may involve emailing or contacting someone you've never met. Engaging in discussion with those who are experts in the field will help you shape your current research and inform future studies.

Box 44.4: Healthcare oriented simulation journals
Advances in Simulation – An open access journal of the Society in Europe for Simulation Applied to Medicine https://advancesinsimulation.biomedcentral.com/

BMJ Simulation & Technology Enhanced Learning – A subscription edition-based journal of the Association for Simulated Practice in healthcare http://stel.bmj.com/

Clinical Simulation in Nursing – A subscription edition-based journal of International Nursing Association for Clinical Simulation and Learning http://www.nursingsimulation.org/

Simulation and Gaming – *Simulation & Gaming (S&G)*: An International Journal of Theory, Practice and Research http://journals.sagepub.com/home/sag

Simulation in Healthcare – A subscription edition-based journal of the Society for Simulation in Healthcare http://journals.lww.com/simulationinhealthcare/pages/default.aspx

TIP #3: Research Questions Take Time and Input from Many People

It is often a surprisingly hard task to narrow down the focus of research and this can sometimes be a struggle for experts as well. We suggest that novice researchers spend time on establishing their research topic. One of the best ways to do this is by sharing your ideas with others, ideally both within and outside of your institution. Before you share your thinking, it's helpful to spend time examining your own interests. What are you passionate about? Why do you want to do this research? What difference might this research make and why? One starting point is to brainstorm all your ideas – either in words or in writing – so that you can sift and prioritise. Next, it is worth taking a step back and pragmatically consider what you would like to produce as a consequence of your research.

Researchers have diverse goals and explicitly identifying what you want to achieve can be very useful. For some, research may simply be information to help make a better simulation, along the lines of scholarship of teaching. For others it might be a publication in a high profile journal; for others again it may be a contribution to the institution or discipline. This motive is important in selecting coming to a research question. For example, if your intent is proving to stakeholders the value of your simulation work, then a rich, theory-based qualitative approach will probably not be of use. On the other hand, if you wish to publish in the education literature, then educational theory becomes increasingly important even before you finalise your research question.

Defining an appropriate research question is of paramount importance. A good research question is one that is feasible, interesting, novel, relevant and ethical [11]. Assessing feasibility involves determining if you have access to: an adequate number of subjects; technical expertise; and appropriate resources. Interesting, novel and relevant research questions will intrigue peers in your community, expand the current knowledge base, and offer potentially meaningful contributions to advancing education and/or clinical care. Lastly, questions must be ethically sound and amenable to approval by local human research ethics boards (See Chap. 34). In all cases, research questions *rarely* spring from the page fully formed. They are influenced by the literature, by what is feasible and by learning the characteristics of a well-formed research question. These standards are easy to articulate but novices find it difficult to ascertain if their research question meets these criteria. This is where interactions with others are vital.

Quality research questions are likely to be those that have been vetted and reviewed by colleagues – be they researchers, educators, clinicians or other stakeholders. Through the process of vetting and review, the research question may be refined or revised based upon a number of factors, including (but not limited to): novelty, relevance/importance, and/or feasibility. Sometimes researchers come up with a great question that is both novel and relevant, but local resources (e.g. human resources, equipment, participants) are insufficient to support execution of the project.

TIP #4: There Are Often Ethical Issues in Researching Students, Trainees and Colleagues

Ethics in educational research is sometimes treated as a 'tick-box' of approval but it is more than this. As with all forms of research, the core ethical research principles of "respect, research merit and integrity, justice, and beneficence"[12] have implications for study designs (See Chap. 34). For example, there has to be very good reasons in order to study participants without their consent, or with deception, or endangering them in some way. The primary reason for research shouldn't be spurious – for promotion or self-betterment – but to have some positive benefit for the community. Healthcare practitioners tend to be alive to ethical issues surround patient care but are less aware of the ethical implications of researching colleagues and students. There is often a power differential and sometimes other conflicts of interests when researching students, trainees and other colleagues. Critically for those researchers who wish to investigate their own program, they may have to take a step back from their teaching role as this may be a real conflict of interest. We strongly encourage researchers to discuss these potential ethical issues with colleagues, collaborators and/or their local ethics board to ensure they don't present a barrier to conducting research.

TIP #5: Your Colleagues May Not Have All the Answers …

A challenge for those beginning in research is that they tended to be very bounded by their immediate experience. So, while we've given advice that the first place that you start is with networking and collaboration, it's important to remember that there may be limitations to this approach, particularly as you are likely to select collaborators that align with and reinforce your own perspectives. At some point, it is worth understanding how your work sits within the entire spectrum of research possibilities.

Seeing the bigger picture is often a very difficult step for novices who, once they have found their 'comfort zone', tend to stay within it. In our experience, this 'comfort zone' sometimes takes the form of sitting on one side of the quantitative/qualitative divide. This may be due to many emerging researchers being simulation educators, who come to research wanting to show that their program makes a difference to clinical practice and ultimately to patient care. This is natural, given health professions' scientific foundations and that most people wish to 'prove' conclusively the value of their work. However, focusing on 'proof', may rule out a whole raft of research approaches.

Research methods and methodologies are very diverse, and it's worth having a working knowledge of those areas outside of your immediate focus. In very general terms, quantitative research emphasises objective measurement while qualitative research generally focusses on the subjective experience. Box 44.5 outlines some useful resources for those interested in quantitative approaches. Box 44.6 outlines some useful resources for those interested in qualitative approaches. In addition, there is evaluation research with its associated newer methodologies such as realist evaluation. Likewise, it's important to explore the theoretical framing of the research across a range of traditions. Theory provides the fundamental principles for educational practice and it is often overlooked or taken for granted, even by experienced researchers. There are many theories which inform healthcare simulation research and Box 44.7 contains accessible references relevant to simulation-based education.

If your curiosity is piqued by your reading of this suggested literature, we suggest expanding your research networks by making contact with individuals who bring skillsets and experience that you lack, or who offer a different perspective. Although there are some general principles that remain constant, expertise in one type of research doesn't always translate to expertise in another. Diversifying your research team is often the best way to learn new skills because you come to see how people approach problems differently. This leads on to the next tip …

Box 44.5 : Resources for designing quantitative simulation research

Norman, G., & Eva, K. W. (2010). Quantitative research methods in medical education. *Understanding medical education: Evidence, theory and practice*, 301-322.

Cheng, A., Kessler, D., Mackinnon, R., Chang, T. P., Nadkarni, V. M., Hunt, E. A., . . . Education Reporting Guidelines, I. (2016). Reporting Guidelines for Health Care Simulation Research: Extensions to the CONSORT and STROBE Statements. Advances in Simulation(1).

Cheng, A., Kessler, D., Mackinnon, R., Chang, T., Nadkarni, V., Hunt, E., . . . Auerbach, M. (2017). Conducting multicenter research in healthcare simulation: Lessons learned from the INSPIRE network. Advances in Simulation, 2(6)

Box 44.6: Resources for designing qualitative research

Some recommended introductory texts are:

- Creswell, J. W., & Poth, C. N. (2017). Qualitative inquiry and research design: Choosing among five approaches. Sage Publications.
- Patton, M. (2002). *Qualitative Research and Evaluation Methods* (Third ed.). Thousand Oaks: Sage Publications Inc.
- Braun, V., & Clarke, V. (2013). *Successful Qualitative Research: A Practical Guide for Beginners*. London: SAGE.

High quality qualitative research is based on often contested notions of rigour, different to quantitative approaches. Introductions to these concepts are found in:

- Tai, J., & Ajjawi, R. (2016). Undertaking and reporting qualitative research. *The Clinical Teacher, 13*, 175-182.
- Varpio, L., Ajjawi, R., Monrouxe, L. V., O'Brien, B. C., & Rees, C. E. (2017). Shedding the cobra effect: problematising thematic emergence, triangulation, saturation and member checking. *Med Educ, 51*(1), 40-50.

Box 44.7: Introductions to theory

Overviews

- Bearman, M., Nestel, D., & McNaughton, N. (2018). Theories informing healthcare simulation practice. In D. Nestel, M. Kelly, B. Jolly, & M. Watson (Eds.), *Healthcare Simulation Education: Evidence, Theory & Practice* (pp. 9-15). Chichester: John Wiley & Sons Ltd.
- Bordage, G. (2009). Conceptual frameworks to illuminate and magnify. *Medical Education, 43*, 312-319.

A series in *Clinical Simulation in Nursing* explored theories that inform simulation education practice.

- Nestel, D., & Bearman, M. (2015). Theory and simulation-based education: definitions, worldviews and applications. *Clinical Simulation in Nursing, 11*, 349-354.
- Husebo, S., O'Regan, S., & Nestel, D. (2015). Reflective practice and its role in simulation. *Clinical Simulation in Nursing*, 368-375.
- Eppich, W., & Cheng, A. (2015). Cultural historical activity theory (CHAT) – informed debriefing for interprofessional teams. *Clinical Simulation in Nursing, 11*, 383-389.
- Reedy, G. (2015). Using cognitive load theory to inform simulation design and practice. *Clinical Simulation in Nursing, 11*, 350-360.
- Smith, C., Gephardt, G., & Nestel, D. (2015). Applying Stanislavski to simulation: Stepping into role. *Clinical Simulation in Nursing*.

Tip #6, You Don't Have to Do Everything

Research in healthcare simulation is not typically an individual endeavour. Many would describe research as a team sport, with various members of the team contributing by offering specific areas of expertise to support different pieces of the project [4]. Your mentors, critical friends and interested colleagues are not the same as your research team; in the team, all members directly contribute to the research study. Again, individual experiences are diverse and context dependent. On the one hand, you may have the good fortune of joining a pre-existing team. On the other, you may be in a situation where you need to work mostly solo. If you can have input into the composition of your research team, choosing collaborators should take into consideration: shared interests; passion for research; personality fit; and availability to meet the needs of the project [13, 14].

In all instances, you don't have to do it all. If you lack training in statistical analysis, then finding a statistician is critical. If you've never conducted a qualitative interview, then it is beneficial to work with someone who has. If you have to write in a language that is unfamiliar, you may seek editorial advice on sentence construction. Within your institution and/or university, you may discover individuals who offer expertise to fill specific gaps. Librarians, for example, offer valuable support into how to search the literature. In addition, working in a team has many other benefits: the quality of the work improves through harnessing multiple perspectives; teamwork allows you to share successes along the way; you may forge lasting scholarly relationships that span many projects.

TIP #7: Work Towards Understanding What Quality Research Looks Like

One of the key capabilities that marks the passage from novice to expert, is coming to understand what makes work 'good' and why [15]. Published reporting standards for quantitative [8, 16, 17] and qualitative [18] studies can be helpful in articulating how rigorous studies should be reported. However, understanding the nuances of different approaches takes some degree of experience. While the references in Boxes 44.1, 44.4, and 44.5 will assist those interested in quality research, standards can be very difficult to understand as abstract concepts, and may be easiest to understand through applying them to already published literature.

'Critical appraisal' – judging the rigour of a study – is a key skill for both practitioners and researchers. This allows you to understand to what extent you can trust the findings of a particular publication. However, once again, it is difficult to learn critical appraisal skills as a solo activity. Discussion and interaction illuminate a more complex view of research, to understand why some compromises may be acceptable but others not. If possible, join a journal club, although simulation specific journal clubs may be far and few between. This is where the online community can really come into its own, particularly when exploring successful articles rather than criticising less rigorous work. For example, Simulcast operates an online journal club [19]. This provides an opportunity to explore different ways of understanding research, and again, to look beyond your research network.

TIP #8: Don't be Shy. Present Your Work at Conferences!

As you become increasingly confident, it is important not to forget the last part of the research equation, dissemination (See Chap. 41). These last three tips revolve around how you promote your work to the wider audience, so that others can take account of your research in their own thinking. While it may be intimidating at first, one of the best places to share your research is at relevant conferences. You may find that presenting your work as a poster or an oral research presentation connects you with others in the field who share similar interests. In some cases, these individuals may become future collaborators, mentees and/or mentors. Presenting your work also provides a forum to receive feedback, which can be used to improve future projects, and an opportunity to brainstorm ideas for next steps within a program of research. There are numerous regional, national and international conferences in healthcare simulation, usually associated with professional associations.

TIP #9: Publication is an Act of Persuasion

A research article is an act of persuasion. The authors have to mount an argument to a particular audience that this work is worthy of sharing with the broader community. It can be helpful to think of publication as slightly different to the research study; research dissemination has a social dimension and in fact can be usefully thought of as "scientific stories"[20]. Like all good stories, there are things you must tell and things you can leave out, and to a certain extent this depends on who your audience is. All this must be done with scientific rigour, honesty, transparency and acknowledgement of limitations (See Chap. 42).

One of the things that novices really struggle with in thinking and writing about their research study is articulating why other people might find the work valuable. We've often heard novices enthuse about a particular point being very interesting, but this is not the same as articulating a rationale in a publication. The 'problem-gap-hook' heuristic is a 'rule of thumb' that can help novices articulate a rationale for their study in the introduction [10]. This heuristic essentially suggests that the authors must answer three questions: what problem is this research addressing? What is the gap in the current research that this study is filling? Why will this particular study make a difference? As these questions illustrate, writing for publication is not just about the act of constructing sentences, it is also about the sequencing of ideas.

TIP #10: Always Begin with the End in Sight: Journals, Reviewers and All That Jazz

Picking a journal for your manuscript can be a daunting task for those who have had little prior experience in publishing. Identifying the most suitable journal for disseminating your work helps to ensure it is shared with the appropriate audience. To identify the most suitable journal, we recommend:

1. Think about appropriate journals BEFORE you start your study, to get a sense of what the final output of your research may look like.
2. Determine the aim and scope of the journal – this can typically be found on the journal website;
3. See what's been published – going online to viewing journal archives can give you a sense of the type of articles and topics of interest to a particular journal;
4. Identify article type – journals have varying different article types; find the article type that best matches the work you would like to disseminate (e.g. innovation, research, methodology, etc.);
5. Is the journal open access? Articles published in open access journals are free and accessible immediately upon publication, which can help improve the dissemination of your work and the authors retain the copyright. However, these journals often charge a fee for publishing;
6. Talk to others – exploring others' experiences with publishing in different journals allows you to understand the review process, including the length of time to receiving reviewer comments, and subsequently to publication. This information may influence your ultimate choice of journal (See Chap. 43).

Finally, once your manuscript has come back with a request for major or minor revisions, we suggest that authors respond to reviewers' comments in a respectful, concise and timely manner. The natural, visceral response to reviewers' comments is often one of defensiveness, and sometimes sprinkled with feelings of frustration, anger, and perhaps even apathy. We highly recommend reading reviewers' comments and then taking a day or two to calm your emotions before drafting the response to reviewers. In crafting a response, create a separate document by cutting and pasting editor and reviewer comments, leaving space to response to each individual comment. We have found that being concise in your responses helps – lengthy, verbose responses run the risk of introducing further criticism while at the same time adding extra work for reviewers. You should do your best to address all reviewers' comments, but sometimes you may not agree. This is okay, but your response should be respectful, and include your rationale for why you chose not to revise the manuscript as suggested. Lastly, respond in a timely manner to ensure the manuscript stays fresh on the minds of reviewers and editors. Putting your response to reviewers at the top of your 'to-do' list ensures you are not the rate-limiting step in the publication process.

Conclusions

This chapter introduces tips that are often not recorded but may be useful to those starting out in healthcare simulation research. Research can seem daunting; in fact, this feeling can remain even when you become more experienced. However, it is also tremendously rewarding. This article does not seek to replicate the excellent educational research methods literature, instead offers practical guidance to help with the social aspects of undertaking healthcare simulation research. We hope that identifying often tacit research processes helps simulation educators and practitioners to take the first step towards research and scholarship. The simulation community is tremendously supportive and we are confident that novices will be welcomed and encouraged.

References

1. Issenberg SB, et al. Setting a research agenda for simulation-based healthcare education: a synthesis of the outcome from an Utstein style meeting. Simul Healthc. 2011;6(3):155–67.
2. Hunt EA, et al. Building consensus for the future of paediatric simulation: a novel 'KJ Reverse-Merlin' methodology. BMJ Simul Technol Enhanc Lear. 2016;2:1–7.
3. Cheng A, et al. Designing and conducting simulation-based research. Pediatrics. 2014;133(6):1091–101.
4. McGaghie WC. Scholarship, publication, and career advancement in health professions education: AMEE Guide No. 43. Med Teach. 2009;31(7):574–90.
5. Cook DA, Beckman TJ. Current concepts in validity and reliability for psychometric instruments: theory and application. Am J Med. 2006;119(2):166 e7–16.
6. Cook DA, Beckman TJ. Reflections on experimental research in medical education. Adv Health Sci Educ Theory Pract. 2010;15(3):455–64.
7. Cook DA, West CP. Perspective: reconsidering the focus on "outcomes research" in medical education: a cautionary note. Acad Med. 2013;88(2):162–7.
8. Cheng A, et al. Conducting multicenter research in healthcare simulation: lessons learned from the INSPIRE network. Adv Simul. 2017;2:6.
9. Boyer EL. Scholarship reconsidered: priorities of the professoriate. Lawrenceville: Princeton University Press; 1990.

10. Lingard L. Joining a conversation: the problem/gap/hook heuristic. Perspect Med Educ. 2015;4(5):252–3.
11. Hulley SB, Cummings SR, Browner WS. Designing clinical research: an epidemiologic approach. 2nd ed. Philadelphia: Lippincott Williams and Wilkins; 2001.
12. National Statement on Ethical Conduct in Human Research 2015.
13. Irving SY, Curley MA. Challenges to conducting multicenter clinical research: ten points to consider. AACN Adv Crit Care. 2008;19(2):164–9.
14. Minnick A, et al. The management of a multisite study. J Prof Nurs. 1996;12(1):7–15.
15. Tai J, et al. Developing evaluative judgement: enabling students to make decisions about the quality of work. Higher Education, 2018;6(3):467–481.
16. Cheng A, et al. Reporting guidelines for health care simulation research extensions to the CONSORT and STROBE statements. Simul Healthc J Soc Simul Healthc. 2016;11(4):238–48.
17. Cheng A, et al. Reporting guidelines for health care simulation research: extensions to the CONSORT and STROBE statements. BMJ Simul Technol Enhanc Learn. 2016;12. bmjstel-2016-000124
18. O'Brien BC, et al. Standards for reporting qualitative research: a synthesis of recommendations. Acad Med. 2014;89(9):1245–51.
19. Simulcast: journal club. n.d. 19/3/2018; Available from: http://simulationpodcast.com/category/journal-club/.
20. Lingard L, Driessen E. How to tell compelling scientific stories: tips for artful use of the research manuscript and presentation genres. In: Researching medical education. Hoboken: Wiley; 2015. p. 259–6 8.

Case Study 1: Joining the Research Conversation: The Romance and Reality

45

Stephanie O'Regan

Joining the Research Conversation: The Romance and the Reality

Romance is a dangerous thing; that overwhelming feeling of excitement and attraction that blinds us to flaws and minimizes difficulties. The day to day slog is glossed over as "not a problem" and timeframes are elastic. I entered the world of research as a self-confessed romantic; blinkered and blissfully ignoring the challenging road ahead. Three years in, I'm now not exactly sure what my goals were, and as a goal driven person this confession is frightening. I should have realized that because I kept my research venture very quiet (a secret from most people) that when the word got out that I was doing my PhD, I brushed it off saying that I wanted to learn to do some research and this seemed to be a good way to go about it. This of course is a total lie; I just wasn't sure I could make it, I wasn't sure the romance would last, that I could do this.

My opening paragraph probably sounds like a call to step away, don't do it; but interestingly it's not. I'm hoping that by sharing my story new researchers, nervous ones, perhaps lacking confidence about research can feel assured that they're not alone, and their feelings are real. Research takes courage, support and time. It also requires a reality check – this is hard work, don't ever believe it isn't. When you start at ground zero like me, the hill is very steep. In this case study, I want to share five lessons I have had, what has worked for me and what has not; but I know that it won't be the same for you. This is an individual journey, some find it easy, some challenging. I share my experiences of joining research conversations in the healthcare simulation community and specifically, joining research conversations on my research topic of interest – of optimizing learning for observers in healthcare simulations.

Lesson One: Know Yourself

I am an emergency nurse, that remains my dominant identity even though I have not worked at the clinical coalface for 15 years. That is part of who I am. Getting to the core of what works and what doesn't was important to my nursing work then and remains so now. As a clinical nurse I read best practice guidelines and tried to apply them. I sought information that was applicable to my day to day work. As an educator this is no different, but I was finding that my education questions couldn't be answered. I searched for best practice guidelines and found there was a concentration on those learners actively involved in a simulation, but very little on learners who watched, the observers. I talked to colleagues who suggested I do some research in this area, but as a novice I knew I needed strong mentorship and some practical knowledge. I completed short courses in qualitative and quantitative research to remind myself of lessons learned long ago and as preparation for a more substantial foray into this world. It was not enough, but still worthwhile. I am at my best when I am true to who I am – a pragmatist who likes the concrete and struggles with esotericism. I want practical answers to questions that I can apply in practice. I am also impatient and like results, and I worried (and still do) that this would not serve me well over the long haul.

The bulk of this book is directed toward a description of the concepts and skills needed to conduct high quality healthcare simulation-based research. Little time, however, has been spent exploring the journey that new researchers take as they attempt to master this field and join in the wider conversation. With this in mind, we offer the following chapter by Stephanie O'Regan, a doctoral student at Monash University, as a window into how this road appeared to her as she has traveled it. Our hope is that it will be an encouragement to other young researchers as they embark on their own travels.

S. O'Regan (✉)
Sydney Clinical Skills and Simulation Centre, Northern Sydney Local Health District, St Leonards, NSW, Australia

Faculty Medicine, Nursing and Health Sciences, Monash University, Clayton, VIC, Australia
e-mail: stephanie.oregan@health.nsw.gov.au

D. Nestel et al. (eds.), *Healthcare Simulation Research*, https://doi.org/10.1007/978-3-030-26837-4_45

Lesson Two: Choose Your Supervisors Well

I was very lucky to avoid a major error that I now realise would have completely derailed my journey. I approached my affiliated university and asked about the research program. They kindly provided me with a list of potential projects and supervisors, each with considerable expertise, but not within my field or particular subject interest. This is not an uncommon approach [1], but I knew this would not work for me. I was doing this to answer my own questions, not join the work of others. I personally knew two people who did not complete primarily due to an incompatible supervisory relationship and I didn't want to fall at the first hurdle.

I have worked in healthcare simulation for a long time and so have a network of smart colleagues. I spoke with several and found that others were interested in the topic and were able to provide me with some direction. In one of these conversations an offer of supervision was made and accepted. I ended up studying with supervisors who were located interstate at a university I hardly ever attend, and speaking with others it seems this is also not uncommon in healthcare simulation. In fact, many pursue studies internationally, so do not despair if you do not have a potential supervisor in your neighbourhood.

Three areas have been identified as important for successful supervisor selection; the supervisor's expertise in the student's preferred topic area, matched interpersonal working patterns and research methodology expertise [1]. This can be satisfied with a co-supervisory model, which was the advice given to me. It has proved valuable as there is a synergy between my supervisors and myself and I feel as if we each have different perspectives and contributions to make. However most important for me is the enthusiasm my supervisors have for my topic matter, their ability to both expand and focus my direction, and the support provided when I had a period of personal issues which impacted on my progress. Lack of support can be a major contributor to program withdrawal [1]. Good supervision promotes the development of higher level graduate attributes, described by Barrie as *transformative* including the *ability to interact with, not just disseminate knowledge* [2]. My supervisors have provided opportunities for development and growth through collaborations, research strategies, article review and academic presentations. This case study is one such opportunity.

Lesson Three: Select a Topic for Study

My initial plan was to investigate strategies which optimize learning for observers of simulations. Partly this was fueled by seeing "learners" on their devices whilst watching a scenario play out starring their colleagues. The cynic in me saw it as distraction although I will acknowledge they may have been referring to clinical guidelines. This was, I thought a simple question and not "worthy" of doctoral studies; I was concerned about how I could make it so. Had I then looked to the 2011 IMSH research consensus summit agenda for simulation I would have seen that one agenda item was learner characteristics and how to better identify and meet learning needs [3]. Learning requirements of the observer would fit within this agenda. Mariani and Doolen [4] also report a need to understand optimal student numbers and roles in simulation. Ideas for study can be found and supported using these research agendas.

As an insider I had an understanding of what the observer role entailed, but interestingly my understanding differed from others. That was a big eye opener. If there were differences in what I thought was a simple fundamental definition, then maybe there was actually more depth to this than I originally considered. The healthcare simulation literature on this topic at the time was not extensive. I recall being worried when I narrowed the systematic review down to only nine papers that it would not be enough. It turned out it was just the beginning. Although there was an effort involved in undertaking the systematic review – of learning search strategies, of identifying quality indicators of manuscripts, of trying to make meaning of findings relative to what educational and other theories espouse – the result has been somewhat gratifying. I was able to identify some key factors that support learning in the observer role, and apply them to my practice, satisfying my need for immediacy [5].

At this point I realized I was not alone in wondering about this question. The conversations quite *literally* started. Colleagues contacted me to both thank me for the paper but also for challenging them to consider the observer in their practice. I was asked to provide feedback on a master's thesis on the topic and advise on simulation program practice change at a number of places. I was actually amazed at the interest and it did a lot to boost my self-esteem, which was being severely challenged in this research process. The article has now had 23 citations! And when I launched the second phase of my study, a survey of simulation educators' practices in supporting observers, I have been heartened by the generosity of people who were prepared, even happy to provide their perspective.

Lesson Four: Document the Journey

This is such good advice and one with which I have struggled. Time passes in a blink of an eye, but memory fails regarding thoughts at the time, decisions made and reasons why. I have a colleague who keeps a handwritten notebook in addition to his files where he writes his ponderings. He says they flow better through a pen than a keyboard. When it comes time to document the journey in the form of a thesis, this richness of detail will be so useful. For myself, I have files of conversations, data and readings. Let's hope they suffice. I also have a moleskin notebook; sadly only two pages even have a mark, one with text and the other a simple drawing.

Lesson Five: Join the Community

As an external student located 1000 kilometers away from the university, and working full time, I have also been neglectful here. We have a research group, led by a doctoral student and there is opportunity to present to the group, listen to others, and regular Pomodoro© [6] writing sessions. At times I have been able to attend virtually, occasionally face to face but often due to work pressures I have missed out. I am lucky to work in an environment where scholarly activities and research are supported and encouraged, which somewhat fills this deficit.

The Road Goes On...

The role of research is to contribute knowledge, building on the work of others, suggesting new ideas and moving the conversation forward. As an emergency nurse I provided front line care, one patient at a time. As an educator I encourage clinicians to reflect and be their best, contributing to patient care from a distance. As a researcher I am starting a conversation, and perhaps challenging others to see how this might fit into their practice. Working in healthcare simulation I discovered an interest in research; the simulation specialty is innovative and growing exponentially. Researching the observer role in simulation allows me to contribute to the conversation in a way that is meaningful, and it is inspiring to be able to do so. I know that what I have contributed so far has been useful for some in healthcare simulation, and that is gratifying. If you have a question, an urge to know, then follow it. If you want to know, so will others. Find a mentor, chart a path and take just one step. You will develop skills and attributes which are translatable to your workplace; you will become better at your job because of the process, discipline and rigor required [2]. Join the research conversation because you will add value, you will contribute, and that is important to remember.

I am not yet at the end of my journey, and I have had a few bumps along the way. It has been difficult, there have been tears both of frustration and joy, and I expect there will be more. I have learned a great deal, and not just academic knowledge, but knowledge about myself, my abilities (which are more than I thought) and my fears. I still suffer from quasi imposter syndrome in this world, but am reassured that I am not alone here, and my contribution is worthwhile. Like a true romance, as the excitement and newness settled a commitment to a longer-term relationship developed. Time will tell whether I succeed, but regardless it will have been worthwhile.

References

1. Ives G, Rowley G. Supervisor selection or allocation and continuity of supervision Ph. D students progress and outcomes. Stud High Educ. 2005;30(5):535–55. https://doi.org/10.1080/03075070500249161.
2. Platow MJ. PhD experience and subsequent outcomes a look at self perceptions of acquired graduate attributes and supervisor support. Stud High Educ. 2012;37(1):103–18. https://doi.org/10.1080/03075079.2010.501104.
3. Issenberg SB, Ringsted C, Ostergaard D, Dieckmann P. Setting a research agenda for simulation-based healthcare education: a synthesis of the outcome from an Utstein style meeting. Simul Healthc. 2011;6(3):155–67. https://doi.org/10.1097/SIH.0b013e3182207c24.
4. Mariani B, Doolen J. Nursing simulation research: what are the perceived gaps? Clin Simul Nurs. 2016;12:30–6. https://doi.org/10.1016/j.ecns.2015.11.004.
5. O'Regan S, Molloy E, Watterson L, Nestel D. Observer roles that optimise learning in healthcare simulation education: a systematic review. Adv Simul. 2016;1(1) https://doi.org/10.1186/s41077-015-0004-8.
6. Cirillo F. The pomodoro technique (the pomodoro). 2007. Retrieved from: http://www.baomee.info/pdf/technique/1.pdf.

46 Case Study 2: Discovering Qualitative Research

Sharon Marie Weldon

Introduction

Whether they are aware of it or not, the philosophical stance (or research paradigm) that a researcher takes is crucial to how the research is conducted. The "stance" or ontological (what is reality?) and epistemological (how can we understand reality?) beliefs taken will direct the entire research approach and in particular the type of theoretical perspective (what approach should we take to acquire knowledge of reality?) selected (positivism, interpretivism, or pragmatism). This in turn directs the research methodology (the procedures used to acquire the knowledge) and methods (the tools used to acquire the knowledge) used. For example, positivists believe that reality can be measured and known, and therefore they tend to generate questions and hypothesis that are testable to provide definitive answers, whereas interpretivists believe that reality is constantly changing and therefore needs to be explored and interpreted within its context, without making prior assumptions. On the other hand, pragmatists believe that some types of reality can be measured and known, and some need to be explored and interpreted depending on the subject under study. They therefore aim to either choose the approach that seems most likely to answer the question at hand or mix research approaches to provide answers to what, why, and how. A more holistic understanding of the subject under study can then be generated. However, positivist and interpretivist purists often don't believe this approach is valid due to their strong beliefs on how reality is known. An awareness of the philosophical assumptions made in research is crucial to ensure the quality of the research and to enable the researcher to be confident in their approach and creative with their methods.

Healthcare research is predominately approached through a positivist/postpositivist lens, an objective approach that relies on their being a reality that exists and is discoverable. This approach aligns well with large-scale quantifiable or experimental research that asks questions relating to effectiveness, sensitivity, specificity, causation, and costs [1]. However, these methods do not always answer the why questions, leaving a gap in our understanding. Interpretative/critical lenses attempt to fill this gap by guiding research that seeks to understand or emancipate. Healthcare research has predominately used positivist/postpositivist lenses to answer questions related to disease, morbidity, and mortality. They have been successful in this approach with the majority of mortality and morbidity rates decreasing since the conception of healthcare research due to a better knowledge of diseases and treatments [2–4]. Nonetheless this approach often ignores issues related to experience and context, subsequently generating gaps in our knowledge and ultimately gaps in approaches to how we conduct healthcare.

As a nurse, I was accustomed to the positivist/postpositivist lens with limited insight into interpretative/critical approaches. The qualitative research I had been exposed to was basic in its conduct and interpretation, which doesn't detract from its meaningful and usefulness but instead highlights the lack of depth and understanding of qualitative research conducted in healthcare. This is of course changing, and the why and how questions are being answered more frequently by interpretative/critical approaches in healthcare, with more and more studies combining approaches through mixed methods (pragmatism) to provide a more holistic view of a subject matter [5–10].

To give an insight into the type of interpretative/critical lens that can be used and what questions it can answer, the following case study of the first truly qualitative study I was involved in will aim to do just that. The case study will describe the study, followed by some diary entries that myself and a social scientist (Dr Terhi Korkiakangas) wrote at the time, and finally the outputs produced.

S. M. Weldon (✉)
Department of Education and Health, University of Greenwich, London, UK
e-mail: S.M.Weldon@greenwich.ac.uk; s.m.weldon@gre.ac.uk

D. Nestel et al. (eds.), *Healthcare Simulation Research*, https://doi.org/10.1007/978-3-030-26837-4_46

Case Study

As a nurse who had worked in operating theatres in the UK, I was always fascinated by the team dynamics and in particular the communication aspect. In an environment where the focus of attention for all healthcare professionals involved is on one patient, good communication is paramount to completing the tasks at hand, safely and effectively.

From my own personal experience, communication in operating theatres had not always been seamless and sometimes could be ineffective and even detrimental to the task at hand. Therefore, when a research study looking into communication in operating theatres through video recordings was advertised, I was intrigued as to how this could be done.

The study I joined was a collaboration between healthcare professionals (namely, surgeons, anesthetists, and nurses) and social scientists (namely, experts in interaction, conversation, and multimodal analysis). It was exploratory in nature (interpretivist), meaning that there was no specific research hypothesis to test but instead a guiding frame of focus – communication between healthcare professionals in operating theatres in the UK.

My role, alongside a social scientist (Terhi), was to conduct the study (gain access to, consent for, and set up cameras and microphones in operating theatres), transcribe and code the audio/video data collected, analyze the results (through the identification of communicative patterns), and write up and disseminate the findings.

As a nurse who had worked in operating theatres, the study set-up component was the easiest for me. What was more difficult was shifting my research lens. Table 46.1 illustrates this difficulty by presenting a page (the first day of

Table 46.1 Diary entries in the operating theatre

Date: Thursday May 17, 2012	
Terhi	Sharon
The first time I was inside an operating theatre I was about 5 and had a tonsillectomy done. This Thursday, it was my second time in an operating theatre. Having dressed in blue scrubs (which are better than pajamas!) and done a fantastic tour around the hospital floor with a senior theatre team member, Sharon and I were allowed in to observe an operation. The patient was a young boy. He was brought into the theatre that was full of people wearing gowns, masks, gloves, and scary looking machines – no wonder he was scared. It was heart breaking to see the boy cry, but his mum was allowed in to hold his hand just until he was asleep. Going to sleep in itself was rather beautiful to watch as it happened very smoothly in the competent hands of an anesthetist The team working on the boy's appendicectomy consisted of two surgeons, a trainee scrub nurse and his mentor, a few circulating nurses, and possibly other trainees or medical students. It was really interesting to observe what actually goes on during surgery and how the team members accomplish a complex task of fixing a person. One thing that struck me was that I struggled to hear what these people wearing masks were saying to each other, even as I moved closer and stood right behind the operating surgeon. Nevertheless, the operation progressed smoothly; I'm fascinated how these interactions unfold, after all surgery is accomplished through communication – whether vocal or nonvocal. I also noted that different levels of experience can affect how the team members attend to the conduct of others which is extremely interesting to me as a social interaction researcher. A wonderful example was the senior theatre member's responsiveness to the eye gaze of the mentor who was standing by the instrument trolley with the trainee scrub nurse. She quickly noted the mentor's "circulating" gaze, swiftly approached her, and asked what she needed. Instances like this are beautiful examples of what we might mean by multimodal communication, how we conduct analyses of each other's gaze, body movement, and gestures with or without talk. I look forward to examining these sorts of things more closely throughout the project.	This was my first introduction to the operating theatres of the hospital. We were met by a member of the theatre team who quickly set us up with the necessary badges, lockers, and scrubs We were then taken on a tour of the operating theatres, including patient waiting rooms, equipment storage, and the recovery area. Apart from a different layout, this was all very familiar to me with only small differences, such as the use of an electronic equipment ordering system and camcorders in the light handles We were asked if we wanted to watch a small operation, to which we responded enthusiastically. The procedure was an open appendicectomy on a child of about 8 years old. This was a procedure I was very familiar with, and memories of my time as a theatre practitioner came flooding back The scrub nurse for the operation was a trainee OPD who was being mentored by a senior nurse. There was also one circulator, the consultant surgeon, and two junior doctor assistants, as well as the anesthetist and another OPD I watched the scrubbed OPD do his instrument and swab count and prepare the instruments and drapes to enable the surgeon to commence the operation. I was struck by how exactly alike this procedure is even in a different hospital with different staffing At first I felt as though I needed to revert back into a nursing role and help to facilitate the operation. It felt alien to stand back and watch and not use the skills I had been trained in. However, once I adjusted to the environment, I was able to see the whole scenario from a completely different perspective and felt privileged at having the opportunity One of the team members was very helpful in talking us through what was happening and in giving us ideas of where would be best to record from and where would not. I was aware that it was my colleague Terhi's first time in an operating theatre and was mindful to check that she was ok; however, she appeared enthralled by the experience As the operation closed, we watched the swab count and got talking to one of the theatre nurses. The environment was very relaxed, and she advised that she would be happy for us to video record her once our research commenced which was extremely encouraging

the study) from the research diaries myself and Terhi had kept independently. Both diary entries were written about the same day that we had both been present at.

These diary entries are a great example of the different lenses we both brought to the operating theatre. As a social scientist researcher, Terhi immediately focused on aspects of communication even before data collection had commenced (interpretivism). As a nurse, with limited experience in social science approaches, I was more concerned with the practicalities and processes relating both to my nursing role and how I thought research should be viewed and conducted (postpositivism).

It took time for me to think differently and realize that from an interpretivist lens, my interpretation of the environment and its context was as crucial to the research as the collecting of the video/audio data itself, as this would help me to identify patterns that could later be confirmed or refuted by the data – a very different approach to research than I had been accustomed to.

Data was collected using two wide-angled camcorders and two wireless microphones (placed under a surgeon and scrub nurses sterile gowns) and captured over a period of 6 months. Different operating theatres, theatre teams, and operation types were filmed. Over 60 hours of video data was captured. Patterns of communication were identified by myself and Terhi and recorded through field notes during the data collection period. The data was then transcribed and coded according to a request/response framework. This enabled us to check if the patterns we had seen were as frequent as we believed, as well as providing concrete examples, and an opportunity to quantify the qualitative data to look at the frequencies of the patterns across our dataset (pragmatism). Once this had been done, the findings (in the form of video clips) were presented to those who we had filmed in the operating theatre to confirm if our interpretations had been correct or not (interpretivism).

Once complete, this study was very successful in identifying areas of communication issues and breakdowns that had not previously been identified or understood. Many of which were mundane (such as whether or not a vocal response was provided to a surgeons request, the positioning of the scrub trolley, or the playing of music); however, accumulatively the mundane instances of communication could cause increased frustrations and a sense of mistrust and inability among team members, prohibiting good seamless communication between each other [11–18].

Prior to this study, research on communication issues in operating theatres had been predetermined (the communication issues decided in advance; postpositivist approach) and assessed through checklists. By using a different lens (interpretivist), this research produced novel findings that could add to our understanding of what the issues were and what could be done to improve communication in operating theatres.

A simulation training program (ViSIOT™ – Video Supported Simulation for Interactions in the Operating Theatre) was then designed that was based on the empirical findings of the study. This was unique from a simulation perspective (based on and driven by research) and even more useful from a research perspective as the simulations could be recorded and coded much in the same way the original research had been. This generated an ongoing (pragmatic) dataset that could employ an iterative approach to problem-solving the communication issues identified and involve the professionals by verifying the original research, comparing it to their own simulated practice, and strategizing solutions. In this format, data was made available from real operating theatres, simulated operating theatres, and simulated potential new ways of working in operating theatres – thus building a body of video-based research across a range of operating theatre (simulated, projected, and real) contexts [19]. A potential next step would be to follow the teams back into real-life practice and qualitatively record any changes in communication patterns, bringing the research full circle.

References

1. Röhrig B, du Prel JB, Wachtlin D, Blettner M. Types of study in medical research: part 3 of a series on evaluation of scientific publications. Dtsch Arztebl Int. 2009;106(15):262–8.
2. Armstrong K. Methods in comparative effectiveness research. J Clin Oncol Off J Am Soc Clin Oncol. 2012;30(34):4208–14.
3. Johnson ML, Crown W, Martin BC, Dormuth CR, Siebert U. Good research practices for comparative effectiveness research: analytic methods to improve causal inference from nonrandomized studies of treatment effects using secondary data sources: *The ISPOR good research practices for retrospective database analysis task force report—Part III*. Value Health. 2009;12(8):1062–73. ISSN 1098-3015. https://doi.org/10.1111/j.1524-4733.2009.00602.x.
4. Francis DR. 2018. Why do death rates decline? The National Bureau of Economic Research. Retrieved from: https://www.nber.org/digest/mar02/w8556.html.
5. Scotland J. Exploring the philosophical underpinnings of research: relating ontology and epistemology to the methodology and methods of the scientific, interpretive, and critical research paradigms. Engl Lang Teach. 2012;5(9):9–16. ISSN 1916-4742
6. Creswell JW. Research design: qualitative and mixed methods approaches. London: SAGE; 2009.
7. Creswell JW, Plano Clark VL. Designing and conducting mixed methods research. Ohio: SAGE Publications, Inc; 2018.
8. Crotty M. The foundations of social research. London: Sage; 1989.
9. Grix J. The foundations of research. London: Palgrave Macmillan; 2004.
10. Groff R. Critical realism, post-positivism and the possibility of knowledge. London/New York: Routledge; 2007.
11. Bezemer J, Korkiakangas T, Weldon S-M, Kress G, Kneebone R. Unsettled teamwork: communication and learning in the operating theatres of an urban hospital. J Adv Nurs. 2015;72(2):361–72. Availabel from: http://onlinelibrary.wiley.com/doi/10.1111/jan.12835/pdf.
12. Korkiakangas T, Weldon SM, Bezemer J, Kneebone R. Nurse–surgeon object transfer: video analysis of communication and situation awareness in the operating theatre. Int J Nurs Stud.

2014;51(9):1195–206. Availabe from. https://doi.org/10.1016/j.ijnurstu.2014.01.007.
13. Korkiakangas T, Weldon SM, Bezemer J, Kneebone R. Video-supported simulation for interactions in the operating theatre (ViSIOT). Clin Simul Nurs. 2015;11(4):203–7. Availabe from. https://doi.org/10.1016/j.ecns.2015.01.006.
14. Korkiakangas T. Mobilising a team for the WHO surgical safety checklist: a qualitative video study. BMJ Qual Saf Published Online First: 29 Feb 2016. https://doi.org/10.1136/bmjqs-2015-004887.
15. Korkiakangas T, Weldon S-M, Bezemer J, Kneebone R. "Coming Up!": why verbal acknowledgement matters in the operating theatre. In: Cartmill SWJ, editor. Communication in surgical practice. Sheffield: Equinox; 2016.
16. Weldon SM, Korkiakangas T, Bezemer J, Kneebone R. Communication in the operating theatre: a systematic literature review of observational research. Br J Surg. 2013;100(13):1677–88. Available from. https://doi.org/10.1002/bjs.9332.
17. Weldon SM, Korkiakangas T, Bezemer J, Kneebone R. Music and communication in the operating theatre. J Adv Nurs. 2015;71(12):2763–74.
18. Weldon SM, Korkiakangas T, Bezemer J, Kneebone R, Nicholson K, Kress G. Transient teams in the operating theatre. Oper Theatre J. 2012;261(1):2.
19. Weldon SM, Korkiakangas T, Bezemer J, Kneebone R. Video analysis of bodily conduct in teamwork within the operating theatre. Int J Qual Methods. 2012;11:895–6.

47 Case Study 3: Application of Quantitative Methodology

Gregory E. Gilbert and Aaron W. Calhoun

Introduction

Perhaps one of the most important conversations occurring in the development of a research study concerns the methodology that will be used. In this vignette we will concentrate on quantitative methodology as an example; however, these principles extend to qualitative approaches as well. Without a valid analysis plan, the likelihood a study will be able to meaningfully comment on the research question it was originally intended to address is quite low. This, in turn, will substantially affect its likelihood of acceptance by a peer-reviewed journal.

While necessary, conversations such as this can be fraught with pitfalls. These often stem from simple differences in vocabulary and perspective between the primary investigator (PI) and statistician, particularly if the primary investigator has little statistical training. While the investigator is primarily concerned with the clinical meaningfulness of the question, and thus has the most intuitive sense of what they are trying to get from their data, the statistician's concerns will center on the structure of the data itself. If this is not recognized by each party, misunderstandings can occur. It is thus important that a clear, transparent initial conversation occurs - bridging this potential gap that sets the study off on the right footing. Whenever feasible, statisticians should be involved from the very beginning of study design.

In order to illustrate the value of such a conversation, we offer the following case study script. In the case below "PI" denotes the researcher/PI and "Stat" denotes the statistician.

G. E. Gilbert
SigmaStats® Consulting, LLC, Charleston, SC, USA

A. W. Calhoun (✉)
Department of Pediatrics, University of Louisville
School of Medicine, Louisville, KY, USA
e-mail: aaron.calhoun@louisville.edu

Case Study

PI: Hi, thanks so much for seeing me today. I want to do a research project on the effects of a new simulation-based intervention I have created to enhance the non-technical skills of transport personnel dealing with an arrest.

Stat: Tell me more. Specifically, what do you mean by, "non-technical skills of transport personnel" and "arrest"?

PI: Well, I was on a flight coming back from IMSH and someone in first class had a heart attack. The flight attendant made an announcement asking for physicians on board so I got up to help.

Stat: What happened? Was the person OK?

PI: Yes, the flight attendants and I were able to get him back after about three minutes of chest compressions. We had to use the defibrillator on board, but we got him stabilized and the pilot was able to make an emergency landing. He was transported to the hospital and I understand made a full recovery. But the whole incident got me thinking of better ways to train non-healthcare personnel in life-saving techniques.

Stat: I think that's a great idea. I can see a lot of applications, including the entire airline industry and the military. Have you checked to see if there is a knowledge gap?

PI: What do you mean by a "knowledge gap"?

Stat: I mean, what research has been done in the area. If it has been "researched to death" then an adequate answer to the research question may already exist. If so, then your work is already done and you can simply use the published intervention. Does this make sense? For example, I know there is a body of literature regarding emergencies on airplanes, and that flight crews are trained in CPR. I have over 20 articles on the subject, I'll send you a few of the most relevant ones. What is going to make your research unique?

D. Nestel et al. (eds.), *Healthcare Simulation Research*, https://doi.org/10.1007/978-3-030-26837-4_47

PI: Well, my concern revolves around the flight crew's non-technical skills specifically. I have looked at this already, and I cannot find any literature about this specific issue.

Stat: OK great, so we do have a knowledge gap to fill. I think we should begin with a pilot project. But I'm still not sure what you mean by "non-technical skills." Did the airline staff perform CPR incorrectly?

PI: No, actually. Their technical skills actually looked pretty good to me. Hmm. Perhaps we need to better define our terms. For physicians, "non-technical skills" refers to how well the individuals in the situation organize into a cohesive team. I can see how that would be confusing. Why don't we use the term "teamwork" instead?

Stat: That makes a little more sense to me. Thanks. So, more specifically, what was your concern with the teamwork?

PI: Well, I was thinking that these people may not know each other and have almost certainly never worked together in this way. I also suspect that, for many of them, it had been some time since their last CPR training. My idea centers on how to get these individuals to work more effectively together in these uncommon situations. Like I said, I already have a simulation-based intervention in mind.

Stat: I see now. Getting flight crews to work together more cohesively as a team could really make a difference. I think the first step would be to demonstrate that your intervention can meaningfully improve teamwork in the classroom (called medical education translational level 1 research). If we can do this, then we can approach an airline to see if the training results in an increase in survivability of the inflight incidents (called medical education translational level 2 research). Check out Drolet & Lorenzi and McGaghie to learn more about simulation-based translational research [1, 2].

PI: Great! When can we start collecting data?

Stat: Well, let's not get ahead of ourselves. The first thing we need to do is to write a clear study protocol for IRB submission so that we have addressed and satisfied any ethical concerns. Also, having a clear protocol with a specifically stated question, design, and data analysis plan will make the study results more transparent and more readily reproducible. We need to provide enough information so that other interested investigators could accurately recreate the entire study if they desired. Remember, other researchers should be able to take our data and get the same results – this is called the reproducible results paradigm.

PI: That makes sense. So it seems as if we should begin by specifying a research question.

Stat: Exactly! A well-defined research question will keep us on track. I like to use the PICO framework for creating a research question. PICO comes from evidence-based medicine and is an acronym that stands for **P**atient, **P**opulation, or **P**roblem; **I**ntervention; **C**omparison or **C**omparison Group; and **O**utcome [3]. Can you define those four components for the pilot study?

PI: Sure.

- The **p**opulation is commercial airline flight attendants.
- The **i**ntervention is my new simulation-based intervention.
- The **c**omparison group would be commercial airline flight attendants who have not been trained using my new simulation-based intervention.
- The **o**utcome would be positive outcomes from a resuscitation in-flight.

How's that?

Stat: Well, I think the first three elements make sense to me, but I worry a bit about the outcome measure. I completely agree actual clinical outcomes would be ideal (Kirkpatrick Level 4 data), but the data suggest that only about 0.1–0.6% of all emergencies are cardiac arrests, so we will be hard pressed to get the needed sample size [4, 5].

PI: What do you mean?

Stat: Statisticians split results into two categories: Statistically significant results and clinically or practically significant results. Statistically significant results are the outcome of a statistical test, but those results are not always meaningful. Clinically significant results (or practically significant results) are meaningful to you as a clinician.

PI: Hmm. I could still use some clarification.

Stat: Here is a simple example. Let's say we do a study with a new antihypertensive medication. We enroll the patients, put them on the medication, after a week measure their mean arterial pressure in mmHg and find a 1 mmHg difference. To you, as a clinician, is 1 mmHg meaningful?

PI: No, of course not!

Stat: True, but that result may be statistically significant. You could also have a difference in mean arterial pressures of 20 mmHg that is not statistically significant, but I think that would be meaningful to you.

PI: Ahhhh, I get it. You might test our results and find a difference that doesn't meaning anything in terms of patient care or vice versa.

Stat: You got it, so for us to be able to detect a statistically significant change, we need to examine a high enough number of events to be able to distinguish variation by random chance from an actual change caused by the intervention. Unless we find an airline with a lot of cardiac arrests (in which case we probably shouldn't fly on it ourselves) we may need to create a simulated environment in which to test the question and select an outcome measure that we can assess in all cases.

PI: That makes sense. OK, how about the following revision of the PICO question:

- The **p**opulation is commercial airline flight attendants.
- The **i**ntervention is my new simulation-based intervention.
- The **c**omparison group would be commercial airline flight attendants who have not been trained using my new simulation-based intervention.
- The **o**utcome would be scores on a teamwork assessment tool measured within a simulated environment.

Stat: Great. That should work better. Now let's make it into a question before we go further. Does this capture what you want to do?

Research Question: In a population of commercial airline flight attendants, does participation in a novel simulation-based intervention focused on teamwork in acute medical crises improve their response to a simulated inflight heart attack when compared to commercial airline flight attendants receiving their usual training?

Does that capture what you want to investigate?

PI: Yes, it does. I've also seen questions like this framed in terms of specific aims. I have had to prepare those for other study applications in the past. I think a good specific aim for this study would be:

Specific Aim: To evaluate the effects of a novel simulation-based educational intervention on airplane crew's teamwork skills when confronted with an in-flight cardiac arrest.

Stat: I like it. Grant applications commonly require those in addition to a research question, so creating specific aims is a good habit to get into.

PI: Excellent! Can we start collecting the data now? I've already developed the intervention!

Stat: Well, not quite yet. I'm still not clear on what our specific outcome measure will be. You mentioned a tool. Which one would you like to use? The answer to that will make a difference in terms of what statistical test we use and how I perform the sample size calculations.

PI: Oh, well, I guess it's better to plan first.

Stat: I agree. Now, I happen to have a number of potential teamwork skills assessment tools in my briefcase: the Team Emergency Assessment Measure (TEAM), the Concise Assessment of Leadership Management (CALM), and the Team Leadership tool [6–10]. From what you describe, I think that one of these three might work. Take a look and see of any of these match what you are actually looking at improving with your intervention. If not, there are a number of others in the literature.

PI: Hmmm. Let me take a look.

A few moments pass…

PI: You know, I really like the TEAM. As I think about my intervention, it seems that what I really want to improve is the overall teamwork and flow of the event. This fits the best.

Stat: Great, now we can figure out how to analyze the data and think about a sample size calculation.

PI: OK. You know, I have performed some non-educational research in the past. Looking at this tool, it seems like we could just average the scores and perform t-tests. Do you think that would work?

Stat: Well, I have some concerns about that approach. The answer is somewhat technical, but concerns the distribution of the data the tool generates. Most tools using Likert scales like the TEAM tool does don't generate bell-curve shaped (i.e., normally distributed) data, and the t-test assumes that the data have this shape. Fortunately, there are a number of tests, called nonparametric or assumption-free tests, we can use that do not make this assumption.

PI: Alright, what type of tests do you suggest?

Stat: Well, let's not get too far ahead. Why don't we talk about the study design first? The study design we choose will play into our decision about which tests are the most appropriate.

PI: Alright, let's discuss the design. It seems as if we will need both an intervention and control group so that we can make sure we are factoring in other important effects that could influence our findings. It also seems wise to measure the baseline teamwork skills of both groups before we do anything else to make sure they are about the same. If we don't, we may miss important differences between the groups affecting the outcomes of the study. We will also need to measure the intervention group's performance after the educational session to see whether it worked or not. But since we aren't giving any interventions to the control group it seems as if their baseline assessment should be enough. What do you think?

Stat: The problem with that is that you will need to use a simulation to stage the assessment itself, right? Well, there is some possibility that they could learn something useful spontaneously during that case, and so the baseline control group assessment isn't strictly comparable to the post-intervention assessment in the intervention group. Plus, it seems as if you are more interested in the effect of simulation on teamwork behavior, rather than just knowledge, correct?

PI: Yes.

Stat: Well, then we may want to consider providing the control group a non simulation-based refresher of their current training, perhaps the refresher they normally experience, that way you can isolate the effects of actually experiencing the simulation (which would affect both behavior and knowledge) from just providing information (which would affect knowledge primarily).

PI: You're right. I am more interested in behaviors. So, to summarize, we assess both groups at baseline. Then, we do the simulation with the intervention group and provide the control group with their usual refresher training and assess both groups again.

Stat: Exactly.

PI: That makes sense to me. So to go back to the prior question, which statistical tests should we use?

Stat: Well, if we were using a one-group pretest-posttest design, I would suggest a Wilcoxon signed-rank test because our data are nonparametric (i.e., not normally distributed). However, with the pretest-posttest control group design we really have two variables we have to account for: 1) how well each participant does in the pretest and 2) the intervention. The way to do this is using analysis of covariance or ANCOVA. You have probably heard of ANOVA. Well, ANCOVA is ANOVA with the addition of a continuous variable. There is one more twist though: because the data are nonparametric we need to use nonparametric ANCOVA.

PI: So what will the analysis look like? It sounds complicated.

Stat: I understand, it could be really complicated, but it doesn't have to be. What I think we should do is carryout the study and, when we go to analyze the data, sort the participants by their posttest score (statisticians call this rank-transforming). Then we can use a "regular" parametric ANCOVA on the data, with the independent variable indicating whether a participant was in the treatment group or not, and the covariate being the pretest score. In this way we do one test instead of four. This approach it is also more powerful from a statistical perspective, so we will have a greater chance of detecting a difference in the control and treatment groups when there really is a difference to be detected.

PI: Sounds good.

Stat: You will, of course, have to draw your conclusions based on the rank-transformed data and also report the medians and interquartile ranges of the original dataset, but I will give you a draft of the Results section and we can fine tune it. Remember the data is not normally distributed, so means and standard deviations aren't the best measures of central tendencies and dispersion.

PI: Alright. That makes sense. So how many subjects do you think we will need to run these tests?

S: Well, to do that we will need to perform a power calculation. To do this we can pretend we are doing a parametric ANCOVA and then multiply by 1.15. It's been demonstrated for reasonable sample sizes (like 20 or 30) that nonparametric tests do not require more than 15% more subjects than equivalent parametric tests. I can get the reference for the manuscript [11]. To begin with I will need information on the difference between the two groups you would consider clinically meaningful, an estimate of the standard deviation, and the expected correlation between the baseline and outcome variable. We can get that from the literature. If we can't find a value for the correlation in the literature we can use a small value such as 0.10 for a conservative estimate of the sample size. A conservative estimate means a larger rather than smaller sample size. We also need to decide on the alpha level and the power of the study. Typically, 80% power is adequate. For an alpha level we can use 0.05.

PI: I'm not sure what you mean about important differences?

Stat: Well, to really answer this you need to state your question again, this time in the form of a hypothesis predicting the magnitude of change in your subjects' performance you expect to result from your intervention.

PI: How do I get a sense of the magnitude?

Stat: That will depend largely on the degree of change that you think would be meaningful. Since we are using the TEAM tool, I think the best way forward would be to carefully examine the tool and ask yourself what difference in overall score you think would really make a difference clinically in terms of the problem you are trying to address. I can't answer the question for you, since the clinical aspects of this are your primary domain.

PI: Let me see.

Time passes…

PI: Well, the tool has 11, 5-point Likert scale items (scored between 0 and 4) that look at individual team behaviors, but then ends with a separate 10-point global score. My thought was to look at the median of the 11 behavioral items as the primary outcome. A 20% improvement in scores seems likely to be clinically significant, and so we would be looking at a 1 point change in median scores. I would also like to compare the global scores as well, but I think we should power the study based on the behavioral scores. They seem more specific and clinically meaningful to me. So given all of this, my hypothesis is as follows:

Hypothesis: We hypothesize the simulation-based intervention will result in a 20% improvement in median behavioral scores obtained from the TEAM tool when compared to a control group not receiving the intervention.

Stat: That sounds excellent. I can work with that. As far as the global scores go, we can run a separate ANCOVA on them as well, but that will not be our primary outcome. At this point, I think that we have everything we need to begin work on a protocol.

PI: Great. How about I start work on the Introduction and Literature Review while you perform the power calculations and write the Methods section.

Stat: It's a good plan. Before we go, though, I want to mention one final set of analyses we will need to do. You see, statistical significance alone does not tell the whole story, but only really tells us whether a difference we observe is likely to be present in the population from which we obtained the sample. To really show the full picture, we also need a measure of magnitude.

PI: Hmm, I'm not sure I understand.

Stat: Well, imagine that you are performing a study examining a new drug to treat cancer. You are able to get a sample size in the thousands, and the results show that the drug significantly lengthens life with a p-value of 0.001. But then you decide to quantify exactly how much longer the intervention group lives. What you find is that the drug lengthened life by an average of two days. This leads to the question of the clinical significance of that drug. You see what I mean? While the p-value was very small, the clinical significance also relates to the absolute magnitude of the change observed. There are a number of different statistical ways to quantify this that we refer to collectively as measures of effect size [12].

PI: Now, I see what you mean. So which of these effect size measures should we use for this study?

Stat: Well, there are a number of effect size statistics we could use for an ANCOVA, but the most commonly and easiest to calculate is the partial eta squared [13]. This number tells us the proportion of variability in our results that can be attributed to our intervention, so higher is better. It's not perfect, but should give us a good estimate of the strength of the intervention if interpreted alongside the actual change in scores [14]. I'll add this to the analysis plan.

PI: Great!

Stat: There is one final thing I would like to bring up. We all lose focus and research projects lose momentum. To be more efficient, and get more publications, I like to harness team's energy at the very beginning of research studies to get the Background and Introduction sections of the manuscript written. We can use the IRB application as a template. I will be writing the Methods section for the IRB application so we can copy and paste that into the paper. However, I also think we should write the Results and Discussion sections.

PI: How can I do that?

Stat: Well writing the results section is easy, just put "XX"s where results should go. If you want tables in the manuscript, go ahead and create them – don't bother to put any numbers in them. Then, write the Discussion section.

PI: The Discussion section?

Stat: Sure, if you have a good Background, you suspect how the outcomes will turn out. Use your knowledge and write the Discussion section. We may have to revise it totally, but you will have a good rough draft. Then I can take what you send me and put statistical computer code in it so that when we get the data all I have to do is literally press a button to get a Microsoft Word document of your entire paper!

PI: Seriously? Like an instant manuscript?

Stat: Yes, if everything goes as expected. If the results are not what you are expecting then, of course, more thought and writing will be needed. Still, it is a good exercise even in those conditions and could really pay off in terms of efficiency. If it works, though, I can send you a final draft of the paper within 5 minutes of receiving the data. Then you email it to colleagues and give them a week to send you suggestions – if they do, great! If not, then you submit without their input. I call it the "B^{2}G Model" for writing a manuscript after the colleagues who introduced me to it.

PI: That is crazy! I could see how productive you could be if you used this model. Excellent. This was a very productive meeting. Thanks for your help.

Stat: And thank you for involving me early. It really makes a difference when we get involved early in a study.

PI: Sounds good. Well, let's go get some data.

Stat: You're forgetting the IRB proposal again.

PI: Right, right…

Conclusion

In summary, the above case study addressed a number of important issues in quantitative research, including the following:

1. Basic study design using the PICO framework
2. Selection of an appropriate assessment methodology
3. Appropriate use of parametric vs nonparametric statistical tests
4. Information needed for an appropriate power calculation
5. Formulation of research questions and hypotheses
6. Complementary value of effect size
7. Use of the B^{2}G Model to leverage momentum at the beginning of the study for better use of time.

In addition to this, it is our hope that the exchange as modeled above clearly demonstrates the value of an early, collegial discussion with a statistician. By taking the needed time to bridge the gap between these professional disciplines at the start of the protocol development process, the primary investigator can attain much needed insight on the design of the study, and the statistician can be oriented to the clinical problems the study is designed to address.

References

1. Drolet BC, Lorenzi NM. Translational research: understanding the continuum from bench to bedside. Transl Res. 2011;157(1):1–5.
2. McGaghie WC. Medical education research as translational science. Sci Transl Med. 2010;2(19):19cm8.
3. da Costa Santos CM, de Mattos Pimenta CA, Nobre MR. The PICO strategy for the research question construction and evidence search. Rev Lat Am Enfermagem. 2007;15(3):508–11.
4. Harden RM, Grant J, Buckley G, Hart IR. BEME guide no. 1: best evidence medical education. Med Teach. 1999;21(6):553–62.
5. Peterson DC, Martin-Gill C, Guyette FX, Tobias AZ, McCarthy CE, Harrington ST, et al. Outcomes of medical emergencies on commercial airline flights. N Engl J Med. 2013;368(22):2075–83.
6. Cant RP, Porter JE, Cooper SJ, Roberts K, Wilson I, Gartside C. Improving the non-technical skills of hospital medical emergency teams: the Team Emergency Assessment Measure (TEAM). Emerg Med Australas. 2016;28(6):641–6.
7. Cooper S, Cant R, Connell C, Sims L, Porter JE, Symmons M, et al. Measuring teamwork performance: validity testing of the Team Emergency Assessment Measure (TEAM) with clinical resuscitation teams. Resuscitation. 2016;101:97–101.
8. Cooper SJ, Cant RP. Measuring non-technical skills of medical emergency teams: an update on the validity and reliability of the Team Emergency Assessment Measure (TEAM). Resuscitation. 2014;85(1):31–3.
9. Nadkarni LD, Roskind CG, Auerbach MA, Calhoun AW, Adler MD, Kessler DO. The development and validation of a concise instrument for formative assessment of team leader performance during simulated pediatric resuscitations. Simul Healthc. 2018;13(2):77–82.
10. Grant EC, Grant VJ, Bhanji F, Duff JP, Cheng A, Lockyer JM. The development and assessment of an evaluation tool for pediatric resident competence in leading simulated pediatric resuscitations. Resuscitation. 2012;83(7):887–93.
11. Cohen J. Statistical power analysis for the behavioral sciences. 2nd ed. Hillsdale: Lawrence Erlbaum Associates Publishers; 1988.
12. Sullivan GM, Feinn R. Using effect size-or why the p value is not enough. J Grad Med Educ. 2012;4(3):279–82.
13. Ialongo C. Understanding the effect size and its measures. Biochem Med. 2016;26(2):150–63.
14. Levine TR, Hullett CR. Eta squared, partial eta squared, and misreporting of effect size in communication research. Hum Commun Res. 2002;28(4):612–25.

48

Case Study 4: Navigating the Peer Review Process in Healthcare Simulation: A Doctoral Student's Perspective

Jessica Stokes-Parish

Introduction

Perhaps one of the more daunting tasks of work for an early career researcher is that of submitting a manuscript for publication. As I reflect on my own journey, I recall a mentor quietly suggesting that my postgraduate coursework would not cut the mustard for the standard of academic writing required to publish. Mildly bruised, I was confused how I could make the leap to quality academic writing. How do you begin the journey of submission, receiving review and revising with the hope to have your work published? In this case study, I share my experience of journeying through manuscript preparation, reviewer feedback, revision and proceeding to publication.

As a way of introduction, I find it useful to reflect on the concept of text being identity shaping [1], as much as it is actual writing. Submitting a manuscript is submitting an identity, a text that represents an individual's position in the scholarly and academic world. Kamler and Thomson [1] propose that writing is in and of itself a process of developing identity, that "writers often experience difficulties and textual struggles because they are negotiating text work and identity work at the same time." I found this text useful in acknowledging the personal tensions that may be present – both from the perspective of testing your knowledge/views on the already established social constructs and paradigms. Submitting a manuscript is worthy of acknowledgement in itself. In doing so, you agree to receive critique on your writing and potentially be challenged on where you situate your worldview. Beyond this, author instructions can be lengthy, and often cover (what feels like) minutia that is seemingly irrelevant. Of course, the guides are necessary for consistency and to encourage a smoother publication process [2, 3]. Additionally, the guides define the objective and goals of the organization to whom the manuscript was submitted. Each journal, organization, etc. has their own mission, objectives and goals.

At its most basic function, a manuscript review is feedback. Clear communication between the author and reviewer/s is fundamental and essential, including beyond the manuscript itself. I sometimes wonder if reviewers would offer their feedback differently it was in person rather than in text. In a manuscript review, we're not granted the introductory rapport-building phase built into face-to-face feedback processes. As the author, it is helpful if the feedback is "framed" in an introduction, with a statement outlining what the reviewer is hoping the author will achieve. As the reviewer, please be clear on what is it that you are delivering feedback – content/theory, writing structure, analysis or contextual framework for which the respective journal has set to achieve.

In Example 1, I share an excellent example of reviewer feedback on a manuscript submitted for publication to a simulation journal. The reviewer identifies immediately that the topic is of relevance and the strengths of the manuscript. The reviewer clearly identifies that he/she is moving on to the areas of issue ("my biggest issue...") and articulates them.

Example 1

> Thank for submitting a paper on a topic with high relevance for the simulation community. The background sections are an informative, concisely written overview of the issues; valuable historical perspectives provide context; the paper is adequately referenced. The authors define terms clearly, including fidelity, realism, and authenticity. In my view, this topic is timely and would add value to the readers of [Journal Title].
>
> My biggest issue with the paper is whether a systematic review is an appropriate methodology given the relative lack of literature. A narrative review, without self-imposed constraints of systematicity, seems to be a better fit. By framing the paper as a systematic review–with a yield of articles in the single digits–the authors appear to diminish the impact of their innovative thinking.

J. Stokes-Parish (✉)
School of Medicine & Public Health, University of Newcastle, Callaghan, NSW, Australia

D. Nestel et al. (eds.), *Healthcare Simulation Research*, https://doi.org/10.1007/978-3-030-26837-4_48

The next reviewer (Example 2) also identifies the topics relevant for study, but isn't so clear on the framing of feedback. There is no rhythm to it and no way for the author, me, to prepare myself for what I am about receive feedback on.

Example 2

Thank you for the opportunity to review your manuscript. This is an important topic and as you know, needs further study. I have provided feedback and suggestions below.

Applying this concept of preparing for feedback in Example 2, the Reviewer could have provided an introductory framing of their approach to feedback and what they set out to achieve through the feedback (*added text marked by underline*). This would allow the author to engage better with the reviewer, almost creating a rapport of sorts – albeit via text.

Revised example 2

Thank you for the opportunity to review your manuscript. This is an important topic and as you know, needs further study. *In reviewing your work, I have spent time providing comments to each line, considering your hypotheses against current evidence and some focus on your grammar and syntax.*

In Example 3 from a different journal (revised manuscript), the reviewer provides a summary of the manuscript and an overarching statement about the issues. However, the reviewer does not provide clarity on the issues discussed in the statement. As the author, it is challenging to consider the value of the feedback if there is no clarity or examples of what the reviewer means. Consider that text can be interpreted differently to what the writer intended; interpretation of the text is shaped by the worldview and prior experiences of the reader [4, 5]. Therefore, reviewer feedback should be specific [6] with demonstrated examples. Ideally, in Example 3, the reviewer could have provided specific comment on where the topics were repeated and how to better organise the paper, that is, suggestions for solutions.

Example 3

This article, which is submitted as a concepts and commentary paper, reviews the topic of moulage and challenges the simulation community to study the impact of moulage. The authors make the argument that use of moulage in simulation is expensive and the impact on realism and effect on the learner has not been studied. The main focus of the article is the minor amount of research dedicated to the validity of moulage. Overall the article is lengthy and could use better organization. Many of the topics and concepts are repeated.

An excellent example of specific feedback is where Reviewer 2 provides specific feedback in the original manuscript to the author [referring to the use of "both"]. "Line 170—there are more health-related disciplines than nursing and medicine. By using the word 'both', you are not acknowledging that. Just eliminate the word 'both'". This is helpful for the author, as there is no ambiguity and it provides perspective that I, the author, may not have considered.

Testing the notion that writing incorporates actual writing and individual identity, it is useful to explore this with regard to being the reviewer. Consider that you are not only providing feedback on the manuscript, but also on the scholar – you are providing feedback as to whether the authors views/concepts fit within the social constraints of the field of discipline. This is evident in Editor in Chief (EIC) comments on the revised version of the original manuscript (Example 4). Fresh eyes from an outsider unattached to the process of developing the theories and subsequent manuscript can be very useful to the author. In this instance, the reviewer takes the time to shape a conversation that engages with the author (me), and as a result, significantly influences the approach taken in the manuscript and scholar directions.

Example 4

I'm not saying that the Dieckmann approach is any better than yours just that there are resonances between this (and likely other things) work and other work on "realism" "reality" "fidelity" etc. of simulation that I don't think you have tapped into fully and that may be as valuable as the relating to the conceptual basis of film.

Another example of shaping the scholar is that of Example 5. The EIC is clear in defining what is done well, what is not in question, with what the trouble spots are. Again, the repetition of specific feedback is helpful. In Example 6 you can see the reviewer identifies the manuscript is also not suitable, however does not provide clarity about what is "less acceptable". Now, in this instance, the manuscript was declined.

Example 5

Thank you for submitting your systematic review… Unfortunately, the manuscript is not suitable for publication … This is not a criticism of … methods but a reflection… of the small number of papers, poor methodological quality of the included papers, and the subsequent inability to draw any meaningful conclusions. However, we believe that you do raise an important topic area for the simulation field ….

Example 6

The re-framing of the systematic review as a narrative review has not worked – the manuscript is in some ways less acceptable now than previously – so the manuscript is not in its current form right for publication - but at heart there is still some really interesting work. So a definite no to acceptance in its current form.

The peer review process of submitting a manuscript is part of a journey in an academic career and necessary for the progression of education and clinical evidence, and to spur further inquiry and discovery [7]. As an author, this can challenge both our identity as a scholar and refine our skillset. Reviewers have an opportunity to shape scholars and add other perspectives to their work. To maximise this potential, it is beneficial to consider the strategies used in the delivery of written feedback. This occurs when reviewers set the scene with clear descriptions of their aims, provide demonstrations of where changes need to occur, and offer examples of changes where possible. As an author, it is productive to receive critique and review of our scholarly work, aiding our development and progress to successful publication [8, 9].

Key Messages

- *To the author*
  1. Nothing is ever perfect. Write and submit. You can only get better with each revision.
  2. Read examples of papers in the journal you wish to submit (from the current editor). This will help to identify if your work fits the style, and hints to what standard the journal is looking for.
- *To the reviewer*
  1. Present a clear framework as a summary for your approach to the review
  2. Provide suggestions for potential solutions
  3. Remember that in reviewing the text you are also shaping the scholar

References

1. Kamler B, Thomson P. Helping doctoral students write: pedagogies for supervision. Hoboken: Taylor & Francis; 2014.
2. Kimmerly-Smith J. 2018. Navigating the peer review process: what you need to know. Retrieved from https://www.scribendi.com/advice/peer_review_process.en.html.
3. Wiley. 2018. The peer review process. Retrieved from https://authorservices.wiley.com/Reviewers/journal-reviewers/what-is-peer-review/the-peer-review-process.html.
4. Janks H. Critical discourse analysis as a research tool. Discourse Stud Cult Polit Educ. 1997;18(3):329–42. https://doi.org/10.1080/0159630970180302.
5. Nestel D, Bearman M. Theory and simulation-based education: definitions, worldviews and applications. Clin Simul Nurs. 2015;11(8):349–54. https://doi.org/10.1016/j.ecns.2015.05.013.
6. Ramani S, Krackov SK. Twelve tips for giving feedback effectively in the clinical environment. Med Teach. 2012;34(10):787–91. https://doi.org/10.3109/0142159X.2012.684916.
7. Yarris LM, Gottlieb M, Scott K, Sampson C, Rose E, Chan TM, Ilgen J. Academic primer series: key papers about peer review. West J Emerg Med. 2017;18(4):721–8. https://doi.org/10.5811/westjem.2017.2.33430.
8. Stokes-Parish JB, Duvivier R, Jolly B. Does appearance matter? Current issues and formulation of a research agenda for moulage in simulation. Simul Healthc J Soc Med Simul. 2016;12(1):47–50.
9. Stokes-Parish JB, Duvivier R, Jolly B. Investigating the impact of moulage on simulation engagement — a systematic review. Nurse Educ Today. 2018;64:49–55. https://doi.org/10.1016/j.nedt.2018.01.003.

Index

D. Nestel et al. (eds.), *Healthcare Simulation Research*, https://doi.org/10.1007/978-3-030-26837-4

T